中国科学院年度报告系列

2022
科学发展报告
Science Development Report

中国科学院

科学出版社

北京

内 容 简 介

本报告是中国科学院发布的年度系列报告《科学发展报告》的第25部，旨在全面综述和分析2021～2022年度国际科学研究前沿进展动态，研判和展望国际重要科学领域研究进展和发展趋势，揭示和洞察科技创新突破及快速应用的重大经济社会影响，观察和综述国际主要科技领域科学研究进展及科技战略规划与研究布局，评述和介绍主要科学奖项的获奖工作，报道我国科学家具有代表性的重要科学研究成果，并向国家决策部门提出有关中国科学的发展战略和科技政策咨询建议，为国家促进科学发展的宏观决策提供重要依据。

本报告对国家各级科技决策部门、科研管理部门等具有连续的重要学术参考价值，可供国家各级科技决策和科研管理人员、科研院所科技研究人员、大专院校师生以及社会公众阅读和参考。

图书在版编目(CIP)数据

2022科学发展报告/中国科学院编．—北京：科学出版社，2023.11
（中国科学院年度报告系列）
ISBN 978-7-03-076168-2

Ⅰ.①2… Ⅱ.①中… Ⅲ.①科学技术-发展战略-研究报告-中国-2022
Ⅳ.①N12②G322

中国国家版本馆CIP数据核字（2023）第152632号

责任编辑：牛 玲 刘巧巧 / 责任校对：何艳萍
责任印制：师艳茹 / 封面设计：有道文化

科学出版社 出版
北京东黄城根北街16号
邮政编码：100717
http://www.sciencep.com
北京中科印刷有限公司 印刷
科学出版社发行 各地新华书店经销
*
2023年11月第 一 版 开本：787×1092 1/16
2023年11月第一次印刷 印张：28 插页：2
字数：450 000

定价：198.00

（如有印装质量问题，我社负责调换）

专家委员会

（按姓氏笔画排序）

丁仲礼　陈凯先　姚建年
郭　雷　曹效业　解思深

总 体 策 划

汪克强　潘教峰

课　题　组

组　长： 张　凤
成　员： 王海霞　叶　京　贾晓琪　裴瑞敏

审 稿 专 家

（按姓氏笔画排序）

习　复　王友德　叶　成　叶小梁　吕厚远
刘文彬　杨茂君　吴乃琴　吴学兵　吴家睿
吴善超　沈电洪　张正斌　赵　刚　聂玉昕
顾兆炎　黄大昉　龚　旭　章静波

前　言

过去的两年，我国重大科技创新成果竞相涌现。中国空间站完成在轨建造并取得重大进展，“天问一号”探测器成功着陆火星，FAST 捕获世界最大快速射电暴样本并精细刻画活跃重复快速射电暴，“祖冲之号”超导量子计算原型机成功构建，先后实现了从二氧化碳到淀粉的人工合成以及以二氧化碳和水为原料合成葡萄糖和脂肪酸，“嫦娥五号”月球样品揭示月球演化奥秘，阐明冠状病毒的跨种识别和分子机制、免疫逃逸机制等，揭示鸟类迁徙路线成因和长距离迁徙的关键基因……

习近平总书记在 2021 年 5 月召开的两院院士大会上指出：“我国广大科技工作者要以与时俱进的精神、革故鼎新的勇气、坚忍不拔的定力，面向世界科技前沿、面向经济主战场、面向国家重大需求、面向人民生命健康，把握大势、抢占先机，直面问题、迎难而上，肩负起时代赋予的重任，努力实现高水平科技自立自强！”① 在 2022 年 10 月 16 日召开的中国共产党第二十次全国代表大会上，习近平总书记再次强调“加快实现高水平科技自立自强”②。因此，发挥战略科学家作用服务国家决策就显得至关重要。

中国科学院作为国家战略科技力量的重要组成部分，作为我国科学技术方面的最高学术机构和国家高端科技智库，把握世界科学技术的整体竞争发展态势和趋势，对科学技术与经济社会的未来发展进行前瞻性思考和布局，促进和提高国家发展决策的科学化水平，是我们义不容辞的使命与责任。

---

① 习近平：加快建设科技强国 实现高水平科技自立自强 . https：//www. gov. cn/xinwen/2022 - 04/30/content _ 5688265. htm.

② 习近平：高举中国特色社会主义伟大旗帜 为全面建设社会主义现代化国家而团结奋斗——在中国共产党第二十次全国代表大会上的报告 . https：//www. gov. cn/xinwen/2022－10/25/content _ 5721685. htm.

1997年9月，中国科学院决定发布年度系列报告《科学发展报告》，按年度连续全景式综述分析国际科学研究进展与发展趋势，向国家最高决策层和社会全面系统地报告世界和中国科学的发展情况，评述科学前沿动态与重大科学问题，报道介绍我国科学家取得的代表性、突破性科研成果。随着国家全面建设创新型国家和推进科技强国建设，《科学发展报告》将进一步系统、全面地观察和揭示国际重要科学领域的研究进展、发展战略和研究布局，服务国家促进科学发展的宏观决策。

《2022科学发展报告》是该系列报告的第25部，主要包括科学展望、科学前沿、2021年中国科研代表性成果、科技领域与科技战略发展观察、重要科学奖项巡礼和中国科学发展建议等六大部分。受篇幅所限，本报告所呈现的内容不一定能体现科学发展的全貌，重点是从当年受关注度最高的科学前沿领域和中外科学家所取得的重大成果中，择要进行介绍与评述。

本报告的撰写与出版是在中国科学院侯建国院长的关心和指导及众多院士专家的参与下完成的，得到了中国科学院发展规划局、中国科学院学部工作局的直接指导和支持。中国科学院科技战略咨询研究院承担本报告的组织、研究与撰写工作。丁仲礼、解思深、陈凯先、姚建年、郭雷、曹效业、汪克强、潘教峰、王友德、聂玉昕、沈电洪、吴学兵、赵刚、习复、叶成、吴善超、龚旭、叶小梁、吴家睿、黄大昉、章静波、杨茂君、张正斌、吕厚远、吴乃琴、顾兆炎、刘文彬等专家参与了本报告的咨询与审稿工作，在此一并致以衷心感谢。

中国科学院《科学发展报告》课题组

2022年12月20日

目　录

CONTENTS

第一章

科学展望

An Outlook on Science

1.1 陆地生态系统在碳中和中的潜力展望

郝天象 徐 丽 于贵瑞

（中国科学院地理科学与资源研究所生态系统网络观测与模拟重点实验室）

气候变化是人类正面临的严峻挑战之一[1]。目前，全球变暖趋势仍在持续，我国是全球气候变化的敏感区。1951～2021 年，中国升温速率（0.26℃/10 a）高于同期全球平均水平（0.15℃/10 a）[2]。积极应对气候变化已然成为全球共识，《巴黎协定》提出了把全球平均气温较工业化前水平的升温控制在 2℃以内，并努力实现升温在 1.5℃以内的目标，形成了 2020 年后全球气候治理总体格局[3]。二氧化碳（CO_2）是大气中最重要的温室气体，约占全球人为温室气体净排放总量的 3/4，其中以化石燃料及工业过程 CO_2 排放为主，占净排放总量的近 2/3[4,5]。美国国家海洋与大气管理局（NOAA）全球监测实验室（Global Monitoring Laboratory，GML）监测数据①显示，1959～2022 年，夏威夷冒纳罗亚（Mauna Loa）大气基线观测站大气中 CO_2 浓度持续攀升。2021 年全球 CO_2 平均浓度达到 415.7 ppm② 的历史新高；2020～2021 年，CO_2 浓度增速高于过去 10 年的平均年增长率[6]。国际社会必须在 2025 年前实现碳达峰[7]，2050 年前实现碳中和，才有可能实现《巴黎协定》的气候治理目标（将 21 世纪全球平均气温上升幅度控制在 2℃以内，并将全球气温上升控制在前工业化时期水平之上 1.5℃以内）[1,4]。

2020 年 9 月 22 日，在第七十五届联合国大会一般性辩论上，国家主席习近平代表我国向世界做出庄严承诺："中国将提高国家自主贡献力度，采取更加有力的政策和措施，二氧化碳排放力争于 2030 年前达到峰值，努力争取 2060 年前实现碳中和。"[8]实现碳达峰碳中和，是以习近平同志为核心的党中央统筹国内国际两个大局做出的重大战略决策，是一场广泛而深刻的经济社会系统性变革[9,10]。我国"双碳"行动具有重大战略意义，是应对世界百年未有之大变局、推动中华民族伟大复兴的宏伟举措，是促进科技进步与绿色转型发展、催生新型生态经济、推进生态文明建设的宏观战略。

① 参见网址：https://gml.noaa.gov/dv/iadv。

② 此处 1 ppm 代表 CO_2 的分子数占空气总分子数的 10^{-6}。

据联合国政府间气候变化专门委员会（IPCC）定义，碳中和是指化石燃料燃烧、工业过程、农业及土地利用活动排放等人为活动导致的碳排放量与通过人为措施产生的生态系统碳吸收量和通过碳捕集、利用与封存（CCUS）等各种技术产生的碳吸收量之间的平衡[11]。尽管能源活动与工业活动是碳排放的主要来源，能源转型和工业减排是实现“双碳”目标的关键途径，但在关键性、颠覆性能源技术还没有取得突破之前，通过人为生态工程建设，巩固和提升生态系统固碳能力，是行之有效、极具规模、绿色安全、经济合理的技术途径，被认为是实现“双碳”行动的“压舱石”及社会经济发展的“稳定器”[12]。

根据“全球碳计划”发布的《2022 年全球碳预算》报告，2012～2021 年，陆地生态系统每年吸收了约 29% 的全球人为碳排放量[13]。2010～2020 年，我国陆地生态系统的碳汇①为每年 10 亿～15 亿 t CO_2，约占我国人为碳排放量的 13%[14]。从《京都议定书》（1997 年）、《哥本哈根协议》（2009 年）到《巴黎协定》（2015 年），生态系统碳保护与增汇都被认为是应对气候变化的重要途径。

综上所述，陆地生态系统碳汇对实现“双碳”目标、应对气候变化的意义重大。围绕陆地生态系统碳循环，国内外学者已开展了大量研究，显著提升了对陆地生态系统碳源汇功能的认识[15-17]。但是，由于调查资料、观测手段、碳计量方法与碳循环模型的局限性，陆地生态系统碳汇功能研究还存在诸多薄弱之处，在陆地生态系统碳收支、碳储量及其时空格局、固碳增汇技术及其潜力预测等方面仍存在很大的不确定性。精准评估陆地生态系统碳通量与储量及其时空动态特征，揭示气候变化和人类活动双重影响下的陆地生态系统碳汇能力变化及其内在机制，构建基于自然生态系统的陆地碳源汇演变过程与增汇措施，整合分析理论和方法学体系，预测未来陆地生态系统碳汇变化和各类措施的固碳潜力等，都是未来陆地生态系统碳汇科研工作所面临的重大挑战。

一、陆地生态系统碳循环

在全球范围内，生物过程是地球系统结构与功能的调控者与塑造者。其中，陆地生态系统无疑是主要的参与者，控制着地球表层系统中大气圈、水圈、土壤圈、岩石圈和生物圈之间的能量流动和物质循环。

大气中的 CO_2 被陆地植物通过光合作用吸收并转化为有机物形态储存在植物中，形成总初级生产力，其中部分以根系分泌物、植物残体和微生物残体等形式迁移到土

① 本文中生态系统碳汇、生态系统碳储量、植被碳储量、土壤有机碳储量、固碳速率等均以 CO_2 的质量进行计算。

壤中。部分光合作用产生的有机碳通过植物自身呼吸作用（自养呼吸，包括地上部呼吸和根系呼吸）和土壤动物与微生物作用（异养呼吸）以 CO_2 形式返回大气；未完全分解的有机质存留在土壤中，或经过漫长的地质年代形成化石燃料储藏在地下，或在人为和自然扰动下以 CO_2 形式释放到大气。这两条途径共同形成了大气—植物—土壤—大气的整个陆地生态系统碳循环过程（图 1）。从 CO_2 吸收和释放过程来看，陆地生态系统碳汇是指植物通过光合作用吸收大气中的 CO_2，并将其固定在植被和土壤中，从而减少大气中 CO_2 浓度的过程[18]；反之，由于人类活动（如土地利用变化、管理措施等）或自然过程（森林火灾、病虫害等）导致生态系统向大气释放 CO_2 的过程则称为碳源。

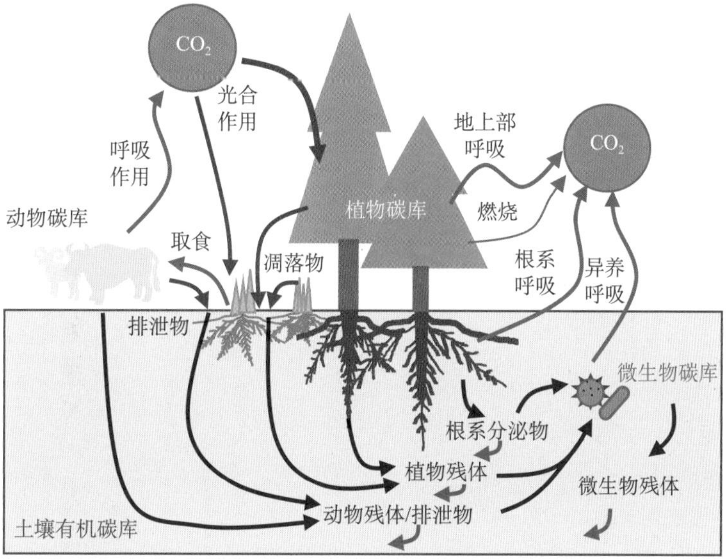

图 1 陆地生态系统碳循环及主要的碳库

箭头表示各个碳库间的主要交换过程。

植物光合作用所形成的有机碳，扣除自养呼吸排放的碳，所剩的这部分有机碳（即净初级生产力），部分以植被的地上生物量（如叶、枝、干）碳库和地下生物量（根系）碳库的形式蓄积在植物碳库中，主要是以淀粉、可溶性糖等非结构性碳水化合物和木质素、纤维素、果胶等结构性碳水化合物等有机碳形态存在；部分被动物和微生物取食用于新陈代谢，在动物碳库中沿着食物链、食物网流动，最终以 CO_2、排泄物或动物残体等形式进入大气或土壤；部分以凋落物、根系分泌物等有机物质形态进入土壤，被土壤动物和微生物分解，未完全分解的残体和分解后形成的含碳有机物储存在土壤有机碳库中。同时，土壤还可以通过无机碳过程直接吸收大气中的 CO_2[19,20]。在土壤高 pH、富钙化环境下，存在 SOC（土壤有机碳，soil organic carbon)- CO_2 - HCO_3^- 体系；在干旱、半干旱地区碱性、富钙化环境下，则为 SOC - CO_2 - HCO_3^- - $CaCO_3$体系。土壤有机质分解释放 CO_2，水解成 HCO_3^- 和 H^+，经过碳酸氢盐的溶解，沉淀形成碳酸盐，产生的 H^+ 以及酸沉降等其他外源 H^+ 输入，会使 $CaCO_3$溶解；而在干旱、CO_2 分压较小和土壤高 pH 的情况下，产生的 H^+ 被土壤 OH^- 中和，可促进 $CaCO_3$的形成。相比于土壤有机碳循环研究，以碳酸盐为主的土壤无机碳循环的定量研究仍较为薄弱，且对土壤有机碳循环与无机碳循环间的周转机制的认识也并不清楚。

二、陆地生态系统碳储量与碳汇能力

（一）国际研究进展

陆地生态系统具有巨大的碳库，其碳库增加或减少都可能导致大气 CO_2 浓度的显著变化，进而对全球碳收支平衡产生深刻影响[13]。相比于大气和海洋，陆地生态系统最为复杂且不确定，主要是由于除了地表存在丰富多样的植被类型外，陆地还包含碳储量巨大的土壤圈，同时与大气和海洋等其他圈层存在多界面和多种过程的碳交换和碳传输[18]。过去几十年，国内外学者利用涡度相关通量观测系统、森林/草地资源清查数据、“应对气候变化的碳收支认证及相关问题”专项调查数据或遥感观测数据等对陆地生态系统固碳速率、碳储量、碳组分分配及其变化规律开展了大量研究，显著提升了对碳循环的认识[15-17]。

2016 年，全球陆地生态系统碳储量为 12.74 万亿 t CO_2（3477.8 Pg C），其中植被碳储量（地上部＋地下部）为 1.62 万亿 t CO_2（441.2 Pg C），土壤有机碳储量（土壤厚度范围为 0～2 m）为 11.13 万亿 t CO_2（3036.5 Pg C）[21]。全球陆地生态系统固碳速率为（125±33）亿 t CO_2/a，约占同期人为活动碳排放量的 29.6%，且过去 60 年全球陆地生态系统碳汇量逐渐增大，从 20 世纪 60 年代的（－7±33）亿 t CO_2/a（弱碳源）增加至 21 世纪头十年的（70±40）亿 t CO_2/a[22]。

（二）我国陆地生态系统碳储量与碳汇研究

据统计，2004～2014 年我国陆地生态系统碳储量为 3652.7 亿 t，其中植被碳储量为 555.1 亿 t，土壤有机碳储量（土壤厚度范围为 0～1 m）为 3097.6 亿 t，以森林和草地生态系统为主[14]。

中国陆地生态系统碳汇监测研究经历了从基于单个生态站的长期定位研究，到多站点层次的观测—研究—示范的初步网络化，在中国陆地生态系统碳汇格局研究方面取得了积极进展。例如，利用不同时期森林清查资料，基于连续生物量换算因子法评估了我国森林生物量的动态变化，发现我国森林生态系统是个显著的碳汇[23]；基于遥感估算、模型模拟以及大气反演等方法，系统评估了我国陆地生态系统碳汇特征，发现我国陆地生态系统碳汇强度为 7.0 亿～9.5 亿 t CO_2/a，占我国化石燃料排放的 28%～37%[24]；2018 年，基于“应对气候变化的碳收支认证及相关问题”专项研究结果，我国科学家在《美国国家科学院院刊》（PNAS）发表“Climate Change, Policy, and Carbon Sequestration in China”专题，全面评估了中国陆地生态系统固碳能力及其形成机制[25-31]。研究发现 2001～2010 年我国陆地生态系统年均固碳约 2 亿 t CO_2，相当于同期中国化石燃料碳排放量的 14.1%[25]；我国的生态修复工程和秸秆还田等农田管理措施的实施，分别贡献了全国总碳汇的 36.8% 和 9.9%[29,30]，其中，生态系统保护和修复重大工程区内生态系统碳储量增加达到 55 亿 t，每年平均碳汇达到 4.84 亿 t，大约抵消了同期我国化石燃料燃烧 CO_2 排放量的 9.4%[29]。尽管由于方法体系、数据来源等差异导致生态系统碳汇能力评估结果存在很大的不确定性，但必须要看到这些工作有力地推动了我国陆地生态系统碳循环研究的快速发展，也为深入认识我国陆地生态系统碳汇现状奠定了坚实的基础。

总的来看，1980～2020 年的 40 年间，采用不同方法评估的我国陆地生态系统碳汇为 9 亿～25 亿 t CO_2/a，并有逐渐增加的趋势；特别是 2010～2020 年，我国陆地生态系统碳汇为 10 亿～15 亿 t CO_2/a，约占 13% 的人为碳排放量，其中森林与灌丛是主要贡献者[14]。

三、陆地增汇途径与潜力

生态系统固碳能力提升是缓解气候变暖的重要途径。从全球来看，陆地生态系统每年的固碳量相当于抵消了化石燃料燃烧碳排放量的 10%～60%[32,33]。提升陆地生态系统的碳汇能力，稳步提升陆地生态系统碳汇增量，高效发挥森林、草地、农田等典型生态系统的固碳作用对实现碳中和目标具有重要意义。

目前，基于陆地生态系统的增汇技术途径主要包括三大类：①传统的农林业减排增汇技术途径；②生态工程增汇途径；③新型生物/生态碳捕集、利用与封存（Bio-CCUS/Eco-CCUS）途径。其中，传统的农林业减排增汇技术途径包括造林和再造林、森林管理、农业保护性耕作、畜牧业减排、草地和湿地管理、滨海生态工程（如“蓝碳”养殖业）等绿色低碳减排或增汇技术措施[34]。生态工程增汇途径包括造林和再造林、退耕还林、天然草地封育等[35]，这些增汇措施需要统筹国土空间绿化与生态环境治理，围绕提升森林、农田、草地、荒漠、内陆湿地、湖泊、滨海湿地等陆地生态系统碳汇功能，挖掘现有成熟技术，整合形成适用于景观、流域到区域的系统化技术模式。Bio-CCUS/Eco-CCUS 是指通过提升陆地生态系统生产力途径来更多地固定大气 CO_2，并将其转换为有机生物质，进而作为能源、化工或建筑材料替代化石产品，或直接埋藏或地质封存，以实现碳的负排放。主要包括利用污染和废弃地等土地资源或利用农林业残余生物量等生物固碳技术[36]。

全球的评估表明，在考虑生物多样性保护和食物及纤维需求的情况下，20 项生态系统固碳增汇途径能提供的最大缓解潜力为 238 亿 t CO_2/a；考虑到成本效益，至少有 113 亿 t CO_2/a 的缓解潜力能以较低的成本实现，这能够抵消 2016～2030 年减排量的 37%，2016～2050 年减排量的 20%，2016～2100 年减排量的 9%[37]。对热带地区的 76 个国家的 12 项增汇途径的研究表明，一半以上的国家（39 个）的固碳增汇潜力可以抵消其工业碳排放量的 50% 以上，超过 1/4 的国家（22 个）的固碳增汇潜力可以超过其工业碳排放量[38]。

通过人为措施巩固与提升生态系统碳汇能力是国际公认的重要碳中和路径。根据美国、德国、日本等主要发达国家和欧盟发布的政策性文件，国际上碳中和路径普遍涵盖了生态减排固碳端（土地利用方式变化、种植业/畜牧业减排、森林保碳/固碳、“蓝碳”、CCUS）及其前沿研究与科技支撑。部分发达国家/组织碳中和路径下的生态系统碳汇增加行动详见表 1。其中，美国、德国和欧盟等在行动方案中不同程度地提及了以森林为主的自然碳汇对应对气候变化的作用。美国政府提出通过加大保护、恢复、持续管理和其他行动进而增加生态碳汇，不仅有利于大气碳移除，同时有利于包括自然生态系统及相关服务和生物多样性保护。德国政府强调了生态碳汇对不可避免的温室气体排放具有重要的约束作用，在《气候变化法案》修正案中明确提出要通过提高生态碳汇对 CO_2 吸收的作用。欧盟在《欧洲气候法案》中提及要促进基于自然的解决方案和提高基于生态系统的适应能力，强调了生态系统对维持、管理和增加生态碳汇、生物多样性保护、应对气候变化的价值，特别是森林作为碳汇、碳库和碳替代不仅有助于减少温室气体，同时还可以提供许多其他服务。

表 1　部分发达国家/组织碳中和路径下的生态系统碳汇增加行动

国家/组织	政策文件（发布时间）	具体行动
美国	《迈向 2050 年净零排放的长期战略》（2021 年 11 月）	农业、林业和土地利用的减排与固碳； 保护森林； 扩大森林面积； 延长林木砍伐周期； 城区/农区造林； 强化气候智能型农业措施（如覆盖作物、循环放牧等）； 提高农业生产力以减少耕地和农业甲烷/氧化亚氮排放（如提高有机肥管理、提高耕地养分管理）； 加大森林保护与管理的投资； 实施基于科学和可持续性努力以减少灾难性森林大火并对火灾林地进行恢复，保护和增加最大的陆地生态系统碳汇； 增加 CO_2 移除（陆地生态系统碳汇与工程措施）

续表

国家/组织	政策文件（发布时间）	具体行动
英国	《净零战略》（2021年10月）	2030年低碳耕作比例达到75%，2035年达到85%； 农业创新实现农业和园艺净零排放，造林率提升两倍，为农民和土地所有者提供净零排放行动资助； 泥炭地恢复，动员私人投资，制定跨政府和行业的政策路线； 垃圾分类与填埋整治，减少含氟气体使用； 推动《环境法案》立法，统筹脱碳与生物多样性等环境目标
澳大利亚	《澳大利亚的长期减排计划》（2021年10月）	降低成本以促进优先技术发展，如碳捕集与封存、土壤碳、新兴技术（如优化牲畜饲料以减少甲烷排放）
德国	《气候变化法案》修正案（2021年6月）	推动有潜力的关键技术研发，发展具有弹性和可持续性农业； 引导健康饮食与可持续消费习惯； 通过发展绿色投资形式来资助和支持净零排放转型，如加强对可持续的林业和农业的支持
日本	《巴黎协定下的长期战略》（2021年10月）	采取直接空气碳捕获与封存（DACCS）技术、生物质能碳捕集与封存（BECCS）技术和森林碳汇措施； 在人口减少、生育率下降和人口老龄化的背景下，在社区与生活领域打造“循环和生态经济”
欧盟	《欧洲气候法案》（2021年7月）	通过更有力的《土地利用、土地利用变化和林业法规》（LULUCF）增强欧盟碳汇

注：上述发达国家/组织以净零排放为目标，即涵盖所有温室气体的综合治理。

我国针对典型森林、草地、农田等生态系统开展了大量增汇相关科学技术的研究[34,39]。例如，针对我国林情（如森林资源分布不均、总量不足、质量低下，林火、病虫害与全球变化等自然干扰强烈，开发利用不合理，森林整体经营管理水平较低等），遵循“分区定制-精准施策”固碳增汇的思路，研发了潜在造林区实施高效固碳造林技术[40,41]，在天然林区实施防灾减排与分类定向管理增汇技术，在人工林区实施全周期结构调控增汇技术，在防护林区实施水分-效益协同增汇技术，针对林业剩余物实施林业生物碳捕集、利用与封存技术[42-45]。针对当前我国农业存在作物生产碳排放总量高、农田土壤碳储量低、种植制度相对单一、施用化肥过量等导致土壤固碳潜力未能充分发挥的问题，探究农作制度调整以及外源有机物料合理配施等绿色低碳种植技术，创新了以植物—土壤—微生物互作增碳提质为核心的农田固碳技术，研发了外源有机物料促进土壤固碳技术[46-48]。针对西南喀斯特区域生态-岩溶碳汇对土地利用变化和植被恢复响应敏感，土少石多、水土俱缺，植被可持续恢复困难等问题，我国学者利用喀斯特区域短期生态恢复、快速固碳的能力，研发了生态修复加速植被恢复、关键带水分养分保障功能提升和地表水体水生生物培养等人工干预技术，显著提高了该地区生态-岩溶过程相互促进的固碳增汇潜力[49-51]。针对我国草地生产力和承载力低，人工草地面积占比较小、集约化程度不够、退化严重的问题，我国学者研发了基于天然草地草畜平衡、高效集约人工种草-区域调配-缓解牧压-草地恢复-草地增汇技术，以及基于围栏禁牧、休牧、划区轮牧-自然恢复＋生态补播＋土壤改良-草地的增汇技术[52,53]。

基于调查数据，我国的科研人员采用了多种技术途径评估和认证了中国陆地生态系统碳源汇功能格局及增汇潜力。结果显示，21 世纪 20 年代中国陆地生态系统固碳速率为 10 亿～15 亿 t CO_2/a[14]，在考虑生态恢复、生态系统管理等情景下，预测到 2060 年将增加至 17 亿～19 亿 t CO_2/a[54]。在陆地各类生态系统中，森林的碳汇强度最大，占比为 68%～71%，其他生态系统（草地、农田、湿地、灌丛等）占 30% 左右。在不同的政策和气候情景下，2030 年和 2060 年，中国陆地生态系统碳汇可分别抵消碳达峰时能源 CO_2 排放量的 12%～15% 和 13%～18%。另一项研究综合不同的人为管理措施，预测了我国陆地生态系统 2020～2060 年的固碳潜力为 10.93 亿～13.18 亿 t CO_2/a（图 2）[14]。其中，由于森林碳汇的估算已包含生态系统保护和修复重大工程和新增造林，为避免重复计量不再加入。根据 2000～2010 年主要林业生态工程碳汇功能评估（2.17 亿 t CO_2/a），估计人为碳汇可贡献生态系统总碳汇潜力的 40% 左右。

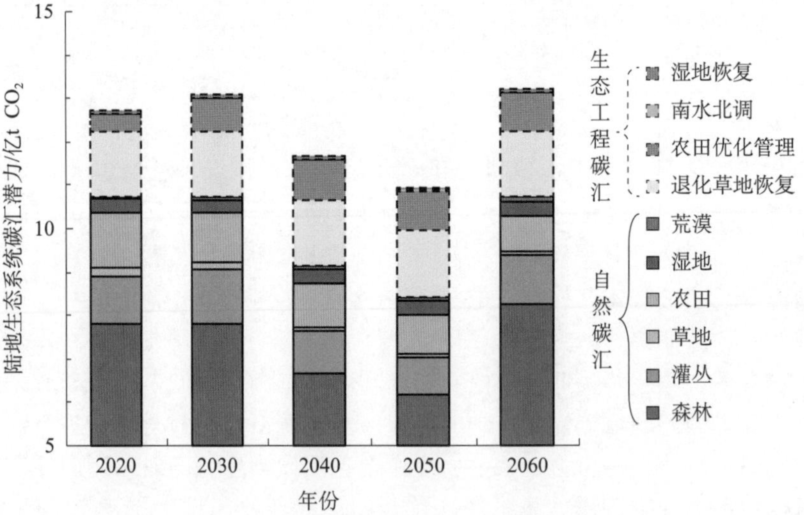

图 2　2020～2060 年我国陆地生态系统固碳潜力预测结果[14]

考虑到陆地生态系统固碳主体为森林生态系统，以下重点介绍有关森林生态系统固碳速率情况。以现有森林为基准，基于逻辑斯谛（Logistic）生长方程的森林固碳模型（forest carbon sequestration model，FCS model），评估 2010～2060 年中国森林生态系统总固碳速率变化情况（表 2）[55]。总体来看，2010～2060 年中国森林生态系统平均固碳速率为（13.12±0.59）亿 t CO_2/a，其中植被和土壤分别为（7.73±0.59）亿 t CO_2/a 和（5.39±0.18）亿 t CO_2/a，占比分别为 59% 和 41%。

表 2　2010～2060 年中国森林生态系统总固碳速率变化情况[55]

单位：亿 t CO_2/a

时间段	植被固碳速率	土壤固碳速率	森林生态系统固碳速率
2010～2015 年	6.56±0.15	1.98±0.04	8.50±0.11
2015～2020 年	7.66±0.26	3.55±0.15	11.21±0.37
2020～2025 年	8.98±0.70	4.76±0.18	13.74±0.88
2025～2030 年	8.57±0.62	5.68±0.26	14.25±0.88
2030～2035 年	8.43±1.65	6.23±0.73	14.66±2.38
2035～2040 年	7.95±0.59	6.34±0.29	14.29±0.88
2040～2045 年	8.24±0.81	7.11±0.55	15.35±1.36
2045～2050 年	6.67±2.13	6.19±1.58	12.86±3.66
2050～2055 年	7.36±2.78	6.96±1.39	14.33±4.18
2055～2060 年	4.07±0.77	5.13±0.66	9.20±1.36
平均	7.73±0.59	5.39±0.18	13.12±0.59

注：固碳速率表示为三种气候情景（RCP 2.6、RCP 4.5 和 RCP 8.5）下的平均值±标准差。

四、展　　望

陆地生态系统固碳及其演化的过程机理复杂，我们对陆地生态系统碳组分的增汇过程、稳定性、潜力和调控机制及其对全球变化的响应与反馈机制方面的认识不足，模型模拟和核算方法也还存在很大不确定性，导致目前采用不同方法估算的我国陆地生态系统碳汇数值之间相差甚大，无法用于准确评估我国陆地生态系统碳汇的潜力及其增量，也难以优先支持“双碳”目标和全球碳盘点[56,57]。因此，我国亟须开展精细化碳通量与储量时空动态特征调查研究，实现全国尺度的全生态系统碳汇动态监测和长期动态的数据积累，精细刻画我国不同类型和不同区域生态系统的碳汇能力及其动态变化，揭示气候变化和人类活动双重影响下的碳汇能力变化及其内在调控机制，从构建精细化的动态观测数据集和建立系统化的“自下而上”碳汇评估方法体系两方面共同推动和引领相关学科的研究工作，服务于中国“双碳”目标的碳核算，为国家制定“双碳”路线图提供高质量的数据支撑。

1. 陆地生态系统碳汇的重大基础科学问题

中国区域的陆地生态系统碳循环规律和特性的认知虽然取得了重要进展，但是应用于指导“双碳”目标的基础理论和方法学体系中依然存在众多不确定性和知识盲点，其核心科学基础是理解陆地生态系统碳循环过程机理及碳汇功能时空变异，以及其与全球气候变化的互馈关系。当前迫切需要回答以下几个与“双碳”目标密切相关的基础科学问题：①碳中和措施的气候效应；②自然碳汇形成与维持机制；③自然和人为碳汇的容量及增汇潜力；④陆地生态系统碳循环与气候变化的互馈机理；⑤多种温室气体间的协同效应；等等。

2. 陆地生态系统碳源汇立体监测与能力评估

及时准确的科技信息是正确决策的基础。一方面，在决策过程中需要及时而准确的信息；另一方面，分析决策行动的实行状态及其行动功效需要科学观测和评估。目前，我国多过程-多要素-多尺度协同的通量观测综合能力以及尺度上推方法研究较为薄弱，生态系统碳汇能力评估缺乏全面、翔实、可靠的地面调查资料（包括植被、土壤、气象等基础数据），尤其是大尺度精准“点对点”的碳储量连续监测数据，导致我国不同类型生态系统碳储量的估算存在很大的不确定性。为此，我国生态碳汇能力提升需要构建陆地生态系统碳循环参数的立体化和网络化的动态观测体系。通过集成生态系统碳通量、碳储量等立体观测数据，形成多源、多尺度、多途径的全国陆地生态系统碳汇核算产品。

3. 增汇技术研发与区域示范

利用生态系统固碳功能，保护植被、土壤及冰冻圈碳库的稳定性，增强自然和人为碳汇功能被认为是最经济有效且最具规模效应的碳中和技术途径。提升生态系统碳汇能力是实现碳中和的重点科技问题与战略任务。目前，已有的生态系统增汇相关技术仍然处于目标单一、零散不成体系的状态，对增汇的效果、可持续性以及与其他生态和经济目标之间的兼容性缺乏系统评估。因此，我国亟须研发、遴选生态系统增汇的单项关键技术，依托野外台站观测研究网络和试验示范平台开展技术增汇效应评估论证；基于长期监测研究形成的生态系统碳循环等理论，遴选各类生态系统关键增汇技术，研发颠覆性生态系统增汇创新技术体系，在增汇技术迭代中打造生态系统增汇与服务功能协同提升范式，为提高区域乃至全国尺度的陆地生态系统碳汇潜力提供科技支撑。

4. 陆地生态系统碳汇模拟与情景预测

准确预测未来情景下的陆地生态系统碳汇，量化全球变化背景下各类增汇措施带来的固碳潜力，对科学评价陆地生态系统在碳中和目标实现中的贡献，以及增汇途径优化提供重要数据支撑。基于多源数据、碳循环原理及其对气候变化与碳汇提升技术的响应规律，构建气候-生态-经济数据驱动的陆地生态系统碳循环模拟系统，降低陆地生态系统碳汇潜力估算与预测的不确定性，准确评估未来气候变化情景及不同管理措施下我国陆地生态系统的固碳能力和增汇潜力，为应对气候变化与实现碳中和的监测、评估、检查、决策提供数据分析平台。

参考文献

[1] IPCC. Summary for policymakers//Masson-Delmotte V, Zhai P, Pirani A, et al. Climate Change 2021: The Physical Science Basis. Contribution of Working Group I to the Sixth Assessment Report of the Intergovernmental Panel on Climate Change. Cambridge: Cambridge University Press, 2021.

[2] 中国气象局气候变化中心．中国气候变化蓝皮书 2022. 北京：科学出版社，2022.

[3] Rogelj J, Den Elzen M, Höhne N, et al. Paris Agreement climate proposals need a boost to keep warming well below 2℃. Nature, 2016, 534(7609): 631-639.

[4] UN. The race to zero emissions, and why the world depends on it. https://news.un.org/en/story/2020/12/1078612[2023-01-17].

[5] IPCC. Summary for policymakers//Shukla P R, Skea J, Slade R, et al. Climate Change 2022: Mitigation of Climate Change. Contribution of Working Group III to the Sixth Assessment Report of the Intergovernmental Panel on Climate Change. Cambridge: Cambridge University Press, 2022.

[6] WMO. Greenhouse Gas Bulletin-No. 18: The state of greenhouse gases in the atmosphere based on global observations through 2021. https://library.wmo.int/doc_num.php?explnum_id=11352[2023-04-15].

[7] 蔡兆男，成里京，李婷婷，等．碳中和目标下的若干地球系统科学和技术问题分析．中国科学院院刊，2021，36(5)：602-613.

[8] 习近平．在第七十五届联合国大会一般性辩论上的讲话．中华人民共和国国务院公报，2020，(28)：5-7.

[9] 中共中央 国务院关于完整准确全面贯彻新发展理念做好碳达峰碳中和工作的意见．中华人民共和国国务院公报，2021，(31)：33-38.

[10] 习近平．高举中国特色社会主义伟大旗帜为全面建设社会主义现代化国家而团结奋斗——在中国共产党第二十次全国代表大会上的报告．中华人民共和国国务院公报，2022，(30)：4-27.

[11] Rogelj J D, Shindell D, Jiang K, et al. Mitigation Pathways Compatible with 1.5℃ in the Context of Sustainable Development//Masson-Delmotte V, Zhai P, Pörtner H O, et al. Global Warming of

1. 5℃. An IPCC Special Report on the Impacts of Global Warming of 1. 5℃ above Pre-industrial Levels and Related Global Greenhouse Gas Emission Pathways, in the Context of Strengthening the Global Response to the Threat of Climate Change, Sustainable Development, and Efforts to Eradicate Poverty. Cambridge: Cambridge University Press, 2018: 93-174.

[12] 于贵瑞,郝天象,朱剑兴．中国碳达峰、碳中和行动方略之探讨．中国科学院院刊,2022,37(4):423-434.

[13] Friedlingstein P, O'Sullivan M, Jones M W, et al. Global carbon budget 2022. Earth System Science Data, 2022, 14(11): 4811-4900.

[14] 丁仲礼,张涛,等．碳中和:逻辑体系与技术需求．北京:科学出版社,2022.

[15] Pan Y, Birdsey R A, Fang J, et al. A large and persistent carbon sink in the world's forests. Science, 2011, 333(6045): 988-993.

[16] Janssens I A, Freibauer A, Ciais P, et al. Europe's terrestrial biosphere absorbs 7 to 12% of European anthropogenic CO_2 emissions. Science, 2003, 300(5625): 1538-1542.

[17] Dixon R K, Solomon A, Brown S, et al. Carbon pools and flux of global forest ecosystems. Science, 1994, 263(5144): 185-190.

[18] Chapin F S, Matson P A, Mooney H A. Principles of Terrestrial Ecosystem Ecology. New York: Springer, 2002.

[19] 李彦,王玉刚,唐立松．重新被"激活"的土壤无机碳研究．土壤学报,2016,53(4):845-849.

[20] 潘根兴．中国干旱性地区土壤发生性碳酸盐及其在陆地系统碳转移上的意义．南京农业大学学报,1999,22(1):51-57.

[21] Walker W S, Gorelik S R, Cook-Patton S C, et al. The global potential for increased storage of carbon on land. Proceedings of the National Academy of Sciences of the United States of America, 2022, 119(23): e2111312119.

[22] Friedlingstein P, O'sullivan M, Jones M W, et al. Global carbon budget 2020. Earth System Science Data, 2020, 12(4): 3269-3340.

[23] Fang J, Chen A, Peng C, et al. Changes in forest biomass carbon storage in China between 1949 and 1998. Science, 2001, 292(5525): 2320-2322.

[24] Piao S, Fang J, Ciais P, et al. The carbon balance of terrestrial ecosystems in China. Nature, 2009, 458(7241): 1009-1013.

[25] Fang J, Yu G, Liu L, et al. Climate change, human impacts, and carbon sequestration in China. Proceedings of the National Academy of Sciences of the United States of America, 2018, 115(16): 4015-4020.

[26] Tang X, Zhao X, Bai Y, et al. Carbon pools in China's terrestrial ecosystems: new estimates based on an intensive field survey. Proceedings of the National Academy of Sciences of the United States of America, 2018, 115(16): 4021-4026.

[27] Chen S, Wang W, Xu W, et al. Plant diversity enhances productivity and soil carbon storage.

Proceedings of the National Academy of Sciences of the United States of America,2018,115(16): 4027-4032.
[28] Tang Z,Xu W,Zhou G,et al. Patterns of plant carbon,nitrogen,and phosphorus concentration in relation to productivity in China's terrestrial ecosystems. Proceedings of the National Academy of Sciences of the United States of America,2018,115(16):4033-4038.
[29] Lu F, Hu H, Sun W, et al. Effects of national ecological restoration projects on carbon sequestration in China from 2001 to 2010. Proceedings of the National Academy of Sciences of the United States of America,2018,115(16):4039-4044.
[30] Zhao Y, Wang M, Hu S, et al. Economics-and policy-driven organic carbon input enhancement dominates soil organic carbon accumulation in Chinese croplands. Proceedings of the National Academy of Sciences of the United States of America,2018,115(16):4045-4050.
[31] Liu H, Mi Z, Lin L, et al. Shifting plant species composition in response to climate change stabilizes grassland primary production. Proceedings of the National Academy of Sciences of the United States of America,2018,115(16):4051-4056.
[32] Solomon S,Qin D,Manning M,et al. Climate Change 2007:The Physical Science Basis—Working Group I Contribution to the Fourth Assessment Report of the IPCC. Cambridge: Cambridge University Press,2007.
[33] Houghton R. Balancing the global carbon budget. Annual Review of Earth and Planetary Sciences, 2007,35(1):313-347.
[34] 何念鹏,王秋凤,刘颖慧,等. 区域尺度陆地生态系统碳增汇途径及其可行性分析. 地理科学进展,2011,30(7):788-794.
[35] 于贵瑞,王秋凤,刘迎春,等. 区域尺度陆地生态系统固碳速率和增汇潜力概念框架及其定量认证科学基础. 地理科学进展,2011,(7):771-787.
[36] Yan P, Xiao C, Xu L, et al. Biomass energy in China's terrestrial ecosystems: insights into the nation's sustainable energy supply. Renewable and Sustainable Energy Reviews, 2020, 127: 109857.
[37] Griscom B W, Adams J, Ellis P W, et al. Natural climate solutions. Proceedings of the National Academy of Sciences of the United States of America,2017,114(44):11645-11650.
[38] Griscom B W,Busch J,Cook-Patton S C,et al. National mitigation potential from natural climate solutions in the tropics. Philosophical Transactions of the Royal Society B, 2020, 375(1794): 20190126.
[39] 于贵瑞,朱剑兴,徐丽,等. 中国生态系统碳汇功能提升的技术途径:基于自然解决方案. 中国科学院院刊,2022,37(4):490-501.
[40] Tong X, Brandt M, Yue Y, et al. Forest management in southern China generates short term extensive carbon sequestration. Nature Communications,2020,11(1):1-10.
[41] Delang C O,Yuan Z. China's Grain for Green Program. Cham:Springer,2015.

［42］ Ji L, Wang Z, Wang X, et al. Forest insect pest management and forest management in China: an overview. Environmental Management, 2011, 48(6): 1107-1121.

［43］ Dai L, Li S, Zhou W, et al. Opportunities and challenges for the protection and ecological functions promotion of natural forests in China. Forest Ecology and Management, 2018, 410: 187-192.

［44］ Wang Y, Xiong W, Gampe S, et al. A water yield-oriented practical approach for multifunctional forest management and its application in dryland regions of China. Journal of the American Water Resources Association, 2015, 51(3): 689-703.

［45］ Gao J, Zhang A, Lam S K, et al. An integrated assessment of the potential of agricultural and forestry residues for energy production in China. GCB Bioenergy, 2016, 8(5): 880-893.

［46］ Xie Z, Liu G, Bei Q, et al. CO_2 mitigation potential in farmland of China by altering current organic matter amendment pattern. Science China Earth Sciences, 2010, 53(9): 1351-1357.

［47］ Ma W, Zhan Y, Chen S, et al. Organic carbon storage potential of cropland topsoils in east China: Indispensable roles of cropping systems and soil managements. Soil and Tillage Research, 2021, 211: 105052.

［48］ Berhane M, Xu M, Liang Z, et al. Effects of long-term straw return on soil organic carbon storage and sequestration rate in north China upland crops: a meta-analysis. Global Change Biology, 2020, 26(4): 2686-2701.

［49］ Xiao K, He T, Chen H, et al. Impacts of vegetation restoration strategies on soil organic carbon and nitrogen dynamics in a karst area, southwest China. Ecological Engineering, 2017, 101: 247-254.

［50］ Hu L, Li Q, Yan J, et al. Vegetation restoration facilitates belowground microbial network complexity and recalcitrant soil organic carbon storage in southwest China karst region. Science of the Total Environment, 2022, 820: 153137.

［51］ Zeng C, Liu Z, Zhao M, et al. Hydrologically-driven variations in the karst-related carbon sink fluxes: Insights from high-resolution monitoring of three karst catchments in southwest China. Journal of Hydrology, 2016, 533: 74-90.

［52］ Xiong D, Shi P, Zhang X, et al. Effects of grazing exclusion on carbon sequestration and plant diversity in grasslands of China—A meta-analysis. Ecological Engineering, 2016, 94: 647-655.

［53］ Wang S, Wilkes A, Zhang Z, et al. Management and land use change effects on soil carbon in northern China's grasslands: a synthesis. Agriculture, Ecosystems & Environment, 2011, 142(3-4): 329-340.

［54］ 杨元合，石岳，孙文娟，等．中国及全球陆地生态系统碳源汇特征及其对碳中和的贡献．中国科学：生命科学，2022，52(4)：534-574.

［55］ Cai W, He N, Li M, et al. Carbon sequestration of Chinese forests from 2010 to 2060: spatiotemporal dynamics and its regulatory strategies. Science Bulletin, 2022, 67(8): 836-843.

[56] 朴世龙,何悦,王旭辉,等. 中国陆地生态系统碳汇估算:方法、进展、展望. 中国科学:地球科学,2022,52(6):1010-1020.
[57] 傅伯杰. 新时代自然地理学发展的思考. 地理科学进展,2018,37(1):1-7.

Perspectives on the Terrestrial Ecosystem in the Context of Carbon Neutralization

Hao Tianxiang, Xu Li, Yu Guirui

China's carbon neutrality goal is a meaningful action to address global climate change and promote the great rejuvenation of the Chinese nation. China's terrestrial ecosystem carbon sink offsets about 13% of the national anthropogenic carbon emission, which is an effective, economical, and large-scale green technology approach in the decarbonization strategy. Therefore, it is logically regarded as the "ballast stone" to guarantee the achievement of carbon neutrality and the "stabilizer" for China's social and economic development. In this context, this paper first introduces the carbon cycles, carbon stock and carbon sink of the terrestrial ecosystem in China, then describes the nature-based solutions and their potential to enhance the terrestrial carbon sink. Finally, we point out four aspects that urgently need to be improved and implemented to effectively consolidate and enhance the carbon sink terrestrial ecosystems towards carbon neutrality.

1.2 21世纪生物医学的三个主要发展趋势①

吴家睿

（中国科学院生物化学与细胞生物学研究所）

20世纪中叶，DNA双螺旋的提出和遗传信息传递的中心法则的建立，标志着分子生物学的诞生，开启了在分子水平上研究生命及其活动的生命科学时代。依靠经验的传统西方医学在现代生命科学的推动下转型成功，成为一门依靠实验科学理论和技术进行疾病诊疗的现代生物医学（biomedicine）。其间不断涌现的“高技术”在人类抗击疾病的过程中扮演了重要的角色，尤其是抗生素、疫苗及化学小分子药物的研发和利用，使人类在全球范围内基本控制了传染病，甚至消灭了天花等恶性传染病。跨入21世纪，为了满足公众不断提高的健康需求，以及抗击慢性病日益增大的健康威胁，现代生物医学在“人类基因组计划”（Human Genome Project，HGP）的推动下正在进入一个新的转型时期，其中有三种主要的发展趋势值得我们关注。

一、从简单性思维的分子生物医学转变到复杂性思维的系统生物医学

20世纪中叶诞生的分子生物学为科研人员提供了这样一种基本的研究范式：利用生物学实验方法以及物理和化学技术等各种研究手段，通过在分子层次上揭示单个基因或蛋白质的结构与功能来阐明生物体的生理或病理活动。美国著名的肿瘤生物学家温伯格（R. Weinberg）对此有很好的总结：“在20世纪，生物学从传统的描述性科学转变成为一门假设驱动的实验科学。与此紧密联系的是还原论占据了统治地位，即对复杂生命系统的理解可以通过将其拆解为组成的零部件并逐个地拿出来进行研究。”[1]

生命科学的进步推动了人类对自身健康和疾病的认识，使依靠经验的传统西方医学转变成为以分子生物学知识和实验方法为基础的分子生物医学（molecular biomedicine）。在生命科学的还原论思维指导下，广为流行着“一个基因一种疾病”的“分子

① 此文曾发表于《生命科学》，2022年34卷第11期，略有修改。

病”观点，即疾病意味着某个基因或蛋白质出了问题，而治疗就是用物理、化学方法去找到并修复这种有问题的分子零件。换句话说，分子生物医学将复杂的病理现象还原为分子层次的个别生物分子之物理或化学功能异常，进而以简单化思维方式去理解疾病并给予诊治。

虽然分子生物医学在抗击传染病方面取得了显著的成绩，但是在抗击肿瘤、代谢性疾病和神经退行性疾病等慢性病方面却面临巨大的挑战。其根本原因在于，慢性病的发生发展过程涉及众多的机体内部因素和外部环境因素，以及这些因素之间存在的复杂的相互作用。显然，在简单化思维指导下的分子生物医学难以认识和处理这类复杂性疾病。温伯格曾经以“一个完整的循环：从无尽的复杂性变为简单性然后又重回复杂性”为题回顾了美国政府于 1971 年启动而以失败告终的“肿瘤战争”，并明确指出：“从事肿瘤研究的科学家见证了这个时期的疯狂转变：从最初面对无数难以理解的病理现象的困惑，到树立了还原论必胜的信念，最近几年再回到重新面对肿瘤这个疾病无尽的复杂性。”[2]

20 世纪末 21 世纪初实施的“人类基因组计划”通过“组学”（omics）整体研究策略从根本上颠覆了这种“碎片化”的科研范式。英国《自然》（*Nature*）期刊曾发表了一篇题为《阻止疾病，现在开始》（*To Thwart Disease*，*Apply Now*）的社论：“似乎在一夜之间就从一个基因、一个蛋白质、一个分子、一次研究一个，转变为所有基因、所有蛋白质、所有分子、一次研究所有。一切都按组学的规模进行。”[3]这种转变不仅仅是将生命体内的研究对象从局部转变为全局，更重要的是对生命的认知从简单性思维转变为复杂性思维。

这种复杂性思维转变之代表是，2000 年之初在生命科学领域兴起了一门交叉学科——系统生物学（systems biology），即整合经典的分子细胞生物学、新兴的各种组学，以及信息科学和数学等非生物学科的研究策略和方法，对生命复杂系统及其生理病理活动进行系统性和整体性的检测和分析。这门新兴学科很快得到了研究人员的接受和重视；《细胞》（*Cell*）期刊在 2011 年 3 月发表了整整一期介绍系统生物学的评论文章，其中一篇文章的标题就是“系统生物学：进化成为主流”（Systems Biology：Evolving into the Mainstream）。

系统生物学因其研究生命复杂系统的能力，很快就被引入医学领域，形成了系统生物医学（systems biomedicine）。美国国立卫生研究院（National Institutes of Health，NIH）在 2003 年发布的 NIH 路线图中，把采用系统生物学的方法和策略开展慢性病研究列为主要任务[4]。此外，欧盟委员会（European Commission）也专门成立了一个“系统生物医学行动协调联盟”（Coordinating Action Systems Medicine Consortium，CASyM），并在 2014 年 6 月发布了系统生物医学的研究规划——

《CASyM 路线图》。该路线图指出，“系统生物医学就是将系统生物学的方法策略应用到医学概念、研究和实践之中……这些活动的开展需要整合不同的学科，包括数学、计算机科学、数据分析、生物学，以及临床医学、伦理和社会实践”[5]。

（一）生命复杂系统的构成：从分子到细胞再到组织器官的相互作用网络

系统生物学最重要的特性之一就是关注生物体内各种元件之间的相互作用。最初的系统生物学研究主要是针对分子层面的相互作用网络，如基因转录调控网络、信号转导网络和代谢调控网络等。2005 年 3 月创立的国际上第一个系统生物学的学术刊物就取名为 *Molecular Systems Biology*。中国生物化学与分子生物学会在 2012 年 7 月成立“分子系统生物学专业委员会”也是基于这样的考虑：由于不同种类的生物分子之间的相互作用是形成生物复杂系统的基础，所以如何形成分子相互作用网络属于系统生物学的核心科学问题。

系统生物学的引入导致了人们对复杂疾病中有关生物分子作用的全新认识。例如，传统肿瘤生物学通常是“孤立”地看待基因或蛋白质产生的突变，认为单个突变可以改变其功能而导致肿瘤的发生发展。2022 年，研究人员利用自动化蛋白质相互作用检测新技术，系统地分析了肿瘤细胞中数以百万计的蛋白质相互作用，发现单个蛋白质突变会改变蛋白质之间的相互作用，进而形成新的蛋白质相互作用网络[6]。也就是说，一个氨基酸残基的改变不仅会影响突变蛋白质本身的功能，还可能产生新的相互作用界面而与其他蛋白质产生新的相互作用，从而形成基于新的蛋白质相互作用网络的功能变异或新功能。

生理或病理活动的复杂性不仅表现在分子层面，还表现在细胞层面和组织器官层面。初期的系统生物学研究技术在细胞层面存在很大的局限性。但随着单细胞 RNA 测序技术的出现，研究人员能够在细胞层面开展系统生物学方面的研究。例如，新加坡的研究人员利用单细胞 RNA 测序技术，比较了胚胎发育过程中人类肝脏细胞以及肝癌细胞的单细胞图谱，并在此基础上发现了一个既可以驱动胚胎肝脏发育又可以促进肝癌细胞免疫抑制的肿瘤-胚胎重编程生态系统[7]。

高分辨空间组学技术的建立和发展则使得研究人员能够更进一步，对三维空间里的组织器官进行系统性研究。例如，耶鲁大学研究人员发展了一种空间组学技术 DBiT-seq——将微流控芯片（microfluidic chip）和条形码（barcoding）技术与单细胞 RNA 测序技术相结合，同时完成组织切片的空间转录组和蛋白质组的测序，其空间分辨率接近单细胞分辨率[8]。又如，深圳华大生命科学研究院联合全球多家研究机构组成的时空组学联盟（The Spatio Temporal Omics Consortium，STOC）2022 年在《细胞》期刊上报告了一项全新的时空组学技术 Stereo-seq，并利用该技术分析了小鼠早期胚胎发育过程，获得了单细胞分辨率水平的小鼠器官形成的时空图谱[9]。

由此可见，当前的系统生物医学不仅有能力揭示基因和蛋白质等生物大分子间的相互作用和功能，而且可以整合生物体不同层次的数据和信息，从而能够更完整地认识人体复杂系统的运行和变化。美国国家癌症研究所（National Cancer Institute，NCI）于2020年启动了“人类肿瘤图谱网络”（The Human Tumor Atlas Network，HTAN）研究计划，拟从分子、细胞、组织器官等多个层次开展肿瘤发生发展机制的研究，并将这些多层次肿瘤生物学数据与患者的临床数据进行整合，形成完整的肿瘤知识网络[10]。

（二）生命复杂系统的运行：基于非线性与动力学的控制

在简单性思维的指导下，生物体内部的运行关系通常被视为线性的。例如，许多研究人员认为mRNA表达水平和其翻译产生的蛋白质丰度之间呈现正相关性，前者高则后者高，反之亦然。但是越来越多的证据表明，在真实的生命复杂系统中，这二者之间存在着复杂的非线性关系。美国的研究人员2020年对各种人类组织中12 000多个基因的表达水平与相应的蛋白质表达水平进行了定量的比较，发现二者的一致性并不是很高，且“组织特有的蛋白质信息能够解释遗传疾病的表型，而仅仅采用转录组信息则做不到这一点”[11]。有研究者对影响mRNA表达水平与蛋白质表达水平的关系进行了系统的总结，认为二者间的数量关系是非线性的，受到细胞内外环境变化、细胞稳态和状态变化，以及mRNA胞内时空分布等各种因素的影响，“转录水平本身在许多情况下不足以用来预测蛋白质表达水平以及解释基因型与表型的关系。因此，在不同层次获取与基因表达水平相关的高质量数据是完全理解生物学过程所必不可少的”[12]。

研究者发现，生物体内不仅各类生物分子的浓度之间存在着非线性关系，而且这些生物分子的行为和功能也有着复杂的表现形态。例如，比利时鲁汶大学的研究人员发现，催化3-磷酸甘油酸合成丝氨酸的磷酸甘油酸脱氢酶（PHGDH）在原发性乳腺癌细胞中通常表现为高表达，从而促进肿瘤细胞的增殖；但位于高度肿瘤血管化区域的肿瘤细胞则往往表现出较低的PHGDH表达水平，而低表达PHGDH能够促进整合素（integrin）$\alpha v\beta 3$的糖基化，进而导致这类乳腺癌细胞具有较强的转移能力[13]。

p53蛋白是目前已知最重要的肿瘤抑制因子，在人类50%以上的肿瘤中都发现过*p53*基因的各种突变。p53蛋白能够通过复杂的调控网络影响众多生理活动，如DNA损伤、细胞周期增殖和细胞凋亡等。南京大学王炜教授运用系统生物学的研究方法，分析了p53网络中调控基因表达调控和信号转导等的动力学机制，发现p53蛋白存在浓度周期性振荡的回路机制：当 DNA 损伤程度较轻时，p53 蛋白浓度的周期性变化可诱导短暂的细胞周期阻断，促进 DNA 修复，并使细胞在完成修复后继续存活；而当 DNA 损伤较为严重时，持续的 p53 蛋白脉冲则诱发细胞凋亡[14]。

复杂生理病理过程普遍存在着一种临界现象，即从一个相对稳定的状态，经过一个临界期后在很短的时间内快速地进入另一个相对稳定的状态。例如，肿瘤或糖尿病等复杂疾病的发生过程中存在一个临界期，在疾病发生前的临界期为可逆阶段，适当的干预可以转归到“正常状态”；但当病变的进展一旦越过临界期，就迅速到达不可逆的“疾病状态”。显然，临界期就是复杂疾病早期监测和干预的关键时间节点。为了预测这种临界期及其关键驱动因子，中国科学院分子细胞科学卓越创新中心的陈洛南研究员及其合作者建立了基于“动态网络标志物”（dynamical network biomarker，DNB）的临界预测方法，即在复杂生物动态演化或疾病发生发展过程中，存在一个可观测的 DNB，它在临界期形成一个分子之间具有强相关并强震荡的奇异分子网络。DNB 不仅可直接用于各种动态生物过程或复杂疾病发生发展的早期诊断，而且可当作复杂疾病发生发展过程的“驱动网络”和关键节点的检测标准[15]。

（三）生命复杂系统的研究：定性分析与定量检测的紧密结合

20 世纪中叶诞生的分子生物学是一门依靠物理和化学方法的实验科学。那个时代的生命科学研究者大多关注定性的研究，以发现新基因或新蛋白质及其结构和功能为主要研究目标。随着后基因组时代的到来，生命科学研究者的定量研究能力已是必需的了。正如曾担任过美国国家科学院院长的分子生物学家阿尔伯特（B. Albert）所说，“对一种蛋白质机器功能的任何一种真正的认识，不仅需要了解它在原子精度的静态结构，而且需要有关它的每个反应中间体的动力学和热力学知识”[16]。而系统生物学正是一门注重定量研究的学科，它不仅注重分子细胞生物学和组学等“湿实验”，而且注重生物信息学和计算生物学等“干实验”。成功的系统生物学研究一定是“干实验”与“湿实验”的紧密结合。

传统的肿瘤靶向治疗是直接针对肿瘤细胞中出现突变且功能异常的靶蛋白。但是，美国哥伦比亚大学系统生物学家卡里法纳（A. Califano）及其合作者认为，这类突变的蛋白质通常处于一个调控网络之中，可以找出该网络的关键调控因子来进行靶向治疗。他们采用调控网络的逆向工程思路，建立了一种算法——VIPER（virtual inference of protein activity by enriched regulon analysis），通过对肿瘤细胞的转录组数据分析去寻找肿瘤异常调控网络中的“瓶颈因子”，并提出相应的靶向治疗方案[17]。卡里法纳等在前期工作的基础上进一步发展了基于网络的新方法——“多组学的主调控因子分析”（multi-omics master-regulator analysis，MOMA），并从近 10 000 个不同肿瘤样本的多组学数据分析中找到了 407 个主调控因子[18]。

系统生物医学通常要面对海量生物分子数据的处理与分析之挑战，尤其是如何把不同种类生物分子的大数据与生物影像以及健康医疗档案等整合起来指导临床实践。2021 年，瑞士科学家提出了一种用来指导肿瘤临床治疗决策的数据整合方案——“肿瘤表型谱”（Tumor Profiling，TuPro），其工作流程大致为：通过不同的分析技术从肿瘤患者样本中获取各种类型的生物分子大数据，然后将这些数据与临床数据进行整合，进而为每位患者生成一份分子研究报告，并提交给多学科医师小组进行讨论，最终制订出一份具体的治疗方案[19]。可以说，系统生物医学的大数据整合工作不过是刚刚开始，其临床运用的可行性和可操作性还有待发展和完善。

二、从基于临床统计研究证据的循证医学转变到关注个体分子特征的精确医学

现代医学的主流是“循证医学”（evidence-based medicine），其诊治方案形成的主要依据是按照各种类型临床研究证据制定的临床指南，以此开展基于统一标准的规范化临床实践活动。临床研究证据的“金标准”是“随机对照试验”（randomized controlled trial，RCT）。这是一种尽可能排除个体差异对研究结果的统计性影响的临床试验，一方面基于临床试验统计学的要求进行试验设计和招募参试者，另一方面对参试者进行试验组和对照组的随机分配，以减少个体差异可能导致的统计学试验偏倚，从而得到具有普遍意义的统计学规律。

由此可见，基于 RCT 等各种临床统计学证据的循证医学的主要特征可以说是看“病”而不是看“人”，即患者仅仅是一个“病例”，而不是一个“病人”。循证医学超越了传统医学那种依靠个人经验的医疗实践模式，能够在科学证据的指导下进行更为客观的医疗实践活动。但是，这种排除个体差异的统计学方式同时导致了循证医学在治疗慢性病患者时的实际疗效往往因人而异，因为患者之间广泛存在着由不同的遗传背景和不同的生活环境而产生的个体差异。显然，循证医学的优点——排除了个体差异并具有统计显著性的治疗方案，对具体的患者来说却成了缺点——治疗不够精确！

“人类基因组计划”在改变循证医学这种“不精确”问题方面同样发挥了重要的作用；正如《科学》（*Science*）期刊在《庆祝基因组》社论中所说的：“基因组草图的完成为一种新的精确医学范式奠定了基础，这种精确医学的目标就是要利用个体独特的基因序列信息去指导治疗和预防疾病的决策。”[20]事实上，美国的研究人员对此很早就有清晰的认识：“与人体有关的分子数据正在爆发性地增长，尤其是那些与患者个

体相关的分子数据；由此带来了巨大的、尚未被开发的机会，即如何利用这些分子数据改善人类的健康状况。”[21] 基于这样的认识，美国政府在 2015 年初正式宣布实施“精确医学计划”（Precision Medicine）；此后，包括中国在内的各国迅速跟进，形成了世界范围的精确医学新潮流。

（一）精确医学的底层逻辑：分子层面的个性与共性之统一

精确医学把主要目标定位在从分子层面认识清楚个体间的遗传差异和表型差异，并相应地把基本任务放在寻找和确定标识个体特征的遗传因子或者表型因子等各种“生物标志物”（biomarker）。例如，欧盟在 2014 年启动的“创新药物先导项目”（Innovative Medicines Initiative 2，IMI2）中明确指出，精确医学的主要任务就是“生物标志物的发现和验证”。此外，NIH 牵头启动的国际癌症基因组项目“癌症基因组图集”（The Cancer Genome Atlas，TCGA）的目标也正是要获取分子层面的信息以进行肿瘤分子分型；截至 2018 年底，该项目已经进行了 33 种不同癌症类型 11 000 名患者的基因组测序和其他种类生物分子数据的采集与分析[22]。2022 年，英国研究人员报道了肿瘤患者样本规模最大的一项全基因组测序研究，他们通过比较 19 种癌症类型 12 222 名患者的全基因组序列，揭示出了 58 种过去未知的肿瘤基因组序列的突变特征，进而为每种癌症类型确定了常见突变特征与罕见突变特征[23]。

越来越多的研究表明，不同个体在分子层面广泛存在着个体间异质性（inter-heterogeneity）。更具有挑战性的是在个体内细胞之间也存在内在异质性（intra-heterogeneity）。例如，研究人员利用特异性结合人体胰岛 β 细胞的膜表面蛋白的抗体技术发现，正常成人胰岛组织中的 β 细胞群体中存在 4 种亚型，这些不同亚型的 β 细胞对葡萄糖的响应有着明显的差别；研究人员还发现，2 型糖尿病患者体内的这 4 种 β 细胞亚型的数量关系发生了明显的改变[24]。

在肿瘤的发生发展过程中，肿瘤细胞的内在异质性更是扮演了重要角色。中国的研究人员在 2015 年通过基因测序等技术分析了一个直径大约为 3.5 cm 的肝癌组织上基因突变情况，推断出这一肝癌组织拥有上亿个突变，且不同肝癌细胞拥有的突变类型和数量是不一样的[25]。2021 年，一支国际研究团队在《细胞》期刊上发表了当时最大规模的肿瘤细胞间异质性的研究工作——通过分析 38 种癌症的 2658 个肿瘤样本的全基因组测序数据，系统地绘制了肿瘤的异质性图谱。研究数据揭示，超过 95% 的肿瘤里都存在代表肿瘤细胞间异质性的亚克隆扩张（subclonal expansion），这些具有不同突变特征的亚克隆扩张驱动着肿瘤的演化[26]。需要强调的是，这种肿瘤细胞间异质性不仅表现在基因组序列的差异上，也表现在基因转录调控和蛋白质表达等各种分子层面上。

研究者在关注研究“个性”的同时通常也需要关注“共性”，这二者就好像一枚钱币的两面是不可分割的。目前国际学术界上采用的“人类参考基因组”是用20多个人的基因组序列拼接成的，其中有大约70%的碱基序列是来自同一个人[27]。2022年，中国、美国等多个国家的研究人员组建了“人类泛基因组参考联盟”（Human Pangenome Reference Consortium，HPRC）。“人类泛基因组”（human pangenome）这个概念，不仅是指一个更高质量和更完整的人类参考基因组，而且是指一个更完整的人类基因组变异框架，涵盖包括重复序列以及单核苷酸多态性等整个基因组范围内的变异信息[27]。换句话说，“人类泛基因组”的提出就是要在分子层面实现个性与共性的整合。

精确医学延续着同样的思路，但它并没有将研究工作局限于寻找和鉴定个体之间的分子差异，而是拓展到对不同个体在分子层面的共性研究，其中最具代表性的就是新的肿瘤类型“泛癌”（pan-caner）概念的提出。为此，在TCGA计划中专门衍生出一个“泛癌图谱计划”（Pan-Cancer Atlas Project），“泛癌图谱计划获得的结果将为下一阶段的工作打下坚实的基础，而后续这类更深入、更广泛和更复杂的工作将有助于实现个体化肿瘤治疗”[28]。这种“泛癌”研究可以超越基于病理特征和解剖形态等传统的宏观疾病分类标准，把不同组织/解剖的肿瘤类型视为一个整体。例如，研究者利用TCGA计划获得的RNA转录组数据，对33种肿瘤类型共9000个样本进行了“增强子表达”（enhancer expression）的共性分析，发现在这些“泛癌”肿瘤样本中，“基因组整体水平的增强子活性与非整倍性（aneuploidy）正相关，而与基因突变的程度则没有相关性”[29]。2019年，荷兰的研究人员从“泛转移瘤”的角度比较了20多种实体瘤的2520对转移性和原发性肿瘤样本的全基因组序列，发现这些实体转移瘤细胞的全基因组扩增（whole genome duplication，WGD）程度比非转移性瘤细胞的要高很多，前者的WGD平均值达到了55.9%，因此研究人员认为WGD是这些不同类型的转移性肿瘤之共同分子特征[30]。

2022年，美国的研究人员在《细胞》期刊上发表了一篇关于不同人脑转移瘤的细胞组成与基因调控模式的研究论文。他们从15个分别患有黑色素瘤、乳腺癌等8种类型的原发性肿瘤的患者体内获得了相应的脑转移瘤样本，对这些样本进行了单细胞转录组测序等各种分析，发现不同患者的脑转移瘤细胞具有高度的异质性，并在这些人脑转移瘤细胞里鉴定出了8种基因调控模式；此外，他们通过对脑转移肿瘤血液-肿瘤界面的研究，发现可以把这些人脑转移瘤细胞分为两种基本类型：一种是增殖型（proliferative），另一种是炎症型（inflammatory）[31]。这一工作很好地反映了精确医学是如何整合对肿瘤的个性与共性研究的。

（二）精确医学的技术路径：理想试验设计与真实世界研究

为了克服 RCT 研究过程中刻意消除个体差异导致的不精确缺点，精确医学发展出了各种理想化的基于个性差异和共性特征的新型临床研究模式。首先是基于“同病异治”思路的“伞形试验”（umbrella trial），即针对单一疾病采用多种药物治疗并评估其效果[32]，如英国目前正在进行一个“肺癌伞形试验”（national lung matrix trial，NLMT），涉及具有 22 个分子标志物的 19 种非小细胞肺癌患者队列和 8 种治疗药物[33]。其次是基于“异病同治”思路的“篮形试验”（basket trial），即按照统一的某个分子标志物把不同类型的疾病患者集中在一起，用来评估某一种药物对这些不同类型疾病的治疗效果[32]，如在 2018 年获得美国食品药品监督管理局（Food and Drug Administration，FDA）批准上市的“原肌球蛋白受体激酶”（tropomyosin receptor kinase，TRK）抑制剂拉罗替尼（Larotrectinib），就是首个依据“篮形试验”结果获批的抗肿瘤药物——不论哪种类型的实体瘤，只要有 TRK 基因融合突变，就可以用此药治疗。

对精确医学而言，理想化的临床研究应该在分子层面实现个性与共性的整合。为此，美国 FDA 提出了一种主方案（master protocols），不仅同时包括了“肺癌伞形试验”和“篮形试验”，而且包括了一种平台试验（platform trial），即在同一个研究平台上平行开展在多个不同分子标志物指导下的单臂药物试验[32]。美国 NIH 下属的国家癌症研究所（National Cancer Institute，NCI）正在开展的“基于分子分析的治疗选择试验”（molecular analysis for therapy choice trial，MATCH）是当前规模最大的一项主方案，从 6000 名肿瘤患者中选出 1000 名分别进入 30 项治疗单臂试验，参与这些试验的患者涉及几乎所有肿瘤类型。

随着大数据时代的到来，生物医学大数据也成为实现精确医学的重要手段。美国国会在 2016 年通过的《21 世纪治愈法案》中提出，日常临床实践中产生的丰富多样的真实世界数据，如电子健康档案和医保数据等构成的“真实世界证据”（real world evidence，RWE）可以作为临床试验证据之外的补充证据。美国 FDA 在 2018 年公布了《真实世界证据方案框架》。中国近年来也逐渐重视 RWE，国家药品监督管理局于 2020 年发布了《真实世界证据支持药物研发与审评的指导原则（试行）》办法；其下属的药品审评中心也在 2021 年 4 月发布了《用于产生真实世界证据的真实世界数据指导原则（试行）》。

获得 RWE 的一个主要途径是真实世界研究（real world study，RWS）。由于 RWS 源于实际医疗场地或家庭社区等场景，可以避免 RCT 那样严格受控实验条件带来的局限性，因此 RWS 也成为临床试验中新的干预型研究手段，如中国国家药品监督管理局药品审评中心在2021年发布的《以临床价值为导向的抗肿瘤药物临床研发指导原则（征求意见稿）》中，就明确给出了临床研究进入“关键研究阶段”时的3种临床试验设计：①随机对照研究；②单臂临床试验；③真实世界研究。

三、从以治病为中心的临床医学转变到以健康为中心的健康医学

随着当今疾病谱从以传染病为主转变为以慢性病为主，医学的理念和形态正在发生着巨大的变化。首先是医学的关注点从“治疗疾病”转变为“维护健康”。慢性病的发生发展通常都要有一段较长的时间；在出现临床症状之前，会先出现亚健康状态或疾病前期状态等各种过渡态，其高危人群数量往往比患病人群要大很多。例如，根据2013年的一篇文章报道：我国国内糖尿病患者数量为1亿，但处于糖尿病前期（prediabetes）的高危人群数量可能已达到5亿左右了[34]。这种疾病演化的“窗口期”给抗击慢性病提供了一个不同于抗击传染病的思路，即将抗击疾病的“关口前移”，加强对人群健康状态的早期监测，并在发现亚健康或疾病前期状态时进行早期干预。

这种“大健康”的思路今天已经上升为中国的基本国策，在2016年国家颁布的《“健康中国2030”规划纲要》中，明确提出其指导思想是：“把健康摆在优先发展的战略地位”，“实现从胎儿到生命终点的全程健康服务和健康保障”。需要指出的是，“关口前移”并非说要忽略对患者的临床诊断和治疗，而是要把维护健康和临床诊治整合为一体，形成“大健康”时代的“健康医学”。也就是说，过去的临床医学主要关注患病人群，而今天的健康医学则拓展到所有个体，正如2019年召开的第72届世界卫生大会倡导的主题——“全民健康覆盖：不遗漏任何一人”。

（一）健康状态的全过程认识：全人群与全数据

健康医学面对的挑战显然要大于临床医学，不仅要处理临床诊治方面的问题，还要解决健康促进和维护方面的问题。显然，健康医学不能按照传统的医学研究模式来选择研究对象，而是要采用把所有个体都纳入研究范围的“全人群”策略。目前NIH正在实施的“全民健康研究项目”（All of Us Research Program），就是这种健康医学“全人群”理念的代表。该项目的前身就是2015年提出的招募美国百万志愿者的“精确医学先导队列项目”。NIH的负责人特别强调：这个项目不关注疾病（disease agnostic），“它不聚焦在某一种疾病、某一种风险因子，或者是某一类人群；反之，它使得研究者可以评估涉及各种疾病的多种风险因子”[34]。因此，该项目特别重视参与者的广泛性和多样性，计划招募的全美百万志愿者不限性别、民族和健康状

态等，并且要覆盖全美各地区和各阶层，包括过去不受重视的族群（underrepresented groups）[35]。

需要指出的是，这种健康医学的“全人群”策略是要在一个“长时程”的维度上展开的。首先，要想有效地统计和分析非特定构成的自然人群中各种常见病的患病率和发病率，不仅需要人群的数量足够大，而且需要连续地对人群进行观察。美国研究者认为，只要美国的自然人群队列人数达到或超过 100 万，在 5～10 年内检测到的多种美国人常见病（如糖尿病、卒中、各种类型肿瘤等）的平均发病数量将超过 2 万例，这些疾病通常伴随着显著的致死致残率[36]。其次，对自然人群队列进行观察和数据收集的时间越长对健康医学就越有价值。例如，在“全民健康研究”项目中，至少要有 3 年的自然人群队列数据才能用于疾病的分类或支撑临床研究，如果有 5 年的自然人群队列数据则效果将更好[37]。

显然，“全人群”研究的目标是获得尽可能完整的人群的健康医学大数据，可以称之为“全数据”，即人群的健康数据越完整、越全面，对健康医学就越有价值。目前在健康医学“全数据”方面最成功的是“英国生物银行”（UK Biobank）项目。英国于 2006 年启动了 UK Biobank 项目，在随后的 5 年时间里收集了 50 万名 40～69 岁英国志愿者的血液、尿液和唾液等生物学样本，以及电子健康档案（EHR）等各种个人信息[38]。要强调的是，UK Biobank 追求的目标正是健康领域的“全数据”。自 2012 年建成至今，UK Biobank 一直在完善其数据的收集工作。例如，初期对 50 万志愿者进行了全外显子测序，2021 年在政府和公司的资助下完成了这些志愿者中 20 万人的全基因组测序，未来将完成其余 30 万人的全基因组测序；此外，在 2022 年 1 月完成了 5 万人的器官成像，下一步将扩大到 10 万人。UK Biobank 的数据量预计在 2025 年将达到 40 PB（1 PB＝10^{15} byte）。由于 UK Biobank 具有这样的健康医学“全数据”，所以被全世界的研究者用于健康医学领域各种各样问题的研究。截至 2022 年，UK Biobank 拥有包括中国和其他国家在内的全球注册用户 28 000 多名，基于这些数据已发表了 6000 多篇研究论文。

（二）健康状态的全过程管理：全方位的生活方式干预

健康医学的基本形态不同于临床医学，前者是人人参与的“主动健康”，后者则主要是依靠医生的“被动健康”。对临床医学而言，大众把诊治自身疾病的职责交给了医生。但是，健康医学面对维护和管理公众全程健康的需求，依靠医生是远远不够的，而是需要每一个人的参与。这种“主动健康”理念已经被纳入 2020 年实施的《中华人民共和国基本医疗卫生与健康促进法》。该法第六十九条清楚地写道：“公民是自己健康的第一责任人，树立和践行对自己健康负责的健康管理理念，主动学习健康知识，提高健康素养，加强健康管理。”

实施健康医学“关口前移”战略的主要举措是早期干预，其中最常用又最简便的是全方位的生活方式干预，包括营养干预、运动干预、心理干预和睡眠调理等。例如，美国癌症协会（American Cancer Society）在2018年提出了未来10年的10项预防肿瘤的措施，其中6项都是生活方式干预，包括戒烟、限酒、健康饮食、防晒、加强运动、控制体重[39]。中国在2019年6月颁布的《国务院关于实施健康中国行动的意见》中也提出了多项具体的生活方式干预行动，如合理膳食行动、全民健身行动、控烟行动和心理健康促进行动等[40]。

饮食对健康的影响从古至今都受到高度重视，中国传统医学很早就提出“药食同源”的观点。美国癌症协会2016年2月在其网站上推荐了一份用于预防肿瘤的“彩虹食谱”：把蔬果按颜色分成5个种类，不同颜色代表不同的植物营养素；人们每天要将不同颜色的蔬果按一定比例搭配进食。此外，许多国家也很重视发布官方的膳食指南，用于指导民众健康饮食，例如，美国营养学家和相关专家组成膳食指南顾问委员会负责颁布《美国居民膳食指南》、中国营养学会负责颁布《中国居民膳食指南》。

随着健康医学的形成和发展，运动干预近年来越来越受到重视。2018年第71届世界卫生大会通过《2018—2030年促进身体活动全球行动计划：加强身体活动，造就健康世界》(Global Action Plan on Physical Activity 2018-2030：More Active People for A Healthier World)，希望到2030年时将缺乏身体活动现象减少15%。2020年世界卫生组织发布了《关于身体活动和久坐行为指南》(WHO Guidelines on Physical Activity and Sedentary Behaviour)，建议所有成年人每周进行150～300分钟中等到剧烈的有氧运动。中国国家体育总局在2017年发布了《全民健身指南》，明确指出：“体育活动已经成为增强国民体质、提高健康水平最积极、最有效、最经济的生活方式。”

生活方式干预不仅被用于预防慢性病的发生，而且还正在被视为治疗疾病的重要手段。2021年，美国心脏协会（American Heart Association）发布了一份声明，建议把运动锻炼作为降低血压和血脂的首选干预措施[41]。2022年初，美国纽约市政府宣布，把生活方式干预纳入大纽约地区的医疗卫生系统，作为针对代谢性疾病等慢性病的一线治疗手段。我国政府和医学界同样很重视生活方式干预的治疗价值，如《中国2型糖尿病防治指南（2017年版）》明确规定，运动和饮食等单纯生活方式干预是血糖控制的首选方法，“生活方式干预是2型糖尿病的基础治疗措施，应贯穿于糖尿病治疗的始终”[42]。由此可以看到，健康医学在疾病治疗手段上形成了“三足鼎立”——药物、手术刀、生活方式干预。

现代医学在科学技术的推动下正在出现新的发展趋势。从以上的分析可以看到主要表现在三个方面（图1）。首先，系统生物医学的形成和发展改变了基于还原论的“碎片化”和“简单化”的分子生物医学，使得人们从生命复杂系统的角度去重新认识生命的生理和病理活动。其次，在系统生物学和生物医学大数据的共同推动下，人们从注重临床研究的统计学证据之循证医学转变成关注个体在分子层面的差异和共性之精确医学。最后，也是最重要的是，对健康的维护不再局限于临床诊断和治疗，而是对生命从正常到异常到临床的全过程监测，以及从营养到运动到治疗的全方位干预。人类社会迎来了一个全新的以健康为中心的健康医学。

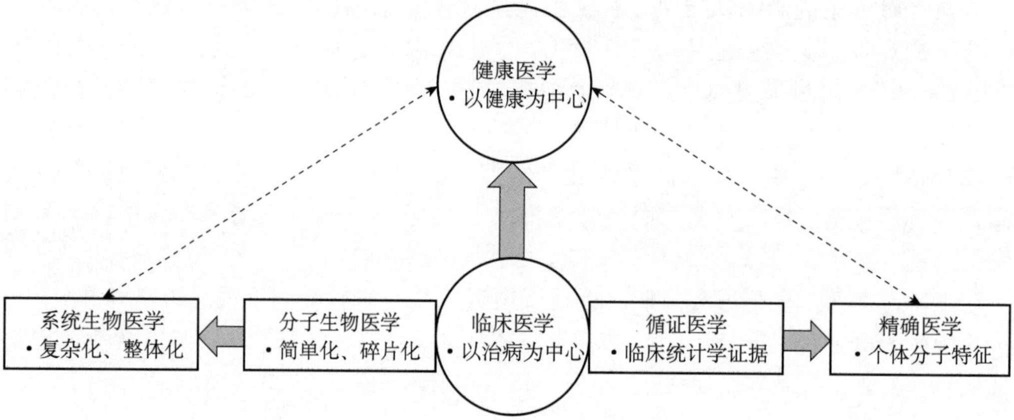

图1　21世纪生物医学发展的三个主要趋势

参考文献

[1] Weinberg R. Point: hypotheses first. Nature, 2010, 464(7289): 678.

[2] Weinberg R. Coming full circle—From endless complexity to simplicity and back again. Cell, 2014, 157(1): 267-271.

[3] Editorial. To thwart disease, apply now. Nature, 2008, 453(7197): 823.

[4] National Institutes of Health. The NIH roadmap. https://www.era.nih.gov/docs/Roadmap_SCCRI_slides_CWG_04-30-03.pdf[2003-04-30].

[5] The CASyM Consortium. The CASyM roadmap: Implementation of systems medicine across Europe. http://www.casym.eu/publications.

[6] Mo X, Niu Q, Ivanov A A, et al. Systematic discovery of mutation-directed neo-protein-protein interactions in cancer. Cell, 2022, 185(11): 1974-1985.

[7] Sharma A, Seow J W, Dutert C A, et al. Onco-fetal reprogramming of endothelial cells drives immuncosuppressive macrophages in hepatocellular carcinoma. Cell, 2020, 183(2): 377-394.

[8] Liu Y, Yang M, Deng Y, et al. High-spatial-resolution multi-omics sequencing via deterministic barcoding in tissue. Cell, 2020, 183(6): 1665-1681.

[9] Chen A, Liao S, Cheng M, et al. Spatiotemporal transcriptomic atlas of mouse organogenesis using DNA nanoball-patterned arrays. Cell, 2022, 185(10): 1777-1792.

[10] Rozenblatt-Rosen O, Regev A, Oberdoerffer P, et al. The human tumor atlas network: charting tumor transitions across space and time at single-cell resolution. Cell, 2020, 181(2): 236-249.

[11] Jiang L, Wang M, Lin S, et al. A quantitative proteome map of the human body. Cell, 2020, 183(1): 1-15.

[12] Liu Y, Beyer A, Aebersold R. On the dependency of cellular protein levels on mRNA abundance. Cell, 2016, 165(3): 535-550.

[13] Rossi M, Altea-Manzano P, Demicco M, et al. PHGDH heterogeneity potentiates cancer cell dissemination and metastasis. Nature, 2022, 605(7911): 747-753.

[14] Zhang X P, Liu F, Wang W. Two-phase dynamics of p53 in the DNA damage response. Proceedings of the National Academy of Sciences of the United States of America, 2011, 108(22): 8990-8995.

[15] Chen L, Liu R, Liu Z P, et al. Detecting early-warning signals for sudden deterioration of complex diseases by dynamical network biomarkers. Scientific Reports, 2012, 2: 342.

[16] Alberts B. The cell as a collection of protein machines: preparing the next generation of molecular biologists. Cell, 1998, 92(3) 291-294.

[17] Califano A, Alvarez J M. The Recurrent architecture of tumour initiation, progression and drug sensitivity. Nature Reviews Cancer, 2017, 17(2): 116-130.

[18] Paull E O, Aytes A, Jones S J, et al. A modular master regulator landscape controls cancer transcriptional identity. Cell, 2021, 184(2): 334-351.

[19] Irmisch A, Bonilla X, Chevrier S, et al. The Tumor Profiler Study: integrated, multi-omic, functional tumor profiling for clinical decision support. Cancer Cell, 2021, 39(3): 1-6.

[20] Fraser C M. A genome to celebrate. Science, 2021, 371(6529): 545.

[21] National Research Council. Toward precision medicine: building a knowledge network for biomedical research and a new taxonomy of disease. http://www. nap. edu/catalog/13284/.

[22] Hutter C, Zenklusen J C. The cancer genome atlas: creating lasting value beyond its data. Cell, 2018, 173(2): 283-285.

[23] Degasperi A, Zou X, Amarante T D, et al. Substitution mutational signatures in whole-genome-sequenced cancers in the UK population. Science, 2022, 376(6591): eabl9283.

[24] Dorrell C, Schug J, Canaday P S, et al. Human islets contain four distinct subtypes of β cells. Nature Communications, 2016, 7: 11756.

[25] Ling S, Hu Z, Yang Z, et al. Extremely high genetic diversity in a single tumor points to prevalence of non-Darwinian cell evolution. Proceedings of the National Academy of Sciences of the United

States of America,2015,112(47):E6496-E6505.

[26] Dentro S C,Leshchiner I,Haase K,et al. Characterizing genetic intra-tumor heterogeneity across 2,658 human cancer genomes. Cell,2021,184(8):2239-2254.

[27] Wang T,Antonacci-Fulton L,Howe K,et al. The human pangenome project:a global resource to map genomic diversity. Nature,2022,604(7906):437-445.

[28] Ding L,Bailey M H,Porta-Pardo E,et al. Perspective on oncogenic processes at the end of the beginning of cancer genomics. Cell,2018,173(2):305-320.

[29] Chen H,Li C,Peng X,et al. A Pan-Cancer analysis of enhancer expression in nearly 9000 patient samples. Cell,2018,173(2):386-399.

[30] Priestley P,Baber J,Lolkema M P,et al. Pan-cancer whole-genome analyses of metastatic solid tumours. Nature,2019,575(7781):210-216.

[31] Gonzalez H,Mei W,Robles I,et al. Cellular architecture of human brain metastases. Cell,2022,185(4):729-745.

[32] Woodcock J,LaVange L M. Master protocols to study multiple therapies,multiple diseases,or both. The New England Journal of Medicine,2017,377(1):62-70.

[33] Middleton G,Fletcher P,Popat S,et al. The National Lung Matrix Trial of personalized therapy in lung cancer. Nature,2020,583:807-812.

[34] Xu Y,Wang L,He J,et al. Prevalence and control of diabetes in Chinese adults. JAMA,2013,310(9):948-958.

[35] NIH. NIH-Wide Strategic Plan for Fiscal Years 2021-2025. https://www.nih.gov/sites/default/files/about-nih/strategic-plan-fy2021-2025-508.pdf[2021-07-31].

[36] Precision Medicine Initiative(PMI) Working Group. The Precision medicine initiative cohort program-building a research foundation for 21st century medicine. https://www.nih.gov/sites/default/files/research-training/initiatives/pmi/pmi-working-group-report-20150917-2.pdf[2015-09-17].

[37] The All of Us Research Program Investigators. The "All of Us" research program. New England Journal of Medicine,2019,381(4):668-676.

[38] Bycroft C,Freeman C,Petkova D,et al. The UK Biobank resource with deep phenotyping and genomic data. Nature,2018,562(7726):203-209.

[39] Siegel R L,Jemal A,Wender R C,et al. An assessment of progress in cancer control. CA:A Cancer Journal for Clinicians,2018,68(5):329-339.

[40] 国务院．关于实施健康中国行动的意见．http://www.gov.cn/zhengce/content/2019-07/15/content_5409492.htm[2019-07-15].

[41] Barone G B,Hivert M F,Jerome G J,et al. Physical activity as a critical component of first-line treatment for elevated blood pressure or cholesterol:who,what,and how?:A scientific statement from the American Heart Association. Hypertension 2021,78(2):e26-e37.

[42] 中华医学会糖尿病学分会．中国2型糖尿病防治指南(2017年版). 中华糖尿病杂志,2018,10:4-67.

Three Main Trends for Biomedicine in 21st Century

Wu Jiarui

In the 21st century, three major trends have emerged from the classical biomedicine. First, the molecular biomedicine, in which biological organisms are viewed as simple systems, has been transformed into systems biomedicine, in which biological organisms are recognized as the complex systems. Second, evidence-based medicine, which is based on clinical statistical data, has been transformed into precision medicine, which focuses on personal characteristics and variations. Third, clinical medicine, which is oriented to disease treatment has been transformed into health medicine, which is oriented to healthcare that includes health-supporting and disease-treatments.

第二章

科学前沿

Frontiers in Sciences

2.1 伽马射线天文学的新时代

曹 臻[1,2,3]

（1. 中国科学院高能物理研究所粒子天体物理重点实验室；
2. 中国科学院大学物理科学学院；3. 天府宇宙线研究中心）

一、引 言

顾名思义，伽马射线天文学①是天文学的一个分支，其研究波段的范围就是伽马射线所覆盖的波长范围。其实，由于明显的粒子特征，在电磁“波”谱上光子能量大于 100 keV 的这个能量范围。人们更习惯使用粒子物理的语言来描述相关的天文观测，原因是这些伽马光子的探测技术就直接来自实验高能物理，而且能量范围非常宽广，直到拍电子伏特（PeV $=10^{15}$ eV）。以我国的“高海拔宇宙线观测站”（LHAASO）在 2021 年发现的最高能量光子（1.4 PeV）为例，能量跨越 10 个量级的范围，占据了整个电磁波段（22 个量级）的近一半，显然不可能用一种统一的探测手段来覆盖。事实上，能量低于 10 GeV（GeV$=10^9$eV）的伽马射线尚可以在卫星探测器上完成测量，而更高能量的伽马光子，就只能借助宇宙线探测技术（即广延空气簇射探测技术），在地面建设大型实验装置才能实现对其探测，并覆盖了能量最高的 5～6 个量级的能量范围，对于单一的探测技术来说也太过于宽广，需要采用截然不同的广延空气簇射探测技术。例如，过去 30 年里取得丰硕成果的切伦科夫望远镜探测技术，就成功地覆盖了直到 10 000 GeV 的伽马光子能区；在更高能量的波段就需要置于高海拔地面的粒子探测器阵列技术，即 LHAASO 采用的技术，才能实现突破性的科学发现。这也就是本文将重点介绍的内容。

① 伽马射线天文学（gamma-ray astronomy），也称作伽马天文学（gamma astronomy），中外论文中这两种叫法混用。

二、伽马射线天文学

1. 伽马射线天文学的两个重要的发展阶段

1967 年 OSO-3① 首次测量到弥散伽马射线，标志着伽马射线天文学的发端。经过超过 40 年的稳定发展，2010 年费米伽马射线卫星大视场望远镜（Fermi-LAT）发布第一个伽马射线源星表，将伽马射线源的数量推高至 1000 个以上，伽马射线天文学进入到鼎盛时期；又经过 1 年多的观测积累，至 2022 年，Fermi-LAT 发布的星表[1]已经包含 6600 多个源，其中不仅包含各种稳定辐射的源，如超新星遗迹等，还包括大量瞬变的源，如脉冲星、活动星系核（active galactic nucleus，AGN）耀发、伽马射线暴（γ-ray burster，GRB）等，甚至包括如费米泡（Fermi bubble）这样扩展到整个银河系尺度的巨大结构。伽马射线天文学产生的众多的科学成果，极大地丰富了人类对宇宙的认知。这些研究也让人们认识到，源的伽马光谱通常都具有一个负幂律函数形式，这也是伽马射线非热辐射的典型特征，谱指数通常介于 2～3，这意味着随能量每升高一个量级，光子流强就要下降到原来的 1/1000～1/100，也就对探测更高能量的伽马光子提出了探测灵敏面积最直接的要求，而卫星探测器的接收面积绝对不可能无限增大！

1989 年，美国 Whipple 望远镜实验采用大气切伦科夫望远镜技术探测到来自蟹状星云（Crab Nebula）的伽马光子[2]，标志着地面探测技术主导的伽马射线天文学时代的到来。大气切伦科夫望远镜阵列的有效探测面积达到 10 000 m^2 的水平，使得人类打开了 1 TeV（TeV=10^{12} eV）的光谱波段——后来人们习惯上称之为甚高能（very high energy，VHE）伽马射线天文学。但是由于大气切伦科夫望远镜阵列有效接收面积的限制，经过 30 多年的观测积累，这种技术将对甚高能伽马射线波段能量探测的上限定格在 100 TeV 以下。在这期间，不但发现了 200 多个甚高能伽马射线源[3]，更重要的是，发现在产生 X 射线和伽马射线的辐射机制——同步辐射之外，还存在更高能量过程主导的光谱分布结构，确立了伽马光子的逆康普顿散射过程辐射机制，并在脉冲星风云（pulsar wind nebula，PWN）等大量源天体的观测中得到证实，有力地推动了多波段天文学（multi-wavelength astronomy）的蓬勃发展。例如，对蟹状星云脉冲星风云的长期精确测量，实现了对覆盖 22 个量级的全电磁波段的光谱（从射电

① OSO 为 orbiting solar observatory 的缩写，中文意思为“轨道太阳观测站”，是指 1962～1975 年美国发射的一组研究太阳和太阳活动的人造卫星，共 8 颗。

波到 100 TeV 伽马光子）近乎完美的物理解释，堪称天体物理研究中的经典之作，如图 1 所示。

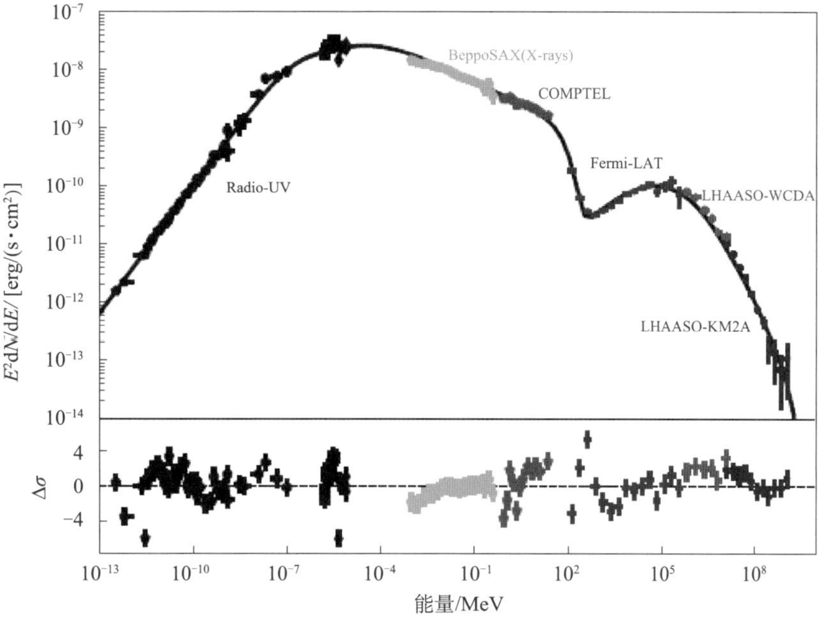

图 1　蟹状星云全波段电磁波能谱分布图[4]

图中所用数据来自射电、红外线、可见光、紫外线的传统天文观测（标注为 Radio-UV，数据点是黑色），以及 X 射线（BeppoSAX 和 COMPTEL 两个卫星探测器，数据点分别是黄色与红色）、伽马射线（Fermi-LAT 空间望远镜实验，数据点是蓝色）和甚高能/超高能（UHE）伽马射线（LHAASO 实验，数据点是浅红色与紫色）各个波段的实验。实线表示的是一个简单的单区电子分布及其辐射的模型计算结果。

2. 高海拔空气簇射探测阵列技术的发展与崛起

几乎与美国 Whipple 望远镜实验同时并行发展的另一种伽马射线探测技术是地面上排布的粒子探测器阵列，以美国犹他州的 CASA-MIA 实验、洛斯阿拉莫斯的 MILAGRO 实验，以及位于我国西藏羊八井的 ASγ 和 ARGO-YBJ 实验为代表，经过了 20 年左右的摸索和积累，人们对这种探测技术在高海拔站址发挥的优势和利用空气簇射中的缪子含量进行光子信号鉴别的能力有了充分的理解。进入 21 世纪，上一代探测器的灵敏度逐渐趋于饱和，欧美国家均提出了各自的换代计划。美国采取了提高站址海拔高度的升级计划（HAWC）；欧洲则提出了宏伟的切伦科夫望远镜阵列（CTA）计划，用超过 100 台望远镜布成阵列，在增大有效面积 10 倍的同时还扩大了能量覆盖范围。我国的计划是建设功能齐全的大型粒子探测器阵列，充分发挥我国拥有世界屋脊上优良高海拔站址和丰富的高山观测经验的优势，在认真总结上一代探测器的经验和教训后，扬长避短，设计出 LHAASO[5]。其具有超群的超高能光子探测灵敏度，覆盖整个甚高能和超高能能量范围，并具有最高巡天灵敏度的扫描望远镜，还具有多变量精确测量宇宙线成分和能谱的综合探测能力[6]。借助于我国强大的基础设施建设能力和配套齐全的现代工业体系，在短短的 4 年里就建成了整个探测装置，虽然比 CTA 晚 5 年提出计划，但在 CTA 仅实现了不到 10% 的设计规模时，LHAASO 就率先投入满负荷科学运行[7]，被业内科学家称为“已经投入观测运行了的未来探测器”。在他们心目中，LHAASO 应该是 CTA 之后的下一代探测器，这一点，在伽马射线天文学领域内所有探测器的灵敏度随能量的分布图[8]（图 2）上也可以清晰地呈现出来。LHAASO 成为国际领先的伽马射线天文学实验装置，不仅使我国在该领域实现跨越式发展，也是地面粒子探测技术在该领域胜出的标志，这种探测器的宽广视场与高灵敏度的结合是史无前例的，正在引发甚高能/超高能伽马射线天文学的革命，就像历史上每一次探测技术上的飞跃总会带来突破性的认知进步一样。

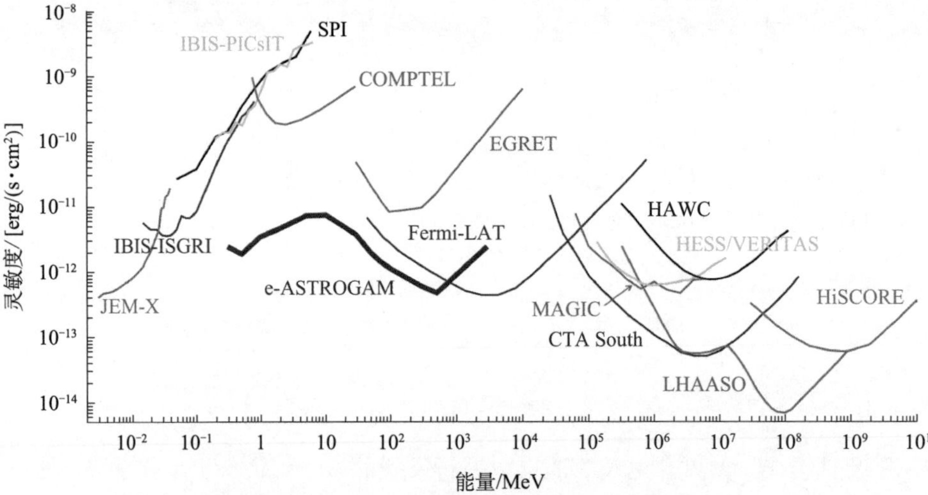

图 2　不同波段上各探测器的伽马射线探测灵敏度[8]

图中列举了现代 X 射线望远镜 JEM-X、IBIS-ISGRI 和上一代较小的 SPI、IBIS-PICsIT 和 COMTEL。可以看到文献［8］介绍的 e-ASTROGAM 是一个雄心勃勃的计划，要覆盖整个硬 X 射线和低能伽马射线波段，这是一个非常重要的波段。以美国主导的上一代 EGRET 和它的后继者 Fermi-LAT 在这个能区有大量和重要的科学发现，对伽马天文产生了重大的影响。在更高的能区，以德国为主的 MAGIC 和 HESS 地面切伦科夫望远镜阵列，与以美国为主的 VERITAS 成为甚高能伽马射线天文学研究中上一代的主力设备，CTA-South 是一个雄心勃勃的下一代望远镜阵列计划，但进度非常缓慢。地面粒子探测技术的上一代实验这里只列出了以美国为主的 HAWC。它完全被已经建成的 LHAASO 所覆盖。在几十 TeV 以上的能区，LHAASO 占据了绝对的领先优势。

三、LHAASO 开启伽马射线天文学新时代

1. 发现“新大陆”

首先是 LHAASO 的高灵敏度引起的突破。LHAASO 在建成一半时，便投入了观测运行，仅 11 个月的观测期，就发现了银河系内 12 个超高能伽马光源[9]，一共记录到 534 个超高能光子，而且光子的最高能量达到前所未有的 1.4 PeV，无论它们是由电子产生，还是由质子产生，都揭示了银河系内存在大量能把粒子加速到 PeV 以上能量的超级加速器，成为一类全新的天体，也就是拍电子伏特宇宙线加速器（PeVatron）。初步的分析发现，PeVatron 的对应天体并不仅仅包括传统上猜测的超新星遗迹一类，也可能是各种其他的天体，如年轻巨质量星团等。这一发现为理解天体物理中粒子加速问题和伽马射线的辐射机制开辟了一个新的分支学科，即“超高能伽马射线天文学”。

2. 进入“无人区”

其次是 LHAASO 的光谱精确测量能力引起的突破。如图 1 中标出的 LHAASO 测量结果，不但覆盖了位于光谱最高端 3.5 个量级的能量范围，更重要的是进入了 0.1 PeV 以上这个此前没有被系统性探索过的超高能区域，这个区域曾经被大家认为是伽马射线不存在的“禁区”——伽马射线光谱“应该”在 0.1 PeV 之前就截断了，原因是理论上认为银河系内的粒子加速源，没有能力把能量提升到 1 PeV 以上。一旦在“禁区”内发现了伽马射线，自然就与貌似非常成功的模型之间存在差别，即使这些差别可能很细微，却往往会引发突破性的发现。例如，对蟹状星云的精确测量，发现这个经典的电子加速源天体模型就不能精确预测观测结果，这种差异也许暗示了蟹状星云可能不仅仅发射出电子，也可能发射出重子宇宙线粒子，从而成为超过“膝”能量的宇宙线粒子的重要候选源天体[10]。通过 LHAASO 今后几年的观测积累，有望回答这一非常重要的问题，这是人类第一次尝试发现“膝”能量以上宇宙线的河内天体起源。

3. 逼近“新极限”

超过 1 PeV 的光子或者其 2.3 PeV 的“父辈”电子，比人类能够在实验室里制备的电子能量高了几万倍，这种极端条件是人工实验室不可能实现的。对它们的观测，可以用来检验基本物理规律在极端条件下是否成立。也可以用来探寻新的甚至是未知

的粒子加速机制，即粒子获得能量变成高能粒子的机制。2021 年的分析表明，LHAASO 的最初观测结果，已经刷新了物理学中最基本的洛伦兹协变性适用范围的高能上限[11]。

4. 开拓“新视野”

LHAASO 的宽广视场，即对伽马射线源邻近区域的暗弱信号的检测能力，也正在引起人们对伽马射线源的认知上的革命性变化。观测表明，LHAASO 不但能够发现明亮的点状伽马射线源，还能测量到源周围的伽马射线“晕”，而这可能是源中心的强大粒子加速器向外发射宇宙线的证据。未来几年，LHAASO 的观测有可能证明这是宇宙线源的标准图像。这是上一代伽马天文窄视场望远镜系统不可能实现的。

5. 新问题驱动新探索

LHAASO 的建成并投入科学运行，以及初步观测的结果，已经逐渐拉开了伽马射线天文学一个全新时代即超高能伽马射线天文学的序幕，大量新的现象不断涌现，刷新了人类对宇宙的认知，尤其是对银河系内天体的高能特征、辐射机制的认知，进而提升了我们对高能天体的演化、银河内气体或尘埃等物质分布的了解。而且，LHAASO 的这些新发现，仅仅是其设计能力几十分之一的积分灵敏度所实现的，因此完全可以预期会有更多、更新奇的现象正等待我们去发现。现在，LHAASO 探测器保持 95% 以上的有效观测时间，运行十分稳定，未来可期。LHAASO 发现的新现象，已经对未来的实验提出了明确的课题，指明了未来研究的方向，甚至于定量设定了未来探测装置的灵敏度需求，例如对未来的超高能中微子探测装置的灵敏度必须达到至少 30 km^3 的水平，才有可能探测到 PeVatron 发出的中微子信号，等等。

四、多学科交叉扩展

LHAASO 不仅仅是一个伽马射线天文望远镜，同时也是精确测量宇宙线各种分布的综合性探测装置，在很多方面具有前所未有的性能。在技术方面，LHAASO 采用了以硅光电管为像素的相机，实现了切伦科夫望远镜阵列的“月盲”观测[12]，与传统的望远镜相比，成倍地提高了观测时间，大大提高了观测效率。在物理研究方面，LHAASO 凭借强大的综合观测能力，在宇宙线空气簇射的地面测量中，首次实现单一宇宙线成分的分离，精确测量其能谱的细致结构，尤其是“膝”的结构。这种实验方法，实际上是对正确测量粒子能谱的回归，是超高能宇宙线研究进入精细测量时代的标志，而这种深刻的转变，在 100 TeV 以下的低能区，已经在 AMS-02、DAMPE

等实验中完成。

将 LHAASO 用于太阳活动的监控，可以提供全新的空间天气预报手段和方法。初步的数据分析显示，其灵敏度可以支撑 1.8 天提前量的大型磁暴预报能力。

五、结　语

伽马射线天文学经历了半个多世纪的发展，仍然是极具活力的研究领域，探测手段及规模呈现清晰的代系更迭，标志着人类探索宇宙不懈的努力。最新一代探测装置 LHAASO 给我们带来了伽马射线天文学研究中革命性的变化，发现“新大陆”、进入“无人区”之后，随之而来的是逼近“新极限”并开拓“新视野”，这将给我们带来观念的转变，以稳健、缜密的思维与方法探索新问题，提升我们对宇宙的认知，为人类的宇宙观增添新的中国元素，加入新的中国思想。

参考文献

[1] Abdollahi S, Acero F, Baldini L, et al. Incremental Fermi large area telescope fourth source catalog. Astrophysical Journal Supplement Series, 2020, 260: 53-77.

[2] Weekes T C, Cawley M F, Fegan D J, et al. Observation of TeV gamma rays from the Crab Nebula using the atmospheric Cerenkov imaging technique. The Astrophysical Journal Letters, 1989, 342: 379.

[3] Wakely S, Horan D. TeV 伽马射线源表. http://tevcat.uchicago.edu.

[4] Nie L, Liu Y, Jiang Z J, et al. Ultra-high-energy gamma-ray radiation from the Crab Pulsar Wind Nebula. The Astrophysical Journal Letters, 2022, 924: 42.

[5] Cao Z, Bi X J, Chan J F, et al. A future project at Tibet: The Large High Altitude Air Shower Observatory (LHAASO). Chin Phys C, 2010, 34: 249-252.

[6] He H H, Collaboration F T L. Design of the LHAASO detectors. Radiat Detect Technol Methods, 2018, 2: 7.

[7] Cao Z. An ultra-high-energy gamma-ray telescope at 4,410 m. Nat Astron, 2021, 5: 849.

[8] de Angelis A, Tatischeff V, Grenier I A, et al. Science with e-ASTROGAM: a space mission for MeV-GeV gamma-ray astrophysics. Journal of High Energy Astrophysics, 2018, 19: 1-106.

[9] Cao Z, Aharonian F A, An Q, et al. Ultrahigh-energy photons up to 1.4 petaelectronvolts from 12 γ-ray Galactic sources. Nature, 2021, 594: 33-36.

[10] Cao Z, Aharonian F, An Q, et al. (LHAASO Coll.), Peta-electron volt gamma-ray emission from the Crab Nebula. Science, 373: 425-430(2021).

[11] Cao Z, Aharonian F, An Q, et al. Exploring Lorentz invariance violation from ultra-high-energy gamma rays observed by LHAASO. Phys Rev Lett, 2021, 126: 051102.

[12] Aharonian F, An Q, Axikegu, et al. Construction and on-site performance of the LHAASO WFCTA camera. Eur Phys J C,2021,81:657.

The New Era of γ-ray Astronomy

Cao Zhen

Over the past 55 years, γ-ray astronomy has evolved from the space borne detectors to ground-based extensive air-shower detection, covering 10 orders of magnitude of energy. With the joining of LHAASO, the coverage is extended beyond 0.1 PeV, and the new era of ultra-high energy γ-ray astronomy is opened up. The initial observations of LHAASO unveiled the existence of sources in our Galaxy with strong radiation of γ-rays above 0.1 PeV, triggering a revolution in γ-ray astronomy. LHAASO enters the new territory where we have discovered the sources of extreme astrophysical processes and have explored the origins of cosmic rays. By updating the record of the highest energy of photons, we have also pushed up the limit by which the fundamental rules are still tested to be correct. We are also motivated to explore new field of astrophysics by the new phenomena found by LHAASO.

2.2 21世纪物理的“乌云”
——“缪子反常磁矩疑难”迎来曙光

李 亮

（上海交通大学）

在1899年岁末，也就是19世纪的最后一天，欧洲著名的科学家欢聚一堂，共贺新年。会上，英国著名物理学家威廉·汤姆生（开尔文男爵）在回顾物理学所取得的伟大成就时说，“十九世纪的物理学大厦已经全部建成，今后物理学家的任务只是修饰和完善这座大厦”。他还对20世纪物理学的前景进行了展望：“物理学美丽而晴朗的天空上还笼罩着两朵乌云：一朵是迈克耳孙-莫雷干涉实验结果，另一朵是麦克斯韦-玻尔兹曼的能量均分定理。”[1]开尔文男爵并没有意识到这两朵“乌云”很快就掀起了物理学的深刻革命：迈克耳孙-莫雷干涉实验结果对以太的否定导致了狭义相对论的诞生，而能量的量子化引发了量子力学的创立。量子力学和狭义相对论开辟了新一代的物理学，造就了20世纪的科学技术的繁荣发展，深远地影响了人类文明的各个方面。

转眼间，一个多世纪过去了，物理学又来到了一个新的十字路口。一方面，量子力学已经发展到了崭新的高度，量子场论成为集大成者。在此基础上，粒子物理作为研究世界万物最基本构成单元以及宇宙起源、演化的学科获得了飞速发展。现在普遍认为物质的最基本单元是三代夸克、三代轻子以及传递彼此之间相互作用的玻色子。人们以此为出发点构建了标准模型理论。该理论在描述粒子的相互作用和一般变化规律时获得了巨大的成功，预言结果与实验观测结果几乎完全符合。另一方面，标准模型远不是完美无缺的，还有若干朵21世纪物理的“乌云”仍然笼罩在标准模型的晴朗天空之上，如暗物质、暗能量、正反物质对称性破缺以及中微子质量来源等。在这些各式各样的“乌云”中，有一朵已经飘浮了20多年，被人们称为“缪子反常磁矩疑难”。这朵“乌云”的背后可能隐藏着超越标准模型的新物理，一场新的物理学革命的序幕即将徐徐拉开。

一、缪子反常磁矩疑难

缪子（muon）又被称为μ子，于1936年由加州理工学院的物理学家卡尔·安德森和塞思·内德迈尔在宇宙射线实验中发现[2]。缪子（μ^-）和其反粒子（μ^+）分别带有负电荷或正电荷，自旋为 1/2，静止质量为 105.658 MeV/c^2，约为电子质量的 207 倍。缪子作为第二代轻子带有一个单位电荷（1.602×10^{-19}C），参与电磁和弱相互作用。由于缪子不直接参与强相互作用，其衰变途径通常不包含比较复杂的强子末态，因此对缪子的测量和计算可以达到相当高的精度。缪子能够通过质子打靶以相对经济的方式大量产生，人们可以由此建造高强度的缪子源，这为精确测量提供了理想的实验条件。

缪子反常磁矩是缪子的基本物理性质，它反映了缪子内禀磁矩与其自旋角动量之间的深刻联系。反常磁矩的表达式为 g-2，其中，g 代表旋磁比（gyromagnetic ratio），为粒子磁矩与自旋角动量之间的比值。对于缪子，g 的实验测量值与经典物理的预期值 2 之间有着 0.1% 左右的差距。这一细微的差距被称为缪子反常磁矩（muon g-2）。

缪子反常磁矩实验是粒子物理最精确的测量之一，同时标准模型能够对缪子反常磁矩给出非常精确的预言。自标准模型诞生以来，人们一直利用缪子反常磁矩的精确测量结果对标准模型进行严格的检验，进而不断改进和发展粒子物理模型。没有缪子反常磁矩实验奠定的实验基础，标准模型就难以获得巨大的成功。

缪子反常磁矩本身具有深刻的物理背景，涉及时空的深层次结构。物理学家们认为宇宙万物所处的空间并不是静态的，即便是所谓的"真空"也并不"空"。真空中一直存在着大量的虚粒子，它们在极短的时间尺度内成对产生又互相湮灭。一般的实验探测手段很难捕捉到虚粒子的蛛丝马迹，但是在缪子附近产生的虚粒子会在极短的时间内作用并改变缪子在磁场作用下的进动频率（图 1）。这些虚粒子种类多种多样，有较轻的电子、光子、π介子，还有较重的 W/Z 粒子和希格斯粒子等，甚至还可能包括一些未知的新粒子。人们可以通过测量缪子反常磁矩探测缪子与周围世界的相互作用，并与标准模型的理论预言进行对比。一旦发现实验测量值与理论预言值不符，就意味着存在有标准模型未曾考虑到的新作用力或者新粒子。这将从根本上动摇标准模型和物理学的根基，引起物理学的下一场革命。

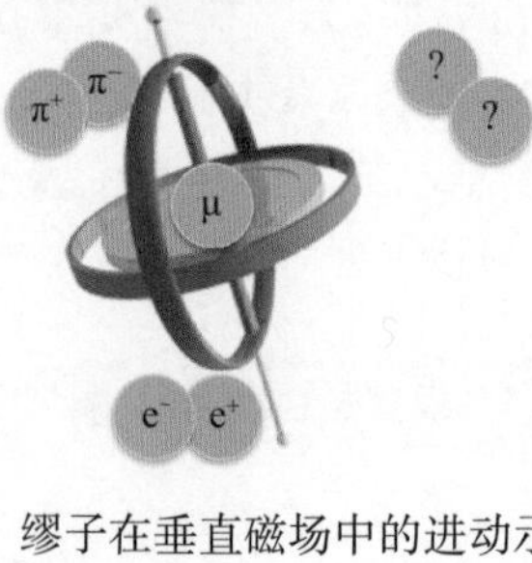

图 1　缪子在垂直磁场中的进动示意图

该进动类似于陀螺的偏轴旋转运动，频率受到附近真空中的虚粒子对的影响而加快或者变慢。

2001 年，美国布鲁克海文国家实验室（BNL）进行的早期缪子反常磁矩实验发现了缪子的一些“反常”迹象[3]：缪子反常磁矩实验值与标准模型的预言值相差 2.7 倍标准差。由于测量和计算精度的限制，这一反常迹象一直无法得到令人满意的解释，人们把该反常迹象称为“缪子反常磁矩疑难”。从那时起，它时刻提醒着人们标准模型可能并不完善，新物理就在前方。

在过去的 20 余年里，为了解答“缪子反常磁矩疑难”，全世界无数科研工作者和工程师们付出了不懈的努力。科学家们采用了一系列新的实验方法不断提高实验测量精度，于 2012 年开始在美国费米国家加速器实验室（FNAL）设计建造新的缪子反常磁矩实验[4]。与此同时，通过计算方法的提高，缪子反常磁矩的理论计算也变得更加精确[5]。2021 年，举世瞩目的缪子反常磁矩实验发布了目前最精确的反常磁矩测量值 $g_\mu=2.002\ 331\ 841\ 22$[6]。该测量结果与标准模型的最新预言值之间的差距扩大到了 3.3 倍标准差。经过更细致的分析，纳入早期的布鲁克海文国家实验室的测量结果后实验测量值与理论预言值之间的差距进一步扩大到了 4.2 倍标准差。这一巨大差距意味着实验值与理论值之间不符的统计学概率达到了惊人的 99.999% 以上，这强烈暗示着新物理的存在。

二、缪子反常磁矩的精确测量

由于缪子是带电粒子，因此缪子在外加垂直均匀磁场的作用下受洛伦兹力的影响将做回旋圆周运动，其运动频率为 ω_c。与此同时，缪子带有自旋，其内禀磁矩与外部磁场的相互作用将使得缪子沿着自旋方向发生进动，其进动频率为 ω_s。在实验室参照系下测得的缪子（反常）进动频率 ω_a 为这两者之差，如图 2（a）所示。

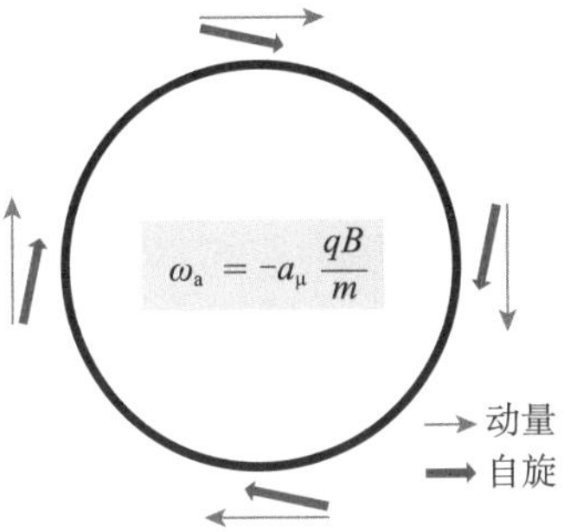

(a)

(b)

图 2 缪子反常磁矩测量装置

（a）缪子在外加垂直磁场中做圆周运动并产生进动；（b）缪子反常磁矩实验所使用的缪子储存环，储存环外部为超导线圈和轭铁以产生垂直均匀磁场，内部为真空。缪子在储存环内部做圆周运动和进动并发生衰变。

$$\vec{\omega}_{a}=\vec{\omega}_{s}-\vec{\omega}_{c}=-\frac{q}{m}\left[a_{\mu}\vec{B}-\left(a_{\mu}-\frac{1}{\gamma^{2}-1}\beta^{2}\times\vec{E}\right)+\cdots\right] \tag{1}$$

其中，B 为磁场，γ、β 为缪子的洛伦兹因子，q、m 为缪子的单位电荷和质量，$a_{\mu}=\frac{g_{\mu}-2}{2}$为缪子反常磁矩。当缪子动量为 3.094 GeV/c 时，式（1）简化为

$$\vec{\omega}_{a}=-a_{\mu}\frac{q\vec{B}}{m},\ a_{\mu}\sim\frac{\vec{\omega}_{a}}{\vec{B}} \tag{2}$$

因此对缪子反常磁矩的测量就转化为对缪子反常进动频率和磁场强度的测量。

在实际测量中，首先将大量自旋极化后的缪子注入缪子储存环中，外加一个稳定均匀的垂直磁场，通过对缪子衰变产生的电子进行精确测量得到缪子反常进动频率。再通过对磁场强度进行精确测量，进而求得它们的比值，最终获得精确的缪子反常磁矩测量值。这种使用储存环测量缪子反常磁矩的实验方法于 20 世纪 70 年代在欧洲核子研究组织被首先使用[7]，在布鲁克海文国家实验室的缪子反常磁矩测量中获得进一步改进[3]，测量精度不断提高。费米国家加速器实验室的缪子反常磁矩实验把该方法发挥到了极致，获得了迄今最精确的反常磁矩测量结果[5]。

费米缪子反常磁矩实验的缪子源通过质子打靶产生。每秒钟打靶 12 次，每次约注入 10^{12} 个具有 8 GeV 能量的质子。质子打靶产生π介子，π介子衰变产生缪子。每次打靶可产生约 10^{4} 个高度极化的缪子，最后通过束流线把缪子引导到 1.45 T 磁场的缪子储存环中进行测量。缪子衰变产生的电子的计数率会随时间发生变化，在排除各种复杂的背景后该变化的频率与缪子反常进动频率 ω_{a} 成正比[8]，通过分析如图 3（a）所示的电子计数率分布能够对 ω_{a} 进行精确的测量。与此同时，通过对磁场的空间位置和强度进行标定和校准[9]，能够获得精确的磁场强度和缪子束团密度分布函数[10]，测量结果如图 3（b）所示。

费米缪子反常磁矩实验从 2017 年夏天开始试运行，第一期物理数据于 2018 年采集完成，大约采集了 80 亿个缪子的数据，占总数据量的 6% 左右。作为 2001 年来缪子反常磁矩测量的首批物理结果，缪子反常磁矩国际合作组在公布最终测量结果之前做了大量的自查和互检，特别是为了排除个人主观因素对实验测量的影响，合作组采用了“双盲分析法”，确保在“揭盲”前没有人能够事先知道或者估算测量结果。经过复杂的数据分析和多轮严密的复查，缪子反常磁矩合作组于 2021 年 4 月 7 日公开了缪子反常磁矩测量的最终测量结果，并和标准模型的预言值进行了比较（图 4）[6]。

该测量结果的精度达到了惊人的 0.46 ppm①，这是迄今最精确的缪子反常磁矩测

① 此处 ppm 表示百万分之一。

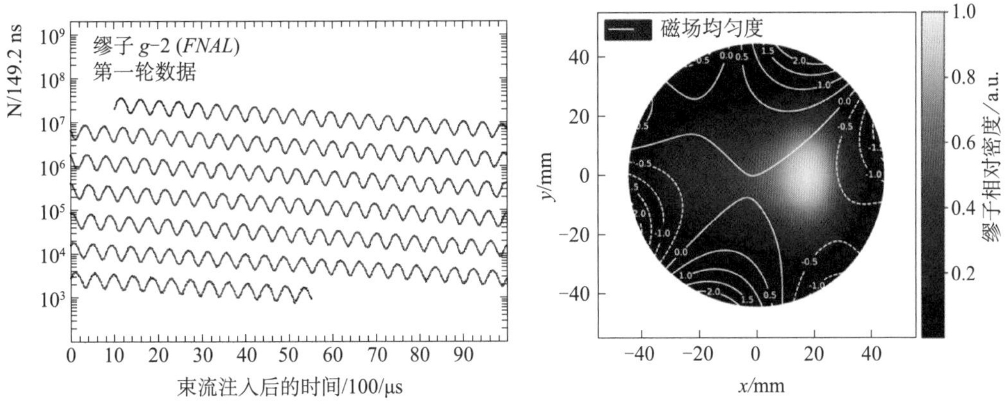

(a)　　　　　　　　　　　　　　　　(b)

图 3　缪子反常进动频率测量结果（a）和缪子空间分布示意图（b）

（a）缪子衰变产生的电子计数率随时间的分布，要求电子能量大于能量阈值 $E_{th}=1.7$ GeV；（b）磁场均匀度在储存环横截面 x-y 平面上的空间分布以及缪子束团的空间相对密度分布函数，右边明亮的部分代表缪子分布密度较大的区域。

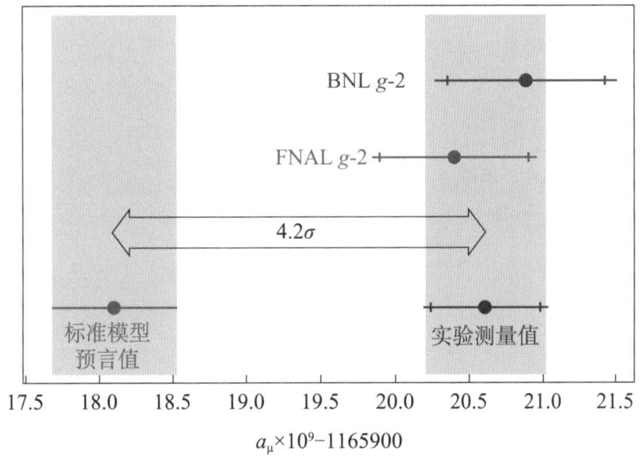

图 4　2021 年缪子反常磁矩测量结果

费米缪子反常磁矩实验测量结果与布鲁克海文国家实验室实验结果以及标准模型预言值之间的比较；绿色宽带为标准模型预言值分布，紫色宽带为实验测量值分布，这两者的中心值之间有 4.2 倍标准差的差距。

量结果，堪称历史性的突破。令科学家感到格外振奋的是，实验测量值与标准模型预言值[5]之间有着 3.3 倍标准差的差距，与布鲁克海文国家实验的结果则完全相符(1 倍标准差以内)。这两个实验的综合测量值与标准模型预言值之间的差距进一步增大到了 4.2 倍标准差，实验家给出的答案与理论家的并不一致。这是非常激动人心的结果，这说明现有理论对缪子的描述很可能并不完全，我们期待新物理对缪子的反常行为做出解释。

实际上有多种新物理模型可以解释缪子反常磁矩疑难[11]，包括超对称物理模型[12,13]、矢量费米子模型[14,15]以及双希格斯子模型[16-18]。还有一种不容忽视的可能性是缪子反常磁矩的理论计算还不够准确，特别是对于比较难以计算的强子真空极化部分。2021 年的一个格点计算结果[19]指出了以往理论计算结果发生偏差的可能，但是这个计算结果又和电弱精确测量结果不符[20,21]。如何理解不同理论计算结果之间的细微差异还需要粒子物理学家们付出更多的时间和努力。

三、缪子反常磁矩实验的未来展望

在激动人心的第一期实验数据测量结果公布后，费米缪子反常磁矩实验第二和第三期实验数据的分析正在紧锣密鼓地进行中。这两期数据量是首批数据量的 4 倍，分析结果预计于 2023 年秋发表，测量精度将提高 1 倍。费米缪子反常磁矩实验预计运行到 2025 年，总共采集约 1 万亿个能量为 3 GeV 左右的缪子，数据量约是早期布鲁克海文国家实验室数据量的 20 倍。

以数量级增长的数据量有助于大大减少统计误差。同时通过不断改进实验方法，缩小系统误差，费米缪子反常磁矩实验的最终测量精度可达千万分之一（0.1 ppm）的水平。该实验的最终测量结果将比第一期测量结果的精度提高 4～5 倍，成为世界上最精确的缪子反常磁矩实验并将保持该记录相当长的时间。在实验测量的中心值不发生较大变化的情况下，即便考虑到近期的格点计算结果与传统理论计算结果之间的差异，精度提高后的实验值与理论值之间的差距仍将超过 5 倍标准差，达到粒子物理界的“黄金判据”标准，缪子反常磁矩疑难将成为一个举世瞩目的重大物理发现。历史上，诺贝尔委员会曾多次关注缪子反常磁矩实验的进展，该实验结果已经成为诺贝尔物理学奖的强有力竞争者。

除了正在进行的费米缪子反常磁矩实验，日本高强度质子加速器实验室（J-PARC）也正在建造一个采取不同实验路线和方法的缪子反常磁矩实验①。J-PARC 缪子反常磁矩实验采用“超冷缪子法”来建造新的缪子源：先通过 28 MeV/c 的表面缪子束流生成缪子素，以二氧化硅气凝胶为载体在真空中将缪子素冷却至 2.3 keV/c，再利用激光电离和激光消融的方法得到低发散度的超冷反缪子，最后再经过一段加速过程将 300 MeV/c 的反缪子注入一个紧凑型的储存环中进行实验测量。该紧凑型储存环为费米实验的储存环的 1/20 大小，磁场强度为 3 T（图 5）。尽管也利用了“缪子储存环”的测量原理，但 J-PARC 实验在缪子源建造和缪子动量、角度和空间测量方面采用了

① 参见：Muon g-2/EDM experiment at J-PARC. https://g-2.kek.jp/.

不同的技术路线和方法[22]。该实验预计 2027 年建成，2028 年正式取数，于 2029 年发表首批实验测量结果。这两个独立进行的缪子反常磁矩实验将会互相验证，使彻底解答缪子反常磁矩疑难更进一步。

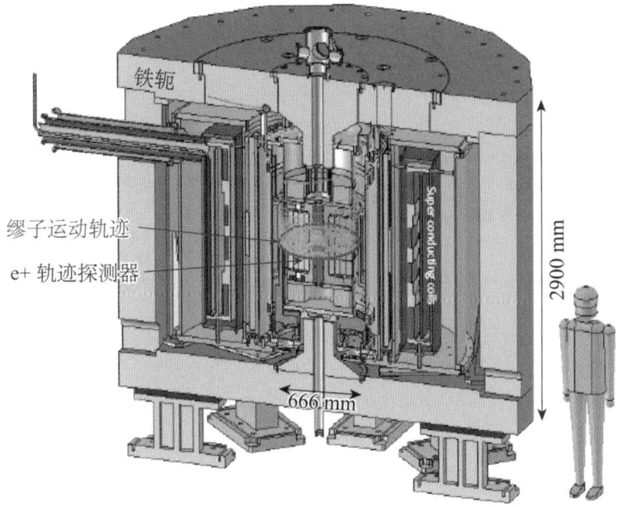

图 5 J-PARC 缪子反常磁矩实验的紧凑型储存环装置示意图

资料来源：Muon g-2/EDM experiment at J-PARC. https：//g-2. kek. jp/.

近年来，我国缪子物理和相关强流加速器的建设也进入了快车道。中国散裂中子源（CSNS）的 100 kW 1.6 GeV 质子加速器已经建成①，作为 CSNS 的升级计划的一部分，我国第一个加速器缪子源设施（EMuS）[23]正在建设中；国家“十二五”重大科学基础设施项目“强流重离子加速器装置”（HIAF）正在建设重离子加速器②；列入“十四五”建设计划的“加速器驱动嬗变系统研究装置”（CiADS）[24]拟建设连续流直线质子加速器；HIAF 的升级计划中缪子束流强度可达到目前缪子反常磁矩实验束流强度的 30 倍。如果能够在这些新型加速器装置的基础上建造下一代高能强流缪子源，必将有力地推动缪子精确测量新高峰的早日到来[25]。

四、总　结

作为 21 世纪物理学天空中最浓厚的“乌云”之一，缪子反常磁矩疑难在推动当

① 参见：中国散裂中子源工程 . http://csns. ihep. cas. cn/.

② 参见：强流重离子加速器装置 . http://hiaf. impcas. ac. cn/.

代物理学发展和检验粒子物理标准模型方面起到了关键性的作用。超级精确的缪子反常磁矩测量已经成为物理学史上的一座丰碑。缪子反常磁矩实验是当今最前沿的科学研究之一，是非常灵敏的新物理探针。它对各种新物理模型有着强大的鉴别能力，将对未来若干年寻找超越标准模型的新物理提供指引[26]。半个多世纪以来，缪子反常磁矩实验一直在不断地改进和提高，我们有充分的理由相信，随着更新、更精确的测量结果的出现，缪子反常磁矩疑难的迷雾终将被拨开，胜利的曙光就在不远的前方。

参考文献

[1] Passon O. Kelvin's clouds. American Journal of Physics, 2021, 89: 1037.

[2] Anderson C D, Neddermeyer S H. Cloud chamber observations of cosmic rays at 4300 meters elevation and near sea-level. Phys Rev, 1936, 50: 263.

[3] Bennett G W, Bousquet B, Brown H N, et al. Final report of the E821 muon anomalous magnetic moment measurement at BNL. Phys Rev D, 2006, 73: 072003.

[4] Fermilab. Muon *g*-2 will better understand the properties of the muon and use them to probe the Standard Model of particle physics. https://muon-g-2.fnal.gov/.

[5] Aoyama T, Asmussen N, Benayoun M, et al. The anomalous magnetic moment of the muon in the Standard Model. Phys Rep, 2020, 887: 1.

[6] Abi B, Albahri T, Al-Kilani S, et al. Measurement of the positive muon anomalous magnetic moment to 0.46 ppm. Phys Rev Lett, 2021, 126(14): 141801.

[7] Bailey J, Borer K, Combley F, et al. Final report on the CERN muon storage ring including the anomalous magnetic moment and the electric dipole moment of the muon, and a direct test of relativistic time dilation. Nucl Phys B, 1979: 150: 1-75.

[8] Albahri T, Anastasi A, Anisenkov A, et al. Measurement of the anomalous precession frequency of the muon in the Fermilab muon *g*-2 experiment. Phys Rev D, 2021, 103: 072002.

[9] Albahri T, Anastasi A, Anisenkov A, et al. Magnetic-field measurement and analysis for the muon *g*-2 experiment at Fermilab. Phys Rev A, 2021, 103: 042208.

[10] Albahri T, Anastasi A, Badgley K, et al. Beam dynamics corrections to the run-1 measurement of the muon anomalous magnetic moment at Fermilab. Phys Rev Accel Beams, 2021, 24: 044002.

[11] Athron P, Balázs C, Jacob D H J, et al. New physics explanations of a_μ in light of the FNAL muon *g*-2 measurement. J of High Energy Phys, 2021, (9): 80.

[12] Everett L L, Kane G L, Rigolin S, et al. Implications of muon *g*-2 for supersymmetry and for discovering superpartners directly. Phys Rev Lett, 2001, 86: 3484.

[13] Ibe M, Yanagida T T, Yokozaki N. Muon *g*-2 and 125 GeV Higgs in split-family supersymmetry. J of High Energy Phys, 2013, (8): 67.

[14] Endo M, Hamaguchi K, Iwamoto S, et al. Higgs mass, muon *g*-2, and LHC prospects in gauge mediation models with vector-like matters. Phys Rev D, 2012, 85: 095012.

[15] Dermisek R, Hermanek K, McGinnis N. Highly enhanced contributions of heavy Higgs bosons and new leptons to muon g-2 and prospects at future colliders. Phys Rev Lett, 2021, 126: 191801.

[16] Lindner M, Platscher M, Queiroz F S. A call for new physics: The muon anomalous magnetic moment and lepton flavor violation. Phys Rep, 2018, 731: 1.

[17] Liu X W, Bian L G, Li X-Q, et al. Type-III two Higgs doublet model plus a pseudoscalar confronted with $h\to\mu\tau$, muon g-2 and dark matter. Nucl Phys B, 2016, 909: 507.

[18] Ferreira P M, Gonçalves B L, Joaquim F R, et al. $(g\text{-}2)_\mu$ in the 2HDM and slightly beyond—An updated view. Phys Rev D, 2021, 104: 053008.

[19] Borsanyi S, Fodor Z, Guenther J N, et al. Leading hadronic contribution to the muon magnetic moment from lattice QCD. Nature, 2021, 593: 51-55.

[20] Keshavarzi A, Marciano W J, Passera M, et al. Muon g-2 and $\Delta\alpha$ connection. Phys Rev D, 2020, 102: 033002.

[21] Crivellin A, Hoferichter M, Manzari C A, et al. Hadronic Vacuum Polarization: $(g\text{-}2)_\mu$ versus Global Electroweak Fits. Phys Rev Lett, 2020, 125: 091801.

[22] Abe M, Bae S, Beer G, et al. A new approach for measuring the muon anomalous magnetic moment and electric dipole moment. Prog Theor Exp Phys, 2019, 2(5): 053C02.

[23] Tang J Y, Ni X J, Ma X Y, et al. EMuS muon facility and its application in the study of magnetism. Quantum Beam Sci, 2018, 2(4): 23.

[24] Xiao G Q, Xu H S, Wang S C. HIAF and CiADS national research facilities: progress and prospect. Nucl Phys Rev, 2017, 34(3): 275-283.

[25] Sun Z Y, Chen L W, Cai H J, et al. Producing high intensity muon, antiproton beams and related physical researches in the HIAF accelerators. Sci Sin Phys Mech & Astro, 2020, 50(11): 112010.

[26] Li L. Precise measurement of muon anomalous magnetic moment. Science & Technology Review, 2022, 40(6): 6-11.

A Cloud over the 21st Century Physics: "Muon Anomalous Magnetic Moment Conundrum" Sees the Dawn

Li Liang

The muon anomalous magnetic moment(a_μ) is a fundamental property of the muon particle and has played a fundamental role in the formation of the Standard Model. The long-standing discrepancy between the experimental value and the theoretical prediction of a_μ is an intriguing conundrum. The first result from the

Fermi National Accelerator Laboratory's muon g-2 experiment strengthens the deviation of the measurement from the Standard Model prediction to a 4.2 standard deviation, which is a strong indication for new physics. Improved measurements and theory calculations may finally shed some light on the over 20 years conundrum.

2.3 蛋白质结构预测：现状和相关问题的未来展望

周耀旗[1] 杨建益[2] 彭珍玲[2]

（1. 深圳湾实验室系统与物理生物学研究所；
2. 山东大学数学与交叉科学研究中心）

一、发展现状

蛋白质结构的从头预测（*de novo* prediction）是指根据蛋白质的一级氨基酸序列来推断、构建它的三级结构。该领域的研究有着 60 多年的历史，被认为是计算生物学领域的一个重大挑战[1]。蛋白质结构预测研究的进展一直非常缓慢，直到 2021 年才有重大突破。基于人工智能（AI）的 AlphaFold2 方法把结构预测的精度提升到了前所未有的高度，对部分蛋白质预测的结构几乎可与实验结构相媲美[2]。

从宏观上来看，AlphaFold2 方法成功的关键因素有三方面：①高质量蛋白质结构大数据多年的积累；②计算机硬件［包括图形处理单元（GPU）、张量处理单元（TPU）等］的发展；③人工智能、深度学习算法最近几年的快速进步。我们认为，这三方面因素缺一不可。

AlphaFold2 是最近几年新兴的全部在人工神经网络（下文简称神经网络）内部进行的“端到端”的方法。这里，我们也称之为“一步走”的方法。相比而言，过去基于人工智能的结构预测方法一直是基于“两步走”方法（图 1）[3]。下面分别简要介绍。

先通过神经网络预测一维主链结构（二级结构）和二维距离接触图，再进行基于能量优化的三级结构建模。相比而言，“端到端”预测方法是“一步走”，所有计算（一维、二维和三级结构）都在神经网络内部进行。

1. “两步走”方法

“两步走”，顾名思义，就是把蛋白质结构预测分成两步：先从一级序列预测一维

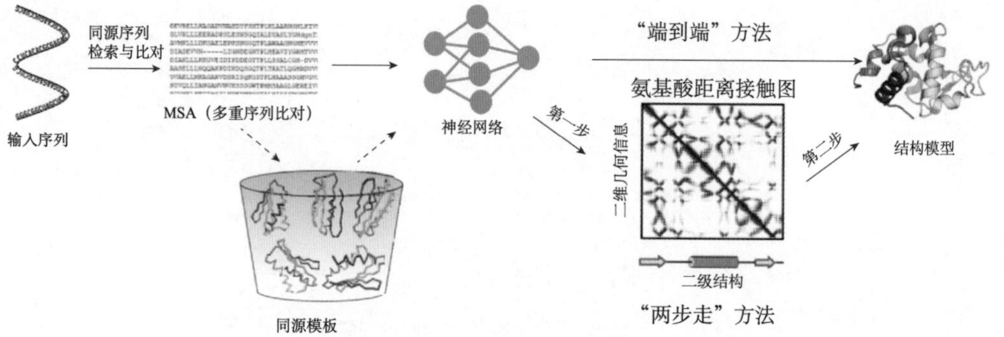

图 1　蛋白质结构预测的"两步走"方法

的主链结构（二级结构）和二维的距离约束信息；再根据约束信息构建三级结构。在AlphaFold2 出现以前，绝大部分方法都是"两步走"方法。它本质上是采用"分而治之"的思想。第一步，降低问题难度，利用人工智能算法求解，即采用在计算机视觉和自然语言处理领域获得成功的算法（从刚开始的浅层网络到不断加深的深度学习），从蛋白质单序列信息到多重序列比对谱出发，预测上述一维和二维几何（边和角）的约束信息。第二步则完成从一维和二维的约束到三级结构的构建过程，用到的基本上是数学中的优化算法，如牛顿迭代法和距离几何方法等，来针对某个能量函数进行优化。"两步走"方法的典型代表包括 RaptorX-Contact[4]、AlphaFold[5]、trRosetta [6,7] 等。

2. "端到端"方法

与"两步走"方法不同，"端到端"方法从蛋白质序列或同源多序列比对出发，利用人工智能算法直接预测蛋白质三级结构。"端到端"方法的最大优点是完全避开由少量参数所构建的、近似的能量函数，可以根据性能及时调整超过百万的网络参数，非常有利于模型训练。第一个成功通过盲测比赛的"端到端"方法是谷歌公司的DeepMind 团队开发的人工智能算法 AlphaFold2。在第 14 届国际蛋白质结构预测竞赛（CASP14）中，AlphaFold2 的性能远远超过其他方法，它所预测结构模型的精确度（GDT-TS，满分 100）的中位数高达 92.4。这是前所未有的精度，AlphaFold2 成为蛋白质结构预测研究领域划时代的方法。为方便生物学家使用和扩大算法影响力，DeepMind 团队把 AlphaFold2 应用于预测人类蛋白质组中 2 万多个蛋白质的结构[8]，后期扩展到现有蛋白质序列数据库 UniProt 中所有 2 亿多个蛋白。

从人工智能算法设计的角度，我们认为 AlphaFold2 算法的成功秘诀可总结为"2＋2"（图 2）。第一个"2"是指"两个表征"：相当于一维的多序列对比表征（MSA representation）和相当于二维成对表征（pair representation），它们通过注意力机制进行信息交互，达到不断改进的目的。第二个“2”是指“两个网络”：处理序列数据的 Evoformer［相当于编码器（encoder）］和处理结构的 IPA（Invariant Point Attention）模块［结构模块，相当于解码器（decoder）］，这两个网络都是基于注意力机制设计的，它们连接在一起，从而达到“端到端”训练的效果。当然，除以上关键要素外，回收（recycle）机制和自蒸馏（self-distillation）等其他方面对改进模型质量有很大的帮助。

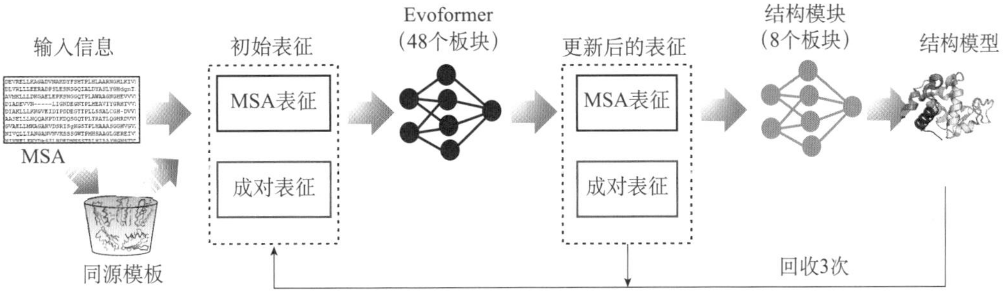

图 2　AlphaFold2 算法的关键要素

Evoformer 是指一种基于注意力机制的神经网络模型。

AlphaFold2 不是第一个“端到端”预测蛋白质结构的方法。第一个方法是循环几何网络（recurrent geometric network，RGN），在 2019 年发表（2018 年出现于 bioRxiv）[9]。该方法的提出基于两项进展：一是 2016 年在神经网络里实现了的任意可微分损失函数的反向传播[10]；二是 2009 年发现的蛋白质主链的二面角可以准确预测，并能用于直接构建合理的三级结构[11,12]而进行可微分学习。但是该方法的硬伤是忽略了支链之间的距离预测。几乎同时出现的“端到端”学习方法神经能量建模及优化（neural energy modeling and optimization，NEMO）[13]和 AlphaFold2 在方法上相似性更加明显。例如，这两个方法都对每个氨基酸残基使用局部参考框架表示，都使用一维和二维作为中间过渡的表征，都考虑旋转和平移不变性（equivariance），都在学习中使用角度、距离的中间表征来优化结构。不同的地方是，NEMO（以及 RGN）使用预测的主链二面角来构建三级坐标及强制主链的约束。相反，AlphaFold2 通过预测独立、氨基酸水平上的仿射变换（affine transformation）来构建主链、生成三级结构，避免了增量误差的传播以及局部最优解。此外，RGN 和 NEMO 利用预处理的一维序列谱，而 AlphaFold2 采用多个同源序列同时输入，可以更加直接地捕获协同进化的距离信息。当然 AlphaFold2 采用了当时所有已知的序列和所有已知的结构来进行最大规模训练，也是其成功的一个原因。值得注意的是，最早“端到端”方法 RNG/NEMO 的准确度并不如“两步走”方法，但从 RGN 到 AlphaFold2 的出现只花了不到两年的时间。相比而言，“两步走”的方法在几十年来一直进展缓慢。这充分表明了方法创新的重要性，即使该类方法一开始的准确度并不高。

在 AlphaFold2 的启发下，华盛顿大学的美国国家科学院院士 David Baker 带领团队开发了基于“三轨”的“端到端”方法 RoseTTAFold [14]。RoseTTAFold 与 AlphaFold2 同一天分别在线发表于《科学》和《自然》上。RoseTTAFold 的精确度比 AlphaFold2 低 10 分左右，但 RoseTTAFold 对算力要求低于 AlphaFold2，而且它可以用于蛋白-蛋白复合物结构预测。目前更多的基于 AlphaFold2 的“端到端”方法在不断涌现[15]。

二、关键问题与发展方向

AlphaFold2 使蛋白质结构预测工作取得了突破性进展，2022 年对 AlphaFold2 预测的结构的分析表明预测的精确结构可以比同源建模多 25% 的氨基酸，对复合物结构预测好于专门工具[16]。但蛋白质结构预测问题尚未彻底解决[17]，我们仍面临诸多挑战[18]。我们认为目前挑战主要集中体现在生物学功能导向的蛋白质结构预测中，简要总结如下[3]。

1. 现有方法严重依赖多重序列比对

自然界中相当一部分蛋白质（如来自病毒），在现有序列数据库中往往找不到同源序列，无法构建多重序列比对（multiple sequence alignment，MSA）。我们把这种蛋白质称作“孤儿”（orphan）蛋白。现有方法对此类蛋白质结构预测的精度普遍偏低（精度在 30～40 分）。因此，如何提升此类蛋白质结构的预测精度，是当前方法面临的一个挑战。利用蛋白质预训练语言模型是一个可能的方向，已有多个团队在此方向开展了一些研究工作[15,19-21]。然而，在评估这些方法时需要特别小心，对于那些能找到同源序列的蛋白质，以单序列作为输入时，性能超越 AlphaFold2 并不能说明方法真正优于 AlphaFold2，这是因为预训练语言模型蕴含了同源序列信息。客观公正的评估应该针对那些没有已知同源序列的蛋白质所预测的蛋白质，例如“孤儿”蛋白和人工设计的蛋白质。在“孤儿”蛋白数据集上，我们的方法 trRosettaX-Single 也仅仅比 AlphaFold2 高 5～6 分，预测结构的精度也只能达到 50 分左右[19]，还有很大的改进空间。一个可能的方法是通过突变和进化产生人工同源序列。但是最终还是需要能够发展不依赖同源序列的甚至不依赖语言模型的深度学习方法，毕竟相当一部分蛋白质的结构，特别是稳定的、单体蛋白质是能够自己从展开的状态折叠成天然状态的，也就是说，单序列的信息应该足够决定三级结构了。

2. 精度与算力的平衡

在当前结构预测领域的“端到端”方法中，AlphaFold2 是最具有代表性的精确方法。然而，AlphaFold2 的训练对算力的要求极高（原版 AlphaFold2 使用 128 个 TPUv3 cores 训练了数周），学术界不太可能拥有如此丰富的计算资源。是否有可能在有限的算力条件下达到或接近 AlphaFold2 的精度？为回答此问题，基于本课题组有限的计算资源（6 张 A100GPU），我们在“两步走”方法 trRosettaX 的基础上，从 AlphaFold2 的“2＋2”借鉴了“2＋1”，即“两个表征”和一个网络 Evoformer，并对其中 Evoformer 做了简化（从 48 个模块降低为 12 个）。用它替换了 trRosettaX 中的第一步，第二步保持不变，我们把这个版本命名为 trRosettaX2。在 CASP14 的数据集上的测试表明[3]，trRosettaX2 相比于 trRosettaX 性能得到了显著提升（指标 TM-score 提升了 8 分），超越 RoseTTAFold，与 AlphaFold2 的差距仅有 5 分。该测试表明，借助 Transformer 等神经网络算法，可在有限算力条件下训练“两步走”方法，并逐步逼近 AlphaFold2 的精度。事实上，第 15 届国际蛋白质结构预测竞赛（CASP15）的最新结果表明，结合 trRosettaX2 与 AlphaFold2，可以较低的成本提升三级结构预测的精度。

3. 动态结构和折叠路径的预测

蛋白质功能与结构的动态特性及折叠路径密切相关，人们往往借助分子动力学模拟对此进行研究。然而，不同蛋白质的模拟参数和设置都不尽相同，因此这些研究工作通常只能针对个例进行开展，从而使得研究结果很难应用于其他蛋白质。最近的研究工作也表明，基于现有最好的结构预测方法也很难得到有意义的折叠路径[22]。由于动态结构的实验数据比较缺乏，研究在短期内也很难有大的突破。利用传统基于物理的分子动力学模拟产生批量数据，再利用深度学习进行“端到端”的训练，可能是解决此问题的一个方向，特别是考虑到蛋白质的折叠路径往往是与结构的天然接触图（native contact map）密切相关的[23]。

4. 蛋白质-配体/药物互作用预测

按理说，蛋白质结构预测的突破对药物研发等领域应该会有较大的促进作用，然而药物设计领域的科学家对此并不是特别乐观[24]。第一个顾虑是预测的结构中的活性位点和侧链的精度并不确定；第二个是如上文所述，蛋白质结构是动态变化的，有的蛋白质在结合配体过程中还会发生显著的构象改变。现有方法先预测蛋白质结构，再通过分子对接的方式预测蛋白质-配体的复合物结构。这种策略或许需要调整，把两步合成一步，借鉴 AlphaFold2 的思想，同步预测蛋白质和配体结构，但是目前的小分子与蛋白质复合物的数据量可能还不足以产生一个可以泛化的模型。

5. 蛋白质-蛋白质复合物预测

与蛋白质-配体复合物结构预测类似，蛋白质-蛋白质复合物的传统预测方法是分子对接预测：先预测蛋白质单体结构，再利用分子对接构建复合物结构。这种方法显然严重依赖于单体结构预测的精度，而且对于蛋白质有较大构象改变的情形并不适用。因此，DeepMind 团队在 AlphaFold2 的基础上，开发了复合物结构预测的版本 AlphaFold-Multimer[25]。虽然它与 AlphaFold2 相比有改进，但性能并不稳定，对不少复合物预测的结构精度几乎等于 0。对于超大复合物，单纯依靠计算方法更是无法构建完整结构。通过算法与部分实验数据的互动（干湿结合），或许是解决复合物结构预测的可行方案[26]。

6. 突变效应预测

由于缺乏大规模高质量数据，突变效应的精准预测始终是一个挑战[27]。这是一个非常重要的科学问题，因为许多致病性突变只需要一个氨基酸的改变。研究表明，结构预测的改进似乎对提高突变效应预测的精准性帮助并不大[28]。主要原因是目前结构预测算法是根据同源序列的进化信息，并假定所有同源序列有同样的结构来获取的，但事实上，高度同源但结构不一样的序列经常可以发生，突变效应的精准预测可能要在解决了无需同源序列就能精准预测蛋白质结构这个问题后才有可能做到。

7. 蛋白质设计

蛋白质设计是蛋白质结构预测的反问题：根据所需的结构骨架来反推能够折叠成这个结构的蛋白质序列。跟蛋白质结构预测一样，蛋白质设计长期以来是基于能量函数的优化来预测蛋白质的序列的[29]。浅层神经网络学习的 SPIN（sequence profiles by integrated neural networks）和深度学习的 SPIN2 方法是最早在神经网络内进行“端到端”尝试的蛋白质设计[30,31]。2018 年后的深度学习算法的大幅度进展，以及 AlphaFold2 的成功，大幅度推动了“端到端”深度学习的蛋白质设计方法的改进[32,33]。天然序列的恢复度从 SPIN 的 30% 和 SPIN2 的 34%，到 2022 年的 40%～50%。但是，天然序列的恢复度显然不是判断设计是否成功的一个好标准，因为几个氨基酸的突变经常就足以破坏蛋白质的结构，而且设计出的蛋白质的可溶性、可表达性也是未知的。但是近十年的快速进展表明蛋白质设计这个问题的解决似乎指日可待[34]。

8. RNA 结构预测

蛋白质结构预测的成功给一个类似的问题——RNA 结构预测提供了新方向。与从头预测蛋白质结构一样，从头预测 RNA 结构长期以来一直也是基于能量函数的优化，进展缓慢[35]。与蛋白质结构不一样，RNA 的二级结构（碱基成对图）比较稳定，先于三级结构的形成。由于存在许多能量相似的二级结构，所以利用能量优化来预测 RNA 结构特别容易陷入局部最优解，因此对能量函数的精确度是一个严峻的考验。RNA 二级结构预测研究长期以来也是基于能量函数的优化展开的，这些方法的精度一直在同一个水平上徘徊不前。我们利用“端到端”深度学习方法对 RNA 的二级结构的预测，成功地打破了 RNA 二级结构预测精度的“天花板”[36]，并利用进化和共进化信息进一步改进了预测的精确度[37]。其他“端到端”深度学习方法以及“两步走”的方法对 RNA 的三级结构预测研究论文也开始出现在预印本网站上[38]。这些文章的结果初步表明，深度学习在 RNA 结构的预测中的参与大幅度改进了二级和三级结构的预测能力，但是仅仅一部分的 RNA 能够获得比较准确的结构。由于 RNA 的结构数据量少，大约只有蛋白质结构数据量的 3%，这些基于小数据量的深度学习方法的泛化能力还需要进一步的求证，特别是考虑到最近相当多的二级结构深度学习预测方法被发现无法预测未见过的 RNA 家族的二级结构[39]。的确，CASP15 的 RNA 结构预测方法评估发现，前四名方法依然是基于能量函数的方法，其中 Alchemy-RNA2（一个基于 RNA-BRiQ 统计势函数[40]的方法）获得了第一名。因此，目前基于 AI 的 RNA 结构预测方法还不能泛化到未见过的结构。

三、结 束 语

AlphaFold2 在蛋白质结构预测精度的革命性进展，打破了蛋白质结构预测不可能达到原子水平的实验精确度的这个预设，进一步激发了 AI 深度学习在生物学研究中应用的热潮。这个进展不是一蹴而就的，它是 60 多年来计算生物学的稳步前进，以及近年来与深度学习算法快速发展交叉融合的结果，特别是任意可微分损失函数的反向传播方法使得用百万甚至千万级别的参数来替代几百/几千参数的能量函数成为可能。当然避免过度训练是深度学习所得到的模型能否泛化、预测未见过的结构类型的关键。AlphaFold2 颠覆性的成功并不意味着不再需要通过实验来测定蛋白质结构，因为还有许多其他问题远远没有解决，尤其是同源序列的依赖性、复合物结构、结构动力学和突变效应等。事实上，AlphaFold2 预测出的蛋白质的精确结构仅仅比同源建模多了 25% 的氨基酸[16]。所以，彻底解决蛋白质结构预测的问题还有待于 AI 深度学习计算方法的进一步发展，特别是在单序列的信息提取和结构预测以及小数据训练的泛化能力上。所以完全可以说："革命尚未成功，同志仍需努力。"

参考文献

[1] Dill K A, MacCallum J L. The protein-folding problem, 50 years on. Science, 2012, 338: 1042-1046.

[2] Jumper J, Evans R, Pritzel A, et al. Highly accurate protein structure prediction with AlphaFold. Nature, 2021, 596: 583-589.

[3] Peng Z, Wang W, Han R, et al. Protein structure prediction in the deep learning era. Current Opinion in Structural Biology, 2022, 77: 102495.

[4] Wang S, Sun S, Li Z, et al. Accurate *de novo* prediction of protein contact map by Ultra-Deep Learning Model. PLoS Comput Biol, 2017, 13: e1005324.

[5] Senior A W, Evans R, Jumper J, et al. Improved protein structure prediction using potentials from deep learning. Nature, 2020, 577: 706-710.

[6] Yang J Y, Anishchenko I, Park H, et al. Improved protein structure prediction using predicted interresidue orientations. Proceedings of the National Academy of Sciences, 2020, 117: 1496.

[7] Du Z Y, Su H, Wang W K, et al. The trRosetta server for fast and accurate protein structure prediction. Nat Protoc, 2021, 16: 5634-5651.

[8] Tunyasuvunakool K, Adler J, Wu Z, et al. Highly accurate protein structure prediction for the human proteome. Nature, 2021, 596: 590-596.

[9] AlQuraishi M. End-to-end differentiable learning of protein structure. Cell Syst, 2019, 8: 292-301.

[10] Wang S, Fidler S, Urtasun R. Proximal deep structured models. Advances in Neural Information Processing Systems, 2016, 29: 865-873.

[11] Faraggi E, Yang Y, Zhang S, et al. Predicting continuous local structure and the effect of its substitution for secondary structure in fragment-free protein structure prediction. Structure, 2009, 17: 1515-1527.

[12] Zhou Y, Duan Y, Yang Y, et al. Trends in template/fragment-free protein structure prediction. Theoretical Chemistry Accounts, 2011, 128: 3-16.

[13] Ingraham J, Riesselman A, Sander C, et al. Learning Protein Structure with a Differentiable Simulator. ICLR, 2019.

[14] Baek M, DiMaio F, Anishchenko I, et al. Accurate prediction of protein structures and interactions using a three-track neural network. Science, 2021, 373: 871-876.

[15] Chowdhury R, Bouatta N, Biswas S, et al. Single-sequence protein structure prediction using a language model and deep learning. Nat Biotechnol, 2022, 40: 1617-1623.

[16] Akdel M, Pires D E V, Pardo E P, et al. A structural biology community assessment of AlphaFold2 applications. Nature Structural & Molecular Biology, 2022, 29(11): 1056-1067.

[17] Moore P B, Hendrickson W A, Henderson R, et al. The protein-folding problem: not yet solved.

Science,2022,375:507.
[18] Jones D T,Thornton J M. The impact of AlphaFold2 one year on. Nat Methods,2022,19:15-20.
[19] Wang W, Peng Z, Yang J. Single-sequence protein structure prediction using supervised transformer protein language models. Nature Computational Science,2022,2(12):804-814.
[20] Singh J,Litfin T,Singh J,et al. SPOT-Contact-LM:improving single-sequence-based prediction of protein contact map using a transformer language model. Bioinformatics,2022,38:1888-1894.
[21] Lin Z,Akin H,Rao R,et al. Evolutionary-scale prediction of atomic-level protein structure with a language model. Science,2023,379(6637):1123-1130.
[22] Outeiral C, Nissley D, Deane C M. Current structure predictors are not learning the physics of protein folding. Bioinformatics,2022,38(7):1881-1887.
[23] Zhou Y, Karplus M. Interpreting the folding kinetics of helical proteins. Nature,1999,401:400-403.
[24] Mullard A. What does AlphaFold mean for drug discovery? Nature Reviews Drug Discovery, 2021,20(10):725-727.
[25] Evans R, O'Neill M, Pritzel A, et al. Protein complex prediction with AlphaFold-Multimer. BioRxiv,2021,https://doi. org/10. 1101/2021. 10. 04. 463034.
[26] Fontana P, Dong Y, Pi X, et al. Structure of cytoplasmic ring of nuclear pore complex by integrative cryo-EM and AlphaFold. Science,2022,376:eabm9326.
[27] Gelman S,Fahlberg S A,Heinzelman P,et al. Neural networks to learn protein sequence-function relationships from deep mutational scanning data. Proc Natl Acad Sci U S A, 2021, 118(48): e2104878118.
[28] Buel G R,Walters K J. Can AlphaFold2 predict the impact of missense mutations on structure? Nat Struct Mol Biol,2022,29:1-2.
[29] Li Z,Yang Y,Zhan J,et al. Energy functions in *de novo* protein design: current challenges and future prospects. Annual Review of Biophysics,2013,42:315-335.
[30] Li Z,Yang Y,Faraggi E,et al. Direct prediction of profiles of sequences compatible with a protein structure by neural networks with fragment-based local and energy-based nonlocal profiles. Proteins:Structure,Function,and Bioinformatics,2014,82:2565-2573.
[31] O'Connell J,Li Z X,Hanson J,et al. SPIN2:predicting sequence profiles from protein structures using deep neural networks. Proteins:Structure,Function,and Bioinformatics,2018,86:629-633.
[32] Anand N,Eguchi R,Mathews I I,et al. Protein sequence design with a learned potential. Nature communications,2022,13:1-11.
[33] Liu Y F,Zhang L,Wang W L,et al. Rotamer-free protein sequence design based on deep learning and self-consistency. Nature Computational Science,2022,2(7):451-462.
[34] 陈志航,季梦麟,戚逸飞. 人工智能蛋白质结构设计算法研究进展. 合成生物学,2023,4(24), doi:10. 12211/2096-8280. 2023-008.

[35] Miao Z C, Adamiak R W, Antczak M, et al. RNA-puzzles round IV: 3D structure predictions of four ribozymes and two aptamers. RNA, 2020, 26: 982-995.

[36] Singh J, Hanson J, Paliwal K, et al. RNA secondary structure prediction using an ensemble of two-dimensional deep neural networks and transfer learning. Nature Communications, 2019, 10: 1-13.

[37] Singh J, Paliwal K, Zhang T C, et al. Improved RNA secondary structure and tertiary base-pairing prediction using evolutionary profile, mutational coupling and two-dimensional transfer learning. Bioinformatics, 2021, 37: 2589-2600.

[38] Feng C, Wang W, Han R, et al. Accurate *de novo* prediction of RNA 3D structure with transformer network. bioRxiv, 2022, doi: https://doi.org/10.1101/2022.10.24.513506.

[39] Szikszai M, Wise M J, Datta A, et al. Deep learning models for RNA secondary structure prediction (probably) do not generalize across families. Bioinformatics, 2022, 38(16): 3892-3899.

[40] Xiong P, Wu R, Zhan J, et al. Pairing a high-resolution statistical potential with a nucleobase-centric sampling algorithm for improving RNA model refinement. Nature Communications, 2021, 12: 2777-2787.

Protein Structure Prediction: Current Status and Future Prospect of Related Problems

Zhou Yaoqi, Yang Jianyi, Peng Zhenling

The breakthrough of AlphaFold2 in protein structure prediction ends the myth that computationally predicted protein structures can never match the atomic resolution of experimental observations. This achievement does not mean that problem of protein structure prediction has been solved, but rather marks the beginning of a new exciting era that will link structure to function, and function to phenotype. The power of deep learning will extend beyond the problems associated with proteins.

2.4 合成生物制造

李 寅

（中国科学院微生物研究所）

当前，全球生物基产品的商业活动高度活跃，生物技术加速渗透和融入传统产业，不断创新、影响和改变各行各业的生产方式，预示着一个以生物制造产业为标志的生物经济时代即将到来。合成生物赋能生物制造，使合成生物制造成为推动新一代生物技术在工业领域应用的主力军[1]。合成生物制造是指基于合成生物学思想和技术，设计、改造和创建合成/工程生物系统（synthetic/engineered biosystems），并利用合成/工程生物系统生产出具有商业价值的产品，用于健康、营养、农业、工业和环境保护等广泛的产业经济领域。合成生物制造所使用的合成/工程生物系统，广义上包括DNA、RNA、蛋白质（酶）、多酶体系、单一物种生物体和多物种生物群体等。由于低碳甚至“净零”碳排放，合成生物制造可以作为面向“碳中和”目标下的重要新技术手段，打开新的产业发展空间，推动区域经济绿色可持续发展做“加法”，实现财富绿色增长，为推进我国制造业向价值链高端攀升、推动生物经济发展、助力实现“双碳”目标、建设美丽中国和促进社会经济可持续发展形成重大科技支撑。

一、发展现状

1. 底层关键技术

合成生物制造的本质简单来说就是以工程化的理念，通过对DNA的精准操控，在中心法则的框架下，获得可预测、可定量的生物学性状，实现高效生物制造产品或开发高效生物工艺的目标。近年来，合成生物学发展飞速，直接原因是DNA测序、DNA合成、DNA编辑、基因表达调控、数字细胞模型、高通量筛选等底层关键技术的快速进步和成本的不断降低，显著提升了对DNA、RNA、蛋白质和细胞表型的计算设计、合成构建、高通量筛选和测试评价的能力，推动了合成生物制造向工程化和智能化发展。

50多年来，DNA测序技术已经从基于双脱氧链终止法为原理的第一代测序技术，发展到高通量、低成本的第二代测序技术，再发展到读长较长、可实现单分子测序与实时测序的第三代测序技术[2]。低成本、高通量的DNA测序技术，为揭示自然界生物多样性的遗传基础、提供合成生物制造的元件来源奠定了大数据基础，也在DNA水平上为合成/工程生物系统的构建提供了低成本的测试和校验平台。

最早基于亚磷酰胺三酯固相合成法的寡核苷酸合成技术，经过30多年的发展，成本已经从最初合成1 kb所需的1万～2万美元下降到100～200美元甚至更低。基于酶促反应的高选择性和高催化活性开发的酶促DNA从头合成技术[3]，有望进一步提高单步合成效率，但在酶对非天然底物的特异性、催化效率以及可控精确合成方面还有很多问题需要解决。未来希望实现每分钟内快速合成kb级长度的寡核苷酸链，合成效率超过99.99%，错误率低于0.01%，快速、高保真、低成本组装Mb级DNA片段[4]，并使工业DNA生产系统更加绿色、低碳、环保[5]。

目前主要应用的三大基因组编辑技术（ZFN、TALEN、CRISPR-Cas），其基本原理均是利用非特异性的核酸内切酶在“向导”引导下切割DNA靶位点，造成DNA双链断裂，从而激发DNA的修复机制，然后利用非同源末端连接（non-homologous end joining，NHEJ）或同源定向修复（homology directed repair，HDR），实现基因编辑的目标。相比NHEJ，HDR效率偏低。一种新的CRISPR方法可以在NHEJ存在的情况下，将HDR的效率提高到67%～100%[6]。基于CRISPR-Cas系统及衍生的基因组编辑和碱基编辑技术已经在大肠杆菌、谷氨酸棒状杆菌、枯草芽孢杆菌、链霉菌、罗氏真养菌、酿酒酵母、嗜热毁丝霉、黑曲霉等重要底盘微生物中得到广泛应用[7]。先导编辑技术、RNA编辑技术和转座酶编辑技术等在真核生物中发展起来的新技术，有望进一步提高合成/工程生物系统的基因编辑效率。

基因组尺度代谢网络模型可以使研究人员在计算机上描述和模拟生命代谢过程[8]。随着更多组学数据、胞内蛋白质和代谢物浓度、酶的催化特性[9]及反应的标准吉布斯自由能（Gibbs free energy）等参数的大量积累，可以通过整合多组学信息构建多约束代谢网络模型[10]，用于模拟基因敲除的影响、预测可行的代谢途径、找出代谢关键瓶颈信息，提高途径设计和靶点预测的精准性。依托海量生物数据，在细胞整体系统水平建立数字细胞模型[11]，将数字细胞技术与基因组测序、合成、编辑等其他底层技术结合起来，将显著提升设计、改造和创造生命的能力。

基因克隆、基因组编辑、编辑序列设计等生物技术的自动化实现，以及流式细胞、液滴微流控、全基因组规模扰动测序等高通量筛选技术的发展，显著提升了合成生物的迭代效率[12]。采用不断迭代的基因编辑、基因表达调控、高通量筛选、机器学习等技术，通过代谢途径设计重构、关键元件进化筛选、底盘形态和性能重塑等研究，氨基酸、有机酸、维生素、乙醇等大宗产品菌种性能不断提升[13]，天然化合物、手性化合物等一批新分子实现了生物合成[14]。

2. 酶与生物催化

高性能工业酶是生物制造产业的核心，2021 年全世界酶制剂市场规模约 70 亿美元，年增速 4%[15]，支撑着下游数十倍甚至数百倍规模的相关产业的发展。对酶蛋白进行分子设计和改造[16]，是创造高性能工业酶、降低生产成本、提升产业竞争力的关键。人工智能的发展显著提升了对酶的设计能力，推动了从新酶设计、新催化反应设计到新生物催化工艺的快速应用。从序列到功能的“端对端”一步设计策略和高度契合酶改造场景的人工智能新算法，在先进算法指导下进行计算机虚拟筛选及从头设计，显著提高了蛋白质改造效率[15]。随着大肠杆菌、芽孢杆菌、酵母、丝状真菌等蛋白表达系统的快速发展，酶的生产成本正在不断降低。多样化、高性能、低成本酶的合成生物制造实现，促进了酶在医药、食品、饲料、纺织、材料、发酵、能源、精细化学品和化学药品制造等重要工业领域中的应用和绿色生物工艺的建立，显著缩短了工艺流程，降低了物耗、能耗与水耗，助力传统加工产业的绿色发展[17]。例如，在机器学习指导下的酶分子改造，大幅度提升了聚对苯二甲酸乙二酯（PET）降解酶的活性，为解决塑料污染开辟了新途径[18]。

利用酶级联反应，将不同的生物转化按顺序反应结合起来合成产品，可以提高合成反应的选择性和减少化学反应的潜在危害，开发出环境友好和成本效益高的药物绿色生物合成新工艺[19]。这种生物合成过程，以简单生物工艺替代数十步的化学合成工艺，形成了绿色、清洁的生产模式，解决了化学合成高污染、高能耗等问题。诸如手性胺、手性醇、天然/非天然氨基酸、甾类化合物、芳香族化合物等医药化学品已实现了绿色生物合成[20]。基于酶分子的智能设计和改造技术，并将生物合成与化学合成紧密结合，从而有望实现更多化学品的绿色合成制造和产业的可持续发展，保障医药化工产业绿色发展。

3. 合成生物制造产品

在合成生物学和代谢工程技术的推动下，以谷氨酸棒状杆菌和大肠杆菌工程菌为代表的氨基酸菌种生产水平不断提升，*L*-谷氨酸、*L*-缬氨酸、*L*-赖氨酸发酵生产水平①可达 220～250 g/L，*L*-苏氨酸、L-丙氨酸发酵生产水平均超过 120 g/L，*L*-亮氨酸、*L*-异亮氨酸、*L*-酪氨酸、*L*-苯丙氨酸、*L*-色氨酸发酵生产水平达到 50～80 g/L[21]。近十

---

① 一般也称为滴度（titer），其单位是 g/L，衡量的是 1 L 发酵液中含有多少 g 目标产品。滴度是反映菌株生产能力的指标，滴度越高，越有利于后提取。

年来，我国氨基酸生产领域一个突出的创新是*L*-丙氨酸的无氧生物制造技术。氨基酸发酵基本都是在有氧状态下进行的，得率（yield）① 最高不超过 0.78 g/g，染菌风险和能耗高。研究人员首先阻断了大肠杆菌中将丙酮酸转化为乳酸、乙酸等有机酸的支路代谢途径，然后引入 NADH 依赖型的*L*-丙氨酸脱氢酶，使丙酮酸转化为*L*-丙氨酸成为无氧条件下再生酵解途径产生 NADH 的唯一途径，并通过在无氧条件下连续进化，提升细胞增殖能力，最后获得的工程菌株，菌种生产*L*-丙氨酸的生产强度（productivity）② 达到 3.9 g/（L·h），得率高达 0.95 g/g。这一技术在安徽华恒生物科技股份有限公司实现了产业化，在甲基甘氨酸二乙酸市场的拉动下，将*L*-丙氨酸从一个年产千吨级的产品提升为一个年产数万吨级的产品，成为国际上第一个在无氧条件下低成本生产氨基酸的范例[22]。

经过工程改造的工程菌解脂耶氏酵母可以高产三羧酸循环的中间产物——柠檬酸、异柠檬酸和α-酮戊二酸，但得率有待提升。在工程菌解脂耶氏酵母中共表达三羧酸线粒体转运蛋白和磷酸脱氢酶，降低胞内腺苷一磷酸（AMP）水平以弱化异柠檬酸脱氢酶活性，柠檬酸的生产水平可达到 97 g/L，得率为 0.5 g/g。而若弱化柠檬酸的转运，同时过表达异柠檬酸转运蛋白，异柠檬酸的生产水平达到 136.7 g/L，得率为 0.74 g/g。在解脂耶氏酵母中过表达其内源的 $NADP^+$ 依赖型异柠檬酸脱氢酶和丙酮酸羧化酶，α-酮戊二酸的发酵生产水平可达到 186 g/L，得率为 0.36 g/g[23]。以葡萄糖为原料生物合成琥珀酸为例，在生物合成过程中还能固定 CO_2，因此得率较高，如大肠杆菌工程菌生产琥珀酸，36 h 产量可达 100 g/L，得率 1.02 g/g。经过工程改造的大肠杆菌工程菌，在无氧条件下*L*-乳酸的发酵生产水平为 150 g/L，得率为 0.95 g/g；*D*-乳酸的发酵生产水平为 100～130 g/L，得率为 0.94～0.97 g/g。大肠杆菌生产琥珀酸和*D*-乳酸均于近年来实现了产业化[24]。

维生素在医药、食品添加剂、饲料添加剂、化妆品等领域具有广泛用途。近年来合成生物学与代谢工程技术的发展，极大地推动了维生素的生物合成不断替代化学合成[25]。维生素 B_2（核黄素）的发酵生产水平可以达到 30 g/L 左右，维生素 B_5（泛酸）可以通过*D*-泛解酸内酯和β-丙氨酸酶法合成获得，其发酵生产水平超过 100 g/L。经工程改造的大肠杆菌也可一步发酵生产泛酸，有望改变泛酸的生产路线。维生素 B_{12}（钴胺素）的发酵生产水平为 200～300 mg/L，已经可以支撑工业生产。维生素 C 是一种大宗维生素，两步发酵法生产水平可以达到 100 g/L 以上。大部分脂溶性维生

① 得率的单位是 g/g，衡量的是从 1 g 原料中能够获得多少 g 目标产品。得率主要影响生物制造的经济性，是最关键的指标。

② 生产强度反映的是单位时间内的发酵生产水平。生产强度越高，越有利于产能的提高。

素还是通过化学合成生产，以微生物发酵合成的β-法尼烯为中间体可以合成异植物醇，进而合成维生素E，实现维生素E的生物+化学合成。

天然产物是药物及其先导化合物的重要来源，近40年获批上市的药物中约有1/4来源于自然界天然产物及其衍生物。目前已知的植物天然产物超过20万种，我国的药用植物在1万种以上。基于合成生物学的原理，设计和创建微生物细胞工厂，发酵生产植物天然产物，可以突破植物资源限制，为植物天然产物的绿色高效合成提供新的路线。目前，青蒿素、紫杉醇、大麻素、莨菪碱等代表性的植物天然产物均已实现了微生物合成，萜类香精、人参皂苷、三萜酸、类胡萝卜素等萜类化合物，灯盏乙素、虎杖苷、根皮素、黄芩素等黄酮类化合物，天麻素、红景天苷、络塞维、迷迭香酸、复杂苯乙醇苷等苯丙素类化合物，以及生物碱类等典型的植物天然产物的生物合成都取得了很大的进展[26]。以解脂耶氏酵母为例，该菌中用于油脂合成的胞质乙酰辅酶A和丙二酰辅酶A，同时也是萜类、黄酮类等天然产物的合成前体；其较高的磷酸戊糖途径代谢通量能够提供大量NADPH供生物合成使用；其丰富的脂滴亚细胞结构，也可以为疏水性天然产物提供储存场所。近年来，经工程改造的解脂耶氏酵母被用于合成一系列萜类化合物，如环状单萜化合物柠檬烯，倍半萜类化合物α-法呢烯、诺卡酮、紫穗槐二烯，三萜类化合物齐墩果酸、桦木酸、原人参二醇、羽扇豆醇，四萜类化合物番茄红素、β-胡萝卜素、虾青素、β-紫罗兰酮等。经过工程改造的解脂耶氏酵母还可以生产100～1000 mg/L的柚皮素、雌黄醇和黄衫素等黄酮类化合物，以及2-苯基乙醇、对香豆酸、白藜芦醇、紫色杆菌素和脱氧紫色杆菌素等莽草酸途径衍生化合物[27]。合成生物制造在生产抗生素、农药/兽药、活性肽、功能蛋白质、功能多糖/寡糖、生物活性材料、生物纳米材料等方面也取得了很多新的进展。

近年来，食品生物制造技术及其产品迅猛发展。一方面，合成生物制造能够利用微生物生产大宗的食品原料，如蛋白质、脂肪等物质，生产出人造肉、人工奶和人工脂肪等产品，或生产出人体不能合成的功能性物质，包括阿洛酮糖等功能糖，重塑传统的食物生产模式，实现食物的绿色可持续供给[28]；另一方面，合成生物制造的飞速发展使高性能菌株的可获得性及获取效率显著提升，对传统低效的发酵优化放大技术提出很大挑战，亟须对发酵优化放大技术进行升级，以满足高通量菌种性能验证及工艺开发能力的需求，人工智能技术特别是数字孪生与知识图谱等技术的应用，将为传统发酵技术的颠覆性发展带来巨大推动力。

4. CO_2 转化和生物质原料利用

将CO_2转化为人类可以利用的有机物是有效固碳途径之一。在自然界中，植物和微生物利用自养固碳途径将CO_2转化为有机物，每年可固定超过7×10^{16} g的无机碳。

利用天然固碳途径来固定 CO_2 生产化学品、对天然固碳途径进行改造来提高固碳效率、创造全新的人工生物固碳途径[29]［如最小酶促碳固定循环 POAP[30] 和人工淀粉合成代谢途径（ASAP）[31]］是当前 CO_2 生物转化的三个主要方向。生物固碳是自然界亿万年进化的结果，生物早已适应了空气中低浓度 CO_2、低密度太阳能的自然条件。为了提高固碳途径的效率，需要进行高效固碳酶的筛选、设计和改造，为创建高效人工固碳途径奠定基础。同时，由于 CO_2 是高度氧化的产物，能量利用效率对 CO_2 人工生物转化的效率有很大的影响。开发生物-化学相组合的技术体系已经成为廉价、高效的 CO_2 还原和固定转化的新方向。具有代表性的例子包括：先通过 CO_2 加氢合成甲醇，再通过人工设计的全新体外多酶催化途径将甲醇转化为淀粉[31]；或通过电催化结合生物合成的方式，将 CO_2 高效还原合成高浓度乙酸，进一步利用微生物合成葡萄糖和脂肪酸[32]；以及电酶催化 CO_2 加氨合成甘氨酸[33]。这种走出植物光合作用模式的新的碳水化合物合成方式，以及不依赖生物体的脂肪酸和氨基酸等物质合成方式，实际上是人类对太阳能利用方式的一种突破。近年来，生物电化学研究发展迅速，不仅模拟海洋生态系统，创建由微生物群落组成的生物太阳能电池，实现生物直接利用太阳能发电[34]，也为利用电能进行生物固碳提供了可能性。未来生物电能利用系统的设计、构建和生物固碳应用方面，还有很大的想象和发展空间，如糖与水反应制氢、糖完全氧化产电，以及利用氢能或电能，将 CO_2 转化为糖的“电—氢—糖循环的新能源理论体系”[35]，以解决电储存和氢经济的“产储运”的难题。

生物质主要由 40%～50% 的纤维素、20%～40% 的半纤维素和 20%～30% 的木质素组成。之前制约纤维素利用的技术瓶颈——纤维素酶生产成本高，在经过多年研发后显著降低，目前每千克酶的成本已经降低到 10～20 美元。但纤维素酶活力的绝对值相比淀粉酶还有较大差距，生物质酶解的效率仍然远低于淀粉的酶解。传统的纤维素利用工艺需要经过生物质原料预处理、纤维素酶发酵和生物质酶解糖化、糖化液发酵生产目标产品三个步骤，而对能够直接降解纤维素的丝状真菌进行代谢工程改造，使其利用纤维素原料生产一步化工产品，生产效率得到了显著提升，因此成为近年来重要的发展方向。目前，经过代谢工程改造的嗜热毁丝霉，以葡萄糖为原料，苹果酸的发酵生产水平达到 226 g/L，得率为 0.88 g/g，生产强度为 1.57 g/（L·h）；直接以纤维素为原料可发酵生产 180 g/L 苹果酸，得率接近 1 g/g；以玉米芯水解液为碳源，也可生产超过 150 g/L 的苹果酸[24]。秸秆制蛋白、秸秆制淀粉[36]也是非常重要的发展方向，其中，有效解决木质素的分离或利用问题是一个核心挑战。

二、关键问题与发展方向

合成生物学技术驱动的生物制造，本质上是创造全新生物系统或改造现有生物系统，开发出新的或者改变现状的产品、工具及工艺，并实现规模化生产和商业化，为社会创造价值。目前，我国一批高等院校、科研机构、企业研发中心和研发型初创公司已建立包括基因元件挖掘、酶设计改造、合成途径优化、发酵放大与分离纯化等环节的合成生物制造全链条研发平台。研发人员已经可以在6～12个月内实现从研究想法到获得酶或菌株，产业部门则可以在12～24个月内实现从酶或菌株到生产出产品，研发和产业化的速度显著提升。一大批化学品、药物、营养品及香精香料等重要产品的合成生物制造已取得突破，正在实现工业化生产的过程中。

从研发角度看，合成生物制造的研发核心包括三个环节：一是高效、专用合成生物（包括酶和菌种）的创建；二是高效、低成本和智能化的生产环节（包括发酵和分离提取过程）；三是产品的应用研发。目前，这三个环节都还有痛点和问题需要解决，需要加强对生物制造底层关键技术的支持，高度重视生物制造产品应用研发，重点是在产业需求引导下，以行业领军企业牵头，找出制约产业发展的底层关键技术方面的真问题，通过生物技术与信息技术的结合，真正解决问题，推动产业发展。

从产业角度看，合成生物制造的产业链亟须延长。目前，生物制造产业链过短，基本是从生物质原料到发酵产品，对经济社会的影响力有限。生物制造从大到强的关键转变是要迅速拓展生物制造的产业链，这不仅需要加强对生物制造核心科技——合成/工程生物系统的研发，还需要现有生物制造产品的应用研发，包括生物制造产品衍生物开发、新应用领域开发、新的解决方案服务。这三个方面的进步将显著延长生物制造的产业链，扩大生物制造在社会经济领域的影响力。

从市场角度看，合成生物制造产业创造新需求的科技动能不足，需创新监管政策，释放科技活力，创造市场需求，带动产业发展。例如，功能明确的重要植物天然产物，以及生物制造未来食品及其组分，其基因重组合成产物市场准入方面还存在限制，是急需监管部门解决的问题。制定科学合理的市场准入、安全评价和规范化管理的政策法规，加快推动合成生物技术成果的产业化应用，是促进我国生物制造产业发展、保障生物产业链安全、参与国际市场竞争的重要课题。创新监管政策，让监管政策助力生物制造产业发展和科技竞争力的提升，有助于进一步释放科技人员的创新活力，将国家巨大的科技投入迅速转换为促进产业发展的新动能。

三、结 束 语

高通量、低成本的DNA测序、合成、编辑技术和基因表达调控技术，正在从根本上重塑以DNA操控为核心的现代生物技术的发展面貌。从核酸元件的设计和改造，到酶和蛋白质的设计和改造，再到细胞工厂和生物体系的设计和改造，合成/工程生物系统的物种和功能多样性已经得到极大的拓展，合成生物制造能力正在不断得到提升，合成生物制造的产业化速度明显加快。随着国家和地方科技资源及社会资本对合成生物学领域支持力度的不断加大，合成生物学和生物制造领域优秀人才的不断汇聚，合成生物制造领域的重大原创成果将会不断涌现。以颠覆性为特征的合成生物学技术，将支撑合成生物制造产业快速成长壮大，助力经济社会的绿色可持续发展。

参考文献

[1] 李寅．合成生物制造．生物工程学报，2022，38(4)：1267-1294.

[2] Karst S M，Ziels R M，Kirkegaard R H，et al. High-accuracy long-read amplicon sequences using unique molecular identifiers with nanopore or PacBio sequencing. Nature Methods，2021，18(2)：165-169.

[3] Lu X Y，Li J L，Li C Y，et al. Enzymatic DNA synthesis by engineering terminal deoxynucleotidyl transferase. ACS Catalysis，2022，12(5)：2988-2997.

[4] Consortium E B R. Engineering biology：a research roadmap for the next-generation bioeconomy. https：//roadmap. ebrc. org[2023-03-01].

[5] 冯淼，王丽娜，汪保卫，等．工业生物技术中DNA合成发展现状及展望．生物工程学报，2022，38(11)：4115-4131.

[6] Ploessl D，Zhao Y，Cao M，et al. A repackaged CRISPR platform increases homology-directed repair for yeast engineering. Nature Chemical Biology，2022，18(1)：38-46.

[7] 杨超，董兴啸，张学礼，等．基因组编辑技术在工业生物领域中的应用现状及展望．生物工程学报，2022，38(11)：4132-4145.

[8] Fang X，Lloyd C J，Palsson B O. Reconstructing organisms in silico：genome-scale models and their emerging applications. Nature Reviews Microbiology，2020，18(12)：731-743.

[9] Li F，Yuan L，Lu H，et al. Deep learning-based k_{cat} prediction enables improved enzyme-constrained model reconstruction. Nature Catalysis，2022，5(8)：662-672.

[10] Lu H，Kerkhoven E J，Nielsen J. Multiscale models quantifying yeast physiology：towards a whole-cell model. Trends Biotechnology，2022，40(3)：291-305.

[11] 袁倩倩，毛志涛，杨雪，等．数字细胞模型的研究及应用．生物工程学报，2022，38(11)：4146-4161.

[12] 涂然，毛雨丰，刘叶，等．工程菌种自动化高通量编辑与筛选研究进展．生物工程学报，2022，38(11)：4162-4179.

[13] 周文娟，付刚，齐显尼，等．发酵工业菌种的迭代创制．生物工程学报，2022，38(11)：4200-4218.

[14] Yu T，Boob A G，Volk M J，et al. Machine learning-enabled retrobiosynthesis of molecules. Nature Catalysis，2023，6(2)：137-151.

[15] 曲戈，袁波，孙周通．工业蛋白质理性设计与应用．生物工程学报，2022，38(11)：4068-4080.

[16] Wang Y J，Xue P，Cao M F，et al. Directed evolution：methodologies and applications. Chemical

Reviews,2021,121(20):12384-12444.

[17] Lu H Y,Diaz D J,Czarnecki N J,et al. Machine learning-aided engineering of hydrolases for PET depolymerization. Nature,2022,604(7907):662-667.

[18] 郑宏臣,徐健勇,杨建花,等. 工业酶与绿色生物工艺的核心技术进展. 生物工程学报,2022,38(11):4219-4239.

[19] Benítez-Mateos A I,Roura Padrosa D,Paradisi F. Multistep enzyme cascades as a route towards green and sustainable pharmaceutical syntheses. Nature Chemistry,2022,14:489-499.

[20] 陈曦,吴凤礼,樊飞宇,等. 手性医药化学品的绿色生物合成. 生物工程学报,2022,38(11):4240-4262.

[21] Wendisch V F. Metabolic engineering advances and prospects for amino acid production. Metabolic Engineering,2020,58:17-34.

[22] 刘萍萍,郭恒华,张冬竹,等. L-丙氨酸厌氧发酵关键技术及产业化. 生物工程学报,2022,38(11):4329-4334.

[23] 荣兰新,刘士琦,朱坤,等. 代谢工程改造解脂耶氏酵母合成羧酸的研究进展. 生物工程学报,2022,38(4):1360-1372.

[24] 李金根,刘倩,刘德飞,等. 秸秆真菌降解转化与可再生化工. 生物工程学报,2022,38(11):4283-4310.

[25] 王岩岩,刘林霞,金朝霞,等. 代谢工程在维生素生产中的应用及研究进展. 生物工程学报,2021,37(5):1748-1770.

[26] 毕慧萍,刘晓楠,李清艳,等. 植物天然产物微生物重组合成研究进展. 生物工程学报,2022,38(11):4263-4282.

[27] 张金宏,崔志勇,祁庆生,等. 解脂耶氏酵母表达调控工具的开发及天然产物合成的研究进展. 生物工程学报,2022,38(2):478-505.

[28] 李德茂,童胜,曾艳,等. 未来食品的低碳生物制造. 生物工程学报,2022,38(11):4311-4328.

[29] 蔡韬,刘玉万,朱蕾蕾,等. 二氧化碳人工生物转化. 生物工程学报,2022,38(11):4101-4114.

[30] Xiao L,Liu G,Gong F,et al. A minimized synthetic carbon fixation cycle. ACS Catalysis,2022,12:799-808.

[31] Cai T,Sun H B,Qiao J,et al. Cell-free chemoenzymatic starch synthesis from carbon dioxide. Science,2021,373(6562):1523-1527.

[32] Zheng T,Zhang M,Wu L,et al. Upcycling CO_2 into energy-rich long-chain compounds via electrochemical and metabolic engineering. Nature Catalysis,2022,5:388-396.

[33] Wu R,Li F,Cui X,et al. Enzymatic electrosynthesis of glycine from CO_2 and NH_3. Angewandte Chemie International Edition,2023,62(14):e202218387.

[34] Zhu H,Xu L,Luan G,et al. A miniaturized bionic ocean-battery mimicking the structure of marine microbial ecosystems. Nature Communications,2022,13:5608.

[35] 宋云洪,吴冉冉,魏欣蕾,等. 电-氢-糖循环的新能源体系研究进展. 生物工程学报,2022,38(11):

4081-4100.
[36] Xu X, Zhang W, You C, et al. Biosynthesis of artificial starch and microbial protein from agricultural residue. Science Bulletin, 2023, 68(2): 214-223.

Biomanufacturing Driven by Synthetic/ Engineered Biosystems

Li Yin

This article summarizesthe recent advances in biomanufacturing driven by synthetic/engineered biosystems. The enabling technologies includeing DNA sequencing, DNA synthesis, and DNA editing as well as in silico cell modeling are highlighted. Subsequently, biomanufacturing of enzymes and biocatalytic products, amino acids, organic acids, vitamins, natural products, and future foods are briefly discussed. Finally, carbon dioxide and biomass utilization technologies are discussed to help the readers gain insight into this rapidly developing field.

第三章

2021年中国科研代表性成果

Representative Achievements of Chinese Scientific Research in 2021

3.1　黎曼流形的极限空间及应用

江文帅

（浙江大学数学科学学院）

自然界中并非处处规整，如宇宙空间的黑洞、地壳中的断层、玻璃破碎的裂缝等，这些特殊现象的研究都是各个相关领域的核心课题，虽然这些现象相对来看似乎不是“很重要或很显眼”，但它们对整体“结构”的影响是巨大的。不规整在数学中被称为“奇异”（singular），很多数学问题的研究都离不开对奇异性的分析。1981 年，著名几何学家格罗莫夫（M. Gromov）[1]引进了格罗莫夫-豪斯多夫距离（Gromov-Hausdorff distance）用于研究空间的几何结构与几何性质。随后，一批数学家通过研究满足一定弯曲程度的空间（称作“黎曼流形”）序列在格罗莫夫-豪斯多夫距离下的收敛极限，研究极限空间的整体结构与奇异结构，证明了一系列奠基性的成果。这些成果在若干重大问题的解决中起到了至关重要的作用，其中七大“千禧年大奖难题”之一的庞加莱猜想（Poincaré conjecture）的解决就非常依赖相关成果，此外，著名的丘（成桐）-田（刚）-唐纳森猜想的证明就基于格罗莫夫-豪斯多夫距离在流形上的进一步发展以及研究。

1997 年，齐格（J. Cheeger）与科尔丁（T. Colding）证明了[2]满足一定弯曲程度的流形（爱因斯坦流形）序列的极限空间由正则部分与奇异部分构成，奇异部分相对很小但是结构较复杂。为了进一步研究极限空间，他们猜测奇异部分具有量化的大小（被称为有限测度猜想），对于这一猜想，我们通过进一步发展极限空间理论，引进颈区域分析、流形分解等技术，并建立一套新方法处理高维奇异性问题，从而完全解决了该猜想[3]。同时通过这套方法，我们还解决了齐格-纳伯（A. Naber）关于爱因斯坦流形的曲率积分猜想，该曲率积分具有重要的理论意义，被国际同行评价为奠基性的成果。

此外，我们发展的方法还具有广泛的应用价值，可以用于研究一般弯曲条件（里奇曲率有下界）的黎曼流形的极限空间。由于一般弯曲条件的情形缺少方程的正则性，研究更加困难，自 1997 年后，关于该极限空间的奇异部分几乎没有任何完整的结果。通过进一步发展我们[3]的方法并证明流形上调和函数更强的衰减估计，我们可以证明极限空间奇异部分具有可求长结构，并且奇异部分的大小具有量化的刻画[4]。该成果被世界闻名的布尔巴基讨论班①组织专题讨论。

我们发展的这套理论，不仅可以应用于研究黎曼流形的极限空间，还可以应用于研究几何其他的一些重要分支[3,4]。目前，该理论已被成功应用于证明杨-米尔斯场的能量恒等式[5]、证明规范场的最优估计[6]、证明流形边界结构定理[7]、解决边界测度猜想[8]等。这些应用都基于我们建立的颈区域分析与流形分解技术。

我们的方法以及极限空间理论的研究由于具有重要的理论应用价值，它的研究远没有结束。从目前的研究可知，它与多个公开的重要问题有着密切的关联，数学家们希望进一步发展极限空间理论来解决几何领域更多的问题，这也将是未来该领域的研究重点之一。

参考文献

[1] Gromov M. Groups of polynomial growth and expanding maps. Publications Mathématiques de l'Institut des Hautes Études Scientifiques，1981，53：53-78.

[2] Cheeger J，Colding T. On the structure of spaces with Ricci curvature bounded below. I. J Differential Geom，1997，46(3)：406-480.

[3] Jiang W S，Naber A. L^2 curvature bounds on manifolds with bounded Ricci curvature. Ann of Math，2021，193(1)：107-222.

[4] Cheeger J，Jiang W S，Naber A. Rectifiability of singular sets of noncollapsed limit spaces with Ricci curvature bounded below. Ann of Math，2021，193(2)：407-538.

[5] Naber A，Valtorta D. Energy identity for stationary Yang Mills. Invent Math，2019，216(3)：847-925.

[6] Wang Y. Sharp estimate of global Coulomb gauge. Comm Pure Appl，2020，73(12)：2556-2633.

[7] Bruè E，Naber A，Semola D. Boundary regularity and stability for spaces with Ricci bounded below. Invent Math，2022，228(2)：777-891.

[8] Bruè E，Mondino A，Semola D. The metric measure boundary of spaces with Ricci curvature bounded below. arXiv：2205. 10609v1[math. DG]，2022.

① 法语名为 Séminaire Nicolas Bourbaki，一个现代数学讨论班系列，1948 年起在巴黎举办，主要活动是出版和每年三次的研究集会。

Gromov-Hausdorff Limit of Riemannian Manifolds and Applications

Jiang Wenshuai

The Gromov-Hausdorff distance is an important tool for studying the geometry of Riemannian manifolds. Consider a sequence of Riemannian manifolds with bounded Ricci curvature and non-collapsed volume. Up to taking a subsequence, there always exists a limit space under Gromov-Hausdorff distance. It is significant to study the structure of this limit space. Based on the neck region analysis and the decomposition theory for manifolds, we can show that the singular set has finite codimension four measure, which solves the Cheeger-Colding conjecture. Furthermore, we can prove a uniform L^2 bound for the Riemann curvature on manifolds with bounded Ricci curvature and non-collapsed volume, which solves Cheeger-Naber L^2 conjecture. We further develop these new techniques and prove a structural theorem for the singular set of limit space with lower Ricci curvature bounds.

3.2 “慧眼”卫星证认快速射电暴起源于磁星 SGR J1935+2154

李承奎

（中国科学院高能物理研究所）

快速射电暴是非常剧烈的天文爆发现象，在无线电波段几毫秒内释放巨大能量，爆发时是宇宙中最亮的射电源。科学家们发现它们是来自遥远宇宙的爆发，有时候不仅会重复发生，而且拥有极高的爆发率。由于一直无法找到准确的对应天体，快速射电暴的起源成为当代天文学的一大谜题。自发现以来的十多年间，科学家们对它的认知集中在直接观测性质，如短时标、高亮度、远距离、大数量、可重复等。

北京时间 2020 年 4 月 28 日 22 时 34 分 24 秒，加拿大 CHIME 天文望远镜和美国 STARE2 射电望远镜分别独立捕捉到同一个快速射电暴 FRB 200428[1, 2]。在 FRB 200428 射电爆发时刻前 8.6 s，我国的硬 X 射线调制望远镜（Insight-HXMT，简称“慧眼”卫星）探测到一例来自 SGR J1935+2154 的 X 射线暴[3]。此时间间隔正好符合通过射电辐射测量的色散延迟的预期。这说明 X 射线爆发和快速射电暴很可能源于同一次天体爆发事件。

接收到 X 射线信号后，“慧眼”卫星以远远高于射电望远镜的精度定位了 X 射线暴，发现跟磁星 SGR J1935+2154 的位置一致（图 1）。并且在 X 射线爆发数据中，发现了两个 X 射线脉冲与快速射电暴的两个脉冲在时间上高度吻合（图 2）。国际上有数台高能望远镜观测到这次爆发，其中“慧眼”卫星提供了最为丰富、精细和准确的时变和能谱信息，并率先指出光变细节上的契合强有力地证实了此 X 射线暴与 FRB 200428 是同一次天体爆发过程产生的。

在观测到与 FRB 200428 相关联的 X 射线暴的高能卫星中，除了“慧眼”卫星以外，其他都不能明确爆发能谱的辐射属性。得益于 1～250 keV 的宽能段覆盖，“慧眼”卫星可以明确此 X 射线爆发源自非热辐射机制（图 3），在磁星 X 射线暴中非常特殊。具体来说，此 X 射线暴的高能截断能量高，低能能谱指数是所有磁星 X 射线暴中最“软”① 的[4]。非热辐射能谱暗示爆发辐射区域粒子数密度较低，散射不充分，可能位于较高

① “软”指能谱中低能端的光子数较多，高能端的光子数较少。

纬度磁极附近。这样的辐射环境也有利于相干辐射的成功释放，产生快速射电暴。

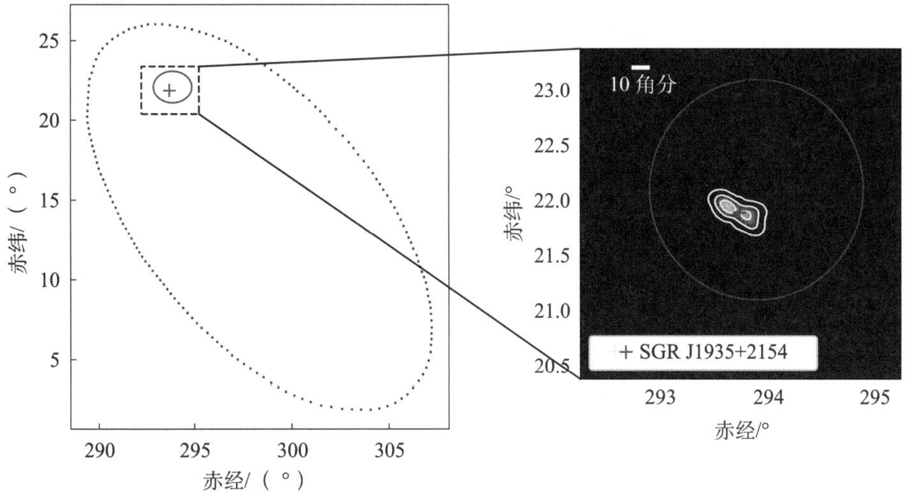

图 1 “慧眼”卫星对与 FRB 200428 相关联 X 射线暴的定位[3]

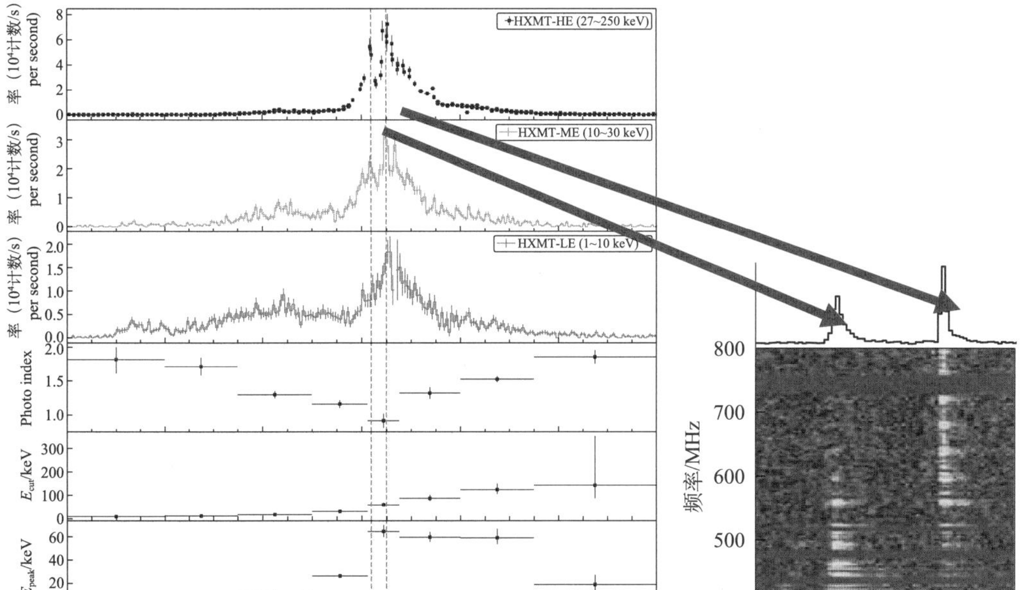

图 2 “慧眼”卫星观测到 X 射线暴中的两个脉冲与 CHIME 天文望远镜观测到 FRB 200428 的两个射电脉冲信号时间高度吻合

资料来源：左图摘自文献［3］，右图摘自文献［1］。

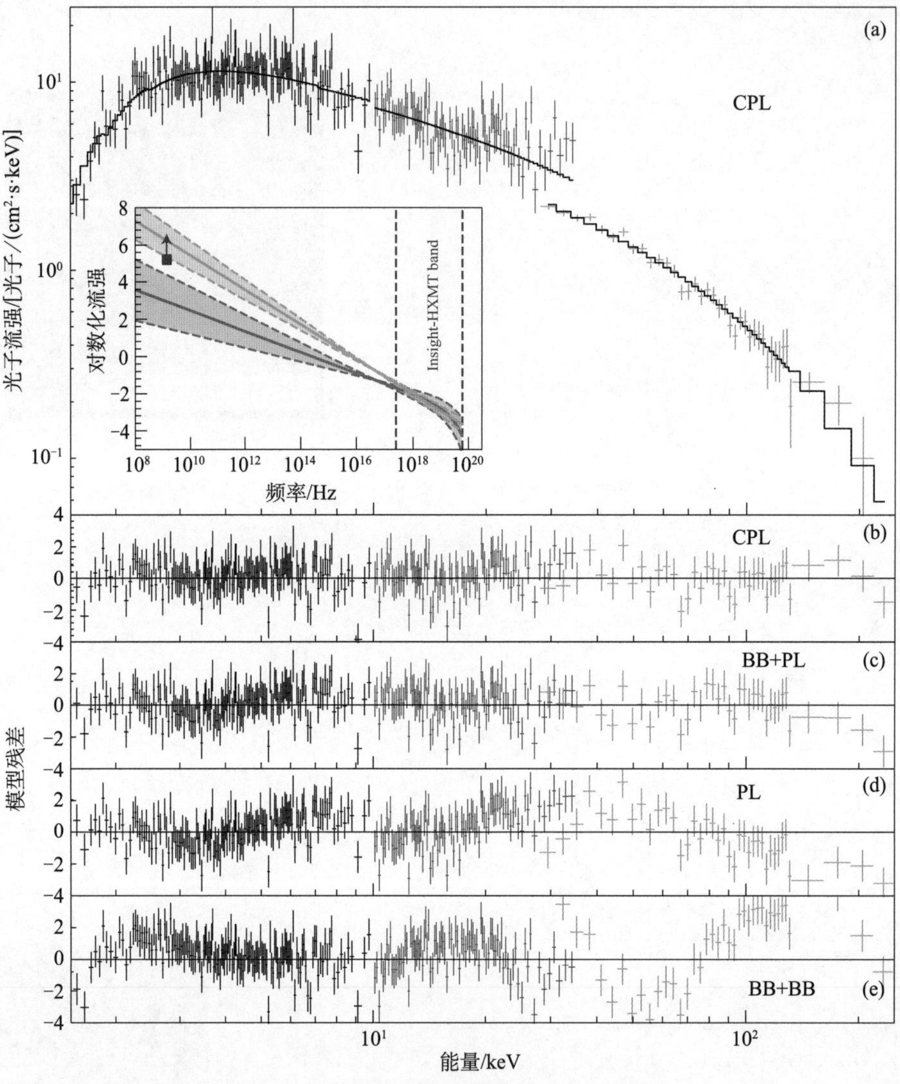

图 3 “慧眼”卫星观测能谱及其拟合结果[3]

(a)“慧眼”能谱及其 X 射线能谱的外推情况；(b) 指数截断幂律谱（CPL）拟合残差；(c) 黑体谱加幂律谱（BB+PL）拟合残差；(d) 幂律谱（PL）拟合残差；(e) 双黑体谱（BB+BB）拟合残差。

快速射电暴起源于银河系内磁星这一发现同时入选《自然》期刊“2020 年度十大科学发现”和《科学》期刊“2020 年度十大科学突破”。这是人类对快速射电暴研究的里程碑。在此次发现中，“慧眼”卫星凭借自身卓越的性能做出了举足轻重的贡献。“慧眼”卫星捕捉到与快速射电暴相关联的 X 射线暴，也是人类首次观测到快速射电暴在其他电磁波段的对应体，通过对 X 射线暴的定位确认快速射电暴起源于磁星，并明确 X 射线暴非热辐射起源的特殊性。“慧眼”卫星的研究成果于 2021 年 2 月 18 日在线发表在《自然-天文》（*Nature Astronomy*）期刊上[3]，截至 2023 年 4 月 19 日，引用次数高达 191 次。

2020 年 4～5 月，“慧眼”卫星对 SGR J1935＋2154 进行了为期一个多月的定点观测，探测到了 75 个 X 射线爆发，并对这些爆发进行了光变和能谱等性质的统计分析[5, 6]。林琳等利用 500 米口径球面射电望远镜（FAST，被誉为“中国天眼”），给出了其中 29 个爆发的射电流量上限[7]，说明大部分磁星 X 射线爆发缺少射电辐射。Westerbork 在 2020 年 SGR J1935＋2154 的爆发活跃期内，发现了两个较弱的射电爆发，但是均未被同时观测的 X 射线卫星观测到 X 射线爆发，说明射电爆发也不一定伴随有 X 射线爆发[8]。

天文学家们构建了许多理论模型来解释观测到的快速射电暴现象，现在已经确定快速射电暴一定是相干辐射。目前主要有两种物理模型用于解释相干辐射的来源：一种是类脉冲星模型，认为相干辐射来自脉冲星的磁层，通过电荷束相干形成的极亮的单脉冲；另一种是类伽马暴模型，认为相干辐射来自脉冲星光速圆柱面以外，通过脉冲星活动触发激波，激波波前与抛射物的壳层发生相互作用，触发脉泽辐射实现高亮度的相干辐射[9]。“慧眼”卫星观测结果表明，不同波段的辐射来自大致相同的区域，这对类伽马暴模型提出了非常苛刻的要求。类脉冲星模型认为，X 射线暴产生于磁层以内，同样产生于磁层内的射电辐射可以自然地解释观测现象。因此，“慧眼”卫星的观测结果倾向于认为 FRB 200428 的辐射来自磁星磁层内的相干辐射。除此之外，“慧眼”卫星确认的 X 射线暴特殊的非热辐射起源也可以进一步限制爆发在磁层中的位置。

2022 年 10 月，SGR J1935＋2154 再次进入爆发活跃期。2022 年 10 月 14 日，我国的引力波暴高能电磁对应体全天监测器卫星（简称 GECAM 卫星，即“怀柔一号”）发现一个跟快速射电暴相关联的 X 射线暴，并确认其来自银河系内的磁星 SGR J1935＋2154[10]。2022 年 10 月 21 日，“慧眼”卫星与中国科学院云南天文台的射电望远镜，再次观测到 SGR 1935＋2154 的一次 X 射线和射电波段的同时爆发[11]。2022 年 11 月 20 日，“怀柔一号”卫星与中国科学院云南天文台和新疆天文台的两台射电望远镜，也同时观测到了 X 射线和射电波段的爆发。SGR J1935＋2154 的这三次新的射电和 X 射线的同时爆发，再次证明磁星 SGR J1935＋2154 是快速射电暴的起源天体，并且可以重复发生。这进一步证明磁星爆发过程可以产生快速射电暴，为深入理解快速射电暴的辐射机制和磁星的爆发机制提供了极为宝贵的数据。

参考文献

[1] CHIME/FRB Collaboration, Andersen B C, Bandura K M, et al. A bright millisecond-duration radio burst from a Galactic magnetar. Nature, 2020, 587: 54.

[2] Bochenek C D, Ravi V, Belov K V, et al. A fast radio burst associated with a Galactic magnetar. Nature, 2020, 587: 59.

[3] Li C K, Lin L, Xiong S L, et al. HXMT identification of a non-thermal X-ray burst from SGR J1935+2154 and with FRB 200428. Nature Astronomy, 2021, 5: 378-384.

[4] Younes G, Baring M G, Kouveliotou C, et al. Broadband X-ray burst spectroscopy of the fast-radio-burst-emitting Galactic magnetar. Nature Astronomy, 2021, 5: 408-413.

[5] Cai C, Xue W C, Li C K, et al. An Insight-HXMT Dedicated 33 day observation of SGR J1935+2154. I. Burst Catalog. ApJS, 2022, 260: 24.

[6] Cai C, Xiong S L, Lin L, et al. An Insight-HXMT Dedicated 33 day observation of SGR J1935+2154. II. Burst Spectral Catalog. ApJS, 2022, 260: 25.

[7] Lin L, Zhang C F, Wang P, et al. No pulsed radio emission during a bursting phase of a Galactic magnetar. Nature, 2020, 587: 63.

[8] Kirsten F, Snelders M, Jenkins, M, et al. Detection of two bright radio bursts from magnetar SGR 1935 + 2154. Nature Astronomy, 2021, 5: 414.

[9] Zhang B. The physical mechanisms of fast radio bursts. Nature, 2020, 587: 45.

[10] Wang C W, Xiong S L, Zhang Y Q, et al. GECAM and HEBS detection of a short X-ray burst from SGR J1935+2154 associated with radio burst. The Astronomer's Telegram, 2022, 15682.

[11] Li X B, Zhang S N, Xiong S L, et al. Insight-HXMT detection of an X-ray burst from SGR J1935+2154 coinciding with the radio burst on 2022-10-21. The Astronomer's Telegram, 2022, 15708.

HXMT Identification of a Non-thermal X-ray Burst from SGR J1935+2154

Li Chengkui

A fast radio burst in our own galaxy was selected as one of the 10 remarkable discoveries by *Nature* and breakthroughs by *Science* in the year 2020. This is a milestone in the research of fast radio bursts. The Insight-HXMT satellite has made a significant contribution to this discovery with its excellent performance. Insight-HXMT captured the X-ray bursts associated with fast radio bursts. This is the first time that the counterpart of fast radio bursts has been confirmed in other electromagnetic wavelengths. Insight-HXMT identified the origin of fast radio bursts from magnetars and clarified the special nature of the non-thermal radiation origin of the X-ray burst. The research result was published online in *Nature Astronomy* on February 18,2021,and has been cited 191 times as of April 19,2023.

3.3 “太极一号”迈出中国空间引力波探测的第一步

罗子人[1,2,3]　张　敏[1,2,4]　胡文瑞[1,3]　王建宇[1,2]　吴岳良[1,2,4]

（1. 中国科学院大学引力波宇宙太极实验室；
2. 中国科学院大学杭州高等研究院浙江省引力波精密测量重点实验室；
3. 中国科学院力学研究所引力波实验中心；
4. 国际理论物理中心（亚太地区），中国科学院大学）

引力波探测“太极计划”是2008年由中国科学院牵头组织论证和规划的中国首个空间引力波探测计划[1-3]。2016年，“太极计划”进一步深化了切实可行的“三步走”发展路线图①[2,3]，最后跟LISA在同一时期发射三颗卫星，实现空间引力波测量。其中规划的第三步——2030年前后发射的探测星组可与美国国家航空航天局（NASA）和欧洲空间局（ESA）合作的空间引力波计划激光干涉空间天线（Laser Interferometer Space Antenna，LISA）形成空间引力波探测的联合编队，能更好地提高空间引力波探测的精度和确定引力波的波源[4]。

由于空间引力波探测所涉及的技术和工程难度极高，其中许多技术是目前人类已掌握的最顶尖的技术，“太极计划”发展路线图首先规划了“太极一号”试验单星和“太极二号”试验双星系统，旨在对所涉及的关键技术进行高精度的在轨验证。由中国科学院空间科学战略性先导科技专项支持的“太极一号”就是试验星系统中的首颗空间引力波探测技术实验卫星。

2019年8月31日，“太极一号”成功发射，“太极一号”主要对星载干涉仪系统和无拖曳控制系统进行了全面的在轨验证[5]（图1）。这两个系统是空间引力波探测最核心的关键技术，由此可以说，“太极一号”的成功为我国“太极计划”奠定了坚实的基础。

① 第一步，发射单颗卫星来验证“太极计划”技术路线图的可行性。“太极计划”的探路者于2019年8月31日成功发射，目前正在运行，它的正式名称是“太极一号”。第二步，2024年发射两颗卫星来验证关键技术，包括在太空中进行甚长基线干涉测量。

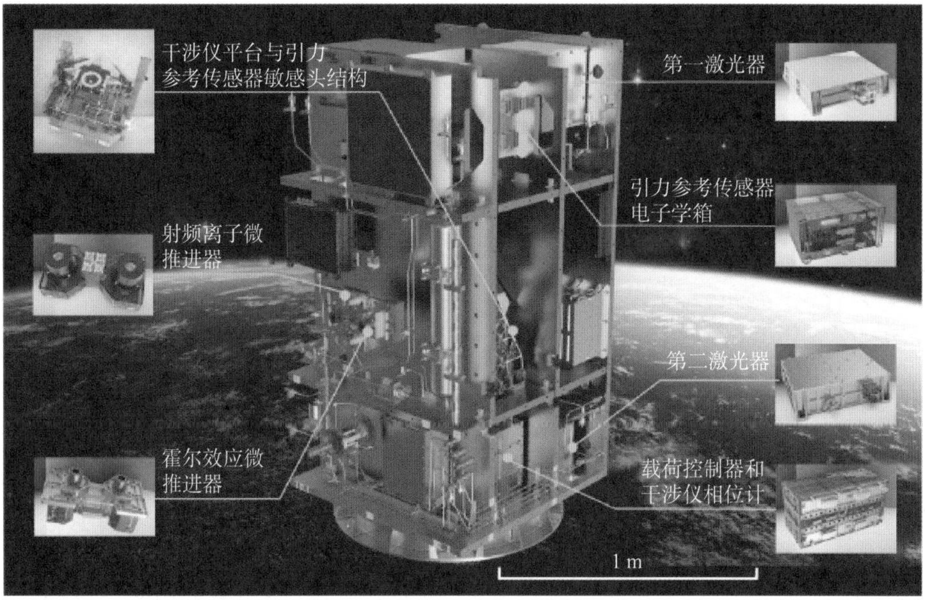

(a) “太极一号”载荷布局图

(b) “太极一号”卫星装配图

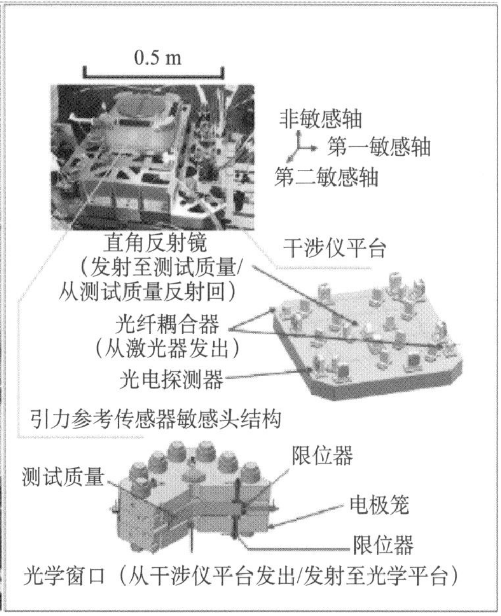

(c) “太极一号”核心测量分系统

图 1 “太极一号”卫星及载荷

“太极一号”于2022年底完成了全部的在轨实验验证。在轨实验验证结果和实测数据表明，“太极一号”取得了干涉仪在轨测量的国内最高精度——在0.01～0.1 Hz达到了100 $pm/Hz^{1/2}$水平，在部分频段可达到25 $pm/Hz^{1/2}$；引力参考传感器在轨实现了$10^{-10}ms^{-2}/Hz^{1/2}$的测量精度，测量精度与动态范围的比率达到$2\times10^{-6}/Hz^{1/2}$，达到国内最高水平。“太极一号”首次在国际上实现了对微牛级射频离子微推进系统和微牛级双模霍尔微推进系统的在轨飞行试验，微推力噪声水平达到了0.15 $\mu N/Hz^{1/2}$；首次实现了我国无拖曳控制的飞行实验，无拖曳控制后的卫星残余加速度水平达到$10^{-9}ms^{-2}/Hz^{1/2}$；对超稳超静卫星平台系统开展了在轨飞行验证，并对卫星平台进行温度控制，在核心测量分系统附近，精密控温达到±2.6 mK，是目前国内最高水平。

“太极一号”实现的重要成果已发表在《自然》期刊子刊《通信-物理》(*Communications Physics*)上[5]。更多有关“太极一号”的研究成果，已由世界科学出版社(World Scientific)以专辑特刊的形式发表在《国际现代物理杂志A》(*International Journal of Modern Physics A*)[6]上。专辑共收录了26篇研究论文，论文作者涉及来自全国30余所科研院校的180余名“太极计划”科学合作组成员；论文内容涉及干涉仪系统、引力参考传感器系统、微推进系统、无拖曳控制技术、超稳超静卫星平台系统、数据处理与仿真技术等，几乎涵盖了空间引力波探测所涉及的全方位技术。

“太极一号”在完成既定的科学任务后，于2021年进入扩展任务模式。扩展任务阶段，“太极一号”持续收集精密定轨、卫星姿态、卫星非保守力等数据。同时，“太极一号”还作为地球重力卫星对全球重力场进行测绘。“太极一号”利用引力参考传感器的数据并结合“北斗”导航的数据，给出了我国首个完全自主的全球重力场模型[7]（图2为“太极一号”数据生成的全球重力场模型前20阶的球谐系数）。

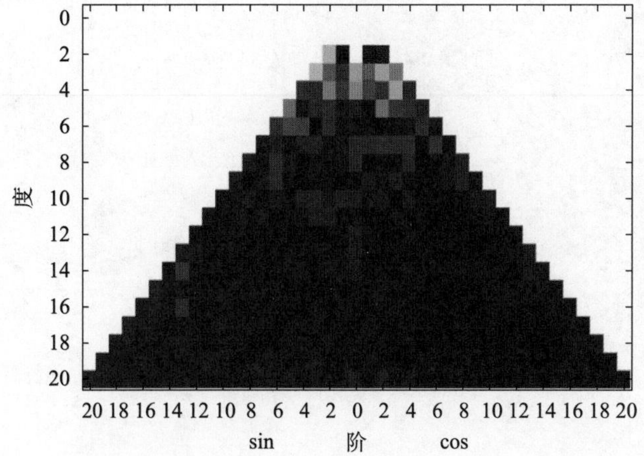

图2 “太极一号”数据生成的全球重力场模型前20阶的球谐系数

“太极一号”不仅迈出了我国空间引力波探测奠基性的第一步，还实现了我国全球重力场测绘零的突破。空间引力波探测的优先部署，不仅将为我国空间科学和基础科学取得重大突破并奠定坚实基础，还将为国民经济发展和国家战略需求提供重要的前瞻性技术支撑。

参考文献

[1] Hu W R, Wu Y L. The Taiji Program in Space for gravitational wave physics and the nature of gravity. Natl Sci Rev, 2017, 4: 685.

[2] Luo Z R, Guo Z K, Jin G, et al. A brief analysis to Taiji: Science and technology. Results in Physics, 2020, 16: 102918.

[3] Luo Z R, Jin G, Wu Y L, et al. The Taiji Program: a concise overview. Progress of Theoretical and Experimental Physics, 2021, (5): 05A108.

[4] Ruan W, Liu C, Guo Z, et al. The LISA-Taiji network. Nat Astron, 2020, 4: 108.

[5] The Taiji Scientific Collaboration. China's first step towards probing the expanding universe and the nature of gravity using a space borne gravitational wave antenna. Communications Physics, 2021, 4: 34.

[6] Wu Y L, Luo Z R, Hu W-R, et al. Taiji program in space for gravitational universe with the first run key technologies test in Taiji-1. Intern J Mod Phys A(special Issue), 2021, 36: Nos. 11 & 12.

[7] Wu L M, Xu P, Zhao S H, et al. Independent global gravity field Model from Taiji-1 observations. Microgravity Science and Technology, 2022, 34: 77.

Taiji-1 Toward the First Step of Chinese Space-Borne Gravitational Wave Detection Mission

Luo Ziren, Zhang Min, Hu Wenrui, Wang Jianyu, Wu Yueliang

The Taiji-1 satellite is the first pilot satellite mission of Taiji program for the space-based gravitational wave detection in China, which is used to verify Taiji's key technology and also to testify the feasibility of Taiji roadmap. Taiji-1 was launched on August 31, 2019, and its designed mission was successfully completed. The in-orbit scientific achievements of the Taiji-1 satellite in the first stage have been published and it was now in the extended task phase. By using the data from Taiji's inertial sensor and the Beidou navigation system, the China's first domestically developed global gravity field model is produced.

3.4 量子计算优越性研究

彭礼超 钟翰森 王 辉 邓宇皓
陈明城 陆朝阳 潘建伟

（1. 中国科学技术大学合肥微尺度物质科学国家研究中心；
2. 中国科学技术大学近代物理系；
3. 中国科学院量子信息与量子科技前沿卓越创新中心）

日常使用的经典计算机采用确定的二进制数字 0/1 比特作为信息处理的基本单元。对应地，量子计算的基本信息单元为量子比特。量子比特除了可以处于 0 或者 1 的状态下，还可以处于量子叠加的状态下，即有效地同时处于 0 和 1 这两种状态下；一个具有 N 个量子比特的系统，可以同时处于 2^N 个不同的状态下。基于这种量子叠加性质，诺贝尔物理学奖获得者费曼（R. P. Feynman）在 1981 年提出了量子计算的初步设想[1]。量子计算机被认为有望完成经典计算机难以高效求解的任务，如加速求解数据库的搜索、质因数分解等具有重要实用价值的问题。

随着对量子比特的制备、操纵以及测量的能力的提升，基于光子的玻色采样[2]和基于超导比特的随机线路采样[3]两个重要的量子计算模型已经在实验上实现了量子计算的第一个里程碑——“量子计算优越性”（即在具有有限的量子比特数目和操纵精度的量子设备上，完成一些经典超级计算机无法高效求解的计算任务[4]）。其中，作者团队构建的 76 个光子 100 个模式量子计算原型机“九章”处理高斯玻色采样的速度比超级计算机快 100 万亿倍[2]。在实验上确凿地证明 40 多年前费曼所提出来的量子计算加速设想后，持续扩大量子计算物理实现的规模、提升计算算力成为世界科技前沿最受关注也是最具挑战的问题之一。

笔者团队专注于高斯玻色采样量子计算。玻色采样最早由美国计算复杂度专家斯科特·阿伦森（Scott Aaronson）和他的学生一起提出[5]。他们证明了对经过随机线性变换后的光子的输出概率分布进行采样是经典计算困难的，这在数学上对应计算相应子矩阵的积和式（permanent），其计算复杂度等级为＃P-hard。实现玻色采样主要需要单光子源制备、光量子线路变换、单光子探测三个步骤。图 1（a）为 3 个光子输入到 6 个模式线性干涉仪的玻色采样示意图。经过一系列的理论研究，人们进一步发展了玻色采样的相关变体模型，这包括随机入口玻色采样[6]、高斯玻色采样[7]等。与单光子态玻色采样不同的是，高斯玻色采样采用压缩真空态作为输入态［图 1（b）］。中大规模的高斯玻色采样实验不仅可用于证明量子计算优越性，同时还被认为在图论、机器学习、量子化学等领域具有潜在应用[8]。

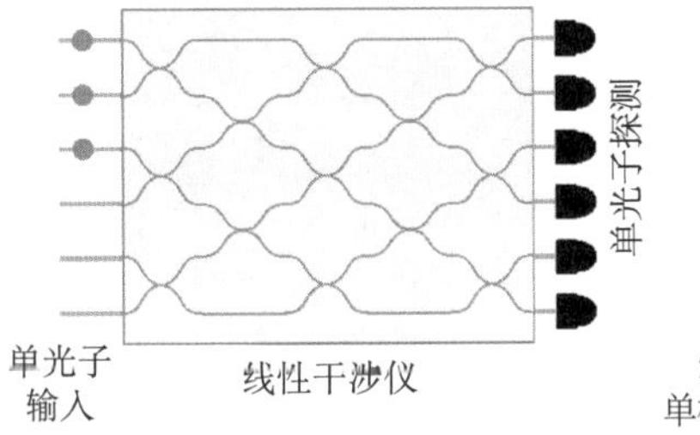

（a）单光子玻色采样示例图

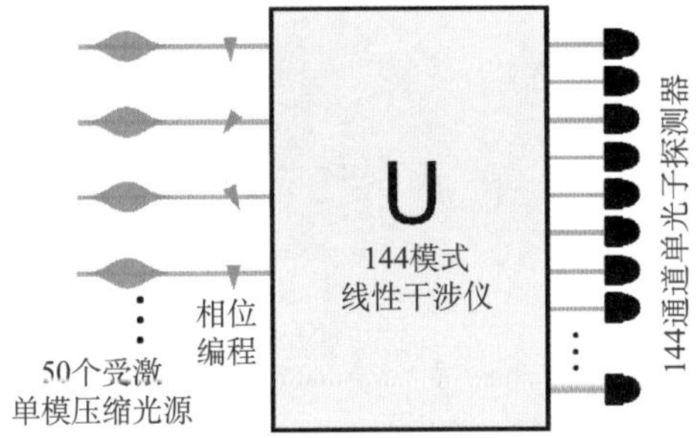

（b）50个单模压缩态输入、144模式变换高斯玻色采样示意图

图 1　玻色采样示意图

拓展高斯玻色采样的规模是极具挑战的，需要从三个方面同时着手：①提高量子光源的产率和品质，从而提高光子的数目；②提高光量子线路的透过效率和模式数，从而提升量子态的维度；③提高相位锁定的精度和动态调节能力，保证相干操纵精度。

2021 年，经过一系列概念和技术创新，笔者团队一方面受到激光——“受激辐射光放大”概念的启发，研究设计并实现了受激双模量子压缩光源，这种量子光源同时具有高产率、高全同性和高收集效率的优点；另一方面，通过三维集成和高效率阵列型光路对准，实现了 144 维度光量子干涉线路；同时，通过动态调节压缩光的相位，实现了对高斯玻色采样矩阵的重新配置，成功构建了可用于求解不同参数数学问题的“九章二号”量子计算原型机（图 2 为其示意图）。在理论上，笔者团队发展了利用高阶关联验证、子系统贝叶斯高效验证等对中大规模高斯玻色采样实验系统输出结果进行检验的理论。在实验中，“九章二号”最高探测到的光子数为 113 个，输出态空间维度达到 10^{43}，根据此前已正式发表的最优化经典算法，其在高斯玻色采样这个问题上的处理速度比最快的超级计算机快 10^{24} 倍。

这一成果于 2021 年 10 月 25 日以“编辑推荐”形式发表在《物理评论快报》（*Physical Review Letters*）上[9]。文章发表后引发了国际学术界的强烈反响和学术讨论。美国物理学会 Physics 网站邀请著名量子物理学家、加拿大卡尔加里大学教授 Barry Sanders 以“量子优势的量子飞跃”为标题撰写长篇评述文章，称赞该工作是

“令人激动的实验杰作”。

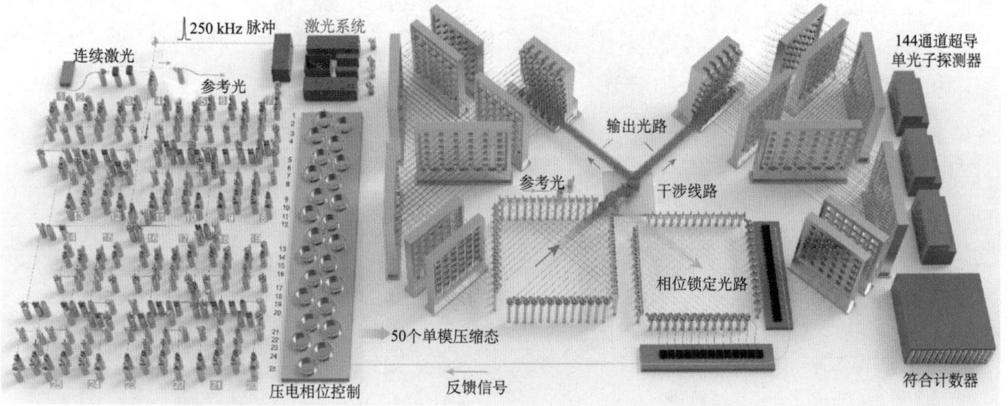

图 2 “九章二号”量子计算原型机整体装置示意图

左上方为激光系统，用于产生高峰值功率飞秒激光脉冲；左方 25 个量子光源通过受激参量下转换过程产生 50 路受激放大单模压缩态输入右方 144 模式光量子干涉网络；最后利用 144 个高效率超导单光子探测器以及符合计数器进行光量子态探测与数据分析。

参考文献

[1] Feynman R P. Simulating physics with computers. Int J Theor Phys, 1982, 21: 467-488.

[2] Zhong H S, Wang H, Deng Y H, et al. Quantum computational advantage using photons. Science, 2020, 370(6523): 1460-1463.

[3] Arute F, Arya K, Babbush R, et al. Quantum supremacy using a programmable superconducting processor. Nature, 2019, 574: 505-510.

[4] Preskill J. Quantum computing in the NISQ era and beyond. Quantum, 2018, 2: 79.

[5] Aaronson S, Arkhipov A. The computational complexity of linear optics. Theory Comput, 2013, 9: 143-252.

[6] Lund A, Laing A, Rahimi-Keshari S, et al. Boson sampling from a Gaussian state. Physical Review Letters, 2014, 113(10): 100502.

[7] Hamilton C S, Kruse R, Sansoni L, et al. Gaussian boson sampling. Physical Review Letters, 2017, 119(17): 170501.

[8] Bharti K, Cervera-Lierta A, Kyaw T H, et al. Noisy intermediate-scale quantum algorithms. Reviews of Modern Physics, 2022, 94: 015004.

[9] Zhong H S, Deng Y H, Qin J, et al. Phase-programmable Gaussian boson sampling using stimulated squeezed light. Physical Review Letters, 2021, 127(18): 180502.

Quantum Computational Advantages

PengLichao, Zhong Hansen, Wang Hui, Deng Yuhao, Chen Mingcheng, Lu Chaoyang, Pan Jianwei

Quantum computers promise to perform certain tasks that are believed to be intractable for classical computers. Boson sampling is such one task and is considered a strong candidate for demonstrating the quantum computational advantage. Here, we report advanced phase-programmable Gaussian Boson Sampling (GBS) that produces up to 113 photon detection events out of a 144-mode photonic circuit. This photonic quantum computer, Jiuzhang 2.0, yields a Hilbert space dimension up to $\sim 10^{43}$, and a sampling rate $\sim 10^{24}$ faster than using brute-force simulation on classical supercomputers.

3.5 科学家实现超导“分段费米面”

郑 浩 贾金锋

(1. 上海交通大学物理与天文学院；2. 李政道研究所)

固体物理学的研究发现，固体材料会有一种叫“能带”的固有性质，固体里的电子会按照能量从低到高的方式填充能带。绝缘体的能带都是被填满的（全满能带），而导体中的能带没有被填满（半满能带）。就像一个挤满了人的房间里，人寸步难行，而一个空房间里，人能够自由走动一样，挤满电子的全满能带不能导电，而半满能带能够导电。另外，就像一瓶半满的水能看到水面一样，一个半满的能带也能观测到“电子面”，物理术语叫“费米面”。具体来说，在固体能带中，能量低于费米面的态都被电子填充，高于费米面的态都空置。所以，一个固体材料有没有费米面会决定它很多物理性质，包括是否导电、是否透光，等等。传统的物态调控都是调控费米面附近态密度，如果能够实现费米面的人工调控，就会给材料物性的调控带来革命性的变化。超导体具有零电阻导电和完全抗磁性等奇特性质，是物理学中一个长盛不衰的研究课题。由于费米能级处超导能隙的存在，超导体均无费米面。

1965年，富尔德（P. Fulde）预言在常规超导体中有足够大的电流时可能导致准粒子谱发生多普勒频移，最终产生零能态[1]。1993年，沃洛维克（G. Volovik）发现零能量的准粒子实际上只占据了正常态费米面垂直于电流的部分[2]。随着电流的增加，这些零能态在动量空间会逐渐扩大，形成类香蕉状的有序结构，被称为“分段费米面”。1993年，实验观测到超导体准粒子谱中的多普勒频移，但动量空间零能有序结构即“分段费米面”尚未被报道。

笔者团队分析认为，前人实验失败的原因可能在于超导能隙闭合所需的超电流通常大于临界电流，所以超电流导致在“分段费米面”形成之前已经令超导体失超。针对这一困难，笔者团队设计制备了拓扑超导体/超导（$Bi_2Te_3/NbSe_2$）异质结。该体系有两个优点：① 超导近邻效应在 Bi_2Te_3 中的诱导出的超导能隙总是小于 $NbSe_2$ 中的超导能隙，所以有机会在不破坏 $NbSe_2$ 的超导状态的前提下，利用多普勒频移来闭合 Bi_2Te_3 中的能隙；②Bi_2Te_3 的拓扑狄拉克表面态中的费米速度比 $NbSe_2$ 要高一个数量级，所以从原理上来说，在相同超电流强度下，Bi_2Te_3 中的多普勒频移将远大于 $NbSe_2$。

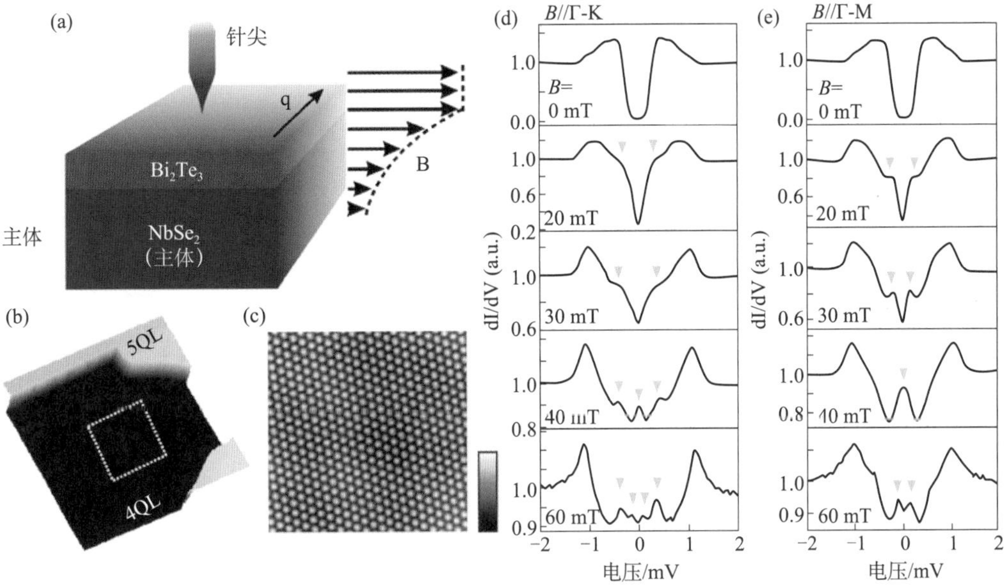

图1　“分段费米面”的实验发现

（a）$Bi_2Te_3/NbSe_2$超导异质结示意图；（b）生长在$NbSe_2$衬底上的高质量Bi_2Te_3薄膜形貌图；（c）Bi_2Te_3薄膜的原子分辨图；（d）沿Γ-K方向不同强度的磁场下的隧道谱；（e）沿Γ-M方向不同强度的磁场下的隧道谱。从（d）、（e）可以看出，在不同大小和方向的面内磁场作用下，隧道谱中来源于准粒子的信号逐渐增加。

笔者团队借助分子束外延技术在$NbSe_2$衬底上生长4 nm厚的Bi_2Te_3薄膜，并施加面内磁场以产生屏蔽超电流；只需要磁通密度（物理符号为B）很低（40 mT）的磁场，就能成功地利用扫描隧道显微谱探测到一个位于零能的峰。更进一步，获得了位于费米能级处的准粒子干涉图像（该图像在零磁场下是完全不存在的）；根据该准粒子干涉图像的形状，可知Bi_2Te_3正常态费米面只有一些分段出现在零能上，而其形状和取向完全由外加磁场的强度和方向控制。笔者团队基于超电流导致的“分段费米面”的理论模拟准确再现了实验上得到的隧道微分电导谱和准粒子干涉图像[3]。

该工作创新性地利用拓扑绝缘体/超导体异质结的特殊性解决了实验中的困难，首次在实验上观察到了50多年前富尔德理论预言的“分段费米面”，并发现不仅可以用磁场方向和大小来调节费米面的形状和大小，还能调控拓扑性，构建新的拓扑超导。该工作开辟了调控物态的新方法。

参考文献

[1] Fulde P. Tunneling density of states for a superconductor carrying a current. Phys Rev, 1965, 137: A783-A787.

[2] Volovik G. Superconductivity with lines of GAP nodes: density of states in the vortex. JETP Lett, 1993, 58: 469-473.
[3] Zhu Z, Papaj M, Nie X A, et al. Discovery of segmented Fermi surface induced by Cooper pair momentum. Science, 2021, 374: 1381-1385.

Scientists Release Segmented Fermi Surface

Zheng Hao, Jia Jinfeng

In 1965, P. Fulde predicted that a sufficiently large supercurrent can close the energy gap in a superconductor and create gapless Bogoliubov quasiparticles through the Doppler shift of the quasiparticle energy due to the Cooper pair momentum. In this gapless superconducting state, zero-energy quasiparticles reside on a segment of the normal state Fermi surface, while the remaining part is still gapped. However, the segmented Fermi surface remains elusive. Here we use scanning tunneling spectroscopy to image field-controlled Fermi surface of Bi_2Te_3 thin films on the superconductor $NbSe_2$. By applying a small in-plane magnetic field, a screening supercurrent is induced, leading to the establishment of a segmented Fermi surface in Bi_2Te_3. Our results pave a new way for the manipulation of superconductivity.

3.6 水气变换反应制氢新进展

马 丁

（北京大学化学与分子工程学院）

氢能经济，指设想以氢气为主要能源的社会状态，被认为是实现社会可持续发展的关键进程之一[1]。氢气（H_2）的高效制备、纯化及储运技术是氢能经济发展的核心。水（H_2O）是自然界中最丰富、易得的氢资源，利用水制备氢的相关技术的突破，对实现清洁、高效、低排放的氢循环具有重要意义[1]。水气变换（water-gas shift，WGS）反应是目前工业制取高纯氢气的关键步骤。

氢燃料电池作为氢能应用的重要技术，可以实现化学能向电能的高效转化。根据美国能源部于 2004 年发布的车载燃料电池发展规划[2]，只有水气变换制氢催化剂的使用成本低于 1 美元/kW 并且催化剂使用量低于 0.11 kg/kW 时，水气变换制氢技术才有望被推广使用。除此之外，氢燃料中少量一氧化碳（CO）会严重毒化电极，制约氢燃料电池的应用，因此，制取高纯氢气时需要除去其中的少量 CO。综上，发展低温、高效、稳定的水气变换制氢催化剂，对上述工业产氢过程和氢能的大规模应用具有重要意义。

在前期研究[3]的基础上，笔者研究团队（包括北京大学化学与分子工程学院马丁课题组、大连理工大学石川课题组以及中国科学院大学周武课题组）创新发展了一种在立方相碳化钼（α-MoC）基底上制备高密度、原子级分散的铂（Pt）物种（单位点 Pt 物种以及亚纳米 Pt 团簇）的有效策略，构建出高效、稳定的 Pt/α-MoC 界面催化结构，直接观察到了水分子在 α-MoC 表面的低温活化解离现象；同时，高密度原子级 Pt 物种的存在有效促进了 CO 的吸附活化，不仅增强了 H_2O 解离产生活性氧物种的快速反应和脱附，同时提出了基于 CO 直接解离步骤的低温协同制氢新路径。上述催化剂在 40～400℃的超宽温度区间实现了高效水气变换制氢，并且大幅度提升了催化剂的反应活性及长效稳定性，在 250℃下实现了寿命周期内每摩尔 Pt 催化产生高达 4 300 000 mol 的氢气，相较于笔者研究团队 2017 年报道的 Au/α-MoC 催化剂[4]产氢量提升了一个数量级。Pt/α-MoC 催化剂的催化性能可与高效的酶催化体系相媲美，这为燃料电池原位供氢提供了新思路。上述结果发表在《自然》期刊上[5]。

以贵金属铂的价格为 6242 美元进行经济衡算，笔者研究团队制备的 Pt/α-MoC 催化剂首次突破了依据美国能源部2004年车载燃料电池发展规划所推算的催化活性限值（图1）。该研究工作不仅为氢能经济的推广提供了新的技术选择，同时也为研究者设计高活性、高稳定性的金属催化剂提供了一种新的思路。

该研究工作被国内外多个科学媒体和权威人士报道并高度评价。美国化学会《化学与工程新闻》（*Chemical & Engineering News*）以“催化剂照亮燃料电池汽车的未来”（Catalyst boosts prospects for fuel-cell vehicles）[6]为题进行了报道，认为这是“一个重要的发现”（a remarkable finding）。

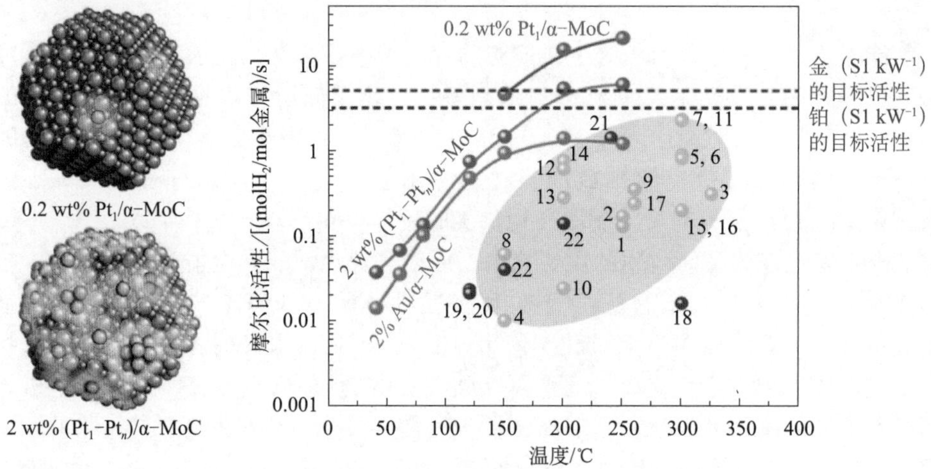

图1　高分散的Pt/α-MoC催化剂及水气变换反应性能

右图中1～22标注了文献报道的典型金属催化剂的摩尔比活性，均低于目标活性，而Pt/α-MoC是唯一超过目标活性的催化剂。

参考文献

[1] Marban G, Valdés-Solís T. Towards the hydrogen economy? International Journal of Hydrogen Energy, 2007, 32: 1625-1637.

[2] Ladebeck J R, Wagner J P. *In* Handbook of Fuel Cells. Vielstich W, Lamm A, Gasteiger H A(eds.). Wiley, 2003.

[3] Steele B C, Heinzel A. Materials for fuel-cell technologies. Nature, 2001, 414: 345-352.

[4] Yao S, Zhang X, Zhou W, et al. Atomic-layered Au clusters on α-MoC as catalysts for the low-temperature water-gas shift reaction. Science, 2017, 357: 389-393.

[5] Zhang X, Zhang M, Deng Y, et al. A stable low-temperature H_2-production catalyst by crowding Pt on α-MoC. Nature, 2021, 589: 396-401.

[6] Chemical & Engineering News. Catalyst boosts prospects for fuel-cell vehicles. https://cen.acs.org/

synthesis/catalysis/Catalyst-boosts-prospects-fuel-cell/99/i3[2021-01-20].

New Progress in H_2-Production by Water-Gas Shift (WGS) Reaction

Ma Ding

The water-gas shift (WGS) reaction is an industrially important source of pure hydrogen(H_2)at the expense of carbon monoxide and water. This reaction is of interest for fuel-cell applications, but requires WGS catalysts that are durable and highly active at low temperatures. To achieve this goal, we demonstrate that the structure Pt/α-MoC catalyses the WGS reaction even at 40℃. We find that it is critical to crowd the α-MoC surface with atomically dispersed Pt and sub-nano Pt cluster species, which prevents oxidation of the support that would cause catalyst deactivation, as seen with gold/α-MoC, and gives our system high stability and a high metal-normalized turnover number of 4,300,000 moles of hydrogen per mole of platinum. We anticipate that the strategy demonstrated here will be pivotal for the design of highly active and stable catalysts for the effective activation of important molecules such as water and carbon monoxide for energy production.

3.7 柔性高分辨 X 射线成像新技术

陈秋水　杨黄浩

（福州大学化学学院）

1895 年伦琴发现 X 射线以来，X 射线成像技术的研究推动了医学影像、工业探伤、安检防爆、高能物理、考古学等领域的科技变革[1]。高灵敏、高分辨、实时动态、柔性的 X 射线成像技术可在较低的辐照剂量下获得更清晰的影像，是当前国际上 X 射线成像技术的一个前沿研究阵地。然而，高端 X 射线探测器长期被国外垄断，X 射线探测器及其元器件被列为制约我国工业发展的 35 项“卡脖子”技术之一。研发高灵敏、高分辨率、柔性的新一代 X 射线成像技术和仪器具有十分重要的科学意义和社会意义，是我国当前的迫切需求。

长期以来，X 射线成像技术存在辐照剂量大、成像分辨率不足、仪器制造成本高等关键科学技术挑战。此外，医用计算机断层扫描（CT）和 X 射线机等设备需要集成大面积高密度像素点的硅基光电传感阵列，这严重限制了 X 射线成像设备的小型化、便携化和柔性化发展[2]。闪烁体材料作为一种可将高能量 X 射线转为可见光的辐射能量转换介质，是制造 X 射线成像设备的关键材料，其光子能量转换效率决定了 X 射线成像设备的灵敏度和分辨率。解决上述这些挑战的关键在于开发闪烁体的 X 射线能量转换新机制以提高辐射发光效率[3]。

为实现上述研究目标，笔者研究团队合成了一类具有 X 射线光学记忆性能的稀土纳米晶闪烁体，由辐射触发的阴离子迁移到宿主晶格中导致被捕获电子的缓慢跃迁，诱导超过 30 天的持续辐射发光（图 1）。这一科学发现为发展非硅基的柔性大面积 X 射线成像设备奠定了基础。笔者研究团队使用一系列可溶液处理的镧系元素掺杂纳米闪烁体开发了一类用于高分辨率 X 射线成像的柔性 X 射线探测器，实现了 20 线对/mm 的三维超高分辨 X 射线成像，辐照剂量比现有 CT 成像低一个数量级（图 2）[4]。该研究为深入了解基于持久电子捕获的 X 射线能量转换机制提供了新思路，为未来放射学研究，包括乳腺摄影、影像引导放射治疗，以及深度学习的可穿戴 X 射线探测器提供了范例。此项研究成果发表在 2021 年 2 月 18 日《自然》[4]上。该项研究成果入选了教育部评选的“2021 年中国高等学校十大科技进展”。

《自然》的同期评述指出“此项研究为医学影像开辟了很有前景的新途径”[5]。英国物理学会主办的《物理世界》(*Physics World*)评论指出“柔性探测器带来了三维高分辨 X 射线成像”。该研究工作颠覆了传统 X 射线成像技术原理，为柔性 X 射线成像技术的研发提供了一种全新的设计思路，希望能够借此有力推进我国高端 X 射线影像装备的国产化，改观我国高端 X 射线影像设备及关键零部件依赖进口的局面。

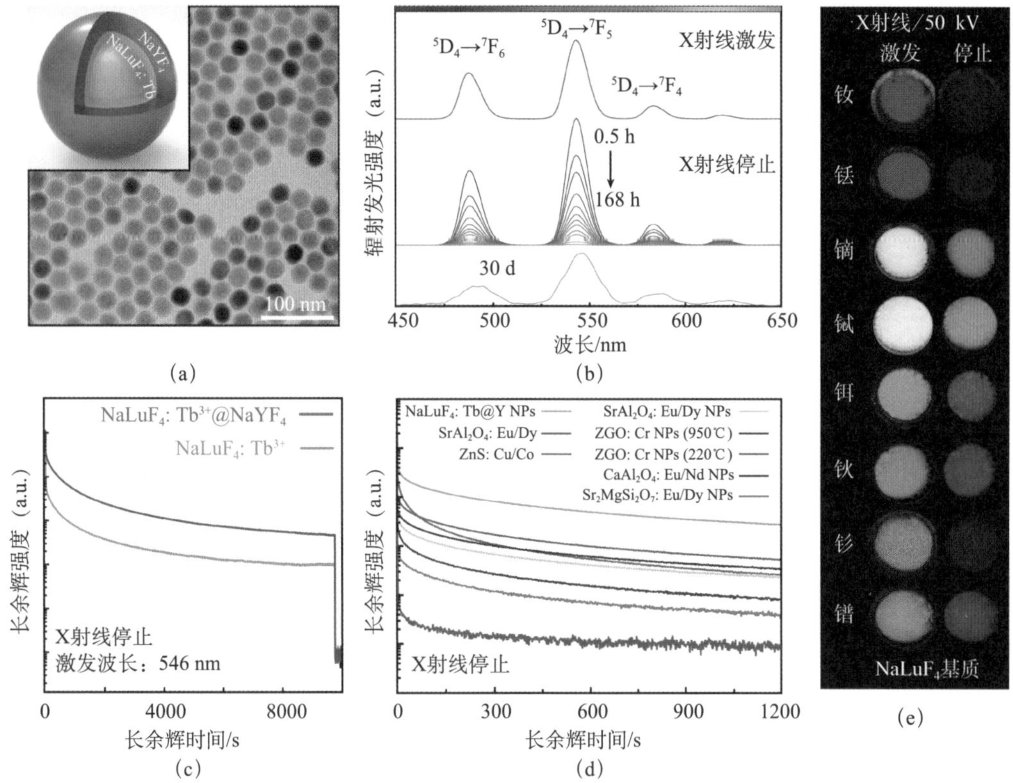

图 1　纳米闪烁体材料的长寿命 X 射线发光[4]

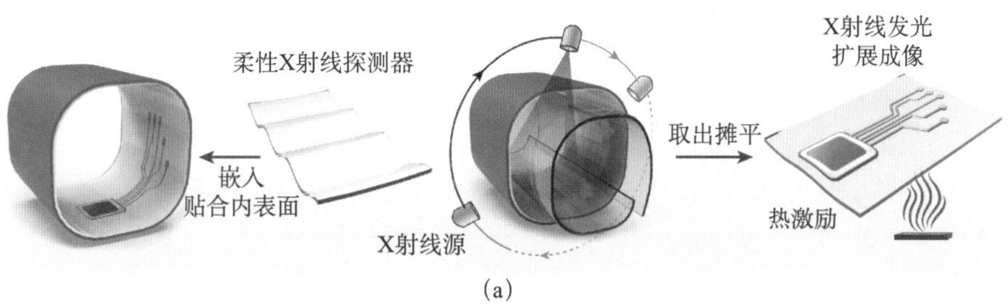

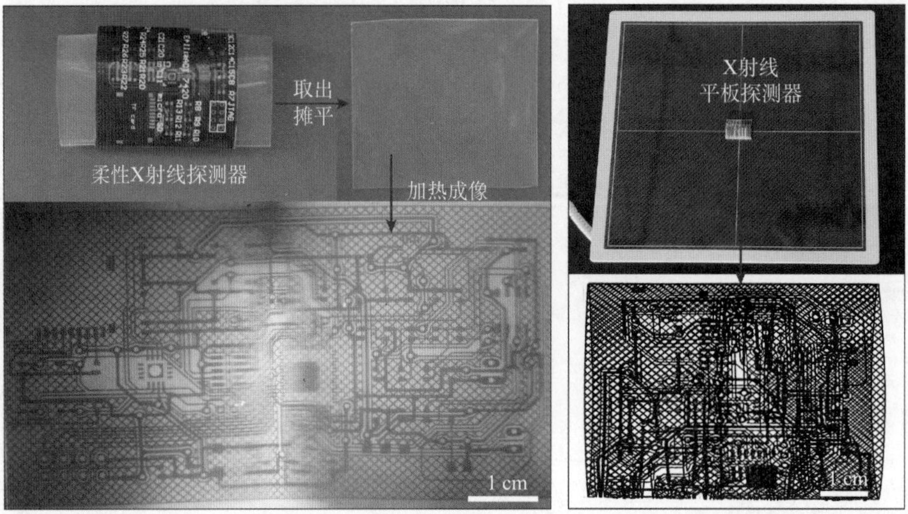

(b) (c)

图 2 柔性高分辨 X 射线成像技术[4]

参考文献

[1] Röntgen W C. On a new kind of rays. Science, 1896, 3(59): 227-231.

[2] Ou X, Chen X, Xu X, et al. Recent development in X-ray imaging technology: future and challenges. Research, 2021: 9892152.

[3] Chen Q, Wu J, Ou X, et al. All-inorganic perovskite nanocrystal scintillators. Nature, 2018, 561(7721): 88-93.

[4] Ou X, Qin X, Huang B, et al. High-resolution X-ray luminescence extension imaging. Nature, 2021, 590(7846): 410-415.

[5] Neto A N C, Malta O L. Glowing nanocrystals enable 3D X-ray imaging. Nature, 2021, 590(7846): 396-397.

A Next-Generation Technology for Flexible High-Resolution X-ray Imaging

Chen Qiushui, Yang Huanghao

X-ray imaging technology has broad applications in various fields, including medical imaging, radiation detection, industrial non-destructive testing, and security and explosion inspection. Currently, clinical CT scanners, X-ray machines, and other instruments generally require integration with large-area silicon-based photoelectric sensor arrays with high-density pixel points, which severely limits the miniaturization, portability, and flexibility of X-ray imaging instruments. In this study, we developed a class of flexible X-ray detectors for high-resolution X-ray imaging using a series of solution processable, lanthanide-doped nanoscintillators. We demonstrated that slow hopping of trapped electrons due to radiation-triggered anionic migration in host lattices can induce persistent radioluminescence for more than 30 days. We also demonstrated X-ray luminescence extension imaging with a resolution of better than 20 line pairs per millimeter. These findings provide a paradigm to motivate future research in wearable X-ray detectors for patient-centered radiography and mammography, imaging-guided therapeutics, high-energy physics and deep learning in radiology.

3.8 分子尺度揭示界面水分子结构在电催化反应中的关键作用

李剑锋

（厦门大学化学化工学院/能源学院）

水是生命之源，覆盖了地球表面约71%的面积，人类从未停止过对水的奥秘的探索。水在各个科学领域都扮演着举足轻重的角色，特别是在表界面科学领域[1]。由于水分子直接参与众多电催化反应，包括析氢、析氧、二氧化碳/氧气/氮气还原等，其结构和行为会直接影响各类催化反应的进行。因此，清楚地认识水分子在固/液界面的物理特性是发展表界面科学、催化科学和能源科学的基础。然而，水分子本身多变，而且界面水处于固/液两相之间，含量远低于体相水，再加上界面环境十分复杂且伴有电催化反应的发生，这些原因极大地增加了对界面水分子原位探测的难度。因此，研究界面水分子的结构及其在电催化反应中的作用，一直是电化学领域长期存在的难题之一[2]。

作为模型催化剂，原子级平整的单晶电极具有明确定义的表面和电场性质，可以克服多晶电极表面状态复杂的缺点，适合在原子水平上揭示电催化反应的构效关系。然而，传统的光谱（如拉曼光谱、红外光谱、和频光谱以及X射线吸收光谱）技术，受仪器和体系的限制，难以用于原位监测单晶/溶液表界面的水分子的物理特性。

对此，利用团队前期发明的壳层隔绝纳米粒子增强拉曼光谱（SHINERS）[3]显示出的极高的表面灵敏度和指纹识别（物质的定性识别）的优势，笔者研究团队原位监测了析氢反应（HER）过程中钯单晶电极/溶液界面水分子的构型及其动态演化过程。直接获取的SHINERS数据表明，在钯单晶电极/溶液界面，除了存在已知的含有氢键的水分子以外，还存在一类与阳离子键合的水分子，例如M·H_2O（M为Li^+、Na^+、K^+、Cs^+、Ca^{2+}等）。随着钯单晶电极电位的改变，与阳离子键合的水分子会在电场调控下向钯单晶电极表面移动，缩短钯原子和水分子中氢原子之间的有效距离，进而提高电极与水分子的电荷转移效率，最终提升HER的性能。

结合理论模拟，笔者研究团队还从热力学和动力学的角度进一步揭示了界面水分子与阳离子协同调控 HER 性能的关键机理。首先，界面水分子与阳离子键合，无序的水分子在有限区域内高效地排列成有序的结构，即熵减少过程，以最大限度地减少 HER 过程所需做的功，并实现最大化的电化学能量转换。其次，通过改变 HER 的反应物和产物的传输路径，即界面阳离子可以作为一种“助催化剂”，在 HER 过程中不断地向电极表面提供水分子，同时将羟基自由基转移到溶液中，且阳离子浓度和价态的增高会进一步增加界面有序水分子的含量，从而加快 HER 的反应速率（图 1）。此外，单晶电极的晶面结构和电子结构也会影响阳离子键合水分子的含量，进而影响 HER 的反应速率。一系列结论均证实了阳离子键合水分子对加速 HER 反应速率具有普适性。

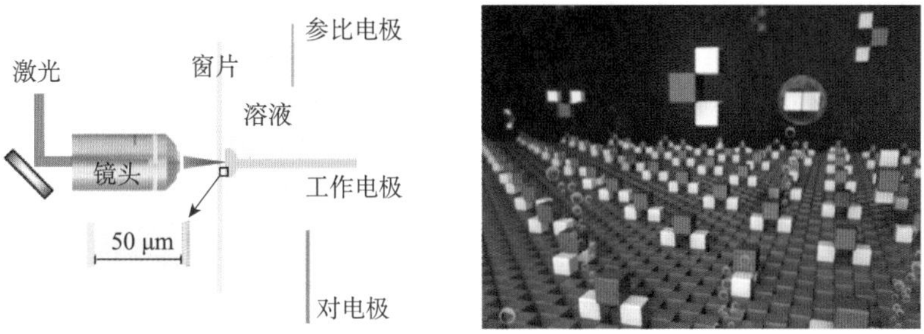

（a）电化学SHINERS测试电解池　　（b）有序界面的水分子结构

图 1　电化学 SHINERS 测试电解池和有序界面水分子结构示意图

该水分子更贴近电极表面，提高了电荷转移效率和 HER 的反应速率。

此项研究从单晶模型体系出发，从分子尺度揭示了界面水分子结构对电催化反应过程的调控机制，解决了长期困扰电化学领域的难题，为提升电催化反应速率提供了一种新的策略。该研究工作结果已发表在国际著名学术期刊《自然》上[4]。

参考文献

[1] Zaera F. Probing liquid/solid interfaces at the molecular level. Chem Rev, 2012, 112(5): 2920-2986.

[2] Bjorneholm O, Hansen M H, Hodgson A, et al. Water at interfaces. Chem Rev, 2016, 116(13): 7698-7726.

[3] Li J F, Huang Y F, Ding Y, et al. Shell-isolated nanoparticle-enhanced Raman spectroscopy. Nature, 2010, 464(7287): 392-395.

[4] Wang Y H, Zheng S, Yang W M, et al. *In situ* Raman spectroscopy reveals the structure and dissociation of interfacial water. Nature, 2021, 600(7887): 81-85.

Revealing the Key Role of Interfacial Water Structure in Electrocatalytic Reactions at the Molecular Scale

Li Jianfeng

Water molecules are directly involved in most electrocatalytic reactions. Its structure and behavior will affect the catalytic reaction progress. Based on the enhanced Raman spectroscopy, we monitor *in-situ* the structure and dissociation of interfacial water at the palladium single crystal electrode/solution interface. The interfacial water molecules are composed of both the hydrogen-bonded water and cation hydrated water. The cations and negative potentials cooperatively impart an ordered structure to the interfacial water. This scenario improves the charge transfer efficiency between the interfacial water and the electrode, thus greatly increasing the HER rate. A higher concentration or valence state of the cations will impose an increased population of ordered interfacial water and further improve the HER performance. The role of cation-hydrated water is also substantiated by the close correlation between the cation-hydrated water population and HER rate on various facets, implying that such a rationale can be generalized to other catalytic systems. These observations have opened up an avenue to explore the mechanism of electrocatalytic reactions and manipulate the water chemistry for targeted application, providing an adaptive strategy to improve the electrocatalytic reaction rate.

3.9　脊椎动物从水生到陆生演化的遗传创新机制

毕旭鹏　张国捷

（浙江大学生命演化研究中心）

在生物漫长的演化史中，脊椎动物登上陆地无疑具有“里程碑”意义，自达尔文以来就备受关注。在从水生到陆生解锁“新地图”的过程中，鱼类的生理功能及表型上出现了许多适应性变化，其中最重要的两个转变分别是，呼吸方式从鳃呼吸转变为肺呼吸，以及运动方式从用鱼鳍游动转变为用四肢行走。这两个功能的转变，奠定了包括人类在内的四足动物在陆地上繁衍的基础。决定脊椎动物从水生到陆生转变的适应性新性状何时出现以及如何出现有两种假说：一种是新性状是从无到有出现的，在此之前并没有相应的遗传基础，是水生到陆生转变过程中全新出现的功能；另一种是祖先群体预先存在相关的基础功能，之后在新环境中物种产生适应性变化。由于早期登陆物种已经灭绝，脊椎动物从水生到陆生的转变过程是如何发生的这个难题一直悬而未决[1,2]。

幸运的是，一些现生水生物种还保留着许多与这些过渡物种类似的生物学特征。塞内加尔多鳍鱼（简称多鳍鱼）又称作“恐龙鳗”或者“恐龙鱼”，与鲟鱼、弓鳍鱼、雀鳝、肺鱼等一样，是最早演化出来的一批硬骨鱼类。由于在地球上生存了几亿年之久，它们的身体结构仍保留着一些原始的形态特征，因此也称作“远古鱼类”。多鳍鱼拥有原始的肺，可以在脱离水环境后继续存活一段时间，并且可以在氧容量极低的环境中通过背部的喷水孔呼吸。另外，与腔棘鱼相似，多鳍鱼也具有肉质柄的胸鳍，这可以帮助它们在水底爬行。这些鱼类活化石所具有的独特生物学特性和演化地位，为我们通过基因组揭示早期陆生脊椎动物的水陆适应性演化提供了可能。

脊椎动物四肢骨骼的解剖结构证实包括人在内的四足动物的“大臂”（肱骨）与远古鱼类胸鳍的后鳍基骨同源。笔者研究团队与西北工业大学和中国科学院水生生物研究所合作，通过对多鳍鱼、匙吻鲟、弓鳍鱼、鳄雀鳝和肺鱼等原始硬骨鱼类进行高质量基因组解析[3,4]（图 1），发现在陆生脊椎动物和原始硬骨鱼类基因组中存在一个相同的、极端保守的增强子，该序列可以作为“开关”调控下游 *Osr2* 基因在滑膜关节处表达；并首次利用多鳍鱼胸鳍再生实验以及原位表达分析证实了 *Osr2* 基因主要在后鳍基骨与鳍条的连接处表达。这一结果提示：滑膜关节使得骨关节之间能够灵活

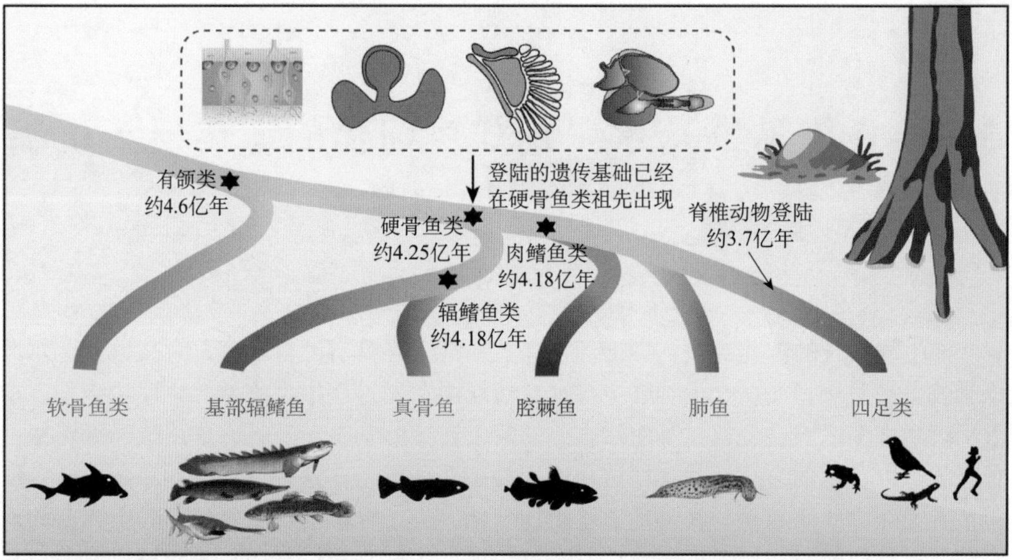

图 1　原始硬骨鱼类高质量基因组解析发现登陆的遗传基础已经在硬骨鱼类祖先出现[3]

脊椎动物从水生到陆生的过程中需要克服众多障碍所需的部分生物学特征在早期硬骨鱼类中就已经产生，而多鳍鱼、鲟鱼、弓鳍鱼、雀鳝等较早起源的原始辐鳍鱼类和腔棘鱼、肺鱼等较早起源的原始肉鳍鱼类里，也保留了这些硬骨鱼祖先的特征。

转动，这为后续鱼鳍到四肢演化的运动灵活性方面提供了结构基础。而所有真骨鱼胸鳍丢失了后鳍基骨元件，因此这很有可能导致这一调控序列在大部分现生鱼类基因组中发生丢失。此外，五指的形成标志着脊椎动物开始真正适应陆地生活。该研究还首次在 *Hoxa11* 基因上游区域发现了一个新起源的调控元件，可能与四足动物的五指形成密切相关（图 2）。与此同时，本研究还发现角质鳍条蛋白编码基因在肺鱼和四足动物中发生渐进式丢失，四足动物中控制后肢发育的 *Hox13b* 基因也发生了特异性的大片段缺失。这些结果在全基因组水平揭示了脊椎动物在从水生到陆生过程中获得陆地性附肢和运动能力的遗传创新机制。

嗅觉感受器的功能主要是感受化学分子刺激再将之转换成嗅神经冲动信息。研究发现，在原始辐鳍鱼和原始肉鳍鱼类（包括多鳍鱼、鲟鱼、弓鳍鱼、雀鳝、腔棘鱼和肺鱼）的嗅觉感受器中同时存在着两种类型的嗅觉受体，除了具有鱼类都拥有的感知水溶性分子的嗅觉受体之外，还具有能够感知空气中分子的嗅觉受体，这与它们能够在空气中呼吸的能力相一致[3]。本研究利用多器官表达谱聚类和系统发生关系，不仅证实了达尔文提出的肺和鱼鳔是同源器官的假说，也证明了现生鱼类的鱼鳔是从早期脊椎动物的肺演化而来的[3]。此外，基因组分析还发现陆生脊椎动物为克服血液与空

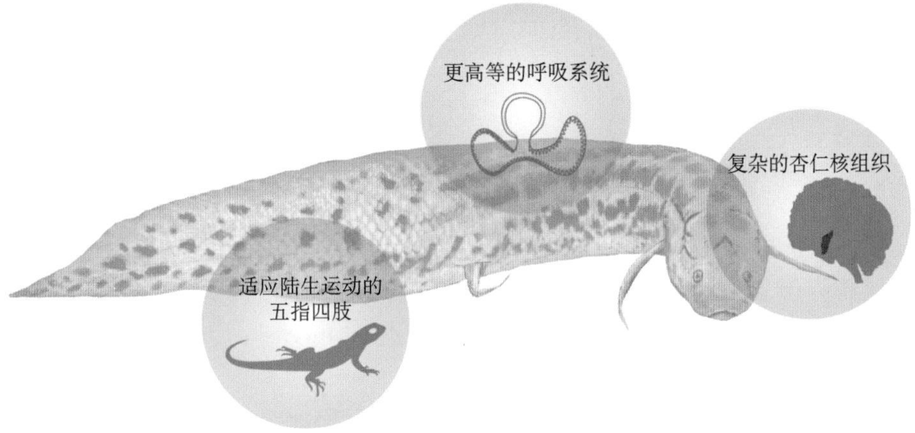

图 2　肺鱼基因组解析揭示了脊椎动物水生向陆生转变过程中的遗传创新[4]

肺鱼高质量超大基因组解析揭示了脊椎动物的登陆可以分为三个阶段性的飞跃：在硬骨鱼类的共同祖先中，出现了初步的空气呼吸能力，在肉鳍鱼的祖先中空气呼吸能力进一步加强，肺表面活性剂蛋白成员基本出现，四足动物祖先则具备了完善的陆地呼吸能力。此外，部分和陆地运动、四肢相关的基因组元件在肺鱼中也已经出现。与抗焦虑密切相关的蛋白神经肽 S 及其受体也是在肺鱼和四足动物的祖先中共同出现，提示大脑的演化对脊椎动物的登陆可能具有十分重要的作用。

气交界面对肺泡造成的表面张力，基因组中的肺表面活性剂蛋白开始不断增多，表明肺所参与的呼吸系统趋于完善及增强。这一开创性成果揭示了陆生脊椎动物肺呼吸能力的遗传创新基础在硬骨鱼类祖先就已经产生。

心脏和呼吸系统的协同演化在脊椎动物演化过程中也发挥着重要作用。动脉圆锥位于心脏流出通道和右心室相接的地方，作为心脏活动的辅助器官，它可以防止血液逆流以及平衡心室血压。笔者研究团队首次鉴定出一个 *Hand2* 基因的保守调控元件，这个调控元件在软骨鱼就已经出现，并在多鳍鱼和其他远古鱼类以及所有陆生脊椎动物里保留了下来，但在其他鱼类演化过程中丢失。这一古老元件参与调控动脉圆锥的形成，继而影响心脏系统功能。在小鼠基因组中靶向删除该调控元件后，发现 *Hand2* 基因在早期胚胎的右心室表达量降低，从而导致小鼠心脏发育不全和因供血不足引起的先天性死亡。这一成果将有助于人类心脏发育缺陷的研究。

基于物种演化过程的基因组比较分析为解析生物复杂性状形成机制提供了重要手段，为生物医学研究提供了有力工具。笔者团队的研究结果不仅揭示了隐藏在现生鱼类中的脊椎动物从水生到陆生演化的遗传奥秘，同时在分子水平揭示了不同类别的脊椎动物重要器官的同源关系。研究说明在祖先物种中存在的基因调控机制为后续新性状的出现提供了重要的功能基础，也为生物跨越式演化提供了可能。相关研究成果在《细胞》期刊上同期发表了两篇文章[3,4]，并成为当期封面故事。2021 年，“脊椎动物从水生到陆生演化的遗传创新机制”研究成果入选由中国科协生命科学学会联合体评选的“2021 年度中国生命科学十大进展”。

参考文献

[1] Zhu M, Zhao W, Jia L, et al. The oldest articulated osteichthyan reveals mosaic gnathostome characters. Nature, 2009, 458(7237): 469-474.

[2] Zhu M, Yu X. Stem sarcopterygians have primitive polybasal fin articulation. Biology Letters, 2009, 5(3): 372-375.

[3] Bi X, Wang K, Yang L, et al. Tracing the genetic footprints of vertebrate landing in non-teleost ray-finned fishes. Cell, 2021, 184(5): 1377-1391.

[4] Wang K, Wang J, Zhu C, et al. African lungfish genome sheds light on the vertebrate water-to-land transition. Cell, 2021, 184(5): 1362-1376.

Genomic Innovations Contribute to the Water to Land Transition of Vertebrates

Bi Xupeng, Zhang Guojie

The water-to-land transition is one of the great leaps in vertebrate evolution. By investigating the genomes of several species from basal lobe-and ray-finned fishes, we found genetic innovations regulating the development of the respiratory system, limbs and heart which underwent anatomical and functional alternations that enabled the vertebrates to move onto land territory. These anatomical structures support their bodies to move flexibly under water without the buoyancy and to breathe oxygen in the air. Our analyses revealed an ancient origin of key regulatory genes underlying these alternations in early vertebrates, but lost in most teleost lineages. Our studies provide molecular evidence that many biological functions associated with terrestrial evolution were present long before the common ancestor of bony fishes, and that these genetic regulatory networks were further refined in the lobe-finned fishes, evolving more sophisticated functions in air breathing, land locomotion, and the nervous system, leading to the subsequent landing event of tetrapods.

3.10 新型冠状病毒 RNA 加帽过程新机制及核苷类似物对其抑制机制的研究

黄羽岑 闫利明 饶子和 娄智勇

（清华大学医学院）

帽（cap）结构是细胞中广泛存在的一种 RNA 修饰方式。细胞中一系列酶分子，通过复杂的催化过程，在 RNA 的 5′端前加上一个鸟苷单磷酸（GMP），形成帽核心结构（cap core，GpppN），再由甲基转移酶对第一、第二、第三个核苷的不同位置进行甲基化，最终形成 cap0、cap1、cap2 结构（图 1）。在高等生物细胞中，cap1 是最主要的帽结构形式。

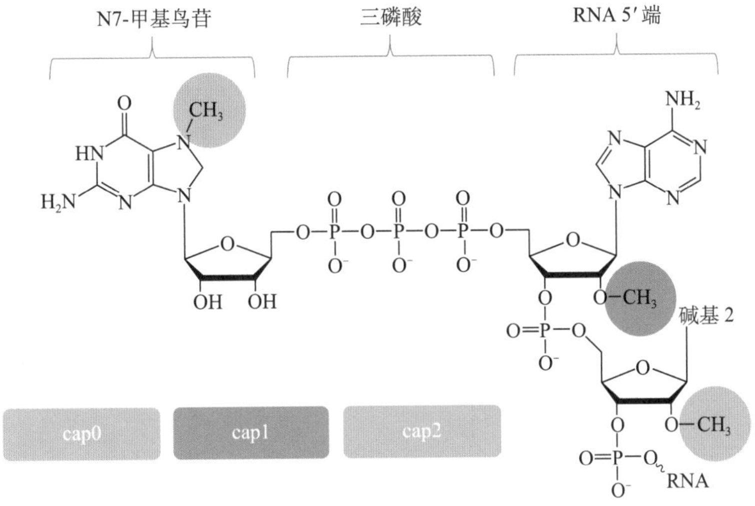

图 1 帽结构

帽结构在细胞生命活动中发挥着关键功能。例如，在高等生物细胞中，帽结构可以维持 mRNA 的稳定性，阻止核酸酶对 mRNA 的破坏，并可通过帽结合蛋白参与 mRNA 与核糖体的识别，保障蛋白质翻译的顺利进行。就必须在宿主细胞中表现生命活性的病毒而言，其在生命活动中，通过宿主或自身编码的蛋白质对自身的 mRNA

进行加帽，其结构与宿主细胞中 RNA 的帽结构完全相同，病毒采用这种加帽方式除了保证病毒蛋白质能够被顺利翻译，还可以使宿主将病毒 mRNA “误认为”是宿主的核酸分子，从而逃避宿主天然免疫的攻击。

新型冠状病毒是目前已知 RNA 病毒中基因组最大的一种病毒（约 30 kb），作为一种正链 RNA 病毒，其基因组 RNA 和 mRNA 的 5′端也具备 cap1 结构。在病毒学传统认知中，冠状病毒核酸的加帽过程通过四步反应过程完成，如图 2（a）所示[1,2]。

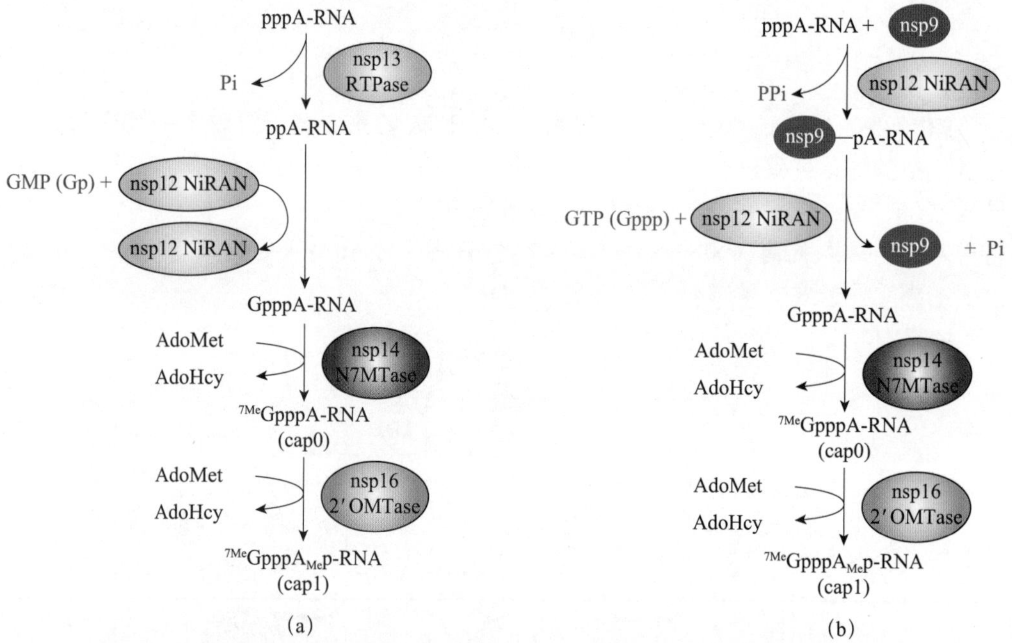

图 2　新型冠状病毒 RNA 加帽的传统机制（a）和本研究工作发现的新机制（b）

pppA-RNA：5′端有 3 个游离磷酸的核糖核酸；RTPase：RNA 三磷酸酶；Pi：磷酸分子；NiRAN：聚合酶相关的核苷转移酶；GMP：鸟苷单磷酸；AdoMet：S-腺苷甲硫氨酸；AdoHcy：S-腺苷同型半胱氨酸；N7MTase：N7 甲基转移酶；2’OMTase：2’O 甲基转移酶；PPi：焦磷酸分子；GTP：鸟苷三磷酸。

自新冠疫情发生后，笔者研究团队围绕新型冠状病毒转录复制过程开展了系统研究，先后阐明了核心转录复制复合体[3]、延伸转录复制复合体[4]、加帽中间态转录复制复合体[1]和 cap0 转录复制复合体[2]的工作机制。在此基础上，笔者研究团队进一步深入探索了新型冠状病毒核酸加帽的机制，发现新型冠状病毒能够利用其转录复制复合体中的单链核酸结合蛋白 nsp9（non-structural protein 9）作为媒介，介导全新的加帽过程，如图 2（b）所示。

该过程分为两个阶段（图 3）。第一阶段，聚合酶 NiRAN 结构域水解新生核酸链 5′三磷酸末端成为单磷酸末端，并将剩余的单磷酸末端与 nsp9 氨基端第一个天冬酰胺的氨基共价连接，形成 RNA-nsp9 中间产物，该过程被命名为“核糖核酸酰化”(RNAylation)。第二阶段，聚合酶 NiRAN 在其“鸟苷口袋”中结合一个三磷酸鸟苷分子，诱导 NiRAN 结构域发生构象变化，将一个我们称之为“断键水分子”的水分子向 RNA 与 nsp9 共价键的距离推进至 3.8 Å，使其发挥亲和攻击能力，导致 RNA 与 nsp9 间的共价键断裂，再由断键形成的高能基团攻击三磷酸鸟苷，最终形成帽核心结构（GpppA）。这一过程的发现，不仅是对新型冠状病毒乃至其他病毒生命过程理解的重要更新，而且也是生物学领域中第一次发现这种由蛋白质作为媒介来介导的 RNA 加帽过程，拓展了生命科学研究中对核酸加工的认识边界，为在人体细胞中发现可能存在的类似现象提供了一个重要的起点。

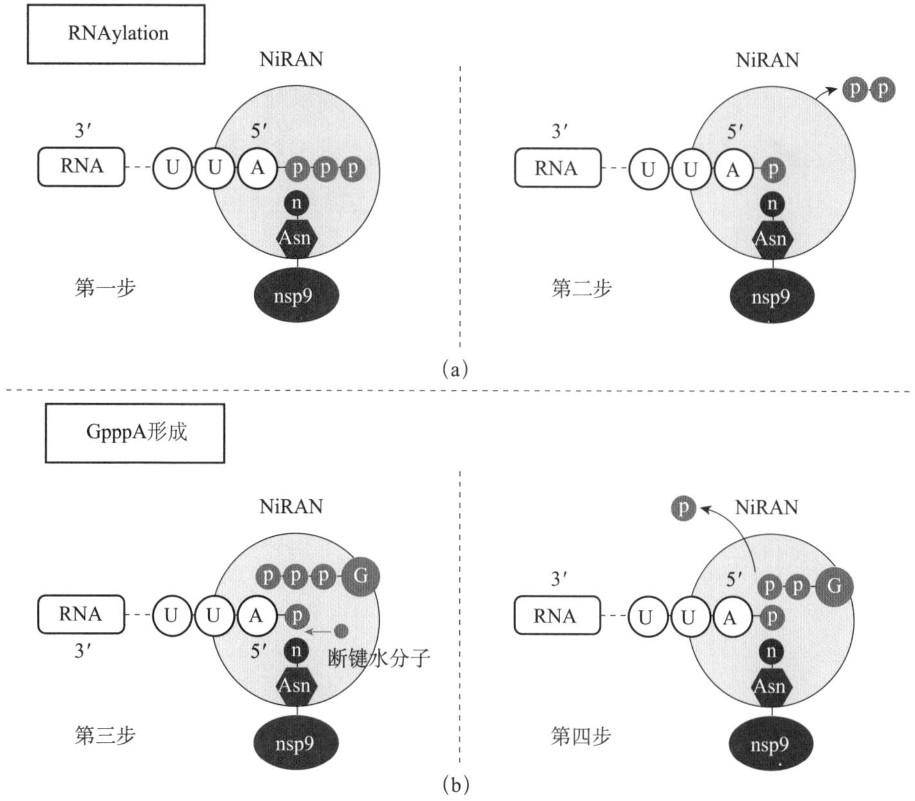

图 3　新型冠状病毒转录复制复合体与 RNA-nsp9 的复合体结构（a）与加帽新机制（b）

笔者研究团队进一步发现，核苷类抗病毒药物能够被聚合酶 NiRAN 结构域通过类似的机制共价连接到 nsp9，进而阻止加帽反应过程。尤其特殊的是，当核苷类抗病毒药物索非布韦（Sofosbuvir）连接到 nsp9 后，药物分子结合在一个被称为“核苷口袋”的位点，并通过其核糖基团上的化学修饰，诱导 NiRAN 结构的“索非布韦环”发生剧烈的构象改变，封闭“鸟苷口袋”中鸟苷三磷酸分子进出的路径（图 4）。这一系列发现，不但为认识核苷类抗病毒药物的作用机制提供了全新视角，还展示了一种“诱导-锁定”的全新的药物设计机制，为进一步研发高亲和力、全新的抗病毒药物提供了一个全新的角度。

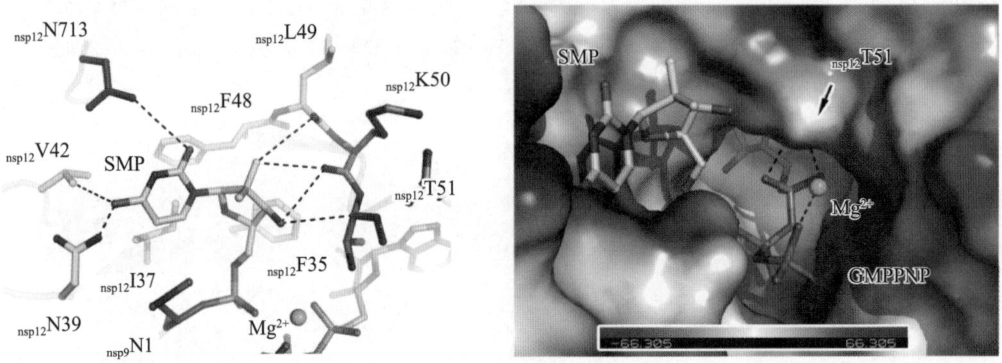

(a) (b)

图 4 索非布韦与索非布韦环的作用（a）和索非布韦环封闭鸟苷口袋的机制（b）
SMP：索非布韦单磷酸；GMPPNP：鸟苷三磷酸非水解类似物。

参考文献

[1] Yan L M, Ge J, Zheng L T, et al. Cryo-EM structure of an extended SARS-CoV-2 replication and transcription complex reveals an intermediate state in cap synthesis. Cell, 2021, 184(1): 184-193.

[2] Yan L M, Yang Y X, Li M Y, et al. Coupling of N7-methyltransferase and 3′-5′exoribonuclease with SARS-CoV-2 polymerase reveals mechanisms for capping and proofreading. Cell, 2021, 184(13): 3474-3485.

[3] Gao Y, Yan L, Huang Y, et al. Structure of the RNA-dependent RNA polymerase from COVID-19 virus. Science, 2020, 368(6492): 779-782.

[4] Yan L M, Zhang Y, Ge J, et al. Architecture of a SARS-CoV-2 mini replication and transcription complex. Nature Communication, 2020, 11(1): 5874.

A Mechanism for SARS-CoV-2 RNA Capping and Its Inhibition by Nucleotide Analogue Inhibitors

Huang Yucen, Yan Liming, Rao Zihe, Lou Zhiyong

The attachment of a cap to mRNA plays essential role in both eukaryotes and viruses. Here we report a mechanism for SARS-CoV-2 RNA capping and document structural details at atomic resolution. The NiRAN domain in the polymerase catalyzes the covalent attachment of the 5′end of the RNA to the first residue of nsp9 (termed as RNAylation), thus providing an intermediate for the formation of the cap core (GpppA) with GTP, again catalyzed by NiRAN. We also reveal that triphosphorylated nucleotide analogue inhibitors can bind to nsp9 and fit into a previously unknown 'Nuc-pocket' in NiRAN, thus inhibiting nsp9 RNAylation and formation of GpppA. The S-loop (residues 50-KTN-52) in NiRAN exhibits a remarkable conformational shift observed in RTC bound to sofosbuvir monophosphate, suggesting an 'induce-and-lock' mechanism for inhibitors design. These findings not only improve the understanding of SARS-CoV-2 RNA capping and the mode of action of NAIs, but also provide a strategy for antiviral drugs design.

3.11 冠状病毒跨种识别的分子机制

齐建勋 王奇慧 刘科芳 高 福

（中国科学院病原微生物与免疫学重点实验室）

冠状病毒严重威胁着人类健康、社会稳定和经济发展，是21世纪以来人类共同面临的重要挑战。在过去的20年中，人类经历了三次严重的冠状病毒感染暴发，即严重急性呼吸综合征冠状病毒（SARS-CoV）、中东呼吸综合征冠状病毒（MERS-CoV）和新型冠状病毒（SARS-CoV-2）[1]。截至2022年11月24日，新冠疫情已经扩散至全球192个国家和地区，造成超过6.3亿人感染和超过660万人死亡①。

病毒与宿主细胞表面受体结合是病毒感染宿主的第一步，也是病毒与宿主共进化的结果。研究病毒和宿主细胞表面受体之间的相互作用，对理解病毒致病的病理机制、病毒的跨物种传播和进化、开发有效的抗病毒药物和疫苗等都具有重要的意义。在冠状病毒编码的所有蛋白中，棘突（spike）蛋白负责与宿主细胞表面受体结合，并诱导病毒囊膜与宿主细胞膜结构（细胞膜或内吞体膜）融合，因此是首选的疫苗设计和抗体筛选的靶蛋白。不同的冠状病毒利用不同的宿主细胞表面受体感染宿主。新冠疫情暴发初期，笔者研究团队首先鉴定出SARS-CoV-2的受体为血管紧张素转换酶2（ACE2），解析了SARS-CoV-2棘突蛋白上的受体结合结构域（RBD）与人源ACE2的2.5 Å分辨率的晶体结构，阐明了SARS-CoV-2侵入宿主细胞的分子机制，为靶向RBD的疫苗开发和抗体治疗奠定了理论基础[1]。此外，研究团队解析了SARS-CoV-2原型（prototype，PT）以及流行变异株Alpha、Beta、Gamma、Delta和Omicron不同亚型BA.1、BA.1.1、BA.2、BA.3的RBD与人源ACE2的复合物结构（图1）[2-4]，揭示了其与人源ACE2结合亲和力变化的结构基础。

疫情之初，笔者研究团队率先提出通过评估SARS-CoV-2 RBD识别不同物种ACE2的能力来为SARS-CoV-2的溯源提供理论基础及评估SARS-CoV-2跨种传播的风险；研究结果显示，SARS-CoV-2具有广泛的受体识别谱，并且具有跨种传播的潜在风险[5]。截至2022年7月31日，已经报道的天然感染SARS-CoV-2的物种有25

① 相关数据参见世界卫生组织网站：https://covid19.who.int/。

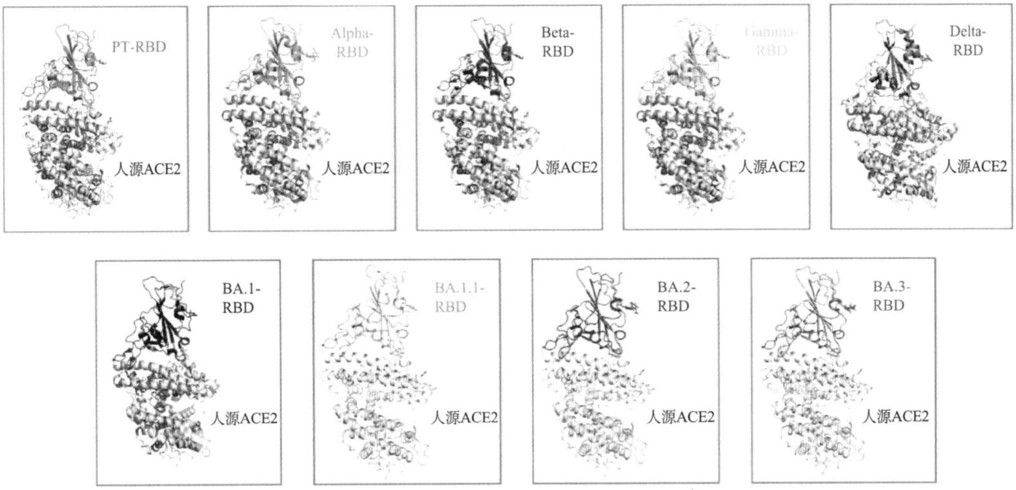

图 1　SARS-CoV-2 不同突变株 RBD 与人源 ACE2 复合物结构

种①。基于此，笔者研究团队聚焦 SARS-CoV-2 跨种传播机制研究，系统阐明了 SARS-CoV-2 RBD 与中华菊头蝠、大耳菊头蝠、猫、狗、马、穿山甲、水貂、小鼠、果子狸、海狮和抹香鲸等哺乳动物 ACE2 相互作用的分子机制（图 2）[5-12]。系统的结构生物学研究表明，SARS-CoV-2 结合人受体的适应性最高[2-5]，提示其在人或具有与人源 ACE2 相似的物种中发生过适应性进化。

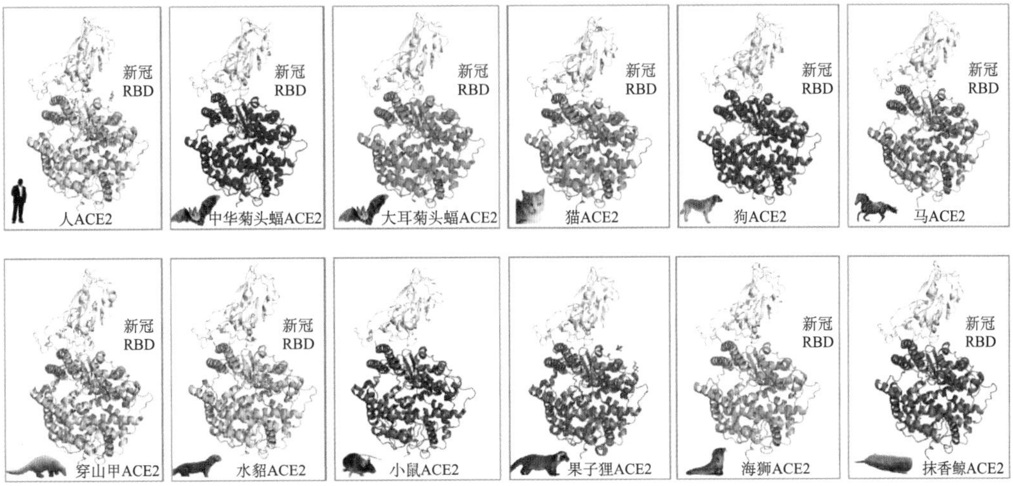

图 2　SARS-CoV-2 RBD 结合不同物种 ACE2 的整体结构[1,5-12]

① 相关数据参见世界动物卫生组织网站：https://www.woah.org/en/document/86934/。

笔者研究团队还聚焦动物携带且与SARS-CoV-2相似的冠状病毒，对蝙蝠来源的冠状病毒RaTG13、RshSTT182和RshSTT200，穿山甲来源的冠状病毒GD/1/2019和GX/P2V/2017的跨种传播潜能进行了系统评估[13,14]。团队发现这5种冠状病毒均可以结合人源ACE2并具有广泛的宿主受体识别谱，且当用假病毒系统评估其感染能力时，这些冠状病毒的假病毒感染表达人源ACE2细胞系的能力均明显弱于SARS-CoV-2假病毒，这表明这些病毒存在跨越种间屏障的潜在风险，但要获得类似SARS-CoV-2的较强跨种传播能力仍需要进一步进化，这也表明需要加强对动物来源的冠状病毒的监测和跨种传播风险评估，提前预警，预防新的冠状病毒跨种传播感染人，并在人际传播、引发疫情。

上述研究工作获得了Faculty Opinions推荐，并入选中国科协生命科学学会联合体评选的“2021年度中国生命科学十大进展”和中国科学院“2021年度科技创新亮点成果”。这些原创性研究成果不仅从分子层面揭示了冠状病毒识别宿主受体及跨种传播的分子机制，还推动笔者团队联合相关企业成功开发了全球首款新型冠状病毒重组蛋白亚单位疫苗（CHO细胞）及新冠肺炎人源治疗性抗体药物埃特司韦单抗(Etesevimab)，为全球抗击新冠疫情提供了科技支撑与防控产品。

参考文献

[1] Wang Q, Zhang Y, Wu L, et al. Structural and functional basis of SARS-CoV-2 entry by using human ACE2. Cell, 2020, 181: 894-904.

[2] Li L, Liao H, Meng Y, et al. Structural basis of human ACE2 higher binding affinity to currently circulating Omicron SARS-CoV-2 sub-variants BA. 2 and BA. 1. 1. Cell, 2022, 185(16): 2952-2960.

[3] Han P, Li L, Liu S, et al. Receptor binding and complex structures of human ACE2 to spike RBD from omicron and delta SARS-CoV-2. Cell, 2022, 185: 630-640.

[4] Han P, Su C, Zhang Y, et al. Molecular insights into receptor binding of recent emerging SARS-CoV-2 variants. Nature Communications, 2021, 12: 6103.

[5] Wu L, Chen Q, Liu K. et al. Broad host range of SARS-CoV-2 and the molecular basis for SARS-CoV-2 binding to cat ACE2. Cell Discovery, 2020, 6: 68.

[6] Tang L, Zhang D, Han P, et al. Structural basis of SARS-CoV-2 and its variants binding to intermediate horseshoe bat ACE2. International Journal of Biological Sciences, 2022, 18: 4658-4668.

[7] Liu K, Tan S, Niu S, et al. Cross-species recognition of SARS-CoV-2 to bat ACE2. Proceedings of the National Academy of Sciences of USA, 2021, 118(1): e2020216118.

[8] Zhang Z, Zhang Y, Liu K, et al. The molecular basis for SARS-CoV-2 binding to dog ACE2. Nature Communications, 2021, 12: 4195.

[9] Xu Z, Kang X, Han P, et al. Binding and structural basis of equine ACE2 to RBDs from SARS-CoV, SARS-CoV-2 and related coronaviruses. Nature Communications, 2022, 13: 3547.

[10] Li L, Han P, Huang B, et al. Broader-species receptor binding and structural bases of Omicron SARS-CoV-2 to both mouse and palm-civet ACE2s. Cell Discovery, 2022, 8: 65.
[11] Li S, Yang R, Zhang D, et al. Cross-species recognition and molecular basis of SARS-CoV-2 and SARS-CoV binding to ACE2s of marine animals. National Science Review, 2022, 9: nwac122.
[12] 仵丽丽，苏佳岐，牛胜，等，新型冠状病毒结合穿山甲新 ACE2 受体的分子机制．科学通报，2021，66(1)：73-84.
[13] Liu K, Pan X, Li L, et al. Binding and molecular basis of the bat coronavirus RaTG13 virus to ACE-2 in humans and other species. Cell, 2021, 184: 3438-3451.
[14] Niu S, Wang J, Bai B, et al. Molecular basis of cross-species ACE2 interactions with SARS-CoV-2-like viruses of pangolin origin. EMBO Journal, 2022, 41: e109962.

Host Range of Coronaviruses and Their Structural Basis

Qi Jianxun, Wang Qihui, Liu Kefang, George F. Gao

Coronaviruses have been a research hotspot since the outbreak of the COVID-19 pandemic. To explore the origin of SARS-CoV-2 and assess the risk of interspecific transmission, our team took the lead in proposing the recognition and evaluation of different species' receptors by the SARS-CoV-2 receptor binding domain (RBD), and clarified the molecular mechanism of SARS-CoV-2's recognition of multiple species' receptors through structural biology methods. In addition, our team systematically evaluated the interspecific transmission potential of five zoonotic coronaviruses, and found that they can bind to human ACE2 and have a broad receptor binding spectrum. These findings underscore the importance of strengthening surveillance and risk assessment of zoonotic coronaviruses to prevent spillover and subsequent epidemics.

3.12 提高中晚期鼻咽癌疗效的高效低毒治疗新模式

马 骏

（中山大学）

鼻咽癌高发于中国，每年新发病例数占到全球的48%，其中华南地区鼻咽癌的发病率是全球平均水平的20倍，居世界首位。鼻咽癌的发病人群以中青年为主，治疗效果较差，严重危害我国人民的生命健康。鼻咽癌的主要治疗方法是放疗，70%左右的鼻咽癌患者可通过放疗得到治愈，而放疗后的全身微小残留肿瘤是治疗失败的主要原因。对于中晚期鼻咽癌患者来说，单纯放疗预后较差，需要辅以全身化疗提高疗效。经典的辅助化疗方案为“大剂量顺铂＋5-FU①”，然而，放疗后患者身体耐受性差，难以耐受高强度的传统化疗（完成率仅为40%～50%），而且严重毒副作用发生率高达42%，这成为制约疗效提高的瓶颈。因此，如何突破传统化疗模式，探讨可有效抑制鼻咽癌转移的新型治疗策略，是临床上亟待解决的一大难题。

近年来，笔者研究团队针对这个难题做了一系列的探索和研究。2006年，笔者研究团队开展了一项前瞻性多中心随机对照临床试验，共纳入476例患者。研究对比了同期放化疗序贯“大剂量顺铂＋5-FU”辅助化疗（辅助化疗组）与单纯同期放化疗（对照组）疗效，发现辅助化疗组与对照组的5年生存率无显著差别，首次证明了经典辅助化疗对鼻咽癌患者无效，并且增加了毒性反应。该研究成果于2012年发表在《柳叶刀-肿瘤学》（*Lancet Oncology*）期刊上[1]，2013年被美国国立综合癌症网络（National Comprehensive Cancer Network）发布的临床实践指南（简称NCCN指南）采纳，改变了国际上沿用了15年的“教科书式”鼻咽癌经典治疗方案，使患者避免了过度治疗。2011年，笔者研究团队开展了第二项前瞻性多中心随机对照临床试验，共纳入480例患者。在经典的“顺铂＋5-FU”两药的基础上，增加了多西他赛，并且根据亚裔人群的特点，将每种药物剂量减少了20%；研究发现，同期放化疗前使用“多西他赛-顺铂-5-FU”（TPF）三药诱导化疗可进一步提高总生存率，该研究成果于

① 5-FU指5-氟尿嘧啶。

2016年发表在《柳叶刀-肿瘤学》[2]，并再次被NCCN指南采纳，使诱导化疗联合同期放化疗成为全世界首选的鼻咽癌治疗方案。然而，临床发现TPF三药诱导化疗方案疗效虽好，但毒性相对较大，严重毒副反应发生率达42%，许多患者无法耐受治疗，不利于该方案在基层医院的推广。2013年，笔者研究团队开展了第三项前瞻性多中心随机对照临床试验，共纳入480例患者。研究发现采用“吉西他滨+顺铂”（GP）双药方案诱导化疗，可以在同期放化疗基础上将3年无瘤生存率提升8.8%，其中GP双药方案的毒副作用较TPF三药方案减少了2/3（GP双药方案5%，TPF三药方案15%），该成果于2019年发布在《新英格兰医学杂志》（*The New England Journal of Medicine*）[3]上，第3次被NCCN指南采纳，成为全世界鼻咽癌首选的治疗方案。

尽管大多数患者在根治性放化疗后可以达到临床完全缓解，但无论是否使用诱导化疗，约30%的患者仍会出现疾病局部区域复发或远处转移。因此，患者迫切需要额外的辅助治疗以减少疾病复发与死亡的风险。2021年，笔者研究团队首次提出了小剂量、长时间口服细胞毒药物卡培他滨的节拍化疗模式。节拍化疗是指，在不延长停药期的情况下，长期高频而规律地给予低剂量化疗药物的抗肿瘤模式，具有毒性低、依从性好等优点，其可通过抗血管生成、杀伤肿瘤干细胞等机制持续抑制肿瘤，同时提高机体耐受性。通过一项多中心、前瞻性临床研究，笔者研究团队发现，在放疗后使用卡培他滨节拍化疗可将患者肿瘤转移风险降低45%，可将高危人群3年生存率提高9%，且严重毒副反应发生率由42%减少至17%，治疗完成率达74%（图1）。同时，卡培他滨口服用药方便可及，易于向基层推广。该研究突破了传统化疗的疗效瓶颈，建立了鼻咽癌国际领先、高效低毒且简单易行的治疗新标准。节拍辅助化疗成为鼻咽癌高危患者高效低毒的新治疗模式。该成果于2021年发布在《柳叶刀》（*Lancet*）杂志上[4]。

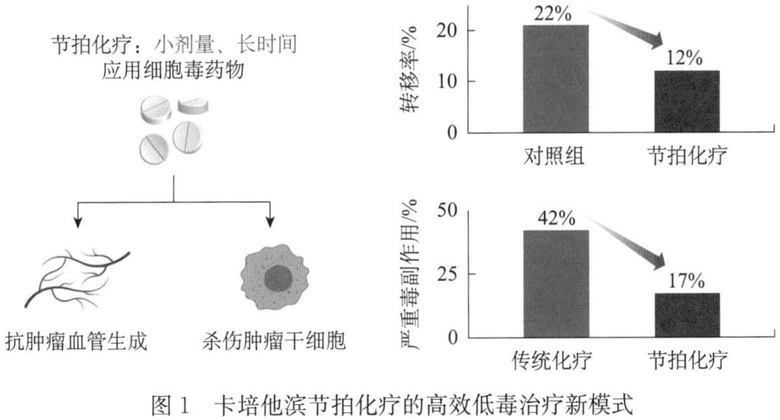

图1　卡培他滨节拍化疗的高效低毒治疗新模式

这一系列研究成果使中晚期鼻咽癌患者5年生存率提高了8%。过去23年，美国

NCCN 指南对鼻咽癌治疗方案做过 7 项重大调整，笔者研究团队为其中的 6 项提供了修订依据，并且牵头制定了《中国-美国临床肿瘤学会鼻咽癌临床诊治国际指南》[5]，这是医学领域内首个由中国学者牵头，联合美国学术组织合作制定的国际循证指南，为国际鼻咽癌诊疗提供了“一种高效低毒的新治疗模式”。

参考文献

[1] Chen L, Hu C S, Chen X Z, et al. Concurrent chemoradiotherapy plus adjuvant chemotherapy versus concurrent chemoradiotherapy alone in patients with locoregionally advanced nasopharyngeal carcinoma: a phase 3 multicentre randomised controlled trial. Lancet Oncol, 2012, 13(2): 163-171.

[2] Sun Y, Li W F, Chen N Y, et al. Induction chemotherapy plus concurrent chemoradiotherapy versus concurrent chemoradiotherapy alone in locoregionally advanced nasopharyngeal carcinoma: a phase 3, multicentre, randomised controlled trial. Lancet Oncol, 2016, 17(11): 1509-1520.

[3] Zhang Y, Chen L, Hu G Q, et al. Gemcitabine and cisplatin induction chemotherapy in nasopharyngeal carcinoma. N Engl J Med, 2019, 381(12): 1124-1135.

[4] Chen Y P, Liu X, Zhou Q, et al. Metronomic capecitabine as adjuvant therapy in locoregionally advanced nasopharyngeal carcinoma: a multicentre, open-label, parallel-group, randomised, controlled, phase 3 trial. Lancet, 2021, 398(10297): 303-313.

[5] Chen Y P, Ismaila N, Chua M L K, et al. Chemotherapy in combination with radiotherapy for definitive-intent treatment of stage II-IVA nasopharyngeal carcinoma: CSCO and ASCO guideline. J Clin Oncol, 2021, 39(7): 840-859.

A New Therapeutic Strategy with High Efficacy and Low Toxicity to Improve Prognosis of Patients with Nasopharyngeal Carcinoma

Ma Jun

Patients withlocoregionally advanced nasopharyngeal carcinoma have a high risk of disease relapse, despite a high proportion of patients achienve complete clinical remission after receiving standard-of-care treatment (i. e. definitive concurrent chemoradiotherapy with or without induction chemotherapy). Additional adjuvant therapies are needed to further reduce the risk of recurrence and death. Metronomic chemotherapy refers to the frequent and regular administration of chemotherapeutic drugs at substantially lower doses over prolonged periods without prolonged drug-free break periods, which has the advantages of low toxicity and superior compliance. The team of Prof. Ma initiated initiated a multicenter, randomized, phase 3 clinical trial, and first proposed the addition of metronomic adjuvant capecitabine to chemoradiotherapy significantly improved failure-free survival in patients with high-risk locoregionally advanced nasopharyngeal carcinoma.

3.13 异源四倍体野生稻快速从头驯化

张静昆 孟祥兵 余 泓 李家洋

（中国科学院遗传与发育生物学研究所植物基因组学国家重点实验室）

未来世界粮食安全仍然面临严峻挑战。一方面，粮食需求随着人口增长而不断增加，据联合国粮食及农业组织（Food and Agriculture Organization of the United Nations，FAO）估计，与2012年相比，2050年粮食产量需要再增加50%才能满足需求[1]；另一方面，近年来全球气候变暖，极端天气发生频率逐年升高，严重威胁粮食生产。水稻是最主要的粮食作物之一，为世界一半以上的人口提供主粮。虽然我国在水稻育种方面取得了辉煌的成就，但近年来水稻单产增速放缓甚至遇到瓶颈，迫切需要新的策略来进一步大幅提升水稻产量及其环境适应能力以保障未来粮食安全。

目前的栽培水稻均为二倍体。除了二倍体栽培稻以外，稻属还有其他25种野生植物，按照基因组特征又可以分成11类，包括6类二倍体基因组和5类四倍体基因组[2]。其中CCDD基因组的异源四倍体野生稻具有生物量大、抗逆性好、环境适应能力强等优势，但同时也具有易落粒、籽粒小等典型的未驯化特征，无法适用于农业生产。

为创制新型多倍体水稻作物突破产量瓶颈，笔者研究团队首次提出异源四倍体野生稻快速从头驯化的新策略，并将其分为四个阶段[3]：第一阶段，收集并筛选综合性状最佳的异源四倍体野生稻底盘种质资源；第二阶段，建立野生稻快速从头驯化技术体系，其中包括三个关键技术，即高质量参考基因组的绘制和基因功能注释、高效遗传转化体系及高效基因组编辑技术体系；第三阶段，品种分子设计与快速驯化，包括控制重要农艺性状基因的挖掘、基于基因组信息的分子设计、多基因编辑聚合优良性状，以及田间综合性状评估；第四阶段，新型水稻作物的示范、推广与应用。

以这一策略为蓝图，笔者研究团队通过对生物量大、抗逆能力强的CCDD型异源四倍体野生稻资源进行筛选，选出一份高秆野生稻（*Oryza alta*）资源，作为后续研究的基础，并将其命名为“PolyPloid Rice 1”（PPR1）。随后团队突破了多倍体水稻组培再生与遗传转化体系、高效精准的基因组编辑技术体系、高质量四倍体野生稻参考基因组等技术瓶颈。在此基础上，进一步注释了驯化基因及农艺性状基因，挖掘鉴定了野生稻从头驯化的候选基因，并对PPR1中控制落粒性、芒长、株高、粒长、茎秆粗度及生育期的同源基因进行了基因组编辑，成功创制了落粒性降低、芒长变短、株高降低、籽粒变长、茎秆变粗、抽穗时间不同程度缩短的各种基因编辑材料（图 1）。这一系列结果最终证明本团队提出的异源四倍体野生稻快速从头驯化策略高度可行，对未来创制新的作物种类、保障粮食安全具有重要意义。

从头驯化全新育种策略的提出，拓宽了育种可用的遗传资源，有望推动新一轮农业革命。该项工作受到国内外同行的高度关注，在《细胞》（*Cell*）、《自然》、《国家科学评论》（*National Science Review*）、《遗传学报（英文版）》（*Journal of Genetics and Genomics*）、《植物学报》和《遗传》等杂志发表专题评述，并入选 2022 年度“中国农业科学重大进展”论文，《细胞》杂志社 2021 中国年度论文。“异源四倍体野生水稻快速从头驯化”入选 2021 年度“中国生命科学十大进展”，及两院院士评选的“2021 年中国十大科技进展新闻”。

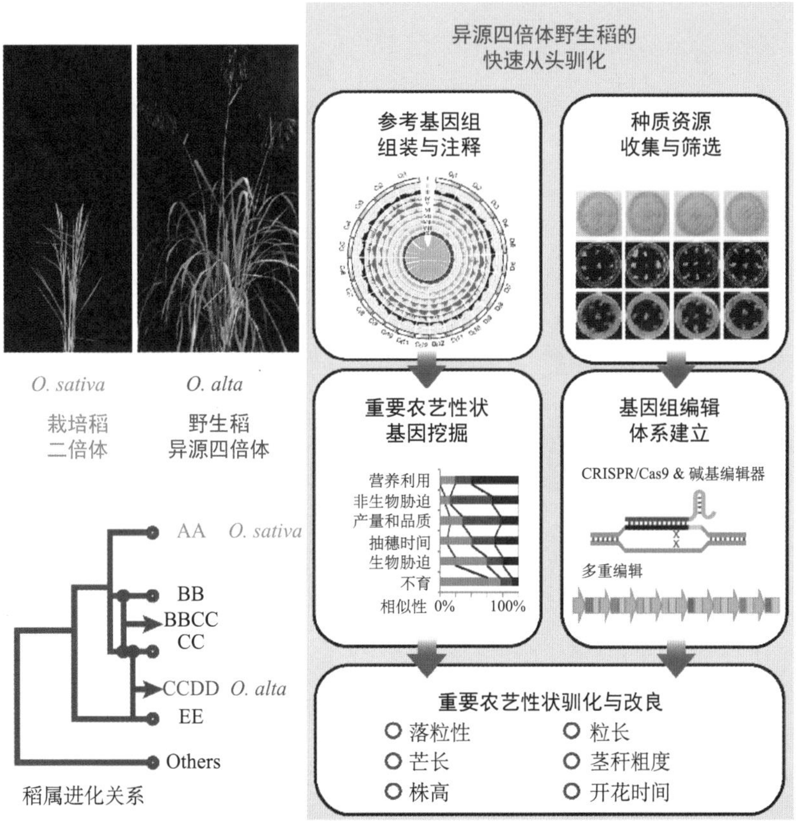

图 1　异源四倍体野生稻的快速从头驯化

参考文献

[1] Food and Agriculture Organization. The Future of Food and Agriculture-Trends and Challenges (Rome: Food and Agriculture Organization). 2017.

[2] Wing R A, Purugganan M D, Zhang Q F. The rice genome revolution: From an ancient grain to green super rice. Nature Review Genetics, 2018, 19(8): 505-517.

[3] Yu H, Lin T, Meng X, et al. A route to *de novo* domestication of wild allotetraploid rice. Cell, 2021, 184(5): 1156-1170.

A Route to *de novo* Domestication of Wild Allotetraploid Rice

Zhang Jingkun, Meng Xiangbing, Yu Hong, Li Jiayang

Certain polyploid rice has remarkable advantages in biomass, resistance, and environmental robustness, making it a candidate wild resource for creating smart crops. Scientists from the Institute of Genetics and developmental Biology, CAS offered a practical strategy, namely *de novo* domestication of wild allotetraploid rice. The strategy can be divided into four steps: (1) collection and selection of wild germplasm; (2) establishment of technical system; (3) design and rapid domestication; (4) application and popularization. By screening allotetraploid rice germplasm, they identified one genotype of *Oryza alta*, and broke three important biotechnological bottlenecks for its *de novo* domestication: (1) an efficient tissue culture, transformation, and (2) genome editing system and (3) a high-quality reference genome and functional gene annotation. Furthermore, they demonstrated that six agronomically important traits could be rapidly improved by editing genes of interest. This groundbreaking work demonstrated that a *de novo* domestication strategy can create new crops and pave a new way to strengthen global food security.

3.14　鸟类迁徙路线成因和长距离迁徙关键基因

谷中如　林婷婷　詹祥江

（中国科学院动物研究所）

“迁徙生物如何发现其迁徙路线?”是 2005 年《科学》期刊公布的 125 个最具挑战性的科学问题之一。迁徙（migration），是指动物个体季节性地离开与返回繁殖地的行为，是动物界的一个普遍现象，在维持地球生物多样性中具有极其重要的作用[1]。鸟类具有独特的飞行特征，它们的迁徙行为具有速度快、距离长、范围广等特点，这使得鸟类成为研究迁徙行为的经典模式类群。

鸟类迁徙的现象早在古代就已经引起了人类的关注，但直到 19 世纪末，人类对鸟类迁徙现象才开始了真正意义上的科学探索。例如，鸟类环志（bird ringing）研究方法的使用，对早期鸟类迁徙研究起到了一定的推动作用。如今，卫星追踪技术凭借其稳定、高精度等优势成为应用最广泛的方法，极大地推动了鸟类迁徙研究的发展[2]。鸟类迁徙研究可以在宏观层次上结合卫星追踪技术和生态学分析，来了解鸟类迁徙路线的成因；但同时也还需要从微观层次借助遗传学、基因组学、细胞生物学等跨学科交叉方法，从迁徙策略量化和迁徙行为进化等多方面进行深入探索。

行为遗传学认为行为是受基因控制的复杂的生物学过程。最近十几年，研究者才逐渐发现一些基因（如 *ADCYAP1*、*Clock*）可能与鸟类的迁徙行为有关，如迁徙冲动、迁徙时间等[3,4]。但这些研究工作，多是采用了候选基因的方法，并没有从全基因组的角度去寻找调控鸟类迁徙行为的基因。由于跨学科整合研究的挑战性，直到最近几年，才有与鸟类迁徙有关的基因组学方面的研究工作发表[5-8]。

为了回答气候和遗传因素如何影响鸟类迁徙路线变化问题，笔者研究团队选择了游隼作为研究物种。游隼作为迁徙鸟类被人们关注由来已久，并且还是世界首例被全基因组测序的猛禽之一[9]，是研究鸟类迁徙基因组学的理想物种。笔者研究团队历时 12 年，在亚欧大陆北极圈内的 6 个繁殖地（科拉、科尔古耶夫、亚马尔、泰梅尔、勒拿、科雷马），为 56 只游隼佩戴了卫星追踪器，并对其中 4 个地区的 35 只游隼进行了高覆盖度的全基因组序列测定，开创性地在大陆尺度上建立了一套北极游隼迁徙研究系统（图 1）。研究发现，游隼主要使用 5 条迁徙路线穿越亚欧大陆，在种群和个体水平上具有非常高的迁徙连接度和重复性；其中，西部 2 群为短距离迁徙，东部 4 群为长距离迁徙。种群动态推断和潜在繁殖以及越冬地重建结果显示，在末次冰盛期到全新世的转换过程中，因冰川消退而带动的繁殖地向北退缩以及越冬地变迁可能是游隼迁徙路线形成的主要历史原因；并且在未来全球变暖日益严峻的情境下，亚欧大陆西部的北极游隼种群可能会面临迁徙策略改变和主要繁殖地退缩的双重风险（图 2）。

笔者研究团队通过聚焦于气候变化下的鸟类迁徙问题，结合遥感卫星追踪、基因组学、神经生物学等新型研究手段和多学科整合分析，阐明了鸟类迁徙路线的过去、现在和未来动态变化及主要驱动因素，揭示了与长期记忆能力相关的 *ADCY8* 基因在长距离迁徙中的遗传改变。该研究成果在 2021 年以封面文章在《自然》期刊上发表[8]，是该期刊首次发表由中国学者主导的鸟类迁徙研究工作。该研究成果还入选了 2021 年度“中国科学十大进展”、“中国生命科学十大进展”、《自然-生态与进化》（*Nature Ecology & Evolution*）期刊评选的 12 篇“年度回顾”工作等。

图 1　北极游隼迁徙研究系统示意图

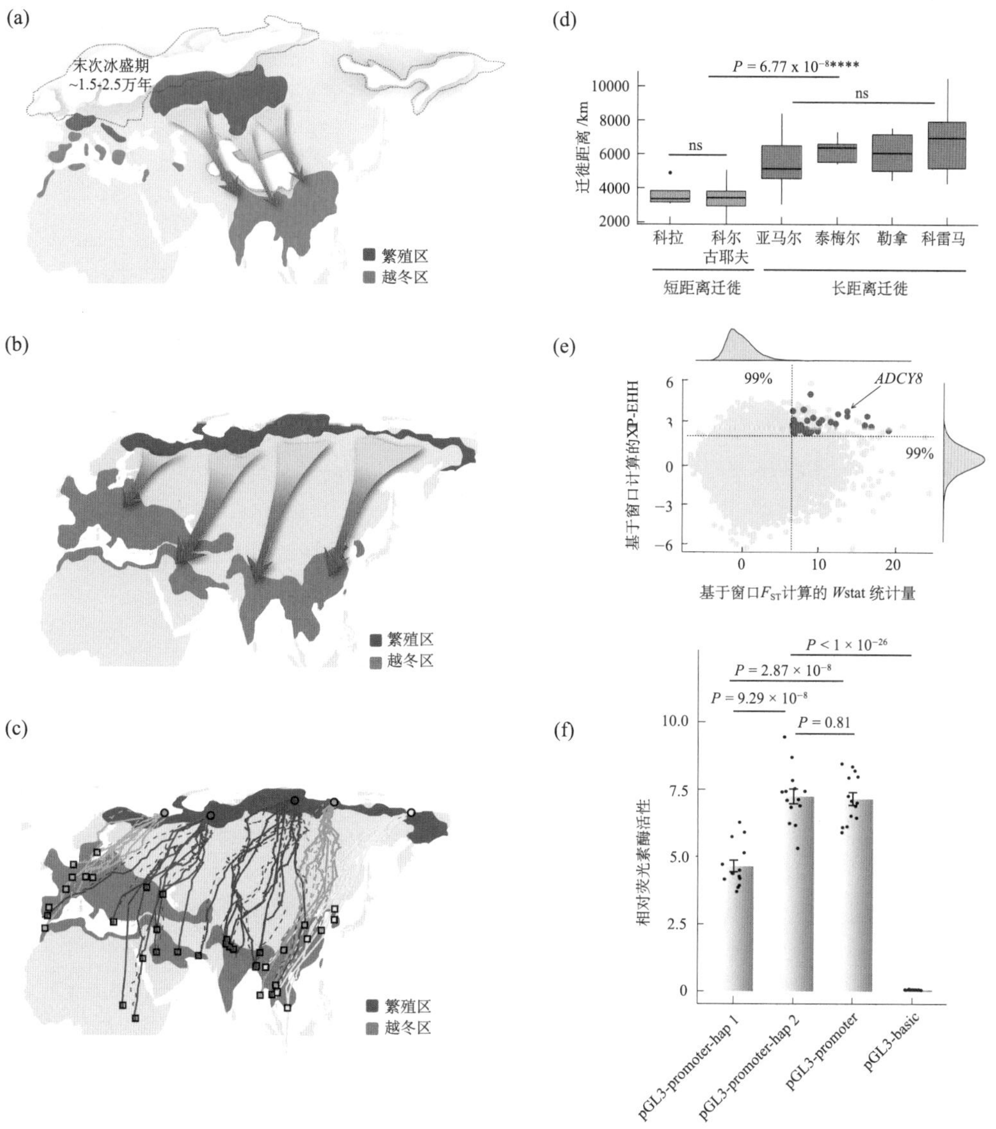

图 2　北极游隼迁徙路线成因与长距离迁徙关键基因

参考文献

[1] Bauer S, Hoye B J. Migratory animals couple biodiversity and ecosystem functioning worldwide. Science, 2014, 344(6179): 1242552.

[2] Kays R, Crofoot M C, Jetz W, et al. Terrestrial animal tracking as an eye on life and planet. Science,

2015,348(6240):aaa2478.
[3] Mueller J C, Pulido F, Kempenaers B. Identification of a gene associated with avian migratory behaviour. Proceedings of the Royal Society B: Biological Sciences, 2011, 278(1719): 2848-2856.
[4] Peterson M P, Abolins-Abols M, Atwell J W, et al. Variation in candidate genes *CLOCK* and *ADCYAP1* does not consistently predict differences in migratory behavior in the songbird genus Junco. F1000Research, 2013, 2: 115.
[5] Toews D P L, Taylor S A, Streby H M, et al. Selection on VPS13A linked to migration in a songbird. Proceedings of the National Academy of Sciences, 2019, 116(37): 18272-18274.
[6] Delmore K, Illera J C, Perez-Tris J, et al. The evolutionary history and genomics of European blackcap migration. Elife, 2020, 9: e54462.
[7] Thorup K, Pedersen L, Da Fonseca R R, et al. Response of an Afro-Palearctic bird migrant to glaciation cycles. Proceedings of the National Academy of Sciences, 2021, 118(52): e2023836118.
[8] Gu Z, Pan S, Lin Z, et al. Climate-driven flyway changes and memory-based long-distance migration. Nature, 2021, 591(7849): 259-264.
[9] Zhan X, Pan S, Wang J, et al. Peregrine and saker falcon genome sequences provide insights into evolution of a predatory lifestyle. Nature Genetics, 2013, 45(5): 563-566.

Formation of Bird Migration Routes and Key Gene for Long Distance Migration

GuZhongru, Lin Tingting, Zhan Xiangjiang

How do migratory animals know where they're going is one of the most challenging 125 scientific questions in *Science*. The research team led by Xiangjiang Zhan from the Institute of Zoology of the Chinese Academy of Sciences, established a continental-scale migration system of peregrine falcons in the Eurasian Arctic. They found that five migratory routes were used across Eurasia, which were likely formed by longitudinal and latitudinal shifts of breeding grounds during the transition from the Last Glacial Maximum to the Holocene epoch. Genomic analysis showed that peregrines migrated longer distance had a dominant genotype of the gene *ADCY8*, which is associated with the development of long-term memory. This suggests that the long-term memory may be an important genetic basis for long-distance migration of birds. The study elucidated the formation of migration routes and the genetic basis of long-distance migration by using multidisciplinary integrated analysis methods, including satellite telemetry, genomics and neurobiology.

3.15　全脑单神经元多样性研究及信息学大数据平台

刘力娟　刘裕峰　赵苏君　彭汉川

（东南大学脑科学与智能技术研究院）

近年来，在哺乳动物脑成像领域，稀疏标记技术[1]和光学成像技术的发展使得在光学显微镜下单个神经元完整形态清晰成像成为可能，同时超大规模的显微镜数据也对大数据管理、图像质量提升、图像可视化、交互等多个层面的技术提出挑战。对此，东南大学脑科学与智能技术研究院组建了跨学科的脑科学研究团队，建立了全脑单神经元多样性研究及信息学大数据平台，涵盖了影像图像处理、图像可视化、高通量神经元形态学标注与重建、跨模态图像配准、大规模计算和分析等多个研究方向（图 1）。

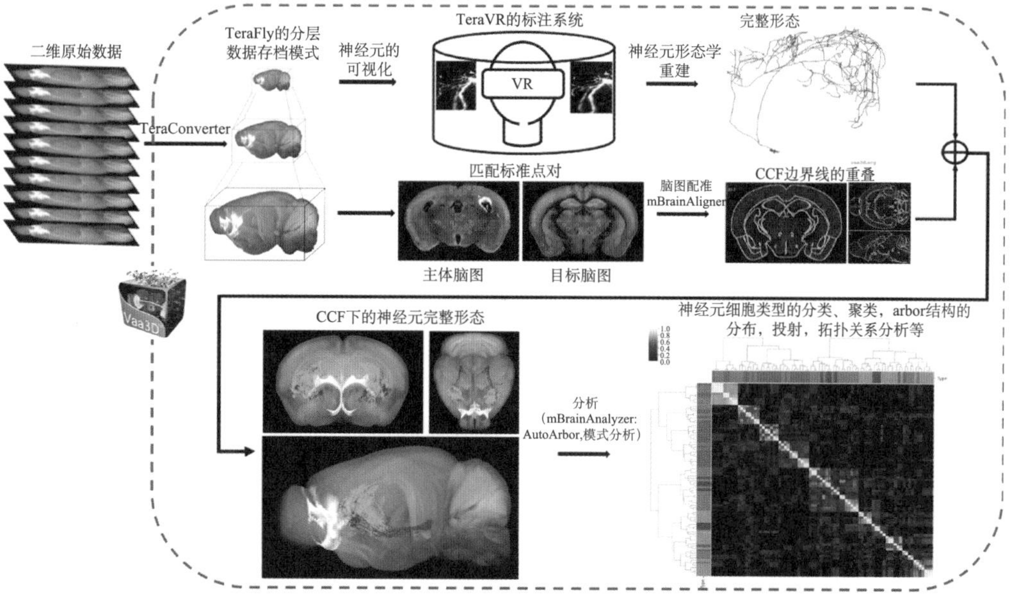

图 1　全脑范围完整神经元形态可视化、重建、配准、分析平台和流程

CCF，全称 Allen Mouse Brain Common Coordinate Framework，艾伦鼠脑公共坐标框架。

Vaa3D[2,3]（3D visualization-assisted analysis），一个跨平台（Windows、Mac 和 Linux）多维度（3D、4D、5D 等）的图像可视化、重建和分析系统，其灵活的插件模式扩展了它的功能和应用范围。以此系统为基础，笔者研究团队搭建了一个完整的神经元形态数据生产和分析的流程。Vaa3D-TeraConverter 将脑影像数据格式转换为具有多重分辨率的 3D 图像块，这种金字塔式的多尺度结构便于在 Vaa3D-TeraFly[4]下快速读取目标区域三维图像，既能从宏观上了解图像全貌，也能锁定感兴趣的脑区，高分辨地查看图像内容。全脑范围形态重建主要分为两个步骤：标注和质量控制。我们将自动算法（APP1/2[5]、tube tracing 等）与手工标注相结合，胞体标记完成后，在 UltraTracer 的框架下，首先对图像信号清晰、干扰较少的神经元树突进行自动追踪，然后借助 Vaa3D-TeraFly[4]模块和 Vaa3D-TeraVR[6]模块标注结构复杂和信号较弱的轴突部分。重建结束后，数据将通过严格的质量控制来确保精准度：①Vaa3D 的神经元数字模型校验插件初步筛查；②标注人员在手机移动平台 Hi5 或者 TeraVR 模块下协同合作，交叉验证后完成数据修改。生产的大量数据集和对应的图像数据将被存储在 MorphoHub[7]——PB 级数据的存储和管理软件平台。我们使用 mBrainAligner[8]将多模态、多维全脑图像和重建数据映射到公共坐标空间——艾伦鼠脑公共坐标框架图谱（Common Coordinate Framework atlas，CCFv3），实现了全面的神经生物学研究，如神经细胞类型鉴定、脑空间定位图谱以及神经联接图谱等。

基于以上平台和方法，笔者研究团队已成功分析了 15 000 个以上的神经元树突，并在 1891 个带有长程投射轴突的神经元中检测到了超 200 万个潜在的突触位点，重建了近 2000 个高精度小鼠神经元。同时实现了对 1891 个单神经元形态结构的特征提取和量化分析，包括分子水平、投射模式、轴突末端模式、脑区特异性和单细胞差异性等，其中鉴定了 11 种主要的长程投射神经元类型[9]。

借助上述全脑单神经元多样性研究及信息学大数据平台，我们会持续研究大脑细胞分型与功能、脑连接环路、全脑大规模模拟、类脑计算、基于生物脑的新型人工智能算法和系统等相关问题，更有效地探索脑科学领域。

参考文献

[1] Thomas R, Smallwood P M, John W, et al. Genetically-directed, cell type-specific sparse labeling for the analysis of neuronal morphology. PLoS One, 2008, 3(12): e4099.

[2] Peng H, Ruan Z, Long F, et al. V3D enables real-time 3D visualization and quantitative analysis of large-scale biological image data sets. Nature Biotechnology, 2010, 28: 348-353.

[3] Peng H, Bria A, Zhou Z, et al. Extensible visualization and analysis for multidimensional images using Vaa3D. Nature Protocols, 2014, 9(1): 193-208.

[4] Bria A, Iannello G, Onofri L, et al. TeraFly: real-time three-dimensional visualization and annotation

of terabytes of multidimensional volumetric images. Nature Methods, 2016, 13: 192-194.

[5] Xiao H, Peng H. APP2: automatic tracing of 3D neuron morphology based on hierarchical pruning of gray-weighted image distance-trees. Bioinformatics, 2013, 29(11): 1448-1454.

[6] Wang Y, Li Q, Liu L, et al. TeraVR empowers precise reconstruction of complete 3-D neuronal morphology in the whole brain. Nature Communications, 2019, 10: 3474.

[7] Jiang S, Wang Y, Liu L, et al. Petabyte-Scale multi-morphometry of single neurons for whole brains. Neuroinformatics, 2022, 20(2): 525-536.

[8] Qu L, Li Y, Xie P, et al. Cross-modal coherent registration of whole mouse brains. Nature Methods, 2022, 19(1): 111-118.

[9] Peng H, Xie P, Liu L, et al. Morphological diversity of single neurons in molecularly defined cell types. Nature, 2021, 598(7879): 174-181.

Diversity of Single Neurons at Whole-Brain Scale Is Effectively Studied Using a Big Data Informatics Platform

Liu Lijuan, Liu Yufeng, Zhao Sujun, Peng Hanchuan

A primate braincontains trillions of interconnected neurons, forming an intricate connectome that drives animal behavior. Our understanding of the diversified morphologies and functions of these circuits remains limited due to the lack of complete morphologies available. To address this issue, we established a comprehensive pipeline consisting of sparse labeling, whole-brain imaging, morphological reconstruction, registration, and analysis, to systematically describe and examine complete single-neuron morphology at the a whole-brain scale. Our framework successfully reconstructed 1891 neurons from different brain regions such as the cortex, thalamus, and striatum, and identified 11 major projection classes with distinct morphological features and transcriptomic identities. By analyzing the diversity and principles of individual cell types embedded in neural circuits, we can better delineate their roles in brain function and comprehensive behaviors.

3.16 全新世温度演变的几点新认识及其不足

张 旭 郝 硕 孙宇辰

（中国科学院青藏高原研究所，
青藏高原地球系统与资源环境全国重点实验室）

全新世开始于大约距今 11 000 年前的“新仙女木”事件结束之后，是最年轻的地质年代。全新世的大气 CO_2 浓度、全球平均海平面、植被等气候边界条件变化相对缓和，与现代较为接近（图 1），加之（特别是工业革命以前）气候变化基本不受人类活动的影响，因此其一直是研究气候变率、检验气候模型性能的重要时窗。理解全新世气候演变的特征与机理将为理解当下全球变暖、预测未来温度变化提供宝贵的参考。

温度变化是衡量气候变化的一项重要指标。然而，全新世全球年平均温度的长期变化趋势是上升还是下降，目前仍存在较大分歧，这一问题被称为“全新世温度谜题”[1,2]。温度代用记录集成的全球年平均温度显示，全新世温度呈现先上升后下降的演变特征——在全新世（早）中期温度达到最高，即“全新世大暖期”，随后至工业革命前逐渐降低[1,3]；然而，气候模式模拟的全球年均温度则呈现出全新世逐步升温的趋势[2]，并得到一些中高纬度地区温度重建的支持[4-6]。一些研究表明，记录的季节性偏差以及气候模型自身的缺陷可能是造成不同记录之间、记录与模拟之间分歧（即“全新世温度谜题”）的主要原因[7-9]。

记录的季节性偏差被认为大都与太阳辐射在不同季节的演变差异有关。为此博瓦（Bova）等在引入一种线性的“季节-年均温度转换”（seasonal to mean annual transformation，SAT）方法剔除记录中由太阳辐射引起的季节偏差后发现，记录集成的全球均温在全新世呈逐渐上升趋势，与模拟结果一致，这证实了记录的季节性偏差或是造成“全新世温度谜题”的主因[7]。但是，SAT 方法应用的合理性受到质疑[10,11]——温度变化对太阳辐射的响应并非一直是线性关系[11]，即使认为其在间冰期是线性响应，SAT 方法也可能低估气候系统内反馈（例如冰消期对间冰期极地气候的影响）对温度的调制作用[10]，因此该结论的可靠性仍需进一步评估。

张文超等基于北半球陆地孢粉记录，定量重建了全新世冬季温度与夏季温度，结

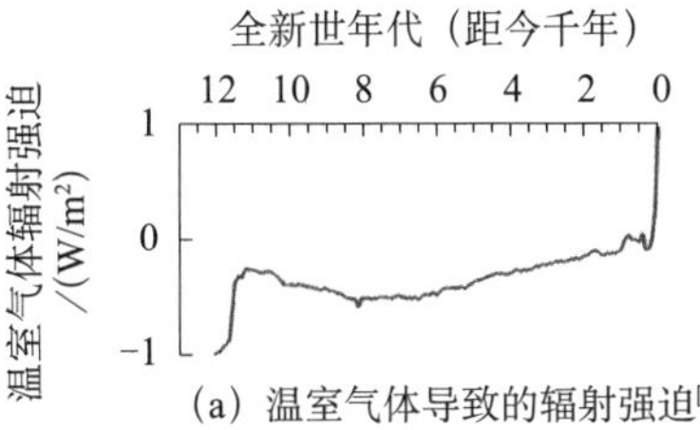

(a) 温室气体导致的辐射强迫[7]

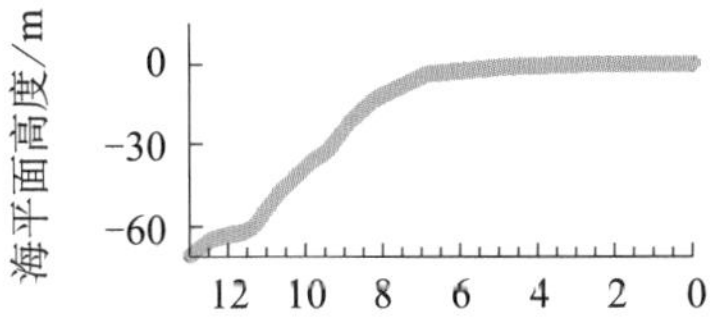

(b) 海平面重建记录[17]

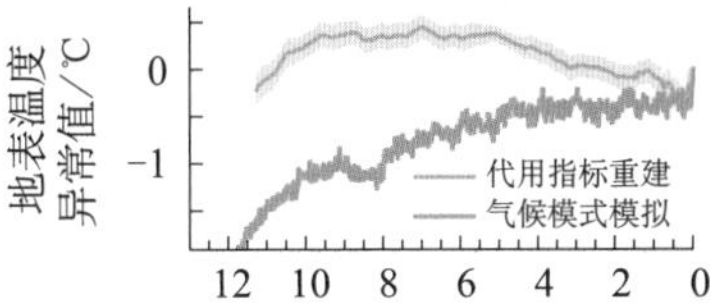

(c) 全球年均温度重建记录（黄线）[1]与数值模拟结果（绿线）[2]

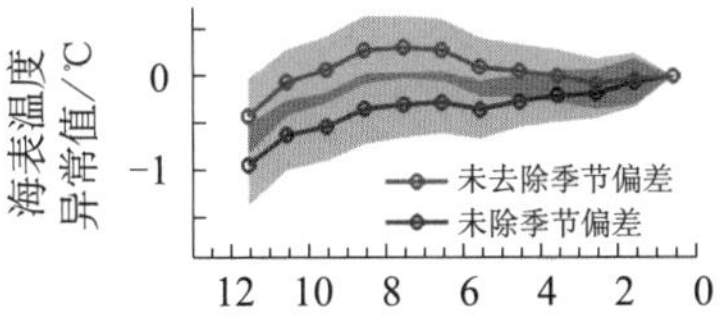

(d) 未经季节校正的全球海表面温度记录（红线）[7]
与经SAT方法计算后得到的全球年均海表面温度（蓝线 ）[7]

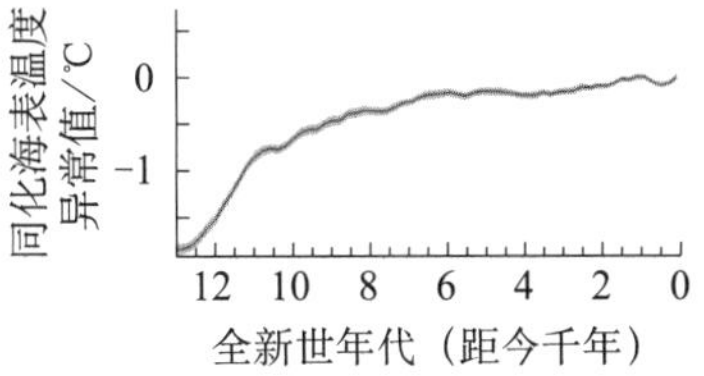

(e) 全球年均海表面温度同化结果 [13]

图 1　全新世气候背景及温度演变

果表明温度代用记录的季节性并非造成“全新世温度谜题”的主要原因，代用记录重建显示中全新世以来冬季温度呈降温趋势，显著不同于模式模拟的增温趋势，这可能源于模式对气候系统内部反馈过程刻画的不足[14]。董亚杰等人基于中国黄土蜗牛记录，定量重建了过去两万年以来的季节温度，结果同样表明代用记录重建不存在显著的季节性偏差，同时支持“全新世大暖期”的存在[15]。这些工作表明，代用指记录否存在季节性偏差目前仍存在较大争议，未来仍需深入研究。

在重建全球均温的记录中，陆相记录主要集中在 40°N～70°N 的北美和欧洲大陆，海洋记录则主要分布在 50°S 以北的海岸带（图 2）[16]，因此严格意义上讲，记录集成结果并不是气候学中的全球均温。虽然用于古气候研究的气候模型缺少对不同圈层互馈过程的刻画[8,9,17]，但可提供有动力约束的地球全域的物理信息。通过古气候数据同化方法，可以将记录信息与模式信息相融合，重建时空全域的全球温度，可在一定程度上弥补记录空间分布不均以及气候模型自身缺陷的问题。最新的同化结果显示，全新世全球年平均温度呈逐步上升趋势（图 1），与气候模式的模拟结果相似[13]。古气候数据同化结果的鲁棒性取决于古记录的丰度和气候模式的先验，奥斯曼（Osman）等用于同化的温度记录，几乎全都来自 50°S 以北的海洋指标，缺少南大洋和陆相温度记录的约束，并且其模式先验仅取自单一气候模式的结果，未进行多模式先验的对比分析，因此其结论仍存在不确定性，亟须未来进一步的研究。

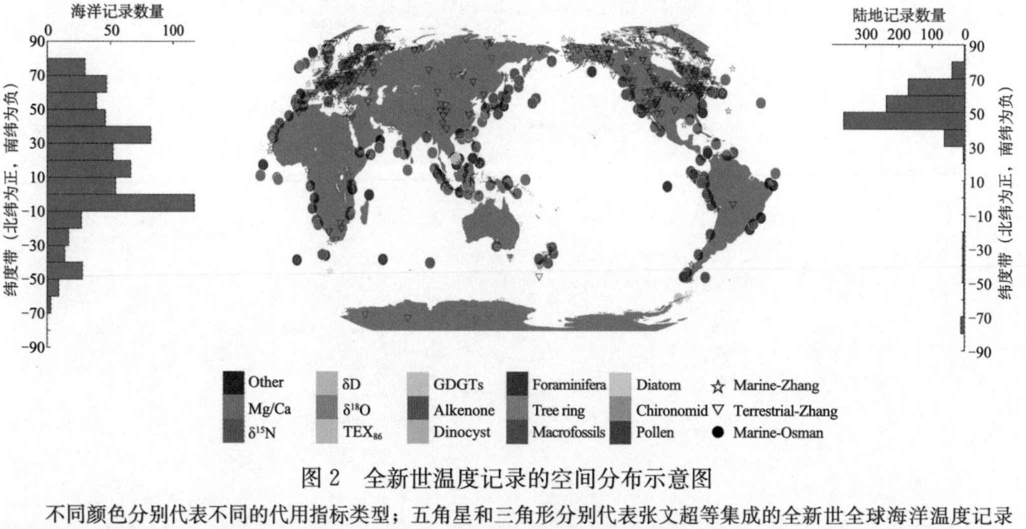

图 2　全新世温度记录的空间分布示意图

不同颜色分别代表不同的代用指标类型；五角星和三角形分别代表张文超等集成的全新世全球海洋温度记录（372 条）和陆地温度记录（931 条）[18]，圆圈代表奥斯曼等在全新世全球温度同化中使用的海洋温度记录[14]（经筛选，仅选择在中全新世期间有值的记录，260 条）。左侧柱状图表示主图中所有海洋记录的纬度分布，右侧柱状图表示主图中所有陆地记录的纬度分布。

参考文献

[1] Marcott S A, Shakun J D, Clark P U, et al. A Reconstruction of regional and global temperature for the past 11,300 years. Science, 2013, 339: 1198-1201.

[2] Liu Z Y, Zhu J, Rosenthal Y, et al. The Holocene temperature conundrum. Proceedings of the National Academy of Sciences of the USA, 2014, 111: E3501-E3505.

[3] Kaufman D, McKay N, Routson C, et al. Holocene global mean surface temperature, a multi-method reconstruction approach. Scientific Data, 2020, 7: 201.

[4] Meyer H, Opel T, Laepple T, et al. Long-term winter warming trend in the Siberian Arctic during the mid-to late Holocene. Nature Geoscience, 2015, 8: 122-125.

[5] Baker J L, Lachniet M S, Chervyatsova O, et al. Holocene warming in western continental Eurasia driven by glacial retreat and greenhouse forcing. Nature Geoscience, 2017, 10: 430-435.

[6] Marsicek J, Shuman B N, Bartlein P J, et al. Reconciling divergent trends and millennial variations in Holocene temperatures. Nature, 2018, 554: 92-96.

[7] Bova S, Rosenthal Y, Liu Z Y, et al. Seasonal origin of the thermal maxima at the Holocene and the last interglacial. Nature, 2021, 589: 548-553.

[8] Park H S, Kim S J, Stewart A L, et al. Mid-Holocene northern hemisphere warming driven by Arctic amplification. Science Advances, 2019, 5: eaax8203.

[9] Thompson A J, Zhu J, Poulsen C J, et al. Northern Hemisphere vegetation change drives a Holocene thermal maximum. Science Advances, 2022, 8: eabi6535.

[10] Zhang X, Chen F H. Non-trivial role of internal climate feedback on interglacial temperature evolution. Nature, 2021, 600: E1-E3.

[11] Laepple T, Shakun J, He F, et al. Concerns of assuming linearity in the reconstruction of thermal maxima. Nature, 2022, 607: E12-E14.

[12] Clark P U, He F, Golledge N R, et al. Oceanic forcing of penultimate deglacial and last interglacial sea-level rise. Nature, 2020, 577: 660-664.

[13] Osman M B, Tierney J E, Zhu J, et al. Globally resolved surface temperatures since the Last Glacial Maximum. Nature, 2021, 599: 239-244.

[14] Zhang W C, Wu H B, Cheng J, et al. Holocene seasonal temperature evolution and spatial variability over the Northern Hemisphere landmass. Nature Communications, 2022, 13(1): 5334.

[15] DongY J, Wu N Q, Li F J, et al. 2022, The Holocene temperature conundrum answered by mollusk records from East Asia. Nature Communications, 2022, 13: 5153.

[16] CartapanisO, Jonkers L, Sanchez P M, et al. Complex Spatio-temporal structure of Holocene Thermal Maximum. Nature Communications, 2022, 13: 5662.

[17] Liu Y G, Zhang M, Liu Z Y, et al. A possible role of dust in resolving the Holocene temperature

conundrum. Scientific Reports, 2018, 8: 4434.
[18] Zhang W C, Wu H B, Geng J Y, et al. Model-data divergence in global seasonal temperature response to astronomical insolation during the Holocene. Science Bulletin, 2022, 67: 25-28.

New Insights of Holocene Temperature Evolution and Their Shortcomings

Zhang Xu, Hao Shuo, Sun Yuchen

The Holocene climate evolution provides key insights into the natural variability and internal feedbacks of Earth system. However, its global mean annual temperature remains controversial change. In particular, proxy reconstructions indicate the existence of Holocene Thermal Maximum (HTM) in the early mid-Holocene, followed by a cooling trend till the pre-industrial period, while climate models simulate a long-term warming trend throughout the Holocene. To date, seasonal biases in temperature proxies and model deficiencies have been considered responsible for this discrepancy. In this context, studies that either remove seasonal biases in the marine proxies via a linear transformation method or produce proxy-constrained global temperature reanalysis datasets via paleoclimate data assimilation consistently demonstrate the long-term warming feature as indicated by climate models, thereby advancing our knowledge of Holocene temperature evolution. Nevertheless, several shortcomings remain that undermine their robustness and thus require further investigation in the future.

3.17　地表物质深循环与原始俯冲在 32.3 亿年前出现

王孝磊　王　迪　李军勇　李林森

（南京大学地球科学与工程学院）

板块构造是地球区别于太阳系其他固体星球的重要特征之一。板块构造理论认为，经过地表水-岩反应的物质很容易经由俯冲作用到达地壳深部或地幔，从而引发水等挥发分物质的深部再循环。然而，地球板块构造的启动时间和早期机制还不甚明确，因而识别地球早期地表物质的深循环对于我们理解古代板块构造在地球早期的演化非常重要。

为深入探索地球早期的动力学机制，笔者研究团队对南非卡普瓦尔（Kaapvaal）克拉通的巴伯顿（Barberton）地区进行了多次野外科考工作，对该地区出露的典型的古太古代—中太古代（约 34.6 亿～32 亿年前）的 TTG 岩石［地球早期特征的花岗质岩石，TTG 是奥长花岗岩（trondhjemite）、英云闪长岩（tonalite）、花岗闪长岩（granodiorite）三个岩类英文首字母的简称］进行系统采样和室内分析，获取了该地区 TTG 岩石的同位素地质年代学和地球化学以及锆石原位微区氧同位素数据。这些新的数据揭示，地球早期地表物质深循环的启动可能与构造体制的转换有关。

锆石离子探针 U-Pb 同位素定年结果表明，南非巴伯顿地区的 TTG 岩石主要形成于 34.6 亿～34.2 亿年前和 32.65 亿～32.05 亿年前两个阶段。选择 U-Pb 体系封闭的锆石颗粒，进行离子探针原位氧同位素分析，发现大约以 32.3 亿年为界，该区 TTG 岩石中的锆石氧同位素组成（以与 SMOW 标准氧同位素的千分差 $\delta^{18}O$ 表示）发生了一次突变——早于 32.3 亿年的 TTG 岩石中的锆石 $\delta^{18}O$ 值都在地幔值的范围（5.3‰±0.6‰）内，平均值分布在 5.07‰～6.02‰；而 32.3 亿年及之后的 TTG 岩石中的锆石 $\delta^{18}O$ 值基本在 6‰以上，在 5.95‰～7.08‰，甚至个别分析点的值接近 8‰［图 1（a）］。而且，笔者研究团队使用锆石激光拉曼光谱分析和累积辐射计算，指出这些氧同位素分析结果基本没有受到锆石放射性衰变的影响。因此可以相信，这种氧同位素变化是可靠的。

2021 年，Smithies 等[1]报道了 29 亿年前的氧同位素变化，而笔者团队此次观察到的氧同位素变化比这还要早。而且在格陵兰、澳大利亚和印度南部，都有类似的 32 亿年前氧同位素变化的痕迹，这意味着这一时期的氧同位素变化应当具有一定规模。

进一步研究发现，该变化与巴伯顿地区页岩的三氧同位素突然下降[2]也一致。

氧同位素的升高是什么源区物质引起的？一种可能是地表的燧石和页岩。在古太古代，这些地表燧石和页岩的 $\delta^{18}O$ 值可达 10‰以上，而这些≤32.3 亿年的 TTG 岩石中的锆石氧同位素 $\delta^{18}O$ 并没有高太多。因此，笔者研究团队认为，玄武质岩浆在海底喷发时与海水之间的低温热液蚀变，是升高其 $\delta^{18}O$ 值的最可能机制。在约 32.3 亿年前，这些玄武质地壳由于全球构造体制的改变［类似于现代俯冲作用的启动；图 1（b）］而被带到地壳深部，在那里发生熔融就可以产生高 $\delta^{18}O$ 的 TTG 岩石。研究团队发现，在这些 TTG 岩石样品中，不只是 $\delta^{18}O$ 发生了突变，全岩微量元素也发生了同步改变。32.3 亿年及之后的 TTG 岩石中稀土元素镝（Dy）与镱（Yb）的比值更高［图 1（c）］，这意味着其源区有更多的石榴子石矿物残留，这说明其岩浆来源比 32.3 亿年之前的 TTG 岩石更深（＞40 km）。本次工作发现的氧同位素升高可能是一个系统性的突变，在其他大陆也有记录［图 1（a）］，且对应于全球铼（Re）亏损模式年龄指示的克拉通化的生长［图 1（d）］，这首次有力地证明了 32.3 亿年前地表物质成规模地深循环到了＞40 km 的深度。

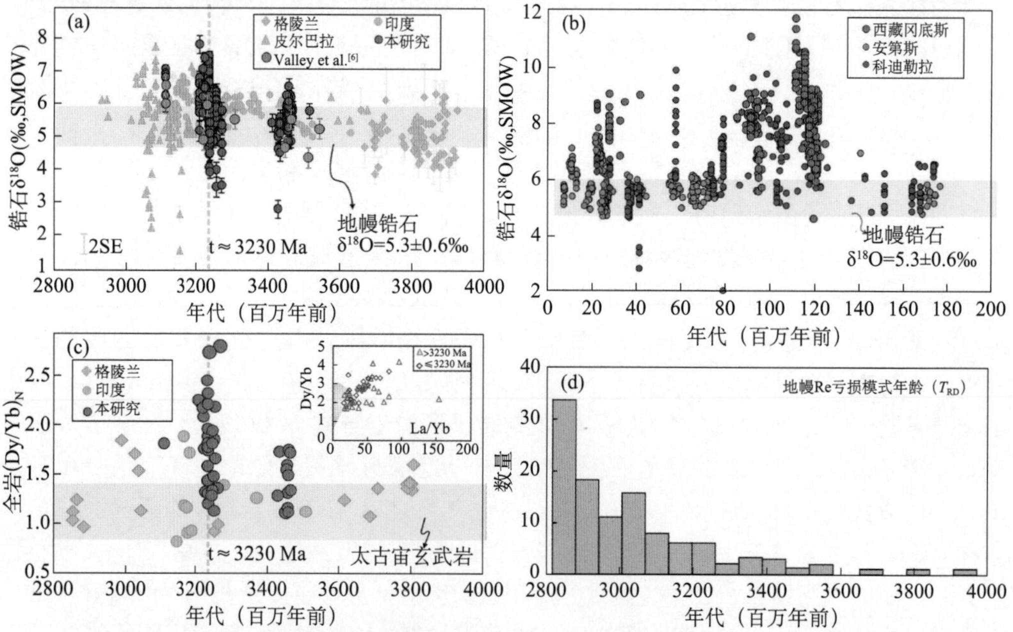

图 1　南非巴伯顿地区的 TTG 地球化学随时间的变化及其与全球数据的对比

（a）目前报道的 TTG 的锆石 $\delta^{18}O$ 随年龄的分布图；（b）显生宙典型大陆弧地区岩浆岩的锆石 $\delta^{18}O$ 值变化；（c）TTG 岩石的全岩 Dy/Yb 随年龄变化图；（d）地幔 Re 亏损模式年龄（T_{RD}）分布图展示在 32.3 亿年前起有一个快速生长。

关于地球早期板块构造的启动时间和机制，还存在较大的争议[3,4]。目前的研究表明，当时的地幔温度较高，古代板块构造如果存在，应是以“暖”俯冲（不同于现今的“冷”的板片俯冲）的形式进行[5]。俯冲的洋壳本身就很厚，然后在汇聚板块边缘进一步加厚（类似于大陆碰撞带）。当加厚到一定程度时，沿加厚的边缘出现重力不稳定，导致根部发生拆沉，结果引起深部低温蚀变玄武质洋壳发生部分熔融，产生高 $\delta^{18}O$ 的 TTG 岩石。因此，TTG 岩石的源岩产生在发散板块边缘的洋中脊，而形成 TTG 岩石的位置是汇聚板块边缘。产生 TTG 岩石的这个两阶段机制可能最早出现在始太古代（约 40 亿年前），并一直持续到晚太古代。

如今，地表物质深循环常发生于俯冲带边缘。但是巴伯顿地区的这一 32.3 亿年前的地表物质深循环与今不同，有着不同的诱发机制，可能是早期的一种特殊形式，可称为“原始俯冲”（图 2）。由于早期地球的地幔温度更高，这种俯冲不能持续太久，可能是间歇的幕式出现的，这可能也是 TTG 岩石时代不连续的一个重要原因。这种原始俯冲可以使得地表的挥发分进入地球内部，并由于水的润滑作用推动正常俯冲的逐渐产生。相关成果发表于《国家科学评论》(*National Science Review*)。

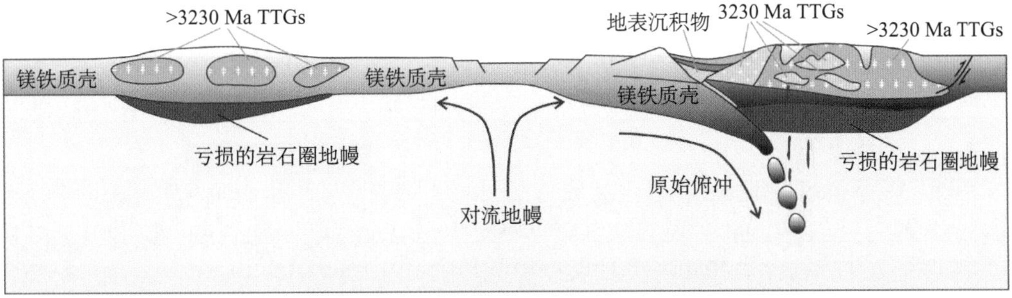

图 2 古太古代和中太古代时期 TTG 岩石形成的模式简图[6]

参考文献

[1] Smithies R H, Lu Y J, Kirkland C L, et al. Oxygen isotopes trace the origins of Earth's earliest continental crust. Nature, 2021, 592: 70-75.

[2] Bindeman I N, Zakharov D O, Palandri J, et al. Rapid emergence of subaerial landmasses and onset of a modern hydrologic cycle 2.5 billion years ago. Nature, 2018, 557: 545-548.

[3] Tang M, Chen K, Rudnick R L. Archean upper crust transition from mafic to felsic marks the onset of plate tectonics. Science, 2016, 351: 372-375.

[4] Deng Z, Chaussidon M, Guitreau M, et al. An oceanic subduction origin for Archaean granitoids revealed by silicon isotopes. Nature Geoscience, 2019, 12: 774-778.

[5] Zheng Y F, Zhao G C. Two styles of plate tectonics in Earth's history. Science Bulletin, 2020, 65: 329-334.

[6] Valley J W, Lackey J S, Cavosie A J, et al. 4.4 billion years of crustal maturation: oxygen isotope ratios of magmatic zircon. Contribution to Mineralogy and Petrology, 2005, 150: 561-580.

Onset of Deep Recycling of Supracrustal Materials and Proto-subduction at 3.23 Billion Years Ago

WangXiaolei, Wang Di, Li Junyong, Li Linsen

On the modern Earth, plate tectonics offers an efficient mechanism for mass transport from the Earth's surface to its interior, but how far back in the Earth's history this mechanism extends is still uncertain. This study presents new zircon oxygen (O) isotopes of Early Archean TTG (tonalite-trondhjemite-granodiorite) rocks from the Kaapvaal Craton, southern Africa. The mean $\delta^{18}O$ values increase abruptly at the Paleo-Mesoarchean boundary (ca. 3.23 billion years ago; Ga), from mantle zircon values of 5‰-6‰ to approaching 7.1‰, and this increase is associated with elevated Dy/Yb ratios. This finding indicates the melting processes in the garnet stability field (>40 km depth) and the onset of enhanced recycling of supracrustal materials in the Mesoarchean. The recycling may also have developed elsewhere in other Archean cratons, reflecting a proto-subduction and a significant change in the tectonic realm during craton formation and stabilization.

3.18　气候模式系统性偏差导致未来强厄尔尼诺发生频次被高估

罗京佳　唐　韬　彭　珂
祁　莉　唐绍磊

（南京信息工程大学气候与应用前沿研究院）

厄尔尼诺是指赤道中东太平洋水域海水温度异常增高的现象。在当前气候状态下，厄尔尼诺现象通常每隔2～7年出现一次。其中，1982～1983年、1997～1998年、2015～2016年发生的超强厄尔尼诺事件，引发了全球范围内极端天气-气候事件，对社会经济发展及生态环境等方面造成了严重影响[1]。因此，准确预估在全球变暖背景下未来超强厄尔尼诺发生频率的变化，对人类社会的可持续发展和防灾减灾政策的制定具有重要意义。

基于耦合模式比较计划第五阶段（CMIP5）提供的模拟数据，此前研究认为“超强厄尔尼诺发生频率在全球变暖背景下将加倍”[2]。然而，这一结论并没有考虑气候模式系统性偏差[3]及其对未来超强厄尔尼诺发生的频率变化预估的可能影响［图1（a）］。进一步分析表明，此前研究中基于Niño3区域①对流全值定义的超强厄尔尼诺[2]，其发生频率加倍的主要原因是Niño3区域对流气候态的增强［图1（b）］，而决定Niño3区域对流气候态变化的主要因素是热带太平洋东-西向海温变化梯度［图1（c）］。因此，校正模式系统性偏差对热带太平洋东—西向海温变化梯度的影响是更准确预估未来超强厄尔尼诺发生的频率变化的前提。

笔者研究团队以过去110年（1901～2010年）中的观测数据为参照，利用观测约束（emergent constraint，又称涌现约束）方法[4]分析了CMIP5模式中13个系统性偏差对热带太平洋东—西向海温变化梯度的影响。这13个系统性偏差包括：①热带太平洋皮叶克尼斯（Bjerknes）正反馈[5]；②历史长期变化趋势[6]；③热带三大洋相

① 赤道东太平洋区域（5°S～5°N，150°W～90°W）。

互作用[7-8]；④热带海温气候态[9]；⑤大气负反馈[10]；⑥厄尔尼诺-南方涛动（ENSO）偏度[11]，等等。

考虑到不同因子之间可能存在显著的线性相关，利用多元线性回归的方法去除上述 13 个系统性偏差对未来热带太平洋海温变化预估的重叠影响后，结果表明：海温增暖的最大值中心从原始预估的热带东太平洋转移到热带西太平洋，热带太平洋信风增强，热带东太平洋区域对流将减弱［图 2（a）～(d)］；基于订正后的 Niño3 区域对流气候态变化，未来超强厄尔尼诺发生的频率相比于历史时期几乎不变［图 2（e)］；另外，基于 Niño3 区域海温异常值定义的超强厄尔尼诺事件，其未来发生频率相比于历史时期也几乎不变。考虑到观测资料的不确定性，笔者研究团队还检验了以 1923～2010 年和 1981～2010 年的观测数据为参照的模式系统性偏差对预估未来超强厄尔尼诺发生频率变化的影响，均与上述结果基本一致。

该研究揭示了当前气候数值模式的系统性偏差对未来气候变化预估的重要影响。因此，提高模式性能，减少系统性偏差是准确预估未来全球变暖背景下强厄尔尼诺发生频率变化的前提。上述成果[12]发表在《国家科学评论》(*National Science Review*)，获得了气候学界的高度关注。

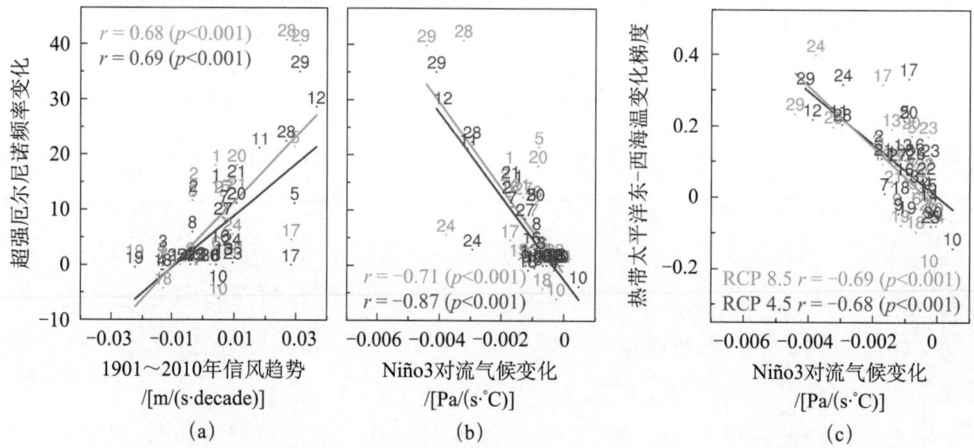

图 1　未来超强厄尔尼诺发生频率的变化与历史模拟偏差以及气候态变化之间的关系

(a) 未来超强厄尔尼诺频率变化与过去百年太平洋信风趋势模拟之间的关系；超强厄尔尼诺频率变化由 Niño3 区域对流的气候态变化 (b) 及热带太平洋东-西海温梯度的变化 (c) 决定。

RCP 8. 5 为高浓度温室气体排放情景；RCP4. 5 为中等浓度温室气体排放情景；PCK 为 representative concentration pathways 的缩写，中文为代表性浓度路径。

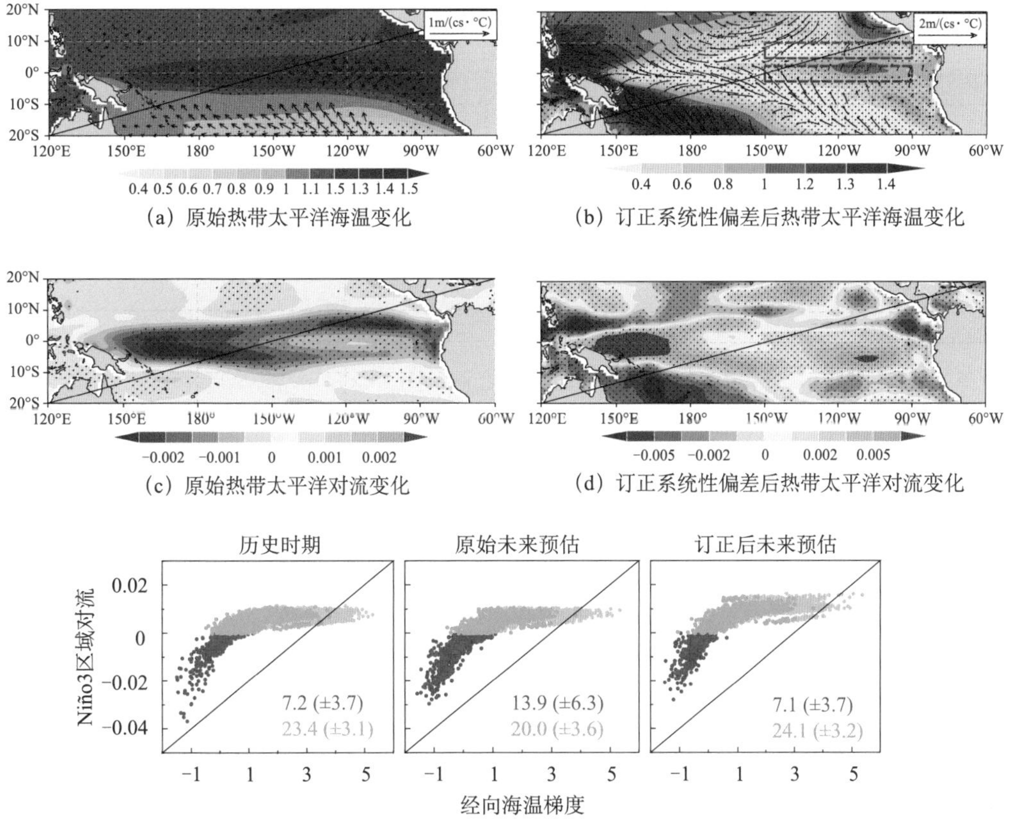

(a) 原始热带太平洋海温变化

(b) 订正系统性偏差后热带太平洋海温变化

(c) 原始热带太平洋对流变化

(d) 订正系统性偏差后热带太平洋对流变化

(e) 厄尔尼诺发生频率

图 2　未来热带太平洋气候态的变化与厄尔尼诺发生频率的变化

(a) 和 (c) 为 CMIP5 多模式平均预估的原始的未来热带太平洋海温（填色）和表面风场（矢量箭头）的变化以及未来热带太平洋对流的变化；(b) 和 (d) 分别对应 (a) 和 (c)，为去除 13 个模式系统性偏差影响后的未来热带太平洋海温、表面风场和对流的变化；(e) 为过去百年厄尔尼诺的发生频率和有无订正模式系统性偏差影响的未来厄尔尼诺发生频率的预估，红色代表超强厄尔尼诺，绿色代表普通厄尔尼诺，括号中数据为不同模式之间的不确定性。

参考文献

[1] McPhaden M J, Zebiak S E, Glantz M H. ENSO as an integrating concept in Earth Science. Science, 2006, 314: 1740-1745.

[2] Cai W, Borlace S, Lengaigne M, et al. Increasing frequency of extreme El Niño events due to greenhouse warming. Nature Climate Change, 2014, 4: 111-116.

[3] Luo J-J, Wang G, Dommenget D. May common model biases reduce CMIP5's ability to simulate the

recent Pacific La Niña-like cooling? Climate Dynamics, 2018, 50: 1335-1351.
[4] Eyring V, Cox P M, Flato G M, et al. Taking climate model evaluation to the next level. Nature Climate Change, 2019, 9: 102-110.
[5] Bjerknes J. Atmospheric teleconnections from the equatorial Pacific. Monthly Weather Review, 1969, 97: 163-172.
[6] Clement A C, Seager R, Cane M A, et al. An ocean dynamical thermostat. Journal of Climate, 1996, 9: 2190-2196.
[7] Luo J-J, Sasaki W, Masumoto Y. Indian Ocean warming modulates Pacific climate change. The Proceedings of the National Academy of Sciences of the USA, 2012, 109: 18701-18706.
[8] McGregor S, Timmermann A, Stuecker M F, et al. Recent Walker circulation strengthening and Pacific cooling amplified by Atlantic warming. Nature Climate Change, 2014, 4: 888-892.
[9] Wang C, Zhang L, Lee S K, et al. A global perspective on CMIP5 climate model biases. Nature Climate Change, 2014, 4: 201-205.
[10] Bellenger H, Guilyardi E, Leloup J, et al. ENSO representation in climate models: from CMIP3 to CMIP5. Climate Dynamics, 2014, 42: 1999-2018.
[11] Kohyama T, Hartmann D L, Battisti D S. La Niña-like mean-state response to global warming and potential oceanic roles. Journal of Climate, 2017, 30: 4207-4225.
[12] Tang T, Luo J-J, Peng K, et al. Over-projected Pacific warming and extreme El Niño frequency due to CMIP5 common biases. National Science Review, 2021, 8(10).

Over-Projected Pacific Warming and Extreme El Niño Frequency due to CMIP5 Common Biases

Luo Jingjia, Tang Tao, Peng Ke, Qi Li, Tang Shaolei

Extreme El Niño events severely disrupt the global climateand cause significant socio-economic losses. A prevailing view is that extreme El Niño events, defined by total precipitation or convection in the Niño3 area, will increase 2-fold in the future. However, this conclusion has been drawn without considering the potential impact of the Coupled Model Intercomparison Project phase 5 (CMIP5) model common biases. Here, we find that the common biases in simulating tropical climate change in the past can reduce the reliability of the projections of the Pacific sea surface temperature (SST) change and the associated extreme El Niño frequency. The projected Pacific SST change, after removing the impacts of 13 common biases, shows a "La Niña-like" rather than an "El Niño-like" change. Consequently, the extreme El Niño frequency, which is strongly linked to the zonal distribution of the Pacific SST change, would keep almost unchanged under the CMIP5 warming scenarios. This finding increases confidence in coping with climate risks associated with global warming.

3.19 “嫦娥五号”月球样品揭示月球演化奥秘

李献华　杨　蔚　胡　森　林杨挺
李秋立　田恒次　王　浩　陈　意
（中国科学院地质与地球物理研究所）

2020年12月17日，我国“嫦娥五号”返回器成功带回了1731g月球样品，我国成为世界上第三个完成月球采样返回的国家。这也是继美国“阿波罗”（Apollo）计划和苏联月球（Luna）计划44年之后，人类再次从月面采样返回地球，翻开了月球探测和研究的新篇章。月球返回样品的科学研究对认识月球的形成和演化至关重要。研究人员通过对“阿波罗”计划和月球计划带回的月球样品进行研究，构建出月球形成和演化的基本框架——月球形成于约45亿年前一颗火星大小的星球和原始地球发生的大撞击，然后从高温状态的“岩浆洋”开始冷却，大约在28亿～30亿年前基本停止了岩浆活动而成为一个地质意义上的“死亡”星球。由于“阿波罗”计划和月球计划的着陆点仅覆盖月球表面的5%～8%，其返回样品非常有限，难以提供月球最早和最晚地质作用的可靠证据，月球形成的早期历史和演化的最终过程仍存在许多争议，成为当今月球研究的“一老”和“一新”两个重要科学问题[1]。

“嫦娥五号”任务着陆于月球正面风暴洋地体北部的吕姆克山脉附近，该地区远离阿波罗计划和月球计划的采样区，富集放射性生热元素。撞击坑年代学研究发现该地区可能存在非常年轻（20亿～10亿年前）的月海玄武岩[2]。“嫦娥五号”任务的主要科学目标就是采集这些可能存在的年轻玄武岩来解密月球最晚期岩浆作用和热演化历史，因而受到国际学术界的高度关注。

笔者研究团队申请获批了3 g“嫦娥五号”样品，均为月球表面铲取的月壤。研究团队首先从微细月壤中分选出玄武岩岩屑，然后利用自主研发的超高空间分辨同位素精确定年技术，对玄武岩岩屑中50余颗直径为3～8 μm细小的富铀矿物（斜锆石、钙钛锆石、静海石等）进行了系统的铅（Pb）同位素分析，确定“嫦娥五号”玄武岩形成于（20.30±0.04）亿年前（图1）[3]，这一结果把月球最年轻的岩浆活动记录更新为20亿年前，将月球的地质寿命“延长”了8亿～9亿年，并为修订月球撞击坑统计年龄函数提供了一颗“金钉子”。

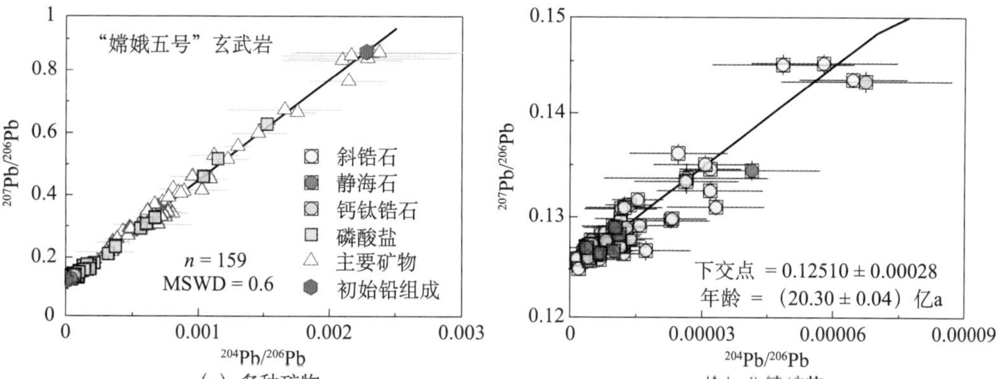

(a) 多种矿物　　(b) 含锆矿物

图 1　“嫦娥五号”玄武岩 Pb/Pb 等时线图

月球的直径约为地球的 1/4，但质量只有地球的 1/81，对于具有如此大的表面积与体积比的一个小星球来说，它应该快速冷却而早早地结束“地质生命”。然而“嫦娥五号”玄武岩定年结果刷新了我们对月球岩浆活动和热演化历史的认知，同时也提出了新的问题——月球的岩浆活动为什么可以持续这么久？

学术界对于月球长时间岩浆活动的原因一般有两种可能的解释：①岩浆源区富含放射性元素以提供形成岩浆的热源；②岩浆源区富含水（或其他挥发分）以降低源区岩石的熔点。笔者研究团队应用超高精度微区原位同位素分析技术研究了“嫦娥五号”玄武岩源区的地球化学性质，发现“嫦娥五号”玄武岩月幔源区的初始锶（Sr）、钕（Nd）、Pb 同位素组成具有亏损月幔的特征（图 2），并不富集铀（U）、钍（Th）、钾（K）等放射性生热元素，而且岩浆在演化过程中也极少卷入富集 K、稀土元素（REE）、磷（P）元素的“克里普物质”［克里普因为富集 K、REE、P（缩写 KREEP，即克里普）而得名。月球形成和演化理论认为，月球形成之初，曾被深达数百公里的岩浆洋覆盖。随着岩浆洋的不断结晶，不相容元素，如 U、Th、K、REE、P 等，在残余熔体中不断富集，形成了克里普，并最终集中在月球的壳-幔之间］[4]，排除了初始岩浆熔融热源来自富集放射性生热元素月幔源区的可能性；同时，应用超低本底纳米离子探针分析技术，测定了“嫦娥五号”玄武岩中熔体包裹体和磷灰石的水含量及氢同位素组成，测得“嫦娥五号”玄武岩月幔源区的水含量仅为 1～5μg/g（图 3）[5]，表明其月幔源区非常“干”，排除了月幔源区因水含量高而降低熔点的猜想。

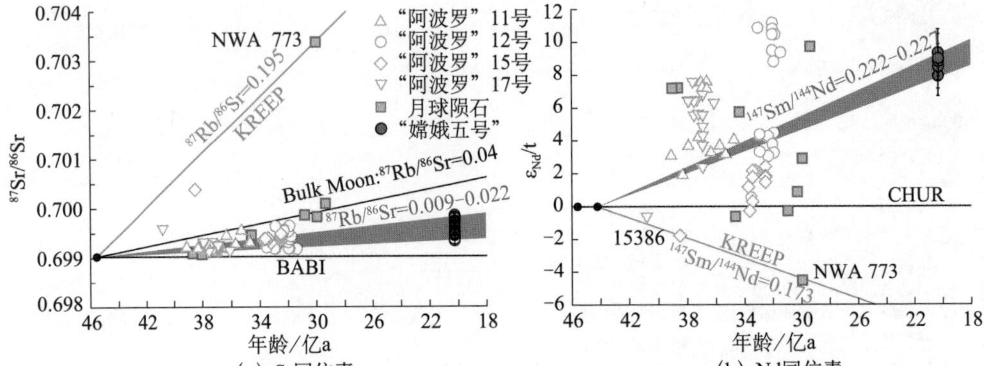

(a) Sr同位素　　(b) Nd同位素

图 2 "嫦娥五号"玄武岩的 Sr 同位素和 Nd 同位素特征[4]

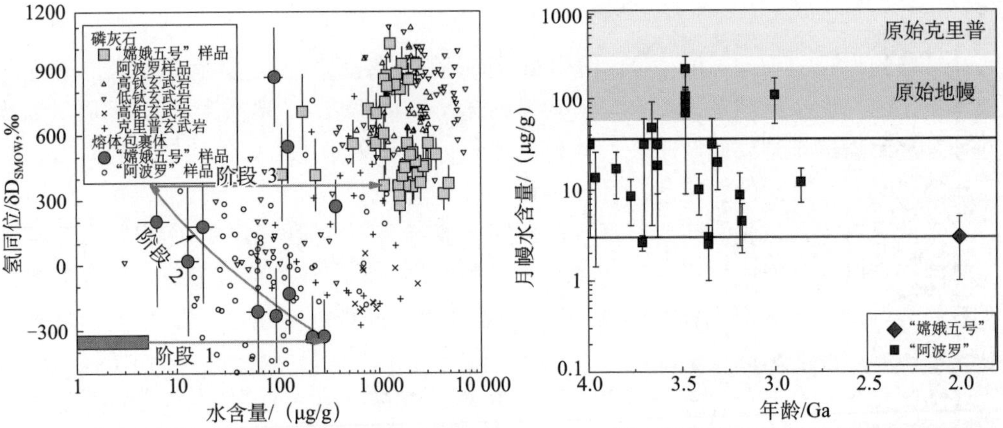

(a)磷灰石和熔体包裹体的水含量和氢同位素组成　　(b) 月海玄武岩源区水含量随时间演化[5]

图 3 "嫦娥五号"玄武岩及其月幔源区水含量

上述"嫦娥五号"月球样品研究成果于 2021 年以三篇论文的形式发表在同一期国际学术期刊《自然》上，入选了科学技术部发布的"2021 年度中国科学十大进展"和两院院士评选的"2021 年中国十大科技进展新闻"，并得到国际同行的广泛关注和好评［三篇文章均入选科学网络数据库（Web of Science）高被引论文］。美国国家科学院院士 Richard Carlson 教授在同期《自然》上发表评述文章指出，"嫦娥五号"任务在过去从未涉足的月球表面，返回了迄今采集到的最年轻的火山岩样品，这些岩石的研究结果表明非常有必要修正已有的月球热演化模型[6]。2022 年 3 月，《自然》再次发表评论文章，指出"中国'嫦娥五号'月球样品点燃了月球研究的热情"[7]。这些"嫦娥五号"月球样品的研究成果不仅刷新了人们对月球岩浆作用和热演化历史的认识，而且为今后的月球研究提出了新的科学问题，并对新一轮月球探测提出了新的方向。

参考文献

[1] 欧阳自远．欧阳自远自传:求索天地间．北京:中国大百科全书出版社,2021.
[2] Hiesinger H, Head J W, Wolf U, et al. Ages and stratigraphy of lunar mare basalts: a synthesis, recent advances and current research Issues in Lunar Stratigraphy. Geological Society of America, 2011, 477: 1-51.
[3] Li Q L, Zhou Q, Liu Y, et al. Two-billion-year-old volcanism on the Moon from Chang'e-5 basalts. Nature, 2021, 600: 54-58.
[4] Tian H C, Wang H, Chen Y, et al. Non-KREEP origin for Chang'e-5 basalts in the Procellarum KREEP Terrane. Nature, 2021, 600: 59-63.
[5] Hu S, He HC, Ji JL, et al. A dry lunar mantle reservoir for young mare basalts of Chang'e-5. Nature, 2021, 600: 49-53.
[6] Carlson R W. Robotic sample return reveals lunar secrets. Nature, 2021, 600: 39-40.
[7] Mallapaty S. China's first moon rocks ignite research bonanza. Nature, 2022, 603: 561-562.

Chang'e-5 Lunar Samples Reveal the Mystery of the Evolution of the Moon

Li Xianhua, Yang Wei, Hu Sen, Lin Yangting, Li Qiuli, Tian Hengci, Wang Hao, Chen Yi

Forty-four years after the Apollo and Luna missions, China's Chang'e-5 mission returned new samples from the Moon, opening a new chapter in lunar exploration and research. The Pb-Pb dating of U-rich minerals in the basalt fragments from the Chang'e-5 lunar soil yields an age of 2 billion years ago, about 800 million years younger than the youngest lunar sample previously found (2.8 billion years ago). This age also provides a "golden spike" for the cratering chronology function. In addition, the basalts are found to originate from a mantle source depleted in both KREEP components and water, as indicated by of Sr-Nd isotope and water content analyses. These results rule out two hypotheses for the long-term volcanic activity on the Moon: that the KREEP-rich materials provided additional heat, and the high water content lowered the melting point. In addition to providing insights into the thermal evolution of the Moon, our research also suggests new directions for future lunar exploration.

3.20 地球极区电离层上空发现“太空台风”

张清和

（山东大学空间科学研究院）

台风或飓风是在地球低层大气中发生的强烈热带气旋。那么，在地球或其他行星的高层大气中是否存在类似台风或飓风的现象呢？研究者一直在寻找它的踪迹。针对这一挑战，山东大学张清和教授、中国科学院国家空间科学中心王赤院士等带领国际团队利用电离层探测卫星等的观测数据，结合磁流体力学数值模拟，开展了系列研究。其中一个成果于2021年发表在《自然-通讯》上[1]，被《自然》选为了研究亮点，并以“新发现的太空台风可倾泻电子‘雨’”（The First Known Space Hurricane Pours Electron “Rain”）为题进行了报道。

该成果首次在地球极区电离层上空发现并命名了“太空台风”（Space Hurricane，图1），因为它伴随有巨型台风气旋状的极光亮斑结构（水平尺度超过1000 km）、环形等离子体对流和速度为零的“台风眼”、降强电子“雨”、电子温度上升、离子上行、环形磁场扰动等与台风或飓风十分相似的特征。利用高分辨率三维磁流体力学数值模拟，团队重现了这一现象及其三维形态（图2）。研究发现，在长时间极端平静地磁条件下，发生在地球高纬度的磁层与太阳风相互作用和演化，使得在北极磁极点上方的电离层与磁层形成了一个巨大的顺时针旋转的漏斗形磁螺旋结构（图3）。该结构形成了太阳风带电粒子进入地球中高层大气、地球带电粒子逃逸至磁层的通道，极大提升了太阳风-磁层能量的耦合效率。

该工作表明，在极端平静地磁条件下，极区仍可能存在堪比超级磁暴的剧烈地磁扰动和能量注入，这给出了地球空间能量传输的新途径，更新了对太阳风-磁层-电离层耦合过程的认识。同时，“太空台风”所造成的极端空间天气环境，会直接影响相关区域的卫星通信导航与超视距雷达探测，也能影响相关区域卫星和火箭的正常运行及空间碎片的轨道稳定等，故而其研究也具有重要的应用价值。

后续研究表明，“太空台风”仅在北半球就年均发生近15次，故而其影响不能忽视。因此，十分有必要进一步开展深入研究，找到“太空台风”的相关规律，并尝试进行建模预报，以合理规避其相关风险和改善通信导航质量。

成果发表后，被美国西弥斯卫星计划（THEMIS）的月球轨道卫星阿特米斯（ARTEMIS）卫星网站选为研究亮点（nugget）长期贴在其主页上；被《中国科学报》、英国 BBC《科学聚焦》、美国《国家地理》杂志、美国地球物理联合会（AGU）会刊 *EOS* 等近千家国内外媒体报道。该成果还成为《自然-通讯》2021 年度七大领域中“物理”和“地球、环境与行星科学”两大领域下载量前 25 的文章，分别位列第 4 和第 5。

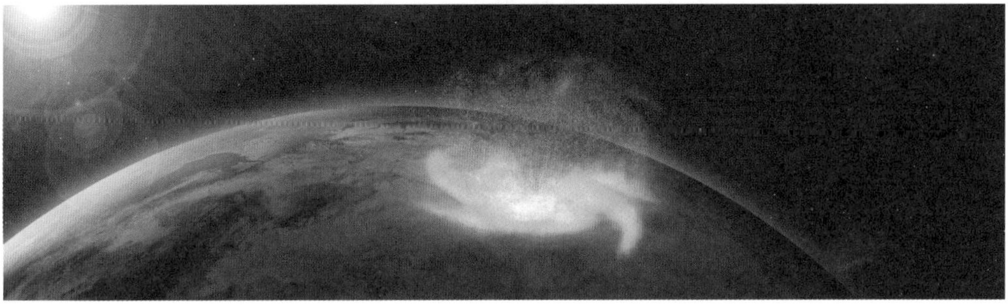

图 1　“太空台风”示意图

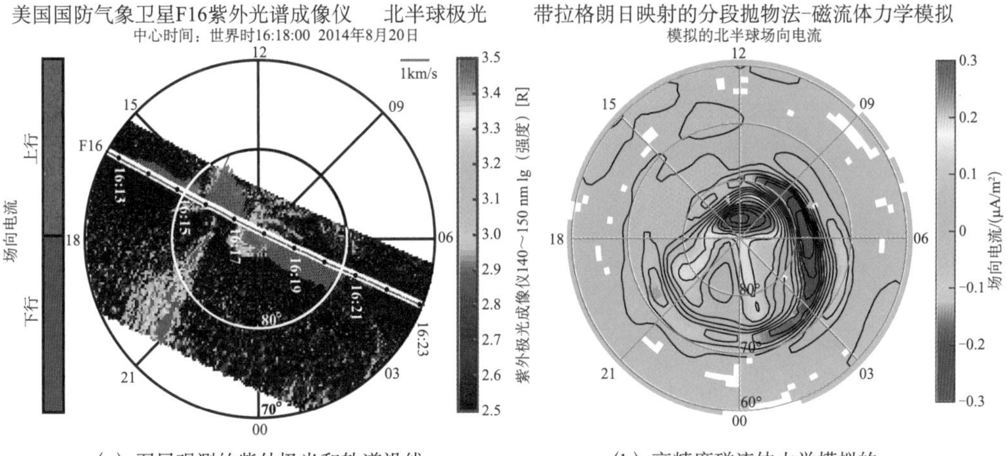

(a) 卫星观测的紫外极光和轨道沿线的水平对流与场向电流观测结果

(b) 高精度磁流体力学模拟的极区电离层场向电流分布情况

图 2　“太空台风”的观测证据与数值模拟

图像位于地磁与磁地方时坐标系中，是从北极上空往下看的结果。

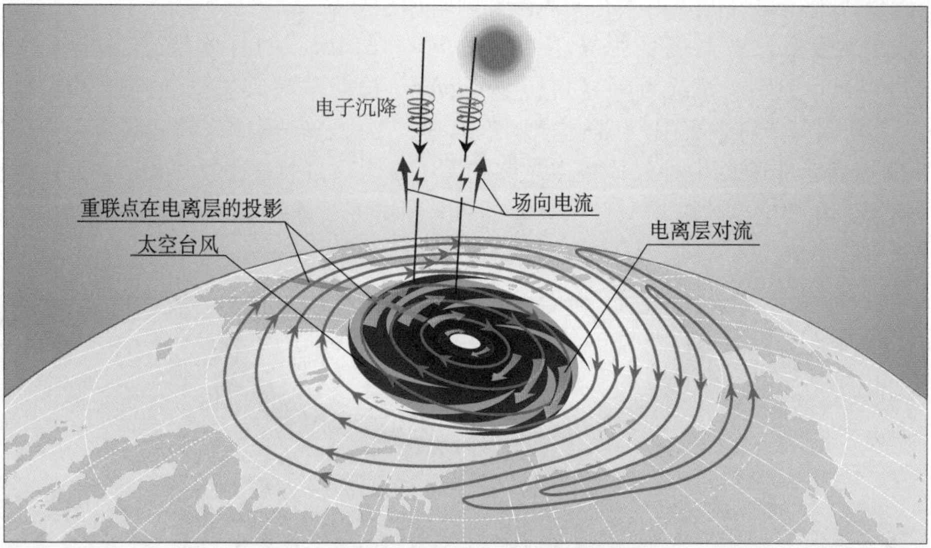

(a)“太空台风”在极区电离层的示意图

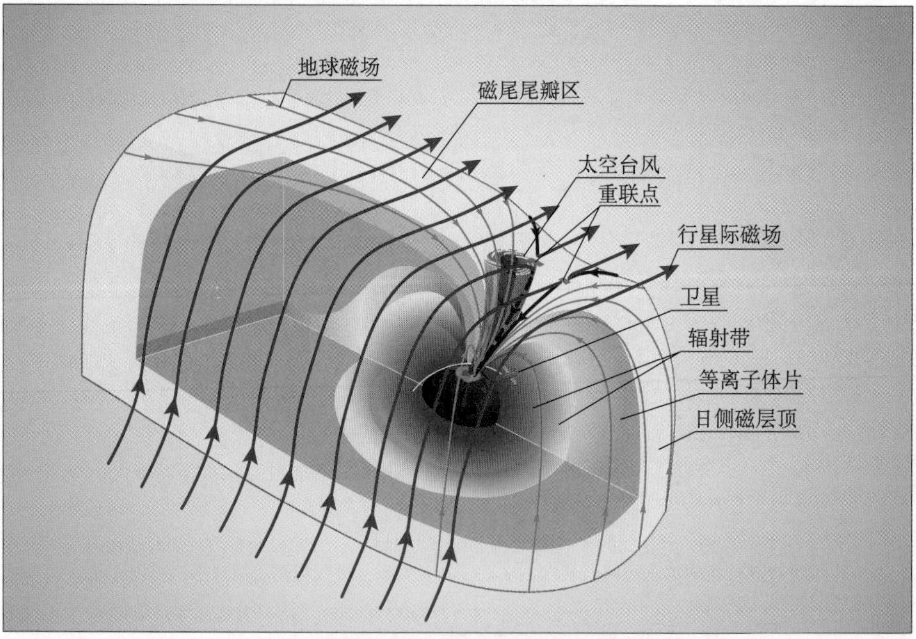

(b)“太空台风”在磁层的示意图

图 3 “太空台风”的形成示意图

参考文献

[1] Zhang Q H, Zhang Y L, Wang C, et al. A space hurricane over the Earth's polar ionosphere. Nature Communications, 2021, 12(1): 1-10.

Space Hurricanes had been Discovered over the Earth's Polar Lonosphere

Zhang Qinghe

A new typhoon-like phenomenon associated with a powerful energy injection, Space Hurricane, has been discovered in the upper atmosphere of the polar region in the Northern Hemisphere. This discovery was made by an international team which was led by Prof. Qing-He Zhang from Shandong University and by Prof. Chi Wang, academician of Chinese Academy of Sciences (CAS), from the National Space Science Center, CAS. The team reveals the formation mechanism of space hurricane through 3D global magnetohydrodynamics (MHD) simulation, which updates the understanding of the solar wind-magnetosphere energy coupling processes.

第四章

科技领域与科技战略发展观察

Observations on Development of Science and Technology

4.1　基础前沿领域发展观察

黄龙光　边文越　张超星　冷伏海
（中国科学院科技战略咨询研究院）

2021～2022年基础前沿领域成果丰硕，粒子物理、量子科技、凝聚态物理和光学等领域突破纷呈，化学合成方法、基础理论不断创新，数据科学与化学结合趋势越发明显，纳米材料基础研究逐渐深化，应用研究遍地开花。世界主要国家持续推进基础前沿领域发展，如俄罗斯加强基础研究布局、世界主要国家和机构陆续推进量子科技战略、美国对国家纳米技术计划做出重要调整等，以进一步提高创新能力，在新一轮国际科技竞争中掌握主导权。

一、重要研究进展

（一）粒子物理、量子科技、凝聚态物理和光学等领域突破纷呈

1. 缪子物理研究的突破吸引全球聚焦粒子物理

美国费米国家加速器实验室研究发现，缪子反常磁矩的测量值与理论预测不符[1]，并把实验值与粒子物理学标准模型预测值之间的差异提高到4.2个标准置信度，接近物理学新发现的5个标准置信度，这表明物理学家可能接近发现新的基本力或粒子。美国费米国家加速器实验室对撞机探测器（Collider Detector at Fermilab，CDF）合作组实现了迄今对W玻色子质量的最精确测量，测量结果与标准模型理论预言值存在7个标准偏差，这或暗示超出标准之外新物理的存在[2]。奇特强子态的突破不断涌现，北京谱仪Ⅲ（BESⅢ）实验发现首个含奇异夸克的隐粲四夸克候选态 Z_{cs}（3985）[3]，欧洲核子研究中心（European Organization for Nuclear Research，CERN）大型强子对撞机上底夸克物理（LHCb）实验合作组发现了4个新型四夸克态[4]［包括 Z_{cs}（4000）、Z_{cs}（4220）、X（4685）和X（4630）］，这些研究成果将有助于更好地理解夸克模型和强相互作用。德国达姆施塔特工业大学等机构研究人员发现迄今最明确的“四中子态”奇异物质存在的证据，这一发现将有助于对核力本质的理论进行微调[5]。

2. 量子科技快速发展势头不减

量子计算方面，中国科学技术大学等机构研究人员构建了 66 bit 可编程超导量子计算原型机“祖冲之二号”[6]，这是我国首次在超导量子体系实现了“量子计算优越性”。同时，该团队还构建了 113 个光子 144 模式的量子计算原型机“九章二号”[7]。IBM公司推出其迄今最大量子比特数的超导量子处理器“鱼鹰”（Osprey），其拥有 433 qbit[8]。芝加哥大学在实验室中创纪录地实现了 512 qbit 的中性原子体系[9]。日本理化研究所[10]、荷兰代尔夫特理工大学[11]、澳大利亚新南威尔士大学[12]等分别独立地验证了双量子比特门保真度 99% 以上的硅量子计算突破。量子通信方面，中国科学技术大学等机构研究人员实现了跨越 4600 km 的星地量子密钥分发[13]，东芝集团欧洲公司等机构研究人员验证实现了 600 km 的双场量子密钥分发[14]，中国科学技术大学等机构研究人员实现了 833 km 的光纤量子密钥分发[15]。这些突破为实现长距离量子安全信息传输奠定了基础。

3. 凝聚态物理持续推进新材料和新物态的发现

三层转角石墨烯引发了研究热潮，美国麻省理工学院等机构和哈佛大学等机构的研究人员分别报道了三层转角石墨烯的超导性[16,17]，美国麻省理工学院等机构的研究人员发现三层转角石墨烯在 10 T 的高磁场下仍显示超导性[18]，美国加州大学圣巴巴拉分校等在菱面三层转角石墨烯中发现了超导性[19]，这些研究成果推动了对转角石墨烯超导机理的进一步探索。上海交通大学等机构的研究人员在超导体中观察到了 50 多年前理论预言的“分段费米面”[20]，为物态调控提供了新的方法。谷歌公司等机构的研究人员通过量子计算机“悬铃木”制造出离散时间晶体[21]，荷兰代尔夫特理工大学等机构的研究人员利用金刚石氮空位中心的核自旋制造出离散时间晶体[22]，这些晶体的出现为研究新物相提供了新机会。德国拜罗伊特大学等机构的研究人员在 100 GPa 高压下开发出一种以前未知的二维材料铍氮烯[23]，美国西北大学等机构的研究人员制造出新的二维材料氢化硼烯[24]，二维材料家族成员得到进一步拓展。美国麻省理工学院[25]、中国科学院国家纳米科学中心[26]等机构研究人员分别独立发现立方砷化硼具有很高的热导率和空穴迁移率，是迄今发现的最好的半导体材料之一。中国科学院物理研究所对铁基超导体 LiFeAs 开展研究，发现应力可以在 LiFeAs 中诱导出的大面积、高度有序和可调控的马约拉纳零能模格点阵列[27]。

4. 光学领域里程碑突破不断涌现

美国国家点火装置实现了 3.15 MJ 的能量产出，首次在可控核聚变实验中实现了核聚变反应的净能量增益，首次证实了惯性核聚变能的基本科学原理和可行性[28]。以色列理工学院等机构的研究人员利用拓扑光子学的概念，开发了一个由 30 台垂直腔面发射激光器组成的阵列[29]，为实现大规模相干激光阵列打下了基础。美国麻省理工学院等[30]和斯坦福大学等[31]机构的研究人员分别利用相变材料，实现了超表面和纳米天线的电可调谐和电切换，推动了可调控光学的进一步发展。中国科学技术大学的研究人员实现了两个吸收型量子存储器之间的可预报量子纠缠，演示了多模式量子中继，为实用化高速量子网络的构建打下了基础[32]。浙江大学等机构的研究人员制备出高质量冰单晶微纳光纤[33]，其可灵活弯曲，也可低损耗传输光，在性能上与玻璃光纤相似。

（二）化学合成方法、基础理论不断创新，数据科学与化学结合趋势越发明显

1. 有机合成方法学不断创新

美国普林斯顿大学的研究人员通过双分子均裂取代机制实现仿生 sp^3-sp^3 交叉偶联，成功构建了一系列 sp^3-季碳中心[34]。英国剑桥大学的研究人员利用光与过渡金属双催化模式，一步合成结构和功能多样的β-芳基叠氮化合物，为构建β-芳基乙胺骨架提供了新策略[35]。美国芝加哥大学的研究人员发展了一种新的骨架编辑反应，通过直接“删除”仲胺中的氮原子，释放出氮气（N_2），从而形成分子内 C—C 键偶联产物[36]。美国科学家通过诺里什-Ⅱ光反应，实现了可见光介导的α-酰化饱和杂环的缩环反应，构建了一系列 *cis*-1,2-二取代环戊烷骨架[37]。中国北京大学的研究人员设计了铜催化的级联活化策略，实现了温和条件下芳环选择性开环裂解[38]。美国匹兹堡大学的研究人员通过室温条件下溶解在四氢呋喃中的锂和乙二胺，改进芳烃化合物的伯奇还原反应[39]。瑞士苏黎世联邦理工学院的研究人员实现了对吲哚骨架的氮原子插入，获得了一系列苯并嘧啶或苯并[b]吡嗪化合物[40]。

2. 对基础理论、基础过程的认识不断深入

美国劳伦斯伯克利国家实验室的研究人员揭示了锿的多项基本化学性质，并与其他锕系元素做了比较[41]。捷克科学家首次观察到卤素原子的不对称电子密度分布，从而证实了 30 年前理论预测的西格玛孔的存在[42]。德国和中国科学家在研究合成氨反应时发现，作为 S 区元素，钙可以像过渡金属一样利用其 d 轨道参与小分子的活化[43]。瑞士和意大利科学家合作发现在水分子与油滴界面发生的 C—H···O 氢键间电荷转移可以使水中油滴更稳定[44]。美国西北大学的研究人员通过使用人工分子机器实

现主动吸附，提出了一种名为机械吸附的新吸附机制[45]。中国浙江大学发明了一种可以直接对溶液中单分子化学反应进行成像的显微镜技术，并实现了超高时空分辨成像[46]。瑞士、西班牙和德国的研究人员借助超高精度扫描隧道显微术首次改变了单个分子内原子之间的键，并在此基础上创造出新键[47]。

3. 化学为可持续发展提供方案

德国康斯坦茨大学的研究人员报道了一种可闭环循环的塑料，由基于可再生原料的单体聚合而成，可在温和条件下降解为单体，由这些单体再聚合得到的材料的性能与原始材料相当[48]。澳大利亚莫纳什大学的研究人员开发了一种基于有机膦盐的高效电化学合成氨方法，有望为实现低碳排放合成氨做出贡献[49]。美国西北大学的研究人员制备了一种串联催化剂 $Pt/Al_2O_3@In_2O_3$，通过耦合丙烷脱氢反应和选择性氢燃烧反应，实现丙烯的高活性、高选择性地合成[50]。美国西北大学的研究人员报道了一种简单方法，可在温和条件下将“永久性化学品”全氟羧酸类物质完全降解[51]。

4. 数据科学在化学领域的应用潜力越发明显

英国深度思维公司的研究人员基于准确的化学数据和分数电子约束训练生成神经网络 DM21，正确描述了电荷离域和相互作用，且在描述主族原子和分子方面比传统密度泛函算法更好[52]。美国科学家开发了一种描述符——最小埋入体积百分比[$\%V_{bur(min)}$]，可以对交叉偶联催化中膦配体的连接状态和反应活性进行二元分类[53]。美国普林斯顿大学的研究人员使用贝叶斯优化算法优化反应条件，在平均优化效率和一致性方面取得优于人类科学家的结果[54]。美国、韩国、加拿大的研究人员使用数据引导的筛选矩阵降维技术、降低不确定性的机器学习算法、自动操作实验和分析结果数据技术，成功探索了铃木-宫浦（Suzuki-Miyaura）偶联反应的通用反应条件。使用该方法得到的反应条件下的平均产率为46%，是之前文献报道条件下产率（21%）的两倍多[55]。

（三）纳米材料基础研究逐渐深化，应用研究遍地开花

1. 纳米材料的制备和调控技术不断升级

南开大学研究人员采用将具有高荧光量子产率的发光基团进行螺旋 π 拓展的分子设计策略，实现了具有优异发光性能的手性纳米石墨烯的制备[56]。耶鲁大学、复旦大学、宾夕法尼亚大学的研究人员通过调整含有侧链的多组分共聚物组成的构建块的各种参数，实现了纳米组装结构在纳米尺度和中尺度上广泛可调，为制备具有多种精确

集成功能的纳米材料提供了方便的合成平台[57]。复旦大学研究人员利用新型的软补丁纳米颗粒及软界面导向的超组装策略制备出高度不对称的中空复合金@银-二氧化硅超胶体，该种胶体在智能纳米机器人、药物封装和递送等方面具有很大的应用潜力[58]。美国麻省理工学院通过逐渐引入与聚合物有不利相互作用能量的非溶剂（如正癸烷），稳定晶格以防止聚合物刷坍塌的方式实现了金纳米超晶格晶体向层次有序的大块固体材料的宏观定制[59]。中国上海交通大学、沙特阿拉伯阿卜杜拉国王科技大学通过剥离由均三甲苯基团稳定的三维层状金属有机骨架，制备具有本征孔隙率的单晶超薄二维镧系 MOF 纳米片，可实现对 17 种手性萜烯和萜类化合物高选择性、高灵敏度的检测[60]。

2. 纳米材料成为实现重大战略目标的重要抓手

天津大学化工学院研究人员通过制备含有级联 DNA 酶和类启动子锌锰铁氧体（ZMF）的 DNA 纳米复合体，实现了生化过程的精确启动和治疗模块的可控顺序释放，提高了基因和化学动力学联合治疗的效率，为实现精准医疗提供了创新路径[61]。南京大学与斯坦福大学研究人员通过分子键合设计和可伸缩耦合试剂（钛酸四丁酯）辅助浸涂方法实现了氧化铝纳米颗粒与蚕丝纤维的耦合，纳米加工的蚕丝具备优异辐射制冷性能，在白天可以低于环境温度约 3.5℃（环境温度约为 35℃），该策略除了可为个人热管理提供可持续的节能方法外，还为开发被动冷却材料和设备以降低能耗提供了新途径[62]。韩国基础科学研究所利用有机半导体聚合物、尺寸可调量子点和弹性基体组合开发出一种纳米复合材料，可用于制备可拉伸、颜色敏感、形状可调的多光谱和多路复用的光电晶体管阵列，在平坦或者变形状态下均能精确检测红色/绿色/蓝色图案，展现出了极具前景的光电探测应用潜力[63]。日本东京大学工程学院制备出一种内壁被氟原子密集覆盖的纳米通道结构，该结构对水有空前的渗透速度，并且具有近乎的完美脱盐能力，有望应用于海水淡化[64]。

3. 纳米技术成为解决新兴技术关键问题的核心技术

韩国基础科学研究所和首尔大学研究人员利用“浮动装配法”实现了橡胶内单层银纳米线导电和弹性纳米膜的制备，为制备具有超薄厚度、类似金属的导电性、高拉伸性和易于图案化可拉伸导体的皮肤电子设备提供了技术途径[65]。法国索邦大学通过构建 CdSe/CdTe/CdSe 异质结结构纳米片，成功实现了绿-红双重光发射，红绿光发射组分的强度不仅可以通过纳米片结构调控，也可以通过激发光强度和电致发光过程中偏压的大小实现调控，这对进一步简化基于量子点的显示技术有重要意义[66]。德国斯图加特大学利用电化学驱动的聚（3,4-乙撑二氧噻吩）：聚苯乙烯磺酸盐的金属-绝缘体相变概念制备了等离子体纳米天线，通过施加仅±1 V的交变电压在高达30 Hz的视频频率下完全关闭和重新打开，完全可以满足全息技术所需的50 000 dpi像素密度的需求[67]。

二、重要战略规划

1. 俄罗斯加强基础研究布局

俄罗斯政府发布《俄罗斯联邦长期基础科学研究计划（2021—2030）》[68]，计划投入2.15万亿卢布（约合人民币1852亿元）支持所有基础科学领域的研究，以获得国家科学技术、社会经济和文化持续发展所必需的，有关人、社会和自然结构、功能和发展基本规律的新知识。该计划包含6个子计划，具体为：①旨在发现重大挑战并完善战略规划体系的分析和预测研究；②旨在获得自然科学、技术、医学、农业、社会和人文科学等重要领域科学知识的基础和探索性研究；③在大型科研装置上进行的基础和探索性研究；④执行《俄罗斯联邦科学技术发展战略》规定优先发展方向的基础和探索性研究；⑤由科学基金和创新活动基金支持的基础和探索性研究；⑥确保国防和国家安全的基础和探索性研究。

2. 主要国家和机构陆续推进量子科技战略

2021年1月，法国发布《法国量子技术国家战略》[69]，计划5年内在量子技术领域投入18.15亿欧元，使法国跻身量子技术国际第一梯队，重点支持量子模拟器和量子加速器、量子计算机、量子传感器、后量子密码学、量子通信、支撑技术等六大方向以及生态环境建设。加拿大政府宣布将在7年内投资3.6亿加元[70]，以启动国家量子战略，旨在增强加拿大在量子研究方面的能力，发展量子技术，培育人才，巩固加拿大在全球量子技术领域的领先地位。德国联邦教育与研究部（Bundesministerium für Bildung und Forschung，BMBF）宣布，通过“量子处理器和量子计算机技术”计划资助8个研究新型量子处理器的大型项目[71]，资助总额为1.1亿欧元；这些项目采用原子阱、离子阱、超导体、半导体和光子等不同技术，构成未来实用型量子计算机的基础。欧盟“地平线欧洲”（Horizon Europe）计划资助5690万欧元开发量子技术[72]，包括下一代量子传感技术、新兴量子计算技术、开发首批大规模量子计算机的框架合作协议、量子技术基础科学、开发大规模量子模拟平台技术的框架合作协议、量子通信技术的框架合作协议、量子传感技术的商业应用、面向量子计算平台的量子软件生态系统等。2021年1月，美国国家科学与技术委员会（National Science and

Technology Council，NSTC）发布《量子网络研究协同路径》报告[73]，提出继续研究量子网络应用场景、优先考虑能产生多重效益的量子网络核心组件、提升现有技术能力以支持量子网络、利用规模适中的量子网络测试平台等 4 项技术建议，以及加强量子网络研发机构间的协同、针对量子网络研发基础设施制定时间表、促进量子网络研发的国际合作等 3 项方案建议，目的是加强美国在量子网络利用方面的知识基础和准备。2022 年 4 月，日本发布新量子技术战略《量子未来社会愿景》[74]，描述了未来社会对量子技术的基本愿景，并提出在量子计算机、量子软件、量子安全网络、量子测量和传感及量子材料等技术领域进行研究和产业开发。2022 年 11 月，欧盟量子旗舰计划（European Quantum Flagship）发布了《战略研究和产业议程》[75]，基于量子计算、量子模拟、量子通信、量子传感与计量等四大技术支柱，描绘了欧盟至 2030 年量子技术发展路线图，并提出了相应的建议。

3. 美国对国家纳米技术计划做出重要调整

不同于此前几版战略规划的一脉相承，美国 2021 年版的《国家纳米技术计划战略规划》对国家纳米技术计划未来 5 年的发展做出重要调整[76]，重新设定了发展愿景和发展目标。新的发展愿景是“通过在纳米尺度理解和控制物质，推动正在发生的技术和产业变革，造福社会”。与 2016 年版相比，2021 年版的《国家纳米技术计划战略规划》发展愿景指出技术和产业变革“已经发生”，反映了美国科技界对科技革命态势的最新判断。在发展目标方面，国家纳米技术计划继续坚持技术研发、商业化、基础设施、负责任地发展四大方向，同时将教育和劳动力方面的内容单列为一个方向，以增加支持力度，反映了国家纳米技术计划对教育、发展劳动力、公众参与对整个纳米技术事业重要性的最新认识。2021 年的《国家纳米技术计划战略规划》核心思想是保持美国纳米技术世界领先，美国不仅要产出原始创新而且要实现成果转化与社会经济效益。

三、启示与建议

1. 系统布局以加强基础研究

基础研究是科技创新的源头，只有持之以恒地加强基础研究，才能不断提高原始创新能力。我国应以重大科学问题为牵引，开展面向世界科技前沿的基础研究，加强物质科学、生命科学等领域的布局，以取得原创性和开拓性的突破；加强面向国民经济主战场、国家重大需求和人民生命健康的应用基础研究，促进重点基础领域技术突破，解决经济社会发展面临的关键科学问题；加大对数学、物理学、化学、天文学、地学、生物学等基础学科的支持力度，大力推进物质科学、生命科学与信息、能源、材料、制造等技术科学的学科交叉和跨学科研究。

2. 积极出台适应数据密集型研发的政策举措

以大数据、人工智能技术为主要特征的数据密集型研发正在越来越多的科研领域证明其大幅加快发现速度、提高研究质量的能力，势必对传统基于经验的试错型研发带来颠覆性影响。为迎接和发展数据密集型研发，建议前瞻性出台一些政策举措，包括：①由资助机构、科研机构、出版商等合作建立数字基础设施，统一数据格式标准，发展高质量的数据资源；②高校设立融合专业知识、数据科学、软件工程课程，培养数据密集型研发所需的人力资源；③把科研伦理放到更加重要的位置，以尊重知识产权和保护隐私的方式共享数据，注意避免技术缺陷带来的风险。

3. 构建纳米材料负面数据的收集和共享平台，促进纳米技术的重大创新

纳米技术和材料在能源、环境、医药、制造、信息等领域发挥了重要作用，并将持续发挥不可替代的作用。我国在纳米材料的制备和合成领域已经处于国际领先地位，发表了大量的原创性论文，同时也积累了丰富的实验数据和制备经验。成功的制备方案和技术多数以论文的形式发表，但制备和合成过程中的失败案例同样值得收集、记录、存储、共享。大量的正数据和负数据将有助于纳米材料开发模型的构建，理解不同类型材料的性质和相互作用，以及学习如何更准确地预测材料的系统性能。为了在纳米材料和技术领域继续取得进展，我国还需构建纳米材料负面数据的收集和共享平台，制订收集、存储和分发材料数据的标准，在节约科研人员实验时间的同时，促进纳米材料领域的重大创新。

致谢：中国科学院化学研究所张建玲研究员对本文初稿进行了审阅并提出了宝贵的修改意见，特致感谢！

参考文献

[1] Abi B, Albahri T, Al-Kilani S, et al. Measurement of the positive muon anomalous magnetic moment to 0.46 ppm. Physical Review Letters, 2021, 126: 141801.

[2] CDF Collaboration. High-precision measurement of the W boson mass with the CDF II detector. Science, 2022, 6589: 170-176.

[3] BES Ⅲ Collaboration. Observation of a near-threshold structure in the K^+ recoil-mass spectra in

$e^+e^- \rightarrow K^+(D_s^- D^{*0} + D_s^{*-} D^0)$. Physical Review Letters, 2021, 126: 102001.

[4] LHCb Collaboration. Observation of new resonances decaying to $J/\psi K^+$ and $J/\psi\phi$. Physical Review Letters, 2021, 127: 082001.

[5] Duer M, Aumann T, Gernhäuser R et al. Observation of a correlated free four-neutron system. Nature, 2022, 606: 678-682

[6] Wu Y L, Bao W S, Cao S R, et al. Strong quantum computational advantage using a superconducting quantum processor. Physical Review Letters, 2021, 127: 180501.

[7] Zhong H S, Deng Y H, Qin J, et al. Phase-programmable Gaussian boson sampling using stimulated squeezed light. Physical Review Letters, 2021, 127: 180502.

[8] Hugh Collins, Chris Nay. IBM unveils 400 qubit-plus quantum processor and next-generation IBM quantum system two. https://newsroom.ibm.com/2022-11-09-IBM-Unveils-400-Qubit-Plus-Quantum-Processor-and-Next-Generation-IBM-Quantum-System-Two. [2022-12-28].

[9] Singh K, Anand S, Pocklington A, et al. Dual-element, two-dimensional atom array with continuous-mode operation. Physical Review X, 2022, 12: 011040.

[10] Noiri A, Takeda K, Nakajima T, et al. Fast universal quantum gate above the fault-tolerance threshold in silicon. Nature, 2022, 601: 338-342.

[11] Xue X, Russ M, Samkharadze N, et al. Quantum logic with spin qubits crossing the surface code threshold. Nature, 2022, 601: 343-347.

[12] Madzik M T, Asaad S, Youssry A, et al. Precision tomography of a three-qubit donor quantum processor in silicon. Nature, 2022, 601: 348-353.

[13] Chen Y A, Zhang Q, Chen T Y, et al. An integrated space-to-ground quantum communication network over 4,600 kilometres. Nature, 2021, 589: 214-219.

[14] Pittaluga M, Minder M, Lucamarini M, et al. 600-km repeater-like quantum communications with dual-band stabilization. Nature Photonics, 2021, 15: 530-535.

[15] Wang S, Yin Z Q, He D Y, et al. Twin-field quantum key distribution over 830-km fibre. Nature Photonics, 2022, 16: 154-161

[15] Park J M, Cao Y, Watanabe K, et al. Tunable strongly coupled superconductivity in magic-angle twisted trilayer graphene. Nature, 2021, 590: 249-255.

[16] Hao Z Y, Zimmerman A M, Ledwith P, et al. Electric field-tunable superconductivity in alternating-twist magic-angle trilayer graphene. Science, 2021, 371: 1133-1138.

[17] Cao Y, Park J M, Watanabe K, et al. Pauli-limit violation and re-entrant superconductivity in moiré graphene. Nature, 2021, 595: 526-531.

[18] Zhou H X, Xie T, Taniguchi T, et al. Superconductivity in rhombohedral trilayer graphene. Nature, 2021, 598: 434-438.

[19] Zhu Z, Papaj M, Nie X A, et al. Discovery of segmented Fermi surface induced by Cooper pair momentum. Science, 2021, (6573): 1381-1385.

[20] Mi X, Ippoliti M, Quintana C, et al. Time-crystalline eigenstate order on a quantum processor. Nature, 2022, 601: 531-536.

[21] Randall J, Bradley C E, van der Gronden F V, et al. Many-body-localized discrete time crystal with a programmable spin-based quantum simulator. Science, 2021, (6574): 1474-1478.

[22] Bykov M, Fedotenko T, Chariton S, et al. High-pressure synthesis of Dirac materials: layered van der Waals bonded BeN4 polymorph. Physical Review Letters, 2021, 126: 175501.

[23] Li Q, Kolluru V S C, Rahn M S, et al. Synthesis of borophane polymorphs through hydrogenation of borophene. Science, 2021, (6534): 1143-1148.

[24] Shin J, Gamage G A, Ding Z W, et al. High ambipolar mobility in cubic boron arsenide. Science, 2022, 6604: 437-440.

[25] Yue S, Tian F, Sui X Y, et al. High ambipolar mobility in cubic boron arsenide revealed by transient reflectivity microscopy. Science, 2022, 6604: 433-436.

[26] Li M, Li G, Cao L, et al. Ordered and tunable Majorana-zero-mode lattice in naturally strained LiFeAs. Nature, 2022, 606: 890-895.

[27] Breanna Bishop. Lawrence Livermore National Laboratory achieves fusion ignition. https://www.llnl.gov/archive/news/lawrence-livermore-national-laboratory-achieves-fusion-ignition [2022-12-28].

[28] Dikopoltsev A, Harder T H, Lustig E, et al. Topological insulator vertical-cavity laser array. Science, 2021, (6562): 1514-1517.

[29] Zhang Y F, Fowler C F, Liang J H, et al. Electrically reconfigurable non-volatile metasurface using low-loss optical phase-change material. Nature Nanotechnology, 2021, 16: 661-666.

[30] Wang Y F, Landreman P, Schoen Da, et al. Electrical tuning of phase-change antennas and metasurfaces. Nature Nanotechnology, 2021, 16: 667-672.

[31] Liu X, Hu J, Li Z F, et al. Heralded entanglement distribution between two absorptive quantum memories. Nature, 2021, 594: 41-45.

[32] Xu P, Cui B, Bu Y, et al. Elastic ice microfibers. Science, 2021, (6551): 187-192.

[33] Liu W, Lavagnino M N, Gould C A, et al. A biomimetic S_H2 cross-coupling mechanism for quaternary sp^3-carbon formation. Science, 2021, 374(6572): 1258-1263.

[34] Bunescu A, Abdelhamid Y, Gaunt M J. Multicomponent alkene azidoarylation by anion-mediated dual catalysis. Nature, 2021, 598(7882): 597-603.

[35] Kennedy S H, Dherange B D, Berger K J, et al. Skeletal editing through direct nitrogen deletion of secondary amines. Nature, 2021, 593(7858): 223-227.

[36] Jurczyk J, Lux M C, Adpressa D, et al. Photomediated ring contraction of saturated heterocycles. Science, 2021, 373(6558): 1004-1012.

[37] Qiu X, Sang Y, Wu H, et al. Cleaving arene rings for acyclic alkenylnitrile synthesis. Nature, 2021, 597(7874): 64-69.

[38] Burrows J, Kamo S, Koide K. Scalable Birch reduction with lithium and ethylenediamine in tetrahydrofuran. Science, 2021, 374(6568): 741-746.

[39] Reisenbauer J C, Green O, Franchino A, et al. Late-stage diversification of indole skeletons through nitrogen atom insertion. Science, 2022, 377(6610): 1104-1109.

[40] Carter K P, Shield K M, Smith K F, et al. Structural and spectroscopic characterization of an einsteinium complex. Nature, 2021, 590(7844): 85-88.

[41] Mallada B, Gallardo A, Lamanec M, et al. Real-space imaging of anisotropic charge of σ-hole by means of Kelvin probe force microscopy. Science, 2021, 374(6569): 863-867.

[42] Rösch B, Gentner T X, Langer J, et al. Dinitrogen complexation and reduction at low-valent calcium. Science, 2021, 371(6534): 1125-1128.

[43] Pullanchery S, Kulik S, Rehl B, et al. Charge transfer across C-HO hydrogen bonds stabilizes oil droplets in water. Science, 2021, 374(6573): 1366-1370.

[44] Feng L, Qiu Y Y, Guo Q-H, et al. Active mechanisorption driven by pumping cassettes. Science, 2021, 374(6572): 1215-1221.

[45] Dong J, Lu Y, Xu Y, et al. Direct imaging of single-molecule electrochemical reactions in solution. Nature, 2021, 596(7871): 244-249.

[46] Albrecht F, Fatayer S, Pozo I, et al. Selectivity in single-molecule reactions by tip-induced redox chemistry. Science, 2022, 377(6603): 298-301.

[47] Häußler M, Eck M, Rothauer D, et al. Closed-loop recycling of polyethylene-like materials. Nature, 2021, 590(7846): 423-427.

[48] Suryanto B H R, Matuszek K, Choi J, et al. Nitrogen reduction to ammonia at high efficiency and rates based on a phosphonium proton shuttle. Science, 2021, 372(6547): 1187-1191.

[49] Yan H, He K, Samek I A, et al. Tandem In_2O_3-Pt/Al_2O_3 catalyst for coupling of propane dehydrogenation to selective H_2 combustion. Science, 2021, 371(6535): 1257-1260.

[50] Trang B, Li Y, Xue X S, et al. Low-temperature mineralization of perfluorocarboxylic acids. Science, 2022, 377(6608): 839-845.

[51] Kirkpatrick J, Mcmorrow B, Turban D H P, et al. Pushing the frontiers of density functionals by solving the fractional electron problem. Science, 2021, 374(6573): 1385-1389.

[52] Newman-Stonebraker S H, Smith S R, Borowski J E, et al. Univariate classification of phosphine ligation state and reactivity in cross-coupling catalysis. Science, 2021, 374(6565): 301-308.

[53] Shields B J, Stevens J, Li J, et al. Bayesian reaction optimization as a tool for chemical synthesis. Nature, 2021, 590(7844): 89-96.

[54] Angello N H, Rathore V, Beker W, et al. Closed-loop optimization of general reaction conditions for heteroaryl Suzuki-Miyaura coupling. Science, 2022, 378(6618): 399-405.

[55] Li J K, Chen X Y, Zhao W L, et al. Synthesis of highly luminescent chiral nanographene. Angewandte Chemie International Edition, 2022, e202215367.

[56] Liang R, Xue Y, Fu X, et al. Hierarchically engineered nanostructures from compositionally anisotropic molecular building blocks. Nature Materials, 2022, 21: 1434-1440.

[57] Yan M, Liu T Y, Li X F, et al. Journal of the American Chemical Society, 2022, 144 (17), 7778-7789.

[58] Santos P J, Gabrys P A, Zornberg L Z, et al. Macroscopic materials assembled from nanoparticle superlattices. Nature, 2021, (7851): 586-591.

[59] Liu Y, Liu L, Chen X, et al. Single-crystalline ultrathin 2D porous nanosheets of chiral metal-organic frameworks. J. Am. Chem. Soc., 2021, 143: 3509-3518.

[60] Yao C, Qi H, Jia X, et al. DNA nanocomplex containing cascade DNAzymes and promoter-like Zn-Mn-Ferrite for combined gene/chemo-dynamic therapy. Angewandte Chemie, 2022, (6): 1-11.

[62] Zhu B, Li W, Zhang Q, et al. Subambient daytime radiative cooling textile based on nanoprocessed silk. Nature Nanotechnology, 2021, (12): 1342-1348.

[63] Song J K, Kim J, Yoon J, et al. Stretchable colour-sensitive quantum dot nanocomposites for shape-tunable multiplexed phototransistor arrays. Nat Nanotech, 2022, 7(8): 849-856.

[64] Itoh Y, Chen S, Hirahara Ry, et. al. Ultrafast water permeation through nanochannels with a densely fluorous interior surface. Science, 2022, 376(6594): 738-743.

[65] Jung D, Lim C, Shim H J, et al. Highly conductive and elastic nanomembrane for skin electronics. Science, 2021, (6558): 1022-1026.

[66] Dabard C, Guilloux V, Gréboval C, et. al. Double-crowned 2D semiconductor nanoplatelets with bicolor power-tunable emission. Nature Communications 2022, https://doi. org/10. 1038/s41467-022-32713-2

[67] Karst J, Floess M, Ubl M, et al. Electrically switchable metallic polymer nanoantennas. Science, 2021, 374: 612-616.

[68] Мишустин М. Михаил Мишустин утвердил программу фундаментальных научных исследований до 2030 года. http://government. ru/docs/41288/[2021-12-28].

[69] Gouvernement Liberté, égalité, Fraternité. Stratégie nationale sur les technologies quantiques. https://www. gouvernement. fr/sites/default/files/contenu/piece-jointe/2021/01/dossier _ de _ presse _ quantique_vfinale. pdf[2021-12-28].

[70] Government of Canada. Developing a National Quantum Strategy. http://www. ic. gc. ca/eic/site/154. nsf/eng/00001. html[2021-12-28].

[71] BMBF. Karliczek: Werkzeugkasten zur Förderung des Quantencomputing vorerst komplett. https://www. bmbf. de/bmbf/shareddocs/pressemitteilungen/de/2021/09/290921-Quantencomputer. html[2021-12-28].

[72] European Commission. Horizon Europe Work Programme 2021-2022: 7. Digital, Industry and Space. https://ec. europa. eu/info/funding-tenders/opportunities/docs/2021-2027/horizon/wp-call/2021-2022/wp-7-digital-industry-and-space_horizon-2021-2022_en. pdf[2021-12-28].

[73] NSTC. A Coordinated Approach to Quantum Networking Research. https://www. quantum. gov/a-coordinated-approach-to-quantum-networking-research/[2021-12-28].

[74] 統合イノベーション戦略推進会議．量子未来社会ビジョン. https://www8. cao. go. jp/cstp/ryoshigijutsu/ryoshimirai_220422. pdf[2022-12-28].

[75] European Quantum Flagship. Strategic Research and Industry Agenda. https://qt. eu/media/pdf/Quantum-Flagship_SRIA_2022_0. pdf [2022-12-28].

[76] National Nanotechnology Initiative. 2021 National Nanotechnology Initiative Strategic Plan. https://www. nano. gov/2021strategicplan[2021-10-08].

Basic Sciences and Frontiers

Huang Longguang, Bian Wenyue, Zhang Chaoxing, Leng Fuhai

In 2021～2022, major achievements have been made in basic sciences and frontier. Major significant breakthroughs are made in particle physics, quantum science and technology, condensed matter physics and optics. Innovations in chemical synthesis methods and fundamental theories are flourish, and the integration of data science and chemistry deepens. Much progress has been made in the field of nanomaterials, and their application is expanding steadily. Major countries in the world continue to promote the development of basic science. Russia have strengthened the layout of basic research. Major countries and institutions have successively promoted their quantum science and technology strategy. The United States has made important adjustments in the National Nanotechnology Initiative. The aim of these actions is to improve the innovation capability and take the lead in the new international science and technology competition.

4.2 生命健康与医药领域发展观察

许 丽 王 玥 李祯祺 施慧琳
杨若南 李 伟 靳晨琦 徐 萍
（中国科学院上海营养与健康研究所/中国科学院上海生命科学信息中心）

生命健康领域是全球科技竞争的焦点之一。大数据与人工智能的融合、学科的深度会聚、技术的突飞猛进、大科学和大团队的组织模式正在推动生命健康科技变革性发展，研究尺度和深度进一步拓宽，全生命周期、全系统研究愈加深入，数字化、系统化、工程化趋势明显，推动全生命周期健康管理广泛融入生活。与此同时，人体处在复杂的环境中，并具有高度的异质性，从个体到群体、个体与环境的研究在新技术的推动下全面开展，全健康（one-health）、精准医学（precision medicine）、群医学（population medicine）等新理念、新路径的提出推动了人们对健康与疾病的全面认识，并研发出更加有效的早发现、早干预与个体化精准治疗方案，有助于全生命周期健康管理目标的实现。

一、重要研究进展

（一）对生命的认识与解析更加深入和系统

生命组学、人工智能、单细胞技术以及超分辨率成像等技术的大发展，推动了分子、细胞、器官、个体等多层次的基础研究向纵深推进，更加全面深入地认识和解析生物体这一复杂系统。

1. 生命组学研究持续开展，为人们更好地认识和解析生命奠定基础

空间组学快速发展，继空间转录组学技术被《自然-方法》（*Nature Methods*）评为2020年年度技术之后，空间多组学技术入选《自然》期刊2022年值得关注的七项年度技术之一。其中，组蛋白修饰分析技术——空间靶向剪切及标记技术（spatial-cleavage under targets and tagmentation，Spatial-CUT&Tag）和染色质可及性分析技术——空间染色质转座酶可及性测序技术（spatial-assay for transposase-accessible chromatin using sequencing，Spatial-ATAC-seq）的开发，突破性实现了发育和疾病相关表观调控的空间映射，新空间蛋白质组技术——基于有机溶剂的器官三维透明成像质谱分析技术（three-dimensional imaging of solvent-cleared organs profiled by mass spectrometry，DISCO-MS）创新推动空间组学分析从二维向三维发展。人类基因组序列绘制持续进行，完整性和准确性不断提高。2022 年4 月，端粒到端粒联盟（the Telomere-to-Telomere Consortium，T2T）报道了有史以来最完整的人类基因组序列 T2T-CHM13。2022 年 5 月，包括中国在内的六国科研人员提出“未充分研究的蛋白质计划”（Understudied Protein Initiative），拟解决蛋白质注释非均一性问题[1]，加快加深蛋白质研究。起始于 2016 年的“人类细胞图谱计划”于 2022 年发布重大成果——绘制了迄今最为全面的人体泛组织细胞图谱，为人类健康和疾病提供生物学新见解[2]。

2. 人工智能深度赋能生命健康科技，引发研究范式变革

“创造性人工智能的快速发展”是《科学》期刊评选的 2022 年度十大突破之一。研究人员现已能够利用人工智能预测或设计出蛋白质已有或全新的结构，应用于疫苗等多个领域。继 AlphaFold2 和 RoseTTAFold 预测蛋白质结构取得突破后，人工智能预测生物大分子结构和功能的工具持续创新，结构预测工具 IgFold、RNA 三维结构预测工具 ARES[3]和 mRNA 序列设计算法 LinearDesign 受到关注。此外，蛋白质功能[4]、序列功能注释[5]，乃至利用人工智能生成新型功能活性蛋白[6]、增强新冠抗体效果[7]等方面的研究也取得了进展，谷歌公司提出的新的深度学习模型将主流蛋白质家族数据库 Pfam 的覆盖范围扩大了 9.5% 以上，超过该库过去十年的数据增加总量[8]。脸书（Facebook）旗下 META AI 公司所开发出的 ESMFold 通过简化训练模型和强化训练参数，高效、准确地预测出 6 亿多种难以表征的宏基因组蛋白[9]。我国基于可预测稳定突变位点的 3D 自监督学习卷积神经网络 utCompute 框架，成功预测出目前最优秀的聚对苯二甲酸乙二醇酯（PET）水解酶[10]。

3. 超分辨率成像向高清、三维、实时、在体观察方向推进，赋能生命认识与解析

单分子及超分辨显微成像技术通过升级超分辨硬件、探针及其标记方法、数据处理算法，推动影像技术向高清、三维、实时、在体观察方向发展。科研人员通过引入量子关联将相干拉曼显微成像的信噪比提升了 35%，有望在克服光损伤的条件下将信噪比和成像速度实现数量级的提高[11]，并研发出 DNA 测序技术与显微镜相结合的原位基因组测序法[12]。我国研究人员提出超分辨卷积神经网络模型，并基于此自主开发

了多模态结构光照明超分辨显微镜[13]。我国首次利用人工智能四维重建技术[14]提升了时间分辨冷冻电镜的分析精度，并提出了深度学习显微成像技术框架[15]，实现了当前国际上速度最快、成像时程最长的活体细胞成像。

（二）改造与利用生物的能力不断增强，造福人类和社会

基因编辑技术、脑-机接口、合成生物学等技术的发展，提高了人类改造、合成、仿生、再生生物的能力，为健康维护提供了更多方案。

1. 基因编辑技术提高了基因“改写”能力，相关临床试验快速推进

基因编辑技术持续迭代，朝着更高效、更安全，操作更简单、更灵活的方向发展。细胞核 DNA 编辑[16,17]、RNA 编辑技术不断优化[18-20]；线粒体 DNA 碱基编辑技术取得突破，开发了新型线粒体 A—G 碱基转换编辑器[21]，扩展了线粒体基因编辑的范围。基于“先导编辑”技术的多项新设计，如双重先导编辑技术 twinPE[22]、PED-AR 系统[23]、PRIME-Del[24]和 MPE 编辑器[25]先后实现了完整的长片段基因的精确插入、置换或删除，甚至多个基因位点的“一步到位”式编辑，这些技术为复杂的生物研究及基因疾病治疗提供了工具与思路。在应用方面，研究人员首次实现了哺乳动物完整染色体的可编程连接，创建出具有全新核型（染色体组型）的小鼠[26]。同时，多项相关的临床试验取得积极结果，截至 2022 年 12 月，全球已开展了 77 项基因编辑疗法临床试验①。其中，CRISPR Therapeutics 公司研发的 CTX001（Exa-cel）成为首个进入Ⅲ期临床试验的基因编辑疗法，用于治疗镰状细胞贫血和输血依赖型 β 地中海贫血，获得美国 FDA 快速通道和孤儿药资格认定。

2. 脑科学、类脑智能及脑-机接口技术迅速突破

脑科学是生命科学研究的热点领域，各国脑科学计划稳步推进实施。英国剑桥大学等主导的“国际人脑图表联盟”分析了全球十万余人脑磁共振数据，构建了覆盖全年龄段的脑发育图谱 BrainChart[27]；我国科研人员也率先实现了全脑单神经元完整形态的重构，构建了国际上最大的小鼠全脑介观神经联接图谱数据库[28]。类脑智能向高效、便携、低功耗发展。例如，英特尔公司发布了第二代神经拟态芯片 Loihi 2，能效提升 1000 倍；瑞士弗雷德里希·米歇尔生物医学研究所等开发的神经拟态芯片能耗仅需当下同类产品的千分之一[29]。脑-机接口技术开始逐步走向成熟，多款通过意念控制数字设备的产品获得突破。例如，美国 Blackrock Neurotech 公司首创的

① 数据来源：ClinicalTrials. gov。

MoveAgain 获得美国 FDA 突破性设备认证；美国 Synchron 公司开发的无需开颅而仅通过血管植入的装置 Stentrode 也已获批开展临床试验。研究人员还利用石墨烯-弹性体复合材料发明了一种用于监测单胺神经递质的柔性传感器 NeuroString，可以安装在大脑或肠道中，而且不会干扰器官的自身功能，标志着下一代传感器向更小、更薄、更灵活的方向发展[30]。

3. 合成生物学与工程生物学提高了设计生物的能力，促进生命健康领域发展

合成生物学作为使能技术突破不断。例如，从头设计模块化蛋白生物传感器[31]、细胞表面蛋白阵列[32]、CO_2 到淀粉分子的全合成[33]等突破奠定了工程生物学的发展基础；Z-基因组生物合成通路[34]使大规模合成 Z DNA 开展噬菌体疗法、DNA 存储等成为可能。在人工生命合成领域，研究人员创造了首个能够进行自我复制的活体机器人[35]。在合成生物学与工程生物学的支撑下，医药产品的生产手段越发丰富。抗癌药物长春碱的前体分子、潜在抗生素黑莫他丁、潜在镇痛药物河豚毒素、抗糖尿病药物阿卡波糖、番木鳖碱等物质的生物合成途径均已构建完成，有望在此基础上建立规模化生产的细胞工厂。我国科学家还通过电催化结合生物合成的方式将 CO_2 的高效还原成高浓度乙酸，并进一步利用微生物将其用于合成葡萄糖和脂肪酸[36]；构建了高效生物合成类胡萝卜素的细胞工厂[37]等。

（三）健康维护向精准化、个体化迈进，疾病防治手段更加多样化，逐步实现全生命周期健康管理

干细胞与再生医学转化进程加速、精准医学研究路径不断推广，创新药物与新型疗法快速突破，以及智能诊疗应用不断拓宽，正逐步推动全生命周期健康管理目标的实现。

1. 干细胞转化进程加速，应用潜力不断扩大

截至 2022 年，全球批准的干细胞产品已累计达到 23 种。基础研究的多项进展推动干细胞走向应用。例如，首次在体外通过诱导人类多能干细胞（human pluripotent stem cell）获得 8 细胞期全能胚胎样细胞（8-cell totipotent embryo-like cell）[38]、利用化学小分子将人类成体细胞诱导编程为多能干细胞[39]等突破，为早期胚胎发育研究及器官再造等奠定了基础。在应用研究方面，保护性封装装置和基因编辑技术的应用，使多能干细胞来源的胰腺细胞在糖尿病患者体内发挥良好的治疗效果[40]，且解决了干细胞移植治疗糖尿病的免疫排斥的问题；研究人员将哺乳动物的心肌细胞重新编

程至具有再生能力的胎儿阶段，实现了受损心肌的再生[41]。类器官领域发展迅速，对器官生理病理的模拟水平不断增强。例如，利用单个类器官重现了心脏和肠道的协同发育和成熟过程[42]。异种器官移植在停滞了近20年后也出现了新一轮研究热潮，2021～2022年，全球开展了2例猪肾脏的人体移植（移植给脑死亡患者）和1例猪心脏人体移植（移植给心力衰竭患者）临床试验，后者患者存活了2个月，这些手术的实施为异种器官移植提供了宝贵经验。

2. 精准医学体系逐渐形成，理念与研究路径广泛应用于疾病研究

精准医学理念与研究路径已广泛应用于疾病防治研究，多个国家已明确将精准医学列入新阶段的规划中，大型人群队列建设广泛开展。在疾病精准分型方面，基于组学特征谱的分子分型改变了传统按照疾病部位进行疾病分类的方式，为疾病精准诊治奠定了基础；一项基于肿瘤免疫环境将不同肿瘤分为12种“免疫原型”的研究提供了疾病精准分型的新方式，为相关免疫疗法开发提供了新思路[43]。在疾病精准早期诊断方面，美国斯坦福大学研究人员开发的一种基于表观遗传学的新液体活检技术——EPIC-seq[44]、希伯来大学等开发的血浆核小体表型分析工具（EPINUC），均能实现高灵敏度、高特异性的肿瘤检测、分型及伴随诊断[45]。

3. 新药创制领域，靶向药物、新型疗法助推个性化精准治疗的实现

基于新靶点、新机制、开辟新适应证的小分子药物不断获批上市，2022年，美国FDA批准了18个首创（first-in-class）小分子药物。以蛋白水解靶向嵌合体（proteolysis targeting chimeras，PROTAC）、分子胶为代表的靶向蛋白质降解技术是小分子药物研发的一个新兴方向，有望靶向目前“不可成药”的约80%的蛋白质，并解决耐药性问题。截至2022年5月，已有17个候选小分子药物进入临床试验阶段，适应证范围从癌症逐步拓宽至自身免疫性疾病。

生物医药处于产业爆发阶段，多个新型疗法持续取得突破。截至2022年7月，全球已有9款免疫细胞治疗药物、14款基因治疗药物，以及15款RNA药物获批上市。免疫细胞疗法在实体瘤中取得突破，2022年，我国科济药业自主研发的CT041成为全球首个进入Ⅱ期临床试验的实体瘤嵌合抗原受体T细胞治疗（chimeric antigen receptor T cell therapy，CAR-T）候选产品。除CAR-T外，其他免疫细胞疗法也迎来重大进展。2023年，美国Iovance公司利用肿瘤浸润淋巴细胞治疗（tumor infiltrating lymphocyte therapy，TIL therapy）研发的黑色素瘤治疗药物Lifileucel上市许可申请已得到美国FDA的优先审评，有望成功上市。

4. 大数据与人工智能深度融合加速智能诊疗落地应用

数据、算法及算力的升级进一步加深人工智能与医疗的融合，助力智能诊疗在多个应用场景落地应用，包括全生命周期健康管理、医院管理、药物研发等。至 2021 年 9 月，美国 FDA 批准的基于人工智能的医疗器械有 300 余个；我国国内有 36 个人工智能辅助诊疗产品获得国家药品监督管理局的三类医疗器械注册证。手术机器人在微创或远程精细操作、获得细胞和分子信息的原位/在体表征，以及以高精度开展靶向治疗等方向快速推进，自主机器人 STAR[46]、远程机器人系统[47]相继被开发出来。人工智能推动新药研发多环节实现降本增效发展，至 2022 年，全球已有 38 个（中国有 4 个）人工智能参与研发的药物候选化合物进入临床试验阶段。

二、重大战略行动

1. 重视关键技术竞争力提升，强化本国科技主权

各国重视关键技术竞争力的提高，以强化本国的科技主权。美国 2022 年新版《关键和新兴技术国家战略》[48]清单关注了生物技术 6 个核心子领域，包括核酸和蛋白质合成，基因组和蛋白质工程，功能表型研究的多组学和其他生物计量学、生物信息学、预测建模和分析工具，多细胞系统工程，病毒和病毒传递系统的工程设计，生物制造和生物加工技术。德国在 2021 年发布的《以技术主权塑造未来》文件中[49]，明确要在疫苗研发、人工智能、数据技术等领域实现技术主权。欧盟 2021 年更新了《欧洲工业战略》[50]，将增强欧盟战略自主权作为优先事项之一，并配套发布了《欧盟的战略依赖与能力》报告，针对原料药等 6 个领域开展审查，识别对外依赖原因并提出相应措施[51]。俄罗斯加大力度发展国产科学仪器以替代外国进口仪器和设备[52]，并向俄罗斯制药公司提供资助建设重要药物标准样品库[53]。韩国发布《国家战略技术培育方案》，将重点培育包括生物主题技术在内的 12 项国家战略技术，以提高技术主权水平。

2. 聚焦热点领域持续布局，推出实施专项规划

各国发布未来发展战略和计划，以及专项规划，对包括数字化、基因组医疗、大规模人群队列、精准医学、癌症、脑科学、新冠疫情、疫苗制造，以及先进疗法开发等领域进行部署。美国国立卫生研究院（National Institutes of Health，NIH）推出“NIH 战略计划（2021—2025）”，提出生物医学的发展目标和优先方向，并在 2023 财

年预算中继续保持对癌症登月计划、精准医学计划、脑科学计划的资助。美国发布《国家生物技术和生物制造计划》，推动生物技术创新；启动《国家生物防御战略和应对生物威胁、加强大流行病防范和实现全球卫生安全的实施计划》[54]，将在五年内拨款约 880 亿美元的强制性资金支持生物防御，以保护美国免受下一轮大流行病及其他生物威胁发生的影响。欧盟"地平线欧洲"计划在 2021～2022 年工作计划中设置了六大健康主题目标[55]，同时启动了"创新健康计划"和"欧洲抗癌计划"[56]；英国政府发布"生命科学愿景"，提出未来 10 年生命科学发展战略[57]，并在 2022 年推进英国基因组计划；日本内阁发布"2021 年综合创新战略"[58]，将生物技术、健康与医药分别确立为战略性基础技术和战略性应用领域，并进一步推进全基因组分析计划的下一阶段路线图和实施方案的制订；加拿大发布了"加拿大生物制造和生命科学战略"[59]；澳大利亚发布了"澳大利亚医学研究与创新战略（2021—2026）"[60]；印度发布了"国家生物技术发展战略（2021～2025）：知识驱动生物经济"[61]。

3. 成立相关研究资助机构，推动科技创新发展

为推动科技创新发展，世界各国纷纷改革科技管理机制，新成立了相关研究资助机构，并已开始运行。美国 NIH 高级健康研究计划局（Advanced Research Projects Agency for Health，ARPA-H）确定了 2023 财年预算资助重点；欧盟委员会创立的欧洲创新理事会（European Innovation Council，EIC）公布了 2021～2022 年资助主题；在欧盟创新药物计划（Innovative Medicine Initiative，IMI）基础上新建的欧盟创新健康计划（Innovative Health Initiative，IHI）发布了"战略研究与创新议程"；法国成立的健康产业战略委员会制定了"法国健康创新 2030 计划"；英国高级研究与发明局（Advanced Research and Invention Agency，ARIA）法案获得了英国议会批准，标志着 ARIA 正式运行。此外，2022 年，韩国生命工学研究所（KRIBB）提出发展先进生物技术的提案，提议设立韩国先进生物技术研究支持小组（ARPA-B），专注于先进生物技术领域的研发。

三、启示与建议

从全球生命健康科技发展趋势与竞争态势来看，我国应"顺大势""长规划""抓关键""补短板"，完成生命健康科技的飞跃发展，实现我国"健康中国 2030"目标。

建议我国顺应数据驱动范式变革的趋势，以大科学计划组织模式，加快布局数据密集型研究项目，以规模化研究促进科学大发现；研判可能影响我国生命健康科技领域发展的关键核心技术和重点领域，并在前沿方向、优势领域进行长周期规划、持续

稳定支持，提高重点领域科技竞争力；加大加快布局数据资源与高端仪器设备等生命科学研究的重要“基石”领域，并细化我国对外依存度较高的产品清单，加大支持力度，提高原始创新能力，实现科技自立自强。

致谢：中国科学院分子细胞科学卓越创新中心吴家睿研究员、惠利健研究员在本文的撰写过程中提出了宝贵的意见和建议，在此谨致谢忱！

参考文献

[1] Kustatscher G, Collins T, Gingras A C, et al. Understudied proteins: opportunities and challenges for functional proteomics. Nature Methods, 2022, 19(7): 774-779.

[2] Liu Z D, Zhang Z M. Mapping cell types across human tissues. Science, 2022, 376(6594): 695-696.

[3] Townshend R J L, Eismann S, Watkins A M, et al. Geometric deep learning of RNA structure. Science, 2021, 373(6558): 1047-1051.

[4] Rives A, Meier J, Sercu T, et al. Biological structure and function emerge from scaling unsupervised learning to 250 million protein sequences. Proceedings of the National Academy of Sciences, 2021, 118(15): e2016239118.

[5] Bileschi M L, Belanger D, Bryant D H, et al. Using deep learning to annotate the protein universe. Nature Biotechnology, 2022, 40(6): 932-937.

[6] Repecka D, Jauniskis V, Karpus L, et al. Expanding functional protein sequence spaces using generative adversarial networks. Nature Machine Intelligence, 2021, 3(4): 324-333.

[7] Shan S, Luo S, Yang Z, et al. Deep learning guided optimization of human antibody against SARS-CoV-2 variants with broad neutralization. Proceedings of the National Academy of Sciences, 2022, 119(11): e2122954119.

[8] Bileschi M L, Belanger D, Bryant D H, et al. Using deep learning to annotate the protein universe. Nature Biotechnology, 2022: 1-6.

[9] Lin Z, Akin H, Rao R, et al. Evolutionary—scale prediction of atomic level protein structure with a language model. bioRxiv, 2022. 07. 20. 500902.

[10] Lu H, Diaz DJ, Czarnecki NJ, et al. Machine learning aided engineering of hydrolases for PET depolymerization. Nature, 2022, 604: 662—7.

[11] Casacio C A, Madsen L S, Terrasson A, et al. Quantum-enhanced nonlinear microscopy. Nature, 2021, 594(7862): 201-206.

[12] Payne A C, Chiang Z D, Reginato P L, et al. In situ genome sequencing resolves DNA sequence and structure in intact biological samples. Science, 2021, 371(6532): eaay3446.

[13] Qiao C, Li D, Guo Y, et al. Evaluation and development of deep neural networks for image super-resolution in optical microscopy. Nature Methods, 2021, 18: 194-202.

[14] Zhang S, Zou S, Yin D, et al. USP14-regulated allostery of the human proteasome by time-resolved

cryo-EM. Nature, 2022, 605: 567-74.

[15] Qiao C, Li D, Liu Y, et al. Rationalized deep learning super-resolution microscopy for sustained live imaging of rapid subcellular processes. Nat Biotechnol, 2022. https://doi.org/10.1038/s41587-022-01471-3.

[16] Jiang T, Zhang X O, Weng Z, et al. Deletion and replacement of long genomic sequences using prime editing. Nature Biotechnology, 2022, 40: 227-234.

[17] Choi J, Chen W, Suiter C C, et al. Precise genomic deletions using paired prime editing. Nature Biotechnology, 2022, 40: 218-226.

[18] Kannan S, Altae-Tran H, Jin X, et al. Compact RNA editors with small Cas13 proteins. Nature Biotechnology, 2022, 40: 194-197.

[19] Özcan A, Krajeski R, Ioannidi E, et al. Programmable RNA targeting with the single-protein CRISPR effector Cas7-11. Nature, 2021, 597: 720-725.

[20] Yi Z, Qu L, Tang H, et al. Engineered circular ADAR-recruiting RNAs increase the efficiency and fidelity of RNA editing in vitro and in vivo. Nature Biotechnology, 2022, 40(6): 946-955.

[21] Cho S I, Lee S, Mok Y G, et al. Targeted A-to-G base editing in human mitochondrial DNA with programmable deaminases. Cell, 2022, 185(10): 1764-1776.

[22] Anzalone A V, Gao X D, Podracky C J, et al. Programmable deletion, replacement, integration and inversion of large DNA sequences with twin prime editing. Nature Biotechnology, 2022, 40(5): 731-740.

[23] Jiang T, Zhang X O, Weng Z, et al. Deletion and replacement of long genomic sequences using prime editing. Nature Biotechnology, 2022, 40(2): 227-234.

[24] Choi J, Chen W, Suiter CC, et al. Precise genomic deletions using paired prime editing. Nature Biotechnology, 2022, 40(2): 218-226.

[25] Yuan Q, Gao X. Multiplex base and prime-editing with drive-and-process CRISPR arrays. Nature Communications, 2022, 13(1): 2771.

[26] Wang L B, Li Z K, Wang L Y, et al. A sustainable mouse karyotype created by programmed chromosome fusion. Science, 2022, 377: 967-75.

[27] Bethlehem R A I, Seidlitz J, White S R, et al. Brain charts for the human lifespan. Nature, 2022, 604: 525-533.

[28] Gao L, Liu S, Gou LF, et al. Single-neuron projectome of mouse prefrontal cortex. Nat Neurosci, 2022, 25: 515-29.

[29] Crame B, Billaudelle S, Kanya S, et al. Surrogate gradients for analog neuromorphic computing. Proceedings of the National Academy of Sciences, 2022, 119(4): e2109194119.

[30] Li J, Liu Y, Yuan L, et al. A tissue-like neurotransmitter sensor for the brain and gut. Nature, 2022, 606: 94-101.

[31] Quijano-Rubio A, Yeh H W, Park J, et al. *De novo* design of modular and tunable protein

biosensors. Nature, 2021, 591(7850): 482-487.

[32] Ben-Sasson A J, Watson J L, Sheffler W, et al. Design of biologically active binary protein 2D materials. Nature, 2021, 589(7842): 468-473.

[33] Cai T, Sun H B, Qiao J, et al. Cell-free chemoenzymatic starch synthesis from carbon dioxide. Science, 2021, 373: 1523-1527.

[34] Zhou Y, Xu X, Wei Y, et al. A widespread pathway for substitution of adenine by diaminopurine in phage genomes. Science, 2021, 372(6541): 512-516.

[35] Kriegman S, Blackiston D, Levin M, et al. Kinematic self-replication in reconfigurable organisms. Proceedings of the National Academy of Sciences, 2021, 118(49): e2112672118.

[36] Zheng T, Zhang M, Wu L, et al. Upcycling CO_2 into energy-rich long-chain compounds via electrochemical and metabolic engineering. Nat Catal, 2022, 5: 388—96.

[37] Ma Y, Liu N, Greisen P, et al. Removal of lycopene substrate inhibition enables high carotenoid productivity in Yarrowia lipolytica. Nat Commun, 2022, 13: 572.

[38] Mazid M A, Ward C, Luo Z, et al. Rolling back human pluripotent stem cells to an eight-cell embryo-like stage. Nature, 2022, 605: 315-324.

[39] Guan J, Wang G, Wang J, et al. Chemical reprogramming of human somatic cells to pluripotent stem cells. Nature, 2022, 605: 325-331.

[40] Dolgin E. Diabetes cell therapies take evasive action. Nature Biotechnology, 2022, 40: 291-295.

[41] Chen Y, Luttmann F F, Schoger E, et al. Reversible reprogramming of cardiomyocytes to a fetal state drives heart regeneration in mice. Science, 2021, 373(6562): 1537-1540.

[42] Sliva A C, Matthys O B, Joy D A, et al. Co-emergence of cardiac and gut tissues promotes cardiomyocyte maturation within human iPSC-derived organoids. Cell Stem Cell, 2021, 28(12): 2137-2152.

[43] Combes A J, Samad B, Tsui J, et al. Discovering dominant tumor immune archetypes in a pan-cancer census. Cell, 2022, 185(1): 184-203.

[44] Esfahani M S, Hamilton E G, Mehrmohamadi M, et al. Inferring gene expression from cell-free DNA fragmentation profiles. Nature Biotechnology, 2022, 40(4): 585-597.

[45] Fedyuk V, Erez N, Furth N et al. Multiplexed, single-molecule, epigenetic analysis of plasma-isolated nucleosomes for cancer diagnostics. Nature Biotechnology, 41, 212-221(2023).

[46] Saeidi H, Opfermann J D, Kam M, et al. Autonomous robotic laparoscopic surgery for intestinal anastomosis. Science Robotics, 2022, 7(62): eabj2908.

[47] Kim Y, Genevriere E, Harker P, et al. Telerobotic neurovascular interventions with magnetic manipulation. Science Robotics, 2022, 7(65): eabg9907.

[48] The White House. Critical and Emerging Technologies List Update. https://www.whitehouse.gov/wp-content/uploads/2022/02/02-2022-Critical-and-Emerging-Technologies-List-Update.pdf [2022-02-08].

[49] Bundesministerium für Bildung und Forschung. Rat für technologische Souveränität nimmt Arbeit auf. https://www.bmbf.de/bmbf/shareddocs/pressemitteilungen/de/2021/08/020921-Rat-technologische-Souveraenitaet.html[2021-09-02].

[50] European Commission. Updating the 2020 New Industrial Strategy: Building a stronger Single Market for Europe's recovery. https://single-market-economy.ec.europa.eu/news/updating-2020-industrial-strategy-towards-stronger-single-market-europes-recovery-2021-05-05_en[2021-05-05].

[51] European Commission. Strategic dependencies and capacities. https://eur-lex.europa.eu/legal-content/EN/TXT/?uri=SWD:2021:352:FIN#:~:text=Strategic%20dependencies%20affect%20the%20EU%E2%80%99s%20core%20interests.%20They,transitions%20at%20the%20core%20of%20the%20EU%E2%80%99s%20priorities[2021-05-05].

[52] Дмитрий Чернышенко: На импортозамещение научного оборудования в этом году будет направлено 8 млрд рублей. http://government.ru/news/45098/[2022-04-09].

[53] Правительство Российской Федерации. В России появится собственная база образцов жизненно важных лекарств. http://government.ru/news/45158/[2022-04-15].

[54] The White House. National Biodefense Strategy and Implementation Plan for Countering Biological Threats, Enhancing Pandemic Preparedness, and Achieving Global Health Security. 2022. https://www.whitehouse.gov/wp-content/uploads/2022/10/National-Biodefense-Strategy-and-Implementation-Plan-Final.pdf.

[55] European Commission. Horizon Europe Work Programme 2021-2022:4. Health. https://ec.europa.eu/info/funding-tenders/opportunities/docs/2021-2027/horizon/wp-call/2021-2022/wp-4-health_horizon-2021-2022_en.pdf[2022-05-10].

[56] European Commission. Europe's Beating Cancer Plan. https://ec.europa.eu/health/sites/health/files/non_communicable_diseases/docs/eu_cancer-plan_en.pdf[2021-12-01].

[57] HM Government. Life Sciences Vision. https://www.gov.uk/government/publications/life-sciences-vision[2021-07-06].

[58] 内閣府．統合イノベーション戦略 2021. https://www8.cao.go.jp/cstp/tougosenryaku/2021.html[2021-06-18].

[59] Government of Canada. Canada's Biomanufacturing and Life Sciences Strategy. https://ised-isde.canada.ca/site/biomanufacturing/en/canadas-biomanufacturing-and-life-sciences-strategy[2021-07-28].

[60] Austrilian Government, Department of Health and Aged Care. Australian Medical Research and Innovation Strategy 2021-2026. https://www.health.gov.au/resources/publications/australian-medical-research-and-innovation-strategy-2021-2026?language=en[2021-11-08].

[61] Department of Biotechnology. Nation Biotechnology Development Strategy(2021-2025). https://dbtindia.gov.in/sites/default/files/NATIONAL%20BIOTECHNOLOGY%20DEVELOPMENT%20STRATEGY_01.04.pdf[2021-01-26].

Life Health and Medicine

Xu Li, Wang Yue, Li Zhenqi, Shi Huilin,
Yang Ruonan, Li Wei, Jin Chenqi, Xu Ping

Life health is one of the focal points of global science and technology competition. In 2021～2022, the integration of big data and artificial intelligence, the deep convergence of disciplines, the rapid progress of technology, and the organizational model of big science and large teams are promoting the transformative development of life and health technology, further broadening the scale and depth of research, deepening the whole life cycle and whole system research, and obvious trends of digitalization, systematization and engineering, promoting the wide integration of life-cycle health management into life. At the same time, the human body is in a complex environment with a high degree of heterogeneity, and the research from individual to group, individual and environment is carried out in an all-round way with the promotion of new technologies. The proposal of new concepts and new paths such as one-health, precision medicine and population medicine has promoted people's comprehensive understanding of health and disease, and developed more effective early detection, early intervention and individualized precision treatment programs. It helps to achieve the goal of life-cycle health management.

4.3 生物科技领域发展观察

丁陈君 陈 方 郑 颖 吴晓燕 宋 琪 邓诗碧

（中国科学院成都文献情报中心）

生物技术作为重要科技创新领域兼具引领性和颠覆性。当前，合成生物学、基因编辑等前沿技术保持快速发展。合成基因组学研究上升到新的高度，面向不同需求的多种新型基因编辑器不断问世。蛋白质结构预测工具迭代更新，预测精度和速度进一步提升。组学技术、生物显微成像技术、单细胞测序和时空组学技术、人工智能等前沿技术的交叉融合使更系统地认识生命成为可能，并推动分子、细胞、组织、器官等重点单元的基础研究向纵深发展，为揭示人类发育和疾病发生机制提供更多新线索。同时，关键核心技术的不断革新也正在持续赋能食品、农业、医药、能源、化工、环保等产业绿色低碳转型。

一、国际重大研究进展与趋势

1. 生物资源与生物多样性保护领域科学问题研究取得多项成果

生物资源与生物多样性保护领域得益于第三代测序技术、合成生物学技术、基因编辑技术等新兴技术的进步，在解析气候变化和人类活动对生态系统、生物多样性造成的影响等重大科学问题，以及生物种质资源创新与利用方面取得了重要成果。美国加州大学研究人员揭示了气候变化对小型哺乳动物群落的物种丰富度影响不大，但鸟类群落受气候变化影响物种丰富度下降明显[1]。法国图卢兹第三大学等机构合作揭示了人类活动对全球淡水鱼类生物多样性的影响严重，许多濒危鱼类面临灭绝的风险，且已危及淡水生态系统的服务功能[2]。德国康斯坦茨大学等机构合作通过确定巨型肺鱼庞大的基因组阐明脊椎动物由水生向陆生生活的演变机制[3]。

美国冷泉港实验室研究人员通过高通量CRISPR技术揭示了古代植物同源盒基因的保守多效性功能，可用于指导作物改良，减少由此引发的不良影响[4]。德国马克斯·普朗克发育生物学研究所研究人员基于细菌色氨酸阻遏物开发了一种生物传感器可用于高时空分辨率的植物生长素梯度体内成像[5]。德国马克斯·普朗克研究所发现二萜类化合物的可控羟基化作用可实现植物化学防御而无自毒性[6]。美国纽约大学医学院与俄罗斯生物初创公司 Gero LLC 等机构相关研究人员合作，发现细菌硫化氢生物发生抑制剂可靶向改善抗生素耐药性和耐受性现象[7]。

2. 生物技术前沿研究领域颠覆性突破层出不穷

合成基因组学研究和合成生物技术应用创新上升到新的高度。英国剑桥医学研究委员会分子生物学实验室研究者首次在全人工合成的大肠杆菌体内删减密码子，使其能够抵御病毒，并利用非天然氨基酸合成聚合物[8]。德国美因茨大学研究人员合成的双膜状细胞器可实现高效正交化真核翻译，并与宿主细胞中其他翻译过程存在空间分离[9]。日本立教大学研究人员通过分裂大肠杆菌 3 Mb 染色体，成功构建了由 3 个 1 Mb 染色体组成的大肠杆菌，且仍能稳定增殖[10]，实现了大规模基因组操纵。作为合成生物学的基础，DNA 合成技术也得到长足发展。酶促 DNA 合成技术加快商业化进程，推动合成生物学向可预测、可定量的工程化方向发展。美国国家标准与技术研究院和克雷格-文特尔研究所等机构合作，通过仅放回 7 个基因，成功对其在 5 年前合成的“最小细胞”进行校正，获得了能够像天然细胞一样正常生长和分裂的 JCVI-syn3A 细胞[11]。此外，基于原核细胞的类真核合成细胞[12]、超过 100 个菌株的合成肠道微生物群落[13]，含重构遗传密码且不与自然生物体发生基因交换的大肠杆菌基因组[14]等人工合成系统也陆续被开发出来。

多种新型基因编辑器不断问世，在前期研究基础上扬长避短大大提升了技术的应用价值。基因编辑技术领域始终保持研发热度，研究人员通过开发新的编辑器来实现多种应用，包括加深理解遗传疾病和实现新的疗法、提升育种水平、加快微生物设计等。美国麻省理工学院麦戈文脑科学研究所研究人员发现了一种新型 RNA 靶向编辑工具，将其命名为 Cas7-11，并通过哺乳动物细胞实验证明其更具精准性且毒性更低[13]。美国博德研究所张锋团队从转座子系统中共发现了 3 种此前未表征的可被非编码 RNA 引导的核酸内切酶，这表明转座子编码的 RNA 引导的核酸酶可能广泛存在[16]。该团队通过结构和机制解析，从功能互作与进化角度加深了对基因编辑系统的理解[17]，为进一步开发和工程化改造奠定了基础。该团队还确定了多蛋白效应复合物类型编辑系统中 Cas7-11 的蛋白底物、结构和作用机制，揭示 CRISPR 系统可同时具有核酸酶和蛋白酶功能[18]。美国加州大学旧金山分校的研究者以 CRISPR 系统为基础，开发出一套持久可遗传且可逆的表观遗传编辑器 CRISPRoff[19]。美国博德研究所刘如谦团队开发了噬菌体辅助的重编程特异性蛋白酶进化系统，为重编程蛋白酶建立了通用平台，可选择性切割具有治疗意义的新靶标[20]；该团队还开发了新的先导编辑系统 PE4 和 PE5，可瞬时表达工程化的错配修复抑制蛋白，暂时抑制 DNA 的错配修复，从而提高细胞编辑效率[21]。

生物成像领域发展迅速，开发了具有划时代意义的新型显微成像技术。显微成像技术高速发展，多次获得诺贝尔奖，已在生命科学等多个领域中得到广泛应用。美国纽约威尔康奈尔医学院开发出定位原子力显微术，借鉴超分辨率荧光显微镜的“定位”技术将原子力显微镜分辨率提高到埃米级[22]。澳大利亚昆士兰大学开发的量子显微镜引入量子科技，利用量子相关性增强非线性显微镜成像[23]。美国国家生物医学成像与生物工程研究所成功开发多视角超分辨率共聚焦显微镜，从而实现活体样本精细结构更高分辨率的三维成像[24]。清华大学戴琼海院士团队开发了扫描光场显微镜，实现了对活体哺乳动物的亚细胞结构的长时间、三维、高速高分辨率的显微观测[25]。中国科学院徐涛院士团队开发的轴向单分子定位成像新技术与新型干涉定位显微镜，使得细胞内单分子定位成像轴向分辨率提升到了纳米尺度[26]。

随着高通量测序技术的不断进步，人类可以对生命体内基因组序列进行非常精准的解析，从而揭示生命密码。但是这些不同类型的细胞通过什么方式构成一个完整的、有功能的器官仍然未知，单细胞技术和时空组学使人类从全新的维度认识人体组织和器官。目前已经有很多研究试图通过空间转录组技术探索组织的结构，在发育生物学、神经生物学和肿瘤学中开展了一系列的研究[27]。将空间转录组数据集与基因组原位高通量成像、组织内组蛋白标记的空间分布相结合将极大地提高我们研究组织复杂性的能力[28]。该项技术的进一步优化和广泛应用，将推动聚焦于组织器官再生、疾病进展和发育等领域的生命科学得到极大的发展。

3. 生物-信息技术深度融合发展加速

人工智能蛋白质结构预测极大地填补了传统观测方法无法实现的结构解析。目前，人工智能已能完成多类型蛋白质的结构预测，且达到近原子水平，未来还将探索如何更好地将这种策略用于特定应用，尽可能接近真实情况地快速预测传统观测方法所不能解析的结构。世界知名人工智能团队“深度思维”（DeepMind）宣布，利用“阿尔法折叠”（AlphaFold）预测了几乎所有人类表达的蛋白质结构，以及其他 20 种生物几乎完整的蛋白质组[29]。美国华盛顿大学研究人员开发出一种准确度更高的预测蛋白质结构新工具“RoseTTAFold”[30]。除了在蛋白质结构预测领域取得突破，人工智能在 RNA 三维结构预测方面也取得重要进展。斯坦福大学生物化学系副教授 Rhiju Das 和计算机科学系 Ron O. Dror 团队，利用机器学习技术开发出 RNA 结构预测模型 ARES（atomic rotationally equivariant scorer），相关研究结果作为封面文章在《科学》上发表[31]。随着计算机算力的提升、生物学相关数据的积累以及 AI 算法的进步，AI 和生物学的融合不断加深，其中以蛋白质合成领域的发展最为迅猛。2022 年，国内外生成式人工智能（artificial intelligence generated content，AIGC）蛋白质设计模型争相爆发。美国 Generate Biomedicines 公司和美国华盛顿大学 David Baker 团队分别发布了基于扩散模型的蛋白质生成平台“Chroma”[32] 和“RoseTTAFold Diffusion”[33]，实现了按照研究人员需求设计出蛋白质结构的突破。同年，中国科学技术大学的刘海燕教授团队开发了一种基于神经网络形式能量项的统计模型（side chain-unknown backbone arrangement，SCUBA）[34]，实现了蛋白质的从头设计，并且设计出不同于已知蛋白质结构的全新结构蛋白质。

计算生物学的研发突破推动智能生物制造快速发展。美国卡内基梅隆大学研究人员开发了一种机器学习算法，通过匹配微生物代谢组和基因组数据，挖掘潜在的天然药物[35]。斯坦福大学研究者开发了一个系统筛选生物分子合成衍生途径的计算工作流[36]。日本可持续资源科学中心研究者通过计算机辅助设计，在大肠杆菌中构建了一条从葡萄糖合成 1，3-丁二烯的人工代谢途径[37]。美国斯坦福大学和加州大学的研究者开发了一种名为高通量微流控酶动力学新技术，这将极大提高对酶结构功能的认识[38]。

脑信号解码和神经传感等多项脑-机接口尖端技术研发取得突破，正在为脑科学研究、医疗康复、教育娱乐和航空军事领域带来重大变革。《自然》以封面文章的形式报道了由斯坦福大学等机构合作的研究成果——首次从脑电信号中解码手写字母的动作，使瘫痪人士意念中的写字动作可以实时转换成屏幕上的文字[39]。加州大学旧金山分校研究团队通过解码从运动皮层发送到声道的大脑信号，可以让具有构音障碍的瘫痪患者恢复交流能力[40]。瑞士洛桑联邦理工学院研究团队开发出一种仅借助人类大脑发出的电信号就能控制机器人的强化学习系统[41]。美国布朗大学等多所高校及公司合作开发了一种能自主执行神经传感和电刺激的无线联网供电的电子微芯片“neurograins”[42]。Blackrock Neurotech 公司宣布其开创性脑-机接口设备 MoveAgain 获美国 FDA“突破性设备”认定。瑞士洛桑联邦理工学院（EPFL）研究团队开发了一项新技术，通过在瘫痪者脊椎处植入电极，利用小电流刺激脊髓神经元，使患者能够重新行走、游泳、骑行。这是全球首例脊髓完全切断的人依靠科技重新自由行走[43]。

4. 高效生物转化技术为绿色低碳循环经济发展提供重要科技创新路径

设计和创建具有高效生物固碳能力的酶、生化途径、工程生物或微生物组，成为合成生物固碳领域的国际研究热点，低碳生物工程技术发展为洁净能源生产和单碳资源高效利用带来重要机遇。美国国家可再生能源实验室设计并组装了一套全新的藻类生物燃料光电合成途径，能量转化效率达到 9%[44]。德国明斯特大学和慕尼黑工业大学研究团队发现 *Hippea maritima* 细菌中存在逆向运行的三羧酸循环[45]，为 CO_2 利用研究提供了新的方向。法国全球生物能源（Global Bioenergies）公司与英国曼彻斯特大学合作开发了一种改良的甲羟戊酸途径，支持微生物体内合成异丁烯[46]。德国蒂宾根大学研究人员通过改变蓝细菌的代谢通路，生产出具有良好生物降解特性的生物塑料替代品——聚羟基丁酸酯（polyhydroxybutyrate，PHB）[47]。中国科学院天津工业生物技术研究所团队在国际上首次在实验室实现 CO_2 到淀粉的从头合成，为以 CO_2 为原料合成复杂分子提供了新技术路线[48]。英国剑桥大学研究人员开发出一种使用阳光、CO_2 和水即可有效产生乙酸盐和氧气的生物-非生物混合系统[49]。美国劳伦斯伯克利国家实验室利用工程化天蓝色链霉菌成功生产出净热值高达 50 MJ/L 的多环丙烷脂肪酸甲酯生物燃料[50]。

5. 新兴技术为公共卫生领域应对挑战和需求提供有力工具

基因编辑、合成生物学、人工智能等新兴技术在新冠病毒研究及新冠疫苗研发中显示出很大的应用潜力。例如，mRNA 疫苗技术可以实现疫苗的快速开发和部署；机器学习和人工智能技术可以用于预测病毒进化；基于 CRISPR 技术的检测试剂可以快速检测病毒；鼻喷疫苗可额外诱发黏膜免疫的建立，具有更广泛的免疫应答，利用合成生物学技术可以简化当前基于 DNA 和 mRNA 疫苗技术的疫苗开发和生产过程。

在病原体检测方面，新开发的基于 CRISPR 的生物传感检测方法、基于生物传感器的检测方法为病原体检测提供了更多选择，例如检测寨卡病毒（Zika virus）、人乳头瘤病毒（human papilloma virus，HPV）等。在疫苗研发方面，研究人员通过开发病毒样颗粒、肽基疫苗和核酸疫苗等现代疫苗技术来取代灭活疫苗或减毒活疫苗。其中，核酸疫苗因其易于设计等优势将在流感病毒、埃博拉病毒、寨卡病毒、新冠病毒等引发的传染性疾病预防方面颇具应用前景。

二、国际重大战略规划和政策措施

1. 各国高度重视生物技术，加强战略布局和整体推进

越来越多的国家和地区将生物技术视为战略高技术，将生物经济发展纳入政策主流。美国国家情报委员会《全球趋势 2040：一个竞争更加激烈的世界》报告指出，到 2040 年，技术将是国家在未来取得发展优势的关键途径，驱动变革的核心技术主要包括人工智能、智能材料制造、生物技术、太空技术、超级互联[51]。日本经济产业省生物小组委员会《生物技术驱动的第五次工业革命》报告指出，生物技术对基础研究、健康医药、食品农业、环境能源、化学产业等产生深远影响，工程生物细胞和人工智能等信息

技术结合，将构成第五次工业革命的核心驱动力[52]。印度发布《2021—2025年国家生物技术发展战略：知识驱动生物经济》，旨在推动"知识和创新驱动的生物经济"发展，目标是到2025年印度跻身全球生物技术前5强，成为公认的世界生物制造中心[53]。

2. 加快生物技术研发与应用，推动传统工业绿色转型

近年来，生物技术创新呈爆发式增长态势，成为新科技产业革命的核心驱动力。美国提出大力发展新能源产业，从研发预算、重要法案、项目资助等方面积极推进生物能源的发展。拜登政府2022年财政预算提出2022～2026年对生物燃料投入10亿美元的支持，并加强生物精炼、可再生化学和生物基产品制造及可持续航空燃料等的支持力度[54]；能源部投入近1亿美元加强以低成本生产低碳生物燃料所需的科学和工程知识体系研发，促进交通部门去碳化[55,56]。欧盟为了实现"至2030年温室气体减排55%"的新目标，在绿色协议框架的指导下开展了多项资助项目[57]，支持绿色科技创新。欧盟生物基产业联盟（Bio-Based Industries Joint Undertaking，BBI JU）向循环生物项目投资1.045亿欧元，加速从有效处理生物质原料供应到将生物基产品推向市场的新可持续价值链的建设[58]；并升级循环生物基欧洲联合企业伙伴关系计划（Circular Bio-based Europe Joint Undertaking，CBE JU），加大对欧盟实现气候目标的贡献[59]。英国于2021年3月发布《工业生物技术报告：标准和法规的战略路线图》，明确了在利用工业生物技术减少碳排放方面具有中短期潜力的农业、生物燃料、精细和特种化学品等领域，并提供了释放工业生物技术潜力的路线图[60]。同期，英国政府启动绿色燃料、绿色天空竞赛，作为绿色工业革命"十点计划"（Ten Point Plan）的一部分，将家庭垃圾、废木材和多余电力转化为可持续航空燃料，帮助提高航空业的可持续性[61]。英国商业、能源和工业战略部宣布为生物质原料创新方案提供资助，以提高英国的生物质生产能力，为构建多样化绿色能源组合提供原料。日本经济产业省启动"开发生物基产品加速碳循环"计划，投入20亿元/年预算以促使依赖野生植物为原料的药品、化妆品和保健品的活性成分的发酵生产工业化，发展聚酰胺和聚丙烯的生物制造技术等，旨在提升日本生物经济的竞争力。

3. 后疫情时代全球生物安全重要性愈加凸显

在全球抗击新冠疫情、改善卫生治理局面的同时进一步发展国民经济、深化国际合作已成为各国面临的重要考题。2021年12月1日，世界卫生大会同意制定一项全球协议或公约，以预防并应对大流行病；同时设立一个政府间谈判机构，分别于2022年2月24日、2022年7月18～21日举行第一次和第二次会议，并于2022年12月5～7日讨论形成了"世卫组织预防、防范和应对大流行公约、协定或其他国际文书"（WHO CA

十）预稿。美国拜登政府对内加大生物安全能力建设，不断完善生物安全风险防控和治理体系，对外加强全球布局，积极争夺生物安全领域的全球治理权和话语权。2021年度美国在生物安全领域实施了一系列举措，包括：重新设立国家安全委员会全球卫生安全和生物防御部门；发布《国家安全战略临时指南》；发布新的650亿美元的大流行病防备计划提案，加快疫苗开发、测试和生产，改变美国应对大流行病的方式。欧盟于2021年9月16日成立了新的卫生应急准备和响应局，计划未来6年投入60亿美元，重点发展新兴生物医学技术，协调全欧洲的疫情监控和应对行动[62]。

三、启示与建议

生物技术发展日新月异，近年创新呈爆发式增长态势，成为新科技产业革命的核心驱动力。基于全球发展态势与竞争格局，就推动我国生物科技创新发展提出如下建议。

1. 加强生物科学基础研究，发挥生物技术创新引领作用

加强未来生物科学前沿领域的精准研判与系统谋划，加强基础研究相关的基础设施平台建设。牢牢把握以生物技术、信息技术等为核心的新技术革命机遇，着力突破传统科研理念和范式，以重大科学问题为牵引，以原创性、引领性基础研究为先导，及时切换新赛道，实现弯道超车，力求在新一轮科技竞争中掌握主动权。

2. 结合国家“十四五”规划的重点任务，加大加快布局底层关键技术研发

聚焦基因操控、合成生物、表达调控、代谢调控等底层基础性技术、共性关键技术、过程工程技术，以及生物安全前沿技术，重点研发突破对外依赖度高的关键技术“堵点”，给予长周期规划、持续稳定支持。顺应数据驱动范式变革的趋势，加强自主开源的生物数据与计算平台建设，推动生物数据、医学数据和人类遗传资源库的有效整合，研发生物大数据的存储、共享技术，强化分析处理算法、软件工具和环境。

3. 强化各学科交叉融合，大力推进生物科技研究范式的转变

生命系统的复杂性，决定了生物技术研发体系的繁复，多组学技术的应用，为生物技术研发积累了海量的数据，生物技术研发从实验科学范式向数据驱动范式转变成为必然趋势。应基于学科交叉融合，尤其面向生物医学和信息计算领域的快速发展，构建新理科和新医科为基础的新型复合型人才培养体系，组建建制化人才队伍；围绕新理科、新工科和新医科建设，探索生物科学与材料、人工智能、药物等交叉融合，

开展技术攻关，以更好地实现更加高效、精准的科学研究，加速成果产业化。

4. 面向人民群众更高层次需求，推动生物技术为核心的生物经济发展

未来应以生物医药、生物农业、生物制造、生物安全等重点领域为抓手，加快出台支撑国家战略的科技路线图规划，推进创新单元设置和研发布局，加快生物技术赋能健康、农业、能源、环保等多个产业；面向人民生命健康，加快推动生物医药领域新技术、新产品的研发，有效应对公共卫生领域的多种挑战和需求；突破技术瓶颈，实现跨领域综合交叉，促进技术产业协同创新，推动生物技术的产业化应用，发挥生物科技创新在实现“双碳”目标中的引领和支撑作用，支撑绿色、循环、可持续生物经济发展。

致谢：中国科学院成都生物研究所谭周亮研究员、福建医科大学沈晓沛教授在本文撰写过程中提出了宝贵的意见和建议，在此表示感谢。

参考文献

[1] Riddell E A, Iknayan K, Hargrove L, et al. Exposure to climate change drives stability or collapse of desert mammal and bird communities. Science, 2012, 371(6529): 633-636.

[2] Su G H, Logez M, Xu J, et al. Human impacts on global freshwater fish biodiversity. Science, 2021, 371(6531): 835-838.

[3] Meyer A, Schloissnig S, Franchini P, et al. Giant lungfish genome elucidates the conquest of land by vertebrates. Nature, 2021, 590: 284-289.

[4] Hendelman A, Zebell S, Rodriguez-Leal D, et al. Conserved pleiotropy of an ancient plant homeobox gene uncovered by cis-regulatory dissection. Cell, 2021, 184(7): 1724-1739.

[5] Herud-Sikimic O, Stiel A, Kolb M, et al. A biosensor for the direct visualization of auxin. Nature, 2021, 592: 768-772.

[6] Li J C, Halitschke P, Li D P, et al. Controlled hydroxylations of diterpenoids allow for plant chemical defense without autotoxicity. Science, 2021, 371(6526): 255-260.

[7] Shatalin K, Nuthanakanti A, Kaushik A, et al. Inhibitors of bacterial H_2S biogenesis targeting antibiotic resistance and tolerance. Science, 2021, 372(6547): 1169-1175.

[8] Robertson W E, Funke L F H, de la Torre D, et al. Sense codon reassignment enables viral resistance and encoded polymer synthesis. Science, 2021, 372(6546): 1057-1062.

[9] Reinkemeier C D, Lemke E A. Dual film-like organelles enable spatial separation of orthogonal eukaryotic translation. Cell, 2021, 184(19): 4886-4903.

[10] Yoneji T, Fujita H, Mukai T, et al. Grand scale genome manipulation via chromosome swapping in Escherichia coli programmed by three one megabase chromosomes. Nucleic Acids Research, 2021,

49(15):8407-8418.

[11] Pelletier J F, Sun L J, Wise K S, et al. Genetic requirements for cell division in a genomically minimal cell. Cell, 2021, 184(9): 2430-2440.

[12] Xu C, Martin N, Li M, Mann S. Living material assembly of bacteriogenic protocells. Nature. 2022; 609(7929): 1029-1037.

[13] Cheng A G, Ho P Y, Aranda-Díaz A, et al. Design, construction, and in vivo augmentation of a complex gut microbiome. Cell. 2022; 185(19): 3617-3636. e19.

[14] Zürcher J F, Robertson W E, Kappes T, et al. Refactored genetic codes enable bidirectional genetic isolation. Science. 2022, 378(6619): 516-523.

[15] Özcan A, Krajeski R, Ioannidi E, et al. Programmable RNA targeting with the single-protein CRISPR effector Cas7-11. Nature, 2021, 597: 720-725.

[16] Tran H A, Kannan S, Demircioglu F E, et al. The widespread IS200/IS605 transposon family encodes diverse programmable RNA-guided endonucleases. Science, 2021, 374(6563): 57-65.

[17] Hirano S, Kappel K, Altae-Tran H, et al. Structure of the OMEGA nickase IsrB in complex with ωRNA and target DNA. Nature, 2022, 610: 575-581.

[18] Strecker J, Demircioglu F E, Li D, et al. RNA-activated protein cleavage with a CRISPR-associated endopeptidase. Science, 2022, 378(6622): 874-881.

[19] Nuñez J K, Chen J, Pommier G C, et al. Genome-wide programmable transcriptional memory by CRISPR-based epigenome editing. Cell, 2021, 184(9): 2503-2519.

[20] Blum T R, Liu H, Packer M S, et al. Phage-assisted evolution of botulinum neurotoxin proteases with reprogrammed specificity. Science, 2021, 371(6531): 803-810.

[21] Chen P J, Hussmann J A, Yan J, et al. Enhanced prime editing systems by manipulating cellular determinants of editing outcomes. Cell, 2021, 184(22): 5635-5652.

[22] Heath G R, Kots E, Robertson J L, et al. Localization atomic force microscopy. Nature, 2021, 594: 385-390.

[23] Casacio C A, Madsen L S, Terrasson A, et al. Quantum-enhanced nonlinear microscopy. Nature, 2021, 594: 201-206.

[25] Wu Y C, Han X F, Su Y J, et al. Multiview confocal super-resolution microscopy. Nature, 2021, 600: 279-284.

[25] Wu J M, Lu Z, Jiang D, et al. Iterative tomography with digital adaptive optics permits hour-long intravital observation of 3D subcellular dynamics at millisecond scale. Cell, 2021, 184(12): 3318-3332.

[26] Gu L S, Li Y Y, Zhang S W, et al. Molecular-scale axial localization by repetitive optical selective exposure. Nature Methods, 2021, 18: 369-373.

[27] Anderson A C, Yanai I, Yates L R, et al. Spatial transcriptomics. Cancer Cell, 40(9): 895-900.

[28] Rao A, Barkley D, França G S, et al. Exploring tissue architecture using spatial transcriptomics. Nature,

2021,596:211-220.

[29] Tunyasuvunakool K, Adler J, Wu Z, et al. Highly accurate protein structure prediction for the human proteome. Nature, 2021, 596: 590-596.

[30] Baek M, DiMaio F, Anishchenko I, et al. Accurate prediction of protein structures and interactions using a three-track neural network. Science, 2021, 373(6557): 871-876.

[31] Townshend R, Eismann S, Watkins A M, et al. Geometric deep learning of RNA structure. Science, 2021, 373(6558): 1047-1051.

[32] Ingraham J, Baranov M, Costello Z, et al. Illuminating protein space with a programmable generative model. BioRxiv , 2022-12.

[33] Watson JL, Juergens D, Bennett NR. , et al. Broadly applicable and accurate protein design by integrating structure prediction networks and diffusion generative models. BioRxiv, 2022-12.

[34] Rowald A, Komi S, Demesmaeker R, et al. Activity-dependent spinal cord neuromodulation rapidly restores trunk and leg motor functions after complete paralysis. Nature Medicine, 2022, 28: 260-271. [35] Huang, Bin, et al. A backbone-centred energy function of neural networks for protein design. Nature, 2022, 602(7897): 523-528.

[36] Behsaz B, Bode E, Gurevich A, et al. Integrating genomics and metabolomics for scalable non-ribosomal peptide discovery. Nature Communications, 2021, 12: 3225.

[37] Hafner J, Payne J, Mohammadi Peyhani H, et al. A computational workflow for the expansion of heterologous biosynthetic pathways to natural product derivatives. Nature Communications, 2021, 12: 1760.

[38] Mori Y, Noda S, Shirai T, et al. Direct 1,3-butadiene biosynthesis in Escherichia coli via a tailored ferulic acid decarboxylase mutant. Nature Communications, 2021, 12: 2195.

[39] Markin C J, Mokhtari D A, Sunden F, et al. Revealing enzyme functional architecture via high-throughput microfluidic enzyme kinetics. Science, 2021, 373(6553): eabf8761.

[40] Willett F R, Avansino D T, Hochberg L R, et al. High-performance brain-to-text communication via handwriting. Nature, 2021, 593: 249-254.

[41] Moses D A, Metzger S L, Liu J R, et al. Neuroprosthesis for decoding speech in a paralyzed person with Anarthria. The New England Journal of Medicine, 2021, 385: 217-227.

[42] Batzianoulis I, Iwane F, Wei S P, et al. Customizing skills for assistive robotic manipulators, an inverse reinforcement learning approach with error-related potentials. Communications Biology, 2021, 4: 1406.

[43] Lee J H, Leung V, Lee A H, et al. Neural recording and stimulation using wireless networks of microimplants. Nature Electronics, 2021, 4: 604-614.

[44] Li Z D, Wu C, Gao X, et al. Exogenous electricity flowing through cyanobacterial photosystem I drives CO_2 valorization with high energy efficiency. Energy & Environmental Science, 2021, 14: 5480-5490.

[45] Steffens L, Pettinato E, Steiner T M, et al. High CO_2 levels drive the TCA cycle backwards towards autotrophy. Nature, 2021, 592: 784-788.

[46] Saaret A, Villiers B, Stricher F, et al. Directed evolution of prenylated FMN-dependent Fdc supports efficient in vivo isobutene production. Nature Communications, 2021, 12: 5300.

[47] Orthwein T, Scholl J, Spat P, et al. The novel PII-interactor PirC identifies phosphoglycerate mutase as key control point of carbon storage metabolism in cyanobacteria. PNAS, 2021, 118(6) e2019988118.

[48] Cai T, Sun H B, Qiao J, et al. Cell-free chemoenzymatic starch synthesis from carbon dioxide. Science, 2021, 373(6562): 1523-1527.

[49] Wang Q, Kalathil S, Pornrungroj C, et al. Bacteria-photocatalyst sheet for sustainable carbon dioxide utilization. Nature Catalysis, 2022, 5(7): 633-641.

[50] Cruz-Morales P, Yin K, Landera A, et al. Biosynthesis of polycyclopropanated high energy biofuels. Joule, 2022, 6(7): 1590-1605.

[51] National Intelligence Council. Global Tends 2040: A More Contested World. https://www.dni.gov/files/ODNI/documents/assessments/GlobalTrends_2040.pdf[2021-03-31].

[52] バイオ小委員会.バイオテクノロジーが拓く「第五次産業革命」. https://www.meti.go.jp/press/2020/02/20210202001/20210202001.html[2021-02-02].

[53] The Department of Biotechnology. National Biotechnology Development Strategy 2021-2025. https://dbtindia.gov.in/sites/default/files/NATIONAL%20BIOTECHNOLOGY%20DEVELOPMENT%20STRATEGY_01.04.pdf[2021-04-01].

[54] Biomass. Biden's 2022 budget includes funds for SAF, biofuels, bioenergy. http://www.biomassmagazine.com/articles/18046/bidenundefineds-2022-budget-includes-funds-for-saf-biofuels-bioenergy[2021-05-28].

[55] DOE. DOE Announces $61.4 Million for Biofuels Research to Reduce Transportation Emissions. https://www.energy.gov/articles/doe-announces-614-million-biofuels-research-reduce-transportation-emissions[2021-04-08].

[56] DOE. DOE Invests $35 Million to Dramatically Reduce Carbon Footprint of Biofuel Production. https://www.energy.gov/articles/doe-invests-35-million-dramatically-reduce-carbon-footprint-biofuel-production[2021-05-14].

[57] SINTEF. Converting CO_2 emissions into products-European Green Deal project PyroCO_2 kicks off. https://www.sintef.no/en/latest-news/2021/converting-co2-emissions-into-products-european-green-deal-project-kicks-off/[2021-10-01].

[58] Bioeconomy BW. BBI JU to invest 104.5 million into circular Bio-Based projects. https://www.biooekonomie-bw.de/en/articles/pm/bbi-ju-invest-eur1045-million-circular-bio-based-projects[2021-05-08].

[59] Bioeconomy BW. Commission gives the green light to the successor of BBI JU. https://www.biooekonomie-bw.de/en/articles/pm/commission-gives-green-light-successor-bbi-ju[2021-02-23].

[60] BSI. Industrial biotechnology roadmap. https://www.bsigroup.com/en-GB/industries-and-sectors/manufacturing-and-processing/industrial-biotechnology-roadmap/[2021-03-30].

[61] Renewable Energy Magazine. UK Government announces eight companies selected for Green Fuels, Green Skies competition. https://www.renewableenergymagazine.com/biofuels/uk-government-announces-eight-companies-selected-for-20210727[2021-07-27].

[62] PubAffairs Bruxelles. European Health Emergency preparedness and Response Authority (HERA): Getting ready for future health emergencies. https://www.pubaffairsbruxelles.eu/eu-institution-news/european-health-emergency-preparedness-and-response-authority-hera-getting-ready-for-future-health-emergencies/[2021-09-16].

Bioscience and Biotechnology

Ding Chenjun, Chen Fang, Zheng Ying,
Wu Xiaoyan, Song Qi, Deng Shibi

Biotechnology is an important area of technological innovation that is both leading and disruptive. At present, synthetic biology, gene editing and other cutting-edge technologies continue to develop rapidly. Synthetic genomics research has reached new heights, and a variety of new genome editing tools for different needs have been developed. As the tool of protein structure prediction has been iterated, and the accuracy and speed of prediction have been further improved. The integration of cutting-edge technologies such as "Omic" technologies, biomicroscopic imaging, single-cell sequencing and spatiotemporal omics, and artificial intelligence has made it possible to understand life more systematically, and has promoted basic research of key units such as molecules, cells, tissues, and organs, providing more new clues to reveal the mechanism of human development and disease occurrence. At the same time, the continuous innovation of core technologies is also continuously enabling the transition to green and low-carbon industries such as food, agriculture, medicine, energy, chemical industry, and environmental protection.

4.4　农业科技领域发展观察

袁建霞　邢　颖

（中国科学院科技战略咨询研究院）

“十四五”时期是开启全面建设社会主义现代化国家新征程、向第二个百年奋斗目标进军的第一个五年，是实现巩固拓展脱贫攻坚成果同乡村振兴有效衔接、加快农业农村现代化的关键时期。农业现代化，关键是农业科技现代化。党的二十大报告强调，要“加快建设农业强国”，“强化农业科技和装备支撑”①。本文在全面监测国际农业科技领域研究动态和战略举措的基础上，总结了国内外重要科学研究进展和国际重要战略行动，为我国把握农业科技发展态势、制定发展战略提供启示和建议。

一、国内外重要科学研究进展

1. 基因编辑技术取得多项重要新进展

通过控制切割时序、调控甲基化开关及开发脱靶评估软件、有效性预测算法和多重基因激活技术等，基因编辑技术的灵活性、有效性和精准度获得显著提升。例如美国伊利诺伊大学芝加哥分校人员开发的 CRISPR 基因编辑新技术，通过设计和导入多个前向导（proGuide）RNA 系统，可随着时间的推移对靶点序列按顺序进行切割或编辑，相关成果于 2021 年 1 月刊发在《分子细胞》（*Molecular Cell*）上[1]。美国麻省理工学院和加州大学旧金山分校研究人员开发出可“开关”的 CRISPR 表观遗传修饰工具，可通过甲基化使特定基因沉默，并可通过去甲基化酶逆转沉默效应，相关成果于 2021 年 4 月发表在《细胞》上[2]。以色列理工学院和巴伊兰大学研究人员开发出一款可检测、评估和量化基因编辑脱靶活动的软件 CRISPECTOR，能精确测量每一个待测位点的脱靶情况，相关成果于 2021 年 5 月发表在《自然-通讯》上[3]。来自丹麦奥胡斯大学、德国哥本哈根大学、中国华大基因研究院和美国哈佛医学院的国际合

① 习近平：高举中国特色社会主义伟大旗帜 为全面建设社会主义现代化国家而团结奋斗——在中国共产党第二十次全国代表大会上的报告．https：//www.gov.cn/xinwen/2022－10/25/content＿5721685.htm.

作团队开发出预测基因编辑向导 RNA（gRNA）有效性的算法，可以预测哪些 gRNA 最有效，相关成果于 2021 年 5 月发表在《自然-通讯》上[4]。中国科学院遗传与发育生物学研究所团队借助全基因组测序技术揭示了引导编辑系统的高特异性[5]，通过使用成对的引导编辑向导 RNA（prime editing guide RNA，pegRNA）和基于解链温度的引物结合位点设计，有效提升了引导编辑在植物中的效率，并搭建了便捷的植物 pegRNA 设计网站 PlantPegDesigner[6]，开发出更加高效的新型引导编辑系统 ePPE，编辑效率较此前开发的 PPE 平均提高了 5.8 倍[7]，相关成果均于 2021～2022 年发表在《自然-生物技术》上。美国马里兰大学研究人员开发出可对多达 7 个基因同时进行高精确度和高效率激活的植物 CRISPR3.0 系统，并在水稻、西红柿和拟南芥中进行了验证，其激活能力是此前 CRISPR 技术的 4～6 倍，相关成果于 2021 年 6 月发表在《自然-植物》上[8]。美国德州大学奥斯汀分校研究人员开发出更加高效安全的 SuperFi-Cas9，脱靶概率降低至原来的 1/4000，相关研究成果于 2022 年 3 月发表在《自然》上[9]。

2. 作物育种研究开辟出重要新方向

生物技术研究取得的突破性进展为作物品种培育开辟了新方向。中国科学院遗传与发育生物学研究所团队首次提出了异源四倍体野生稻快速从头驯化的新策略，并突破多倍体野生稻参考基因组绘制、遗传转化及基因组编辑等技术瓶颈，建立了从头驯化技术体系，创制了性状得到改良的不同类型的四倍体水稻新材料，为培育新型多倍体水稻作物提供了一条全新的路径，相关成果于 2021 年 3 月发表在《细胞》上[10]。来自北京大学及芝加哥大学、贵州大学等机构的研究团队发现，在水稻和马铃薯中引入去甲基化酶 FTO，可实现针对 RNA 修饰 m6A 去甲基化，使水稻和马铃薯的产量和生物量显著增加约 50%，同时还会显著提高抗旱能力，该研究开发了一种革新的、具有普适性的表观遗传修饰育种技术，相关成果于 2021 年 7 月发表在《自然-生物技术》上[11]。中国农业科学院深圳农业基因组研究所团队联合云南大学等单位运用“基因组设计”的理论和方法体系培育出杂交马铃薯，用二倍体育种替代了四倍体育种，并用杂交种子繁殖替代了薯块繁殖，这一成果是对马铃薯产业的颠覆性创新，相关成果于 2021 年 6 月发表在《细胞》上[12]。与此同时，作物育种向智能化方向发展。2022 年 4 月 14 日，荷兰政府宣布将投入 4200 万欧元用于开发新一代智能育种工具，以加快新作物品种培育[13]。

3. 作物病害防控机理研究取得新突破

中国农业科学院蔬菜花卉研究所研究人员首次发现植物和动物之间存在功能性水平基因转移现象——超级害虫烟粉虱通过“偷窃”寄主植物解毒基因 *PMaT1* 来代谢

植物中广泛存在的抗虫次生代谢物酚糖，从而巧妙利用“以子之矛，攻子之盾”的方式产生广泛的寄主适应性。该成果为探索昆虫适应性进化规律开辟了新的视角，为新一代靶标基因导向的烟粉虱田间精准绿色防控技术研发提供了全新思路，相关成果以封面文章于 2021 年 5 月发表在《细胞》上[14]。中国科学院遗传与发育生物学研究所与清华大学研究人员合作揭示了一种全新的植物免疫受体作用机制，发现 ZAR1 抗病小体具有钙离子通道的功能，在钙信号与植物细胞死亡之间建立了联系。该发现革新了既往对效应物触发免疫（effector-triggered immunity，ETI）触发的细胞死亡的看法，为人工设计广谱、持久的新型抗病蛋白带来新启示[15]，相关成果于 2021 年 5 月发表在《细胞》上[16]。

4. 基因编辑动物研究和商品化取得重大进展

2021 年 10～11 月，两种基因编辑鱼“Madai”红鲷鱼和“22-seiki fugu”虎豚相继在日本获准商业销售。“Madai”红鲷鱼是首批获得日本政府批准的基因编辑动物食品，由区域鱼类公司、京都大学、近畿大学及厚生劳动省和农林水产省共同开发。研究人员利用 CRISPR 基因编辑技术敲除了“Madai”红鲷鱼的一种肌肉生长抑制素基因，其可食用部分是常规鱼类的 1.2～1.6 倍，饲料利用效率提高约 14%。“22-seiki fugu”虎豚由区域鱼类公司与京都大学、近畿大学共同开发。研究人员利用 CRISPR 基因编辑技术敲除了“22-seiki fugu”虎豚的 4 个控制食欲的瘦素受体基因，提高了虎豚的食欲和生长速度，体重是同一养殖时期的传统河豚的 1.9 倍[17]。英国爱丁堡大学罗斯林研究所人员利用基因编辑技术培育出对猪繁殖与呼吸综合征（porcine reproductive and respiratory syndrome，PRRS，又称猪蓝耳病）具有抗性的基因编辑猪，并与动物遗传育种公司 Genus 签署了合作开发协议。PRRS 是对全球生猪产业造成损失最大的传染病之一，每年仅在美国和欧洲给养猪业造成的损失达 25 亿美元[18]。2022 年 3 月 7 日，美国 FDA 宣布，两种通过基因编辑技术改变了皮毛特征（这一性状可以遗传）的肉牛（被称为 PRLR-SLICK 牛）和其后代产品可以安全食用[19]。

5. 智能农机装备研发取得重大进展

世界农机巨头企业重视智能农机装备发展，持续推出相关产品，加快新技术研发和成果转化。2021 年日本久保田与荷兰的初创企业 Aurea Imaging 合作开展果树栽培系统自动化研究，将无人机和物联网传感器获得的数据与人工智能相结合，提供果园等收成预测和绘制土壤图等服务，并将感应技术与农业机械融合，验证了可以为果树栽培领域提供的综合解决方案[20]。美国爱科（AGCO）集团也在 2021 年推出了 23 款新的智能农机产品，包括利用视觉系统、人工智能和机器学习技术开发的定向喷涂产品[21]。跨国企业 Priva 在 2021 年 9 月宣布，与荷兰主要种植者、技术合作伙伴和专家合作开发出市场上第一个可以与员工一起在温室内独立移动的机器人 Kompano，其可由智能设备轻松控制，根据用户的喜好和要求进行调整，全天候为番茄植株脱叶，提供了一种经济可行的番茄作物人工脱叶的替代方案，其智能算法和受专利保护的末端执行器可以确保每周 1 公顷范围内的脱叶效率超过 85%[22]。2022 年 1 月，美国约翰迪尔公司在新闻发布会上宣布研发成功首款商业化无人驾驶拖拉机[23]。

6. 农业食品生产方式取得变革性突破

淀粉的人工合成及细胞培养肉将对传统农业生产方式产生变革性影响。中国科学院天津工业生物技术研究所联合大连化学物理研究所等机构的研究人员成功运用合成生物学技术，首次在实验室实现了从 CO_2 到淀粉分子的人工全合成，从头设计创建了由 11 步核心反应组成的人工淀粉合成途径，合成效率约为传统农业生产淀粉（如玉米作物淀粉合成）的 8.5 倍，理论上 1 m^3 大小的生物反应器年产淀粉量相当于我国 5 亩①玉米地的年产淀粉量[24]。该研究突破了自然光合作用局限，使工业化车间制造淀粉成为可能，相关成果于 2021 年 9 月发表在《科学》上[25]。在细胞培养肉方面，以色列细胞肉公司 Future Meat Technologies（FMT）开发了在反应器中培养动物细胞来生产肉食的技术，且开创了培养基过滤再生专利技术。由于该技术支持高细胞密度，成功克服了成本难题，实验室培养鸡肉的价格从 2019 年的 150 美元/磅②降到了 2021 年的 3.9 美元/磅，使得 FMT 成为业内少有的、利用培养基过滤再生技术实现规模量产的公司[26,27]。2021 年 12 月，英国和日本的研究人员首次从特定化学条件下生长的家畜胚胎中获得干细胞并且开发出新的细胞系，对未来肉类生产具有重大价值和意义。[28]

二、国际重要战略行动

1. 美国通过已有计划支持大量农业科技研发创新项目

2021 年 4 月 5 日，美国农业部国家食品与农业研究所（National Institute of Food and Agriculture，NIFA）宣布[29]通过两项关键计划，即“食品与农业网络信息学与工具”（Food and Agriculture Cyberinformatics and Tools，FACT）和“农业与食品研究计划”（Agriculture and Food Research Initiative，AFRI），共投资 1550 万美元支持 30 个新项目，包括 18 项有关大数据分析、机器学习、人工智能和预测技术的研究，

① 1 亩≈666.7 m^2。

② 1 磅≈0.4536 kg。

以及12项基于纳米技术的解决方案研究。4月28日，NIFA又宣布[30]通过AFRI投资超过865万美元资助22个植物育种研究项目，包括：加速测试、评估和发布公开完成的小麦作物品种；开发具有竞争力的适合机械收获的辣椒品种；开发改良的南方玉米杂交种；评估使用无人机提高决策速度和准确性；培育可有效抗晚疫病和耐热的马铃薯品种等。10月6日，NIFA可持续农业系统计划第三期宣布，投资1.46亿美元支持15个可持续农业研究项目[31]，以改进稳健、有弹性和气候智能型食品和农业系统。

2. 英国启动农业创新计划及未来10年植物科学战略

英国研究与创新署和英国环境、食品与农村事务部于2021年2月宣布，共同发起“农业创新路径”研发计划[32]，拟投入1200万英镑。其中，500万英镑支持可行性研究项目，用于解决农场挑战的早期想法或创新发展技术和商业可行方案；700万英镑支持行业研究项目，用于开发针对农业实际问题的具有高潜力的创新性解决方案；支持重点包括机器人、自动化，如视觉引导机器人除草系统、垂直农业等新型食品生产系统等。同时，两部门还合作启动了农业创新计划，拟投入1750万英镑支持可提高农业生产力和环境可持续性的研究项目，重点方向包括牲畜、植物、新型食品生产系统、生物经济和农林业。英国研究与创新署于2021年3月15日发布了面向未来10年的植物科学研究战略[33]，提出了植物研究创新的5个战略目标：①可持续地平衡农业、生物多样性、碳封存、能源生产和洪水管理的需求；②部署先进的植物育种和作物管理策略，建立能可持续生产安全且营养食品的弹性农业系统；③通过使用生物投入品替代化学投入品、更好地管理植物-土壤相互作用、部署替代性耕作系统等，大幅减少农业碳排放；④通过遥感、生物干预和持久的植物免疫，建立植物病虫害监测、控制和阻止机制；⑤建立全新的植物基食品生产系统，生产植物基疫苗、蛋白质原料和高价值化学品等新产品。

3. 法国启动涉农未来投资计划

2021年5月28日，法国宣布启动第四期未来投资计划[34]，拟在2021～2025年投入200亿欧元支持高等教育、研究与创新。其重点是启动若干国家战略，其中包括两项农业食品发展加速战略，一是可持续农业系统和用于生态转型的农机设备加速战略，二是可持续和健康食品加速战略；并于11月5日宣布为这两项战略投入8.8亿欧元[35]。可持续农业系统和用于生态转型的农机设备加速战略，支持以选种和农业机器人为主的优先研究与设备计划，以及支持农业机器人开发、生物控制与生物刺激素两个“重大挑战”颠覆性技术创新项目。可持续和健康食品加速战略以促进食品工业转型、更好理解食品和健康的关系、为消费者提供高质量本土食品研发为主要支持方向；重点实施“食品系统、微生物和健康”优先研究与设备计划，启动“未来发酵”“重大挑战”颠覆性技术创新项目，支持植物蛋白和多元蛋白质来源开发项目。2021年10月12日，法国公布“法国2030”投资计划，将投资健康、可持续和可追溯食品，加速农业数字化作为十大目标之一[36]。

4. 澳大利亚实施“农业2030”一揽子措施及涉农计划

澳大利亚联邦科学与工业研究组织于2021年5月12日宣布，将在5年内投资超过8.5亿澳元实施“农业2030”一揽子措施[37]，以支持农业部门到2030年实现1000亿澳元产业的目标，支持澳大利亚的土壤和生物多样性管理，并最大限度地发挥澳大利亚渔业和林业资源的效益。“农业2030”一揽子措施具体包括：①加强抵御外来病虫害能力的生物安全新一揽子计划；②支持国家土壤战略；③资助避免食物浪费的有关措施；④鼓励农业用地的管理和生物多样性；⑤支持农产品市场的贸易和生产；⑥林业和渔业的发展，改善农业就业机会；⑦增加对创新农业实践和数字技术的采用。9月，澳大利亚联邦科工组织宣布澳大利亚联邦科工组织“任务”计划将启动3个新的主题（分别是提高抗旱能力、降低农产品出口成本、开创未来食用蛋白质）[38]，与澳大利亚政府和企业界共同出资1.5亿澳元，帮助澳大利亚的农业食品行业到2030年实现200亿澳元的产值。同时，澳大利亚政府重视数字农业平台建设。2022年4月1日澳大利亚政府发布《农业数字基础战略》[39]，规划了数字技术在农业、林业和渔业的发展路径，提出将投入3000万澳元建设国家数字农业中心，投资40万美元建设澳大利亚农场数据代码等。

5. 日本提出绿色食品战略和发展可持续食品供应产业

2021年6月18日，日本政府发布“统合创新战略2021”[40]，作为2021年日本科技创新工作的年度指导。该战略提出将制定并发布绿色食品战略，以提高粮食、农林水产业的生产效率；同时推动战略基础技术在食品和农林水产等重要领域的应用。9月28日，日本综合科学技术创新会议召开[41]，对此前提出的面向高风险、高挑战的大型科研计划——登月计划的发展目标进行完善，提出将在2050年前，充分发掘尚未充分利用的生物功能，在全球范围构建无浪费、可持续的食品生产供应产业，并将其作为登月计划的九大目标之一。

6. 欧盟“地平线欧洲”计划部署农业研发投资领域

2021年3月15日，欧盟委员会通过“地平线欧洲”计划的第一个战略计划[42]，确定了2021～2024年研究与创新投资的战略方向，其中包括加速农业系统转型的农业生态生活实验室和研究基础设施、动物健康、农业数据、安全和可持续粮食系统等领域。6月16日，欧盟委员会发布“地平线欧洲”计划2021～2022年主要工作内容[43]，包括在农业食品领域部署可持续农业、渔业和水产养殖，实现健康、可持续和包容性的粮食系统，以及在农业生产中进行生物多样性管理研究等。9月29日，欧盟在“地平线欧洲”计划下启动了5个任务导向的“欧盟使命”计划[44]，其中土壤健康为使命计划之一。

7. 国际农业研究磋商小组发布《2030研究与创新战略》

国际农业研究磋商小组（Consultative Group on International Agricultural Research，CGIAR）于2021年1月发布《2030研究与创新战略》[45]，提出将在系统改造、弹性农业食品系统和基因创新3个领域开展创新行动，旨在促进粮食、土地和水系统的转变以应对气候危机。其研究和创新的优先事项包括：①推动建立可持续的土地和水资源利用、生计和健康饮食的食品系统，加强数字创新的作用，通过创新能负担得起的健康可持续饮食来解决营养问题；②研究探索生产系统的多种路径，包括利用生态系统功能的农业生态学方法、优化小规模生产者获得和使用现代投入品的技术性方法等，提高农产品系统的复原力；③聚焦基因库和作物育种，开展基因创新。

三、启示与建议

基于上述国内外重要农业科学研究进展，以及发达国家和农业组织在发展农业科技方面采取的战略行动，得出以下启示。

一是在农业科学研究前沿方面，农业生物领域发展最为迅速，尤其是在作物生物育种、病虫害生物防控、食品科学与工程等方向上表现突出，不断出现新突破，开辟新方向，包括野生作物从头驯化、表观遗传修饰、病虫害入侵和植物免疫新机制及淀粉的人工合成和细胞培养肉生产等，为农业育种、病害防控和食物生产等提供了全新的思路和实现路径。与此同时，新型基因编辑技术作为一种前沿生物技术，一直备受关注，持续取得重要新进展，不断优化完善，为农业育种提供了重要技术工具。尤其是近年来，以基因编辑技术为代表的新一代生物技术与信息技术/人工智能技术结合紧密，有可能引发新一轮产业变革，值得关注和进行跨学科布局。智能农机装备是未来农业发展的重要机会之一，具有巨大的市场价值，全球农机企业尤其是巨头企业纷纷投入大量资源进行布局，已取得显著成效，未来该领域的国际竞争将持续加大。

二是在农业科技发展目标方面，主要农业发达国家逐渐将农业科技发展目标定位于促进提高农业产值、发展可持续农业，实现农业系统转型，而不仅是提高农业产量。因此，其农业科技战略和研发支持方向也逐渐增加了相关研究支撑的内容，包括机器人、数字技术等在农业领域的创新应用；新型食品（如食用和饲用替代蛋白质）和生物制品研发；经济作物、高价值农产品的育种、栽培和养殖技术研发；促进农业系统绿色、低碳转型和提高气候适应的多学科技术等。

三是面向未来，我国发展农业科技需面向世界农业科技前沿和我国农业现实战略需求，大力推动农业生物技术、农业自动化智能化技术的研发，前瞻部署农业智能化技术、低碳农业技术、高价值农产品、农业替代食品等的研发，促进农业转型和可持续发展。

致谢：中国科学院遗传与发育生物学研究所的高彩霞研究员和田志喜研究员对本文进行审阅并提出宝贵修改意见，特致谢忱！

参考文献

[1] Clarke R, Terry A R, Pennington H, et al. Sequential activation of guide RNAs to enable successive CRISPR-Cas9 activities. Molecular Cell, 2021, 81(2): 226-238.

[2] Nuñez J K, Chen J, Pommier G C, et al. Genome-wide programmable transcriptional memory by CRISPR-based epigenome editing. https://www.cell.com/cell/fulltext/S0092-8674(21)00353-6 [2021-04-29].

[3] Amit I, Iancu O, Levy-Jurgenson A, et al. CRISPECTOR provides accurate estimation of genome editing translocation and off-target activity from comparative NGS data. https://www.nature.com/articles/s41467-021-22417-4[2021-05-24].

[4] Xiang X, Corsi G I, Anthon C, et al. Enhancing CRISPR-Cas9 gRNA efficiency prediction by data integration and deep learning. https://www.nature.com/articles/s41467-021-23576-0[2021-05-28].

[5] Jin S, Lin Q, Luo Y, et al. Genome-wide specificity of prime editors in plants. https://www.nature.com/articles/s41587-021-00891-x[2021-04-15].

[6] Lin Q, Jin S, Zong Y, et al. High-efficiency prime editing with optimized, paired pegRNAs in plants. https://www.nature.com/articles/s41587-021-00868-w[2021-03-25].

[7] Zong Y, Liu Y J, Xue C X. An engineered prime editor with enhanced editing efficiency in plants. https://www.nature.com/articles/s41587-022-01254-w[2022-03-24].

[8] Pan C, Wu X, Markel K, et al. CRISPR-Act3.0 for highly efficient multiplexed gene activation in plants. https://www.nature.com/articles/s41477-021-00953-7[2021-06-24].

[9] Bravo J P K, Liu M S, Hibshman G N, et al. Structural basis for mismatch surveillance by CRISPR-Cas9. https://www.nature.com/articles/s41586-022-04470-1[2022-03-02].

[10] Yu H, Lin T, Meng X, et al. A route to *de novo* domestication of wild allotetraploid rice. https://www.cell.com/cell/fulltext/S0092-8674(21)00013-1[2021-03-04].

[11] Yu Q, Liu S, Yu L, et al. RNA demethylation increases the yield and biomass of rice and potato plants in field trials. Nature Biotechnology, 2021, 39: 1581-1588.

[12] Zhang C, Yang Z, Tang D, et al. Genome design of hybrid potato. https://www.cell.com/cell/fulltext/S0092-8674(21)00707-8[2021-06-24].

[13] Wageningen University & Research. Dutch cabinet invests 42 million in CROP-XR institute for faster development of resilient agricultural crops. https://www.wur.nl/en/news-wur/Show/Dutch-cabinet-invests-42-million-in-CROP-XR-institute-for-faster-development-of-resilient-agricultural-crops.htm [2022-04-14].

[14] Xia J, Guo Z, Yang Z, et al. Whitefly hijacks a plant detoxification gene that neutralizes plant toxins. https://doi.org/10.1016/j.cell.2021.02.014[2021-04-01].

[15] 中国科普网. 2022 中国农业科学十大重大进展发布. http://www.kepu.gov.cn/www/article/dtxw/01b9422d5a144cf5aba824df6f4359f2[2022-12-17].

[16] Bi G, Su M, Li N, et al. The ZAR1 resistosome is a calcium-permeable channel triggering plant immune signaling. https://doi.org/10.1016/j.cell.2021.05.003[2021-05-12].

[17] ISAAA. Japan's Three Genome-Edited Food Products Reach Consumers. https://www.isaaa.org/blog/entry/default.asp? BlogDate=1/19/2022[2022-01-19].

[18] The University of Edinburgh. Agreement targets disease-resistant pigs. https://edinburgh-innovations.ed.ac.uk/news/genus-disease-resistant-pigs #: ~: text = Agreement% 20targets% 20disease-resistant% 20gene-edited% 20pigs% 20Animals% 20% 26% 20Agriculture, resistant% 20to% 20a% 20respiratory% 20disease% 20affecting% 20livestock% 20worldwide[2021-07-06].

[19] FDA Makes Low-Risk Determination for Marketing of Products from Genome-Edited Beef Cattle After Safety Review. https://www.fda.gov/news-events/press-announcements/fda-makes-low-risk-determination-marketing-products-genome-edited-beef-cattle-after-safety-review. [2022-03-07].

[20] U.S. Food & Drug Administration. FDA Makes Low-Risk Determination for Marketing of Products from Genome-Edited Beef Cattle After Safety Review. https://www.fda.gov/news-events/press-announcements/fda-makes-low-risk-determination-marketing-products-genome-edited-beef-cattle-after-safety-review [2022-03-07].

[21] 株式会社久保田. KUBOTA REPORT 2021. https://www.kubota.com/ir/financial/integrated/data/cn/digest2021.pdf[2021-08-30].

[22] 国科农研院. 国际智能农机装备企业科技创新动态. https://mp.weixin.qq.com/s? __biz=MzU2NzI1NjkzNw==&mid=2247572739&idx=3&sn=864ade8d7a1006563fc78e37a9a1cdaa&chksm = fc9c6f05cbebe613be6b769df9519d3a902c4412d791c69db04e4421a2d744116963ad873bb0&scene=27[2022-09-30].

[23] SeedQuest. Priva introduces the world's first fully automated leaf-cutting robot for tomato crops. https://www.seedquest.com/news.php? type= news&id_article = 132756&id_region = &id_category=2409&id_crop=[2021-09-28].

[24] John Deere. John Deere Reveals Fully Autonomous Tractor at CES 2022. https://www. deere. com/en/news/all-news/autonomous-tractor-reveal/ [2022-01-04].

[25] 董瑞丰，王井怀．我国科学家突破二氧化碳人工合成淀粉技术．http://www. gov. cn/xinwen/2021-09/24/content_5638987. htm[2021-09-24].

[26] Cai T, Sun H, Qiao J, et al. Cell-free chemoenzymatic starch synthesis from carbon dioxide. https://www. science. org/doi/10. 1126/science. abh4049[2021-09-23].

[27] DeepTech 深科技．世界首家细胞培育肉工厂诞生，单份细胞培育鸡胸肉 3. 9 美元，有望明年进入消费市场 | 专访．https://zhuanlan. zhihu. com/p/387211687 3[2021-07-06].

[28] Stem cell study paves way for manufacturing cultured meat. Stem cell study paves way for manufacturing cultured meat-FarmingUK News. [2021-1208].

[29] 李万，钱娅妮．2021 年世界科技进展 100 项．https://mp. weixin. qq. com/s/_ xwxYDEupoWGaDXKP5c08w[2022 04 12].

[30] FarmingUK Team. Stem cell study paves way for manufacturing cultured meat. https://www. farminguk. com/news/stem-cell-study-paves-way-for-manufacturing-cultured-meat_59483. html [2021-12-08].

[31] USDA. NIFA Invests $15. 5M in Food and Agriculture Cyberinformatics Tools to Boost Agricultural Production. https://nifa. usda. gov/press-release/nifa-invests-155m-food-and-agriculture- cyberinformatics-tools-boost-agricultural[2021-04-05].

[32] USDA. NIFA Invests Over $8. 65M for Plant Breeding. https://nifa. usda. gov/press-release/nifa-invests-over-865m-plant-breeding[2021-04-28].

[33] USDA. USDA/NIFA announces more than $146M investment in sustainable agricultural research. https://www. usda. gov/media/press-releases/2021/10/06/usda-announces-more-146m-investment-sustainable-agricultural[2021-10-06].

[34] UK Research and Innovation. UKRI and Defra to launch Farming Innovation Pathways competition. https://www. ukri. org/news/ukri-and-defra-to-launch-farming-innovation-pathways-competition/[2021-02-10].

[35] UK Research and Innovation. UK plant science research strategy a green roadmap for the next ten years. https://www. ukri. org/wp-content/uploads/2021/03/BBSRC-120321-PlantScienceStrategy. pdf [2021-03-15].

[36] Ministère de l'enseignement supérieur de la recherche. Lancement du 4e programme d'investissements d'avenir en janvier 2021: 20 Md € dans la recherche et l'innovation en faveur des générations futures. https://www. enseignementsup-recherche. gouv. fr/cid156296/4e-programme-d-investissements-d-avenir-20-md-dans-la-recherche-et-l-innovation-en-faveur-des-generations-futures. html[2021-05-28].

[37] Ministère de l'enseignement supérieur de la recherche. PIA4: près de 880 M au service de la 3e révolution agricole et de l'alimentation santé. https://www. enseignementsup-recherche. gouv. fr/fr/pia4-pres-de-880-meu-au-service-de-la-3e-revolution-agricole-et-de-l-alimentation-sante-81841 [2021-

11-05].

[38] ELYSEE. Le Président de la République Emmanuel Macron a présenté le plan d'investissement 《France 2030》. https://www.elysee.fr/france2030#publication-list; https://www.gouvernement.fr/france-2030-un-plan-d-investissement-pour-la-france-de-demain[2021-10-12].

[39] CSIRO. 2021-22 Federal Budget Statement. https://www.csiro.au/en/news/News-releases/2021/2021-22-Federal-Budget-Statement[2021-05-12].

[40] CSIRO. $150 million missions to boost Australian agriculture and food sectors. https://www.csiro.au/en/news/News-releases/2021/$150-million-missions-to-boost-Australian-agriculture-and-food-sectors[2021-09-09].

[41] Australian Government Department of Agriculture, Water and the Environment. Digital Foundations for Agriculture Strategy. https://www.dcceew.gov.au/sites/default/files/documents/digital-foundations-agriculture-strategy.pdf [2022-04-01].

[42] 内閣府. 統合イノベーション戦略 2021. https://www8.cao.go.jp/cstp/tougosenryaku/togo2021_honbun.pdf[2021-06-18].

[43] 内閣府. 日本内閣府総合科学技術・イノベーション会議:ムーンショット型研究開発制度の目標設定について、. https://www8.cao.go.jp/cstp/siryo/haihui057/haihu-057.html[2021-09-28].

[44] European Commission. Horizon Europe's first strategic plan 2021-2024: Commission sets research and innovation priorities for a sustainable future. https://ec.europa.eu/commission/presscorner/detail/en/ip_21_1122[2021-03-15].

[45] European Commission. Horizon Europe Work Programme 2021-2022: 9. Food, Bioeconomy, Natural Resources, Agriculture and Environment. https://ec.europa.eu/info/funding-tenders/opportunities/docs/2021-2027/horizon/wp-call/2021-2022/wp-9-food-bioeconomy-natural-resources-agriculture-and-environment_horizon-2021-2022_en.pdf[2021-06-16].

[46] European Commission. Commission launches EU missions to tackle major challenges. https://ec.europa.eu/commission/presscorner/detail/en/IP_21_4747[2021-09-29].

[47] CGIAR. CGIAR 2030 Research and Innovation Strategy: Transforming food, land, and water systems in a climate crisis. https://cgspace.cgiar.org/handle/10568/110918[2021-01-31].

Agricultural Science and Technology

Yuan Jianxia, Xing Ying

We summarized the important scientific research progress and international strategic actions. Globally, agricultural science and technology achieved great progress and breakthrough in agricultural biotechnology and intelligent agriculture, including gene editing technology, crop biological breeding, mechanism of crop disease prevention and control, gene editing of food animals, artificial synthesis and culture of food, and intelligent agricultural machinery equipment. In addition, there were several significant events that deserve attention: the United States supported a large number of agricultural science and technology R&D and innovation projects through existing plans, the United Kingdom launched the agricultural innovation plan and the plant science research strategy for the next ten years, France has launched the agriculture-related future investment plan, Australia has implemented the "Agriculture 2030" package of measures and agriculture-related plans, Japan has proposed the green food strategy and developed the sustainable food supply industry, and the "Horizon Europe" deployed the agricultural R&D investment, and CGIAR released the "2030 Research and Innovation Strategy". Finally, we proposed inspiration and suggestions from two aspects, including global agricultural science research frontier and agricultural science and technology goals.

4.5 生态环境领域发展观察

廖 琴[1] 曲建升[2] 曾静静[1] 裴惠娟[1]
董利苹[1] 刘燕飞[1] 刘莉娜[1]
（1. 中国科学院西北生态环境资源研究院；
2. 中国科学院成都文献情报中心）

2021～2022年，陆地生态系统碳汇、生物多样性、地下水储量、陆地和大气微塑料、气候变化与空气污染协同治理等成为全球生态环境科学领域关注的核心议题，科学界围绕这些领域取得了一系列重要研究进展。同时，在碳中和目标下，世界主要国家和国际组织积极制定碳中和相关战略行动，推进科技创新研发，并在土壤保护与碳汇、生物多样性恢复、水资源管理和研究，以及塑料污染治理等方面也部署了相关战略和行动计划。

一、重要研究进展

1. 全球聚焦于陆地生态系统碳汇潜力及其影响机制研究

在碳中和背景下，陆地生态系统碳储量和固碳潜力及其影响机制成为当前国际研究的焦点。联合国环境规划署（United Nations Environment Programme，UNEP）研究指出，到2050年，在所有生态系统中实施基于自然的解决方案每年可至少减少100亿 t CO_2 排放[1]。英国和澳大利亚科学家研究发现，生态系统的恢复可在2100年额外吸收930亿 t 碳[2]。美国科学家提供了全球陆地碳储存潜力数据集，发现未实现的碳储存潜力达2870亿 t 碳[3]。中国科学家研究发现，中国陆地生态系统固碳能力巨大，2010～2016年，中国陆地生态系统每年吸收的碳相当于人为碳排放的45%，这在以往研究中被低估[4]。然而，多项研究也表明，全球变暖可能会触发陆地生态系统从碳汇向碳源转变[5]。中国科学家发现，1982～2015年，全球陆地生态系统 CO_2 施肥效应呈现显著下降趋势[6]。美国科学家研究指出，2010～2019年，巴西亚马孙地区向大气释放的 CO_2 比吸收的量多了近20%[7]。联合国教育、科学及文化组织（United Nations Educational Scientific and Cultural Organization，UNESCO）等机构研究指出，全球 10 座世界遗产地森林的碳排放量超过碳吸收量[8]。美国和英国科学家预计，全球泥炭地在 2020～2100 年可能释放超 1000 亿 t 的碳[9]。

2. 生物多样性丧失趋势仍未扭转但恶化的速率有所放缓

多项研究表明，全球生物多样性丧失趋势仍然没有得到根本性扭转，但《生物多样性公约》取得长期进展。美国和法国科学家研究估计，从公元 1500 年左右到现在，地球上有 7.5% ～13% 的已知物种灭绝，并且地球可能正在经历第六次物种大灭绝的开始[10]。世界自然基金会（WWF）发布的《地球生命力报告 2022》指出，自 1970 年以来，全球野生动物种群数量平均减少了 69%[11]。英国发布的《英国的生物多样性：繁荣或衰败》报告显示，英国生物多样性大幅下降，优先保护物种减少约 60%，约 15% 的物种面临灭绝的威胁[12]。中国科学家评估了《生物多样性公约》保护目标的长期进展，发现在人类不断的努力下，生物多样性恶化的速率有所放缓[13]。美国和德国科学家评估了全球 3.1 万种陆生脊椎动物的时空分布特征及其趋势，填补了全球生物多样性地图中的许多空白[14]。奥地利和英国等多国研究人员指出，如果优先考虑生物多样性，对全球 30% 的重要区域进行保护，可以实现 81.3% 的陆地植物与脊椎动物保护目标[15]。

3. 地下水资源储量变化及监测方法取得新进展

研究发现，全球超过一半的主要含水层正在枯竭，部分地下水位下降速度惊人[16]。《2022 年联合国世界水发展报告》指出，全球地下水总储量枯竭严重，仅在 21 世纪初，每年就有 1000 亿～2000 亿 m^3 的地下水超采[17]。美国科学家发现，截至 2021 年，世界上有 20% 的地下水开采井面临枯竭的风险[18]。英国和印度等国科学家研究发现，印度和巴基斯坦地下水在 20 世纪的净蓄积量至少为 420 km^3，而在 21 世纪的前 10 年，大约有 70 km^3 的地下水的水位在以平均每年 2.8 cm 的速度下降[19]。德国科学家发现，受气候变化的影响，直到 2100 年，德国地下水位都可能会持续下降[20]。在地下水储量监测方面，美国和德国科学家发现一种更精确地监测地下水储量变化的新方法，他们将重力测量与气候实验（gravity recovery and climate experiment，GRACE）卫星及其第二代卫星（GRACE-Follow On）的重力场数据与其他测量方法进行比较，通过结合不同的方法将大尺度的 GRACE 数据缩小到更小区域，以获得可靠的地下水数据[21]。美国科学家使用干涉合成孔径雷达的表面位移测量和来自 GRACE 卫星的监测数据对陆地储水量的重力估计来表征区域的水文动态，利用新方法有望改善地下水管理[22]。

4. 微塑料在不同环境介质中的赋存及健康影响获重要发现

微塑料研究从海洋环境逐步向其在陆地生态系统中的赋存、环境影响和人类健康风险方向扩展。在陆地和大气微塑料研究方面，美国科学家发现，大气中 84% 的微塑料来自道路交通，其余 11% 来自海洋，5% 来自土壤[23]。新西兰科学家首次对大气微塑料的全球直接气候影响进行评估，认为大气微塑料或能通过反射阳光辐射对气候有微小的冷却效果，而且随着未来塑料持续在地球环境中累积，可能会呈现更强的气候效应[24]。中国科学家创新性地提出了利用稀土配合物掺杂标记对亚微米级塑料颗粒在作物中的吸收转运进行量化和可视化的方法[25]。在微塑料对人体健康影响方面，意大利科学家采用显微拉曼光谱法，首次在人类胎盘中检测到微塑料[26]。中国科学家建立了微塑料表面亚微米尺度化学变化表征方法，发现硅橡胶奶嘴消毒产生的颗粒物是儿童体内和环境中微（纳）塑料的重要来源[27]。荷兰科学家首次在人类血液中检测到微塑料[28]。英国科学家首次从活人肺部组织内检测到微塑料成分[29]。中国科学家研究发现，尺寸为纳米级的微塑料可进入肝细胞和肺并破坏其正常生理功能，对器官造成不良影响[30]。

5. 减污降碳协同治理及健康效益研究成重要趋势

空气污染和气候变化是当前人类健康面临的最大环境威胁[31]。在碳达峰与碳中和目标下，减污降碳协同治理是减少空气污染健康损失的必由之路。中国科学家发现：2002～2017 年，中国 $PM_{2.5}$ 污染相关的死亡人数增加了 23%；污染末端治理政策的实施避免了 87 万人死亡[32]。清华大学等机构的研究指出，2030 年中国实现碳达峰的同时，全国人群 $PM_{2.5}$ 年均暴露水平可从 2015 年的 55 $\mu g/m^3$ 下降到 28 $\mu g/m^3$[33]；与参考情景相比，碳中和路径下 2060 年中国的人均预期寿命可能增加 0.88～2.80 年[34]。南京大学等机构的研究指出，在绿色发展路径和 1.5℃温控目标下，中国在 2030 年和 2050 年可分别避免约 11.8 万人和 61.4 万人的 $PM_{2.5}$ 归因死亡[35]。美国未来资源研究所（Resources for the Future，RFF）的研究指出，美国在 2030 年实现气候目标将创造超过 330 亿美元的健康效益[36]。

6. 气候变化造成极端事件频发并可能继续增加

2021～2022 年，全球频发的极端天气事件打破了长期以来的纪录。联合国政府间气候变化专门委员会（Intergovernmental Panel on Climate Change，IPCC）发布的第六次评估报告第一工作组报告指出：2011～2020 年，全球地表平均温度比 1850～1900 年高 1.09℃，2016～2020 年是自 1850 年有记录以来最热的 5 年[37]；全球升温可能在未来 20 年达到或超过 1.5℃，如果升温达到 2℃，过去每 50 年发生一次的极端温度事件将每隔几年发生一次。该研究成果入选《自然》评选的 2021 年度全球十大科学新闻。瑞士科学家研究发现：气候变化导致破纪录热浪的发生概率增加，其增加程度取决于变暖速度；在高排放情景下，2021～2050 年破纪录极端热浪的发生概率是过去 30 年的 2～7 倍，2051～2080 年为 3～21 倍[38]。美国科学家预测，相较于过去的 300 万年，21 世纪热带气旋可能会在更大的纬度范围内发生，中纬度地区可以发现更多的热带气旋[39]；到 2100 年，热带气旋导致美国东北部联合极端事件发生频率增加 30～195 倍[40]。

二、重大战略行动

1. 主要国家加强碳中和战略行动制定和科技布局

美国于 2021 年 1 月宣布重返《巴黎协定》，提出到 2050 年实现 100% 清洁能源和净零排放的最新目标[41]，并签署《应对国内外气候危机的行政命令》[42]；11 月发布长期战略[43]，系统阐述了美国 2050 年实现净零排放的中长期目标和技术路径；2022 年 9 月发布《工业脱碳路线图》，提出了到 2050 年实现工业净零排放所需的 5 个关键子行业（化工、炼油、钢铁、水泥食品）的脱碳路径[44]。英国于 2021 年 3 月启动投资额为 10 亿英镑的净零创新投资组合计划，重点聚焦海上风电、先进模块化反应堆、储能、氢能、生物质能、工业燃料转换，以及碳捕集、利用与封存（carbon capture，utilization and storage，CCUS）等领域的低碳技术研发[45]；3 月发布《工业脱碳战略》，计划拨款超 10 亿英镑用于降低工业与公共建筑的排放[46]；7 月发布交通脱碳计划，为交通行业到 2050 年实现净零排放制定了路线图[47]；10 月制定了《净零战略》[48]和《英国净零研究与创新框架》[49]，提出到 2030 年启动投资金额为 900 亿英镑的净零投资计划，并确定了未来 5～10 年英国在电力、工业、交通、建筑和供暖等方面的净零研究和创新挑战及需求。德国于 2021 年通过《气候变化法》修正案[50]，规定德国到 2045 年实现气候中和。韩国于 2021 年 3 月发布《碳中和技术创新推进战略》[51]，确定了实现碳中和的关键技术。2022 年 3 月，加拿大发布《2030 年减排计划》，提出投资 91 亿加元，以确保加拿大在 2030 年实现比 2005 年碳排放水平减排 40%～45%的目标，并推动在 2050 年实现净零排放[52]。

2. 多国制定土壤保护和碳汇相关战略及其行动计划

2021 年 5 月，澳大利亚政府发布《国家土壤战略》及其临时行动计划[53,54]，将建立“国家土壤监测计划”，制定土壤数据信息生成、管理与交换的国家标准，并在 2021～2022 年投入 1.02 亿澳元实施“土壤监测与激励试点计划”，2021～2024 年投入 2090 万澳元实施“土壤科学挑战”资助计划。2021 年 5 月，英国发布《2021～

2024年英格兰树木行动计划》[55]和《英格兰泥炭行动计划》[56]，前者计划在2020～2025年通过“自然应对气候基金”投资超5亿英镑用于支持植树造林，后者计划到2025年投资超5000万英镑用于恢复至少3.5万公顷的泥炭地。2021年11月，欧盟发布《2030年土壤战略》[57]，提出到2023年提交土壤健康的专门立法提案，恢复退化的土壤和修复受污染的场地，通过构建评估荒漠化和土地退化程度的方法和相关指标来防治荒漠化，增加土壤研究和监测等关键行动。2022年4月，美国总统拜登签署旨在加强美国森林保护的行政命令[58]，主要行动包括降低野火风险、加强森林建设、遏制全球森林砍伐、部署基于自然的气候解决方案。2022年12月，韩国发布《第四次湿地保护基本计划（2023～2027）》[59]，提出了以科学为基础的湿地调查及评价，以及湿地的有效保护和管理等战略任务，以提高湿地的生态系统价值和碳汇功能。

3. 国际社会采取系列行动助力生物多样性恢复

国际组织通过发布《2020年后全球生物多样性框架》（第一份详细草案）[60]、2020年联合国生物多样性大会《昆明宣言》、《人权与生物多样性》[61]等促进全球生物多样性保护，旨在在2030年前使生物多样性走上恢复之路。英国通过《环境法》修正案，为“在2030年实现英国的自然和生物多样性恢复”设定法律约束，以期逆转物种数量缩减的态势[62]。澳大利亚发布《2021～2031年濒危物种战略》，针对受威胁物种建立了国家优先框架，确定了保护受威胁物种的关键行动领域[63]。欧盟“地平线欧洲”计划公布2021～2022年的生物领域工作计划，资助净额为19亿欧元，其中3.76亿欧元用于生物多样性与生态系统服务研究[64]。欧盟“环境与气候行动计划”（Programme for the Environment and Climate Action，LIFE计划）投资近4亿欧元，支持欧盟《2030年生物多样性战略》，促使欧洲的生物多样性走上恢复的道路[65]。2022年1月，英国公布了“地方自然恢复计划”和“景观恢复计划”两项土地管理计划，计划到2042年恢复30万hm^2的栖息地，以阻止物种减少[66]。5月，全球环境基金（Global Environment Facility，GEF）提出，将向139个发展中国家提供4300万美元资金，以期在全球生物多样性协议达成之前快速采取行动阻止物种损失[67]。12月，《生物多样性公约》第十五次缔约方大会（COP15）第二阶段会议通过《昆明-蒙特利尔全球生物多样性框架》，设定了全球生物多样性保护的长期目标及在2030年前采取的行动[68]。

4. 国际组织和美国确定水资源管理和研究的优先领域

2021年10月，世界气象特别大会批准包括“水与气候联盟”在内的《水宣言》，以加快联合国可持续发展目标6（SDG 6）的实施，并批准了新的水文学愿景和战略以及相关行动计划[69]。11月，美国地质调查局（United States Geological Survey，

USGS）发布《2020—2030 年水资源研究行动计划》[70]，确定了 7 个重点方向及其优先领域，具体包括：①水资源短缺和可用性的驱动因素及结果；②了解和应对极端水文事件和气候变化的影响；③解决水质问题，确保安全和可获得的高质量水；④用于水资源综合管理和治理的水政策、规划和社会经济情况；⑤流域生态系统功能的保护；⑥水技术研发及创新应用；⑦解决劳动力发展和水知识普及问题。2022 年 3 月，UNEP 发布《2022—2025 年淡水战略优先事项》[71]，重点关注淡水与气候行动、淡水与自然行动、淡水与化学品和污染行动，旨在促进全球对水资源的紧迫理解和优先行动。2022 年，UNESCO 发布《2022—2029 年政府间水文计划第九阶段战略》[72]，确定了五大水优先领域，将从科研创新、水教育、数据共享、水资源综合管理、科学水治理方面应对变化环境下的水资源挑战 。

5. 全球推进塑料监测研究及污染治理进入新阶段

2021 年 3 月，澳大利亚发布《2021 年国家塑料计划》[73]，提出将在塑料污染源头预防、塑料回收利用、海洋和水道中的塑料应对以及研究、创新和数据方面采取系列行动举措。5 月，北极监测与评估计划（Arctic Monitoring and Assessment Programme，AMAP）发布《垃圾和微塑料监测计划》[74]，提出有关监测类型、监测指标和监测方案实施的建议。9 月，新西兰发布《国家塑料行动计划》[75]，提出将在 2021～2024 年逐步淘汰一次性和难以回收的塑料，并启动 5000 万美元的塑料创新基金等行动。9 月，印度启动《印度塑料公约》[76]，提出了到 2030 年的行动目标，包括 100% 的塑料包装可重复使用或回收、50% 的塑料包装得到有效回收等。2022 年 3 月，第五届联合国环境大会通过《终结塑料污染：迈向具有国际法律约束力的文书》[77]的决议，推动全球塑料污染治理进入新的阶段。

6. 世界卫生组织和欧盟加严空气质量目标

世界卫生组织（World Health Organization，WHO）的空气质量指南为空气污染治理提供了全球普遍参考的目标框架。自 1987 年首次发布欧洲空气质量指南以来，WHO 定期更新和发布基于健康风险评估的空气质量指南。2021 年 9 月，WHO 发布最新修订的《全球空气质量指南》[78]，更新了 $PM_{2.5}$、PM_{10}、臭氧（O_3）、二氧化氮（NO_2）、二氧化硫（SO_2）和一氧化碳（CO）6 种主要空气污染物的指导值和过渡期阶段目标值。该指南修订的重要变化包括：①将 $PM_{2.5}$ 年均指导值由 10 μg/m³ 收紧为 5 μg/m³；②PM_{10} 年均指导值由 20 μg/m³ 收紧为 10 μg/m³；③NO_2 的年均指导值从 40 μg/m³ 收紧为 10 μg/m³；④增设臭氧浓度高峰季平均值（60 μg/m³）；等等。2022 年 10 月，欧盟委员会通过欧盟《环境空气质量指令》修订案，制定了 2030 年欧盟空气质量中期目标，收紧主要空气污染物 $PM_{2.5}$ 的年度限值[79]。

三、启示与建议

1. 强化碳中和科技布局及关键技术研发

科技创新是实现碳达峰碳中和目标的关键支撑。主要发达国家已对碳中和科技创新优先领域进行了相关部署。2021 年 10 月以来，我国发布了《2030 年前碳达峰行动方案》和部分领域实施方案等政策文件。面向碳中和重大科技需求，我国需要加快布局创新的科技研发体系，推进碳中和关键核心技术研发与应用示范，包括可再生能源、生物质能、氢能、储能、建筑能效、工业脱碳以及生态固碳增汇、CCUS 等低碳、零碳、负碳重点技术的突破[80]。

2. 加强生态系统碳汇潜力及其增强途径研究

生态系统碳汇大小及增汇途径是实现碳中和目标的关键因素。在此背景下，我国陆地生态系统碳汇潜力研究引起了科学界前所未有的关注并取得了一些创新性成果。由于陆地生态系统的强烈异质性，陆地碳汇估算仍然存在较大的不确定性，因而，仍需进一步加强陆地碳汇领域相关研究，包括：我国陆地生态系统碳储量监测体系；生态系统碳循环过程机理和碳汇功能时空变异，以及与全球气候变化的相互影响；我国陆地生态系统固碳速率及其不确定性和稳定性；气候变化和人为活动对生态系统增汇潜力的贡献等。

3. 推进大气污染治理与碳减排协同路径研究

协同推进减污降碳是促进我国全面绿色转型的必然选择。因此，气候变化和大气污染协同治理将是重点的研究方向，未来应加强以下领域的研究：大气污染防治措施等相关政策的健康影响效益分析；碳达峰与碳中和目标下的 $PM_{2.5}$ 和臭氧污染协同治理；空气污染暴露特征及其对我国人群健康影响的本地化科学研究；空气质量标准和目标体系评估；大气污染与气候变化协同路径对健康和经济的影响；碳中和与清洁空气协同的大气复合污染防治技术体系。

4. 开展陆地微塑料通量与人体健康影响研究

当前我国围绕陆地土壤和大气微塑料的研究也取得了一定进展，但总体而言还处于起步阶段[81]，同时微塑料污染对人类健康的潜在影响尚未得到系统评估。我国亟须开展陆地生态系统中微塑料的相关研究，并采取相应措施控制微塑料污染：在微塑料的赋存和迁移方面，建立不同环境介质中微塑料检测和量化分析的统一标准体系，规范微塑料含量的表达方法；探讨环境微塑料的来源以及影响其分布和迁移的因素，对微塑料中吸附的污染物进行识别和量化。在对人体和生物潜在毒性方面，完善微塑料毒性的测定方法，研究微塑料和纳米塑料颗粒的毒性机制；评估不同类型微塑料在人类不同接触途径中的累积。

致谢：中国科学院院士/中国科学院城市环境研究所朱永官研究员、中国科学院南京土壤研究所骆永明研究员、南京农业大学农业资源与生态环境研究所潘根兴教授等审阅了本文并提出了宝贵的修改意见，中国科学院西北生态环境资源研究院吴秀平、牛艺博、李恒吉、魏艳红、王金平、秦冰雪等对本文的资料收集和分析工作亦有贡献，在此一并表示感谢。

参考文献

[1] UNEP. Nature-based Solutions for Climate Change Mitigation. https://www.unep.org/resources/report/nature-based-solutions-climate-change-mitigation[2021-11-04].

[2] Littleton E W, Dooley K, Webb G, et al. Dynamic modelling shows substantial contribution of ecosystem restoration to climate change mitigation. Environmental Research Letters, 2021, 16: 124061.

[3] Walker W S, Gorelik S R, Patton S C, et al. The global potential for increased storage of carbon on land. PNAS, 2022, 119(23): e2111312119.

[4] Wang J, Feng L, Palmer P I, et al. Large Chinese land carbon sink estimated from atmospheric carbon dioxide data. Nature, 2020, 586: 720-723.

[5] Duffy K A, Schwalm C R, Arcus V L, et al. How close are we to the temperature tipping point of the terrestrial biosphere? Science Advances, 2021, 7(3): eaay1052.

[6] Wang S H, Zhang Y G, Ju W M, et al. Recent global decline of CO_2 fertilization effects on vegetation photosynthesis. Science, 2020, 370(6522): 1295-1300.

[7] Qin Y, Xiao X, Wigneron J P, et al. Carbon loss from forest degradation exceeds that from deforestation in the Brazilian Amazon. Nature Climate Change, 2021, 11: 442-448.

[8] UNESCO. World Heritage Forests: Carbon Sinks under Pressure. https://unesdoc.unesco.org/ark:/48223/pf0000379527/PDF/379527eng.pdf.multi[2021-10-28].

[9] Loisel J, Gallego-Sala A V, Amesbury, M J, et al. Expert assessment of future vulnerability of the global peatland carbon sink. Nature Climate Change, 2021, 11: 70-77.

[10] Cowie R H, Bouchet P, Fontaine B. The sixth mass extinction: fact, fiction or speculation? Biological Reviews, 2022, 97(2): 640-663.

[11] World Wildlife Fund (WWF). Living Planet Report 2022. https://www.wwf.org.uk/sites/default/files/2022-10/lpr_2022_full_report.pdf[2022-10-13].

[12] House of Commons. Biodiversity in the UK: Bloom or Bust? https://committees.parliament.uk/publications/6498/documents/70656/default/[2021-06-30].

[13] Hu Y S, Wang M, Ma T X, et al. Integrated index-based assessment reveals long-term conservation progress in implementation of Convention on Biological Diversity. Science Advances, 2022, 8(1): eabj8093.

[14] Oliver R Y, Meyer C, Ranipeta A, et al. Global and national trends, gaps, and opportunities in documenting and monitoring species distributions. PLoS Biology, 2021, 19(8): e3001336.

[15] Jung M, Arnell A, de Lamo X, et al. Areas of global importance for conserving terrestrial biodiversity, carbon and water. Nature Ecology Evolution, 2021, 5: 1499-1509.

[16] Famiglietti J S, Ferguson G. The hidden crisis beneath our feet. Science, 2021, 372(6540): 344-345.

[17] UN WATER. UN World Water Development Report 2022. https://www.unwater.org/publications/un-world-water-development-report-2022/[2022-03-21].

[18] Jasechko S, Perrone D. Global groundwater wells at risk of running dry. Science, 2021, 372(6540): 418-421.

[19] MacAllister D J, Krishan G, Basharat M, et al. A century of groundwater accumulation in Pakistan and northwest India. Nature Geoscience. 2022, 15: 390-396.

[20] Wunsch A, Liesch T, Broda S. Deep learning shows declining groundwater levels in Germany until 2100 due to climate change. Nature Communications, 2022, 13: 1221.

[21] Schmidt A H, Lüdtke S, Andermann C. Multiple measures of monsoon-controlled water storage in Asia. Earth and Planetary Science Letters, 2020, 546: 116415.

[22] Vasco D W, Kim K H, Farr T G, et al. Using Sentinel-1 and GRACE satellite data to monitor the hydrological variations within the Tulare Basin, California. Scientific Reports, 2022, 12: 3867.

[23] Brahney J, Mahowald N, Prank M, et al. Constraining the atmospheric limb of the plastic cycle. PNAS, 2021, 118(16): e2020719118.

[24] Revell L E, Kuma P, Le Ru E C, et al. Direct radiative effects of airborne microplastics. Nature, 2021, 598: 462-467.

[25] Luo Y, Li L, Feng Y, et al. Quantitative tracing of uptake and transport of submicrometre plastics in crop plants using lanthanide chelates as a dual-functional tracer. Nature Nanotechnology, 2022, 17: 424-431.

[26] Ragusa A, Svelato A, Santacroce C, et al. Plasticenta: First evidence of microplastics in human placenta. Environment International, 2021, 146: 106274.

[27] Su Y, Hu X, Tang H J, et al. Steam disinfection releases micro(nano)plastics from silicone-rubber baby teats as examined by optical photothermal infrared microspectroscopy. Nature Nanotechnology,

2022,17:76-85.

[28] Leslie H A, van Velzen M J M, Brandsma S H, et al. Discovery and quantification of plastic particle pollution in human blood. Environment International, 2022, 163:107199.

[29] Jenner L C, Rotchell J M, Bennett R T, et al. Detection of microplastics in human lung tissue using μFTIR spectroscopy. Science of the Total Environment, 2022, 831:154907.

[30] Lin S Y, Zhang H N, Wang C, et al. Metabolomics reveal nanoplastic-induced mitochondrial damage in human liver and lung cells. Environmental Science & Technology, 2022, 56(17):12483.

[31] WHO. COP26 special report on climate change and health: the health argument for climate action. https://www.who.int/publications/i/item/9789240036727[2021-10-11].

[32] Geng G, Zheng Y, Zhang Q, et al. Drivers of $PM_{2.5}$ air pollution deaths in China 2002-2017. Nature Geoscience, 2021, 14:645-650.

[33] Cheng J, Tong D, Zhang Q, et al. Pathways of China's $PM_{2.5}$ air quality 2015-2060 in the context of carbon neutrality. National Science Review, 2021, 8(12):nwab078.

[34] Zhang S H, An K X, Li J, et al. Incorporating health co-benefits into technology pathways to achieve China's 2060 carbon neutrality goal: a modelling study. The Lancet Planetary Health, 2021, 5(11):e808-e817.

[35] Tang R, Zhao J, Liu Y, et al. Air quality and health co-benefits of China's carbon dioxide emissions peaking before 2030. Nature Communications, 2022, 13:1008.

[36] Resources for the Future. The Distribution of Air Quality Health Benefits from Meeting US 2030 Climate Goals. https://www.rff.org/publications/reports/the-distribution-of-air-quality-health-benefits-from-meeting-us-2030-climate-goals/[2022-01-03].

[37] IPCC. Climate Change 2021: The Physical Science Basis. https://report.ipcc.ch/ar6/wg1/IPCC_AR6_WGI_FullReport.pdf[2021-08-09].

[38] Fischer E M, Sippel S, Knutti R. Increasing probability of record-shattering climate extremes. Nature Climate Change, 2021, 11:689-695.

[39] Studholme J, Fedorov A V, Gulev S K, et al. Poleward expansion of tropical cyclone latitudes in warming climates. Nature Geoscience, 2022, 15:14-28.

[40] Gori A, Lin N, Xi D Z, et al. Tropical cyclone climatology change greatly exacerbates US extreme rainfall-surge hazard. Nature Climate Change, 2022, 12:171-178.

[41] The White House. Paris Climate Agreement. https://www.whitehouse.gov/briefing-room/statements-releases/2021/01/20/paris-climate-agreement/[2021-01-20].

[42] The White House. Executive Order on Tackling the Climate Crisis at Home and Abroad. https://www.whitehouse.gov/briefing-room/presidential-on-tackling-the-climate-crisis-at-home-and-abroad/[2021-01-27].

[43] The White House. The Long-term Strategy of the United States: Pathways to Net-Zero Greenhouse Gas Emissions by 2050. https://www.whitehouse.gov/wp-content/uploads/2021/10/US-Long-Term-

Strategy. pdf[2021-11-01].

[44] Department of Energy. Industrial Decarbonization Roadmap. https://www. energy. gov/eere/doe-industrial-decarbonization-roadmap[2022-09-07].

[45] Department for Business, Energy & Industrial Strategy(BEIS). Net Zero Innovation Portfolio. https://www. gov. uk/government/collections/net-zero-innovation-portfolio[2021-03-03].

[46] Department for Business, Energy & Industrial Strategy(BEIS). Industrial Decarbonisation Strategy. https://www. gov. uk/government/publications/industrial-decarbonisation-strategy[2021-03-17].

[47] Department for Transport. Decarbonising Transport: A Better, Greener Britain. https://www. gov. uk/government/publications/transport-decarbonisation-plan[2021-07-14].

[48] Department for Business, Energy & Industrial Strategy (BEIS). Net Zero Strategy: Build Back Greener. https://assets. publishing. service. gov. uk/government/uploads/system/uploads/attachment_data/file/1026655/net-zero-strategy. pdf[2021-10-19].

[49] Department for Business, Energy & Industrial Strategy (BEIS). UK Net Zero Research and Innovation Framework. https://www. gov. uk/government/publications/net-zero-research-and-innovation-framework [2021-10-19].

[50] Federal Ministry for the Eironment, Nature Conservation, Nuclear Safety and Consumer Protection. Revised Climate Change Act Sets out Binding Trajectory Towards Climate Neutrality by 2045. https://www. bmuv. de/en/pressrelease/revised-climate-change-act-sets-out-binding-trajectory-towards-climate-neutrality-by-2045[2021-05-12].

[51] Ministry of Science and ICT. Strategy for Technology Innovation for Carbon Neutral. https://english. msit. go. kr/eng/bbs/view. do? sCode=eng&mId=4&mPid=2&pageIndex=&bbsSeqNo=42&nttSeqNo=492&searchOpt=ALL&searchTxt=[2021-08-27].

[52] Environment and Climate Change Canada. 2030 Emissions Reduction Plan: Canada's Next Steps for Clean Air and a Strong Economy. https://www. canada. ca/content/dam/eccc/documents/pdf/climate-change/erp/Canada-2030-Emissions-Reduction-Plan-eng. pdf[2022-03-29].

[53] Department of Agriculture, Water and the Environment. National Soil Strategy. https://www. agriculture. gov. au/sites/default/files/documents/national-soil-strategy. pdf[2021-05-11].

[54] Department of Agriculture, Water and the Environment. Commonwealth Interim Action Plan National Soil Strategy. https://www. agriculture. gov. au/sites/default/files/documents/commonwealth-interim-action-plan-national-soil-strategy. pdf[2021-05-11].

[55] UK Government. The England Trees Action Plan 2021-2024. https://assets. publishing. service. gov. uk/government/uploads/system/uploads/attachment_data/file/987432/england-trees-action-plan. pdf[2021-05-18].

[56] UK Government. England Peat Action Plan. https://assets. publishing. service. gov. uk/government/uploads/system/uploads/attachment_data/file/987060/england-peat-action-plan. pdf[2021-05-18].

[57] European Commission. Soil strategy for 2030. https://ec. europa. eu/environment/strategy/soil-

strategy_en[2021-11-17].

[58] Executive Office of the President. Strengthening the Nation's Forests, Communities, and Local Economies. https://www. federalregister. gov/documents/2022/04/27/2022-09138/strengthening-the-nations-forests-communities-and-local-economies[2022-04-22].

[59] 정책브리핑（KOREA. KR）. 친환경수소전기열차개발성공,철도분야탄소중립을위한첫걸음내딛다 . https://www. gov. kr/portal/gvrnPolicy/view/H2212000000956951? policyType=G00301&srchTxt=%ED%83%84%EC%86%8C%EC%A4%91%EB%A6%BD[2022-12-28].

[60] UNEP. 1st Draft of the Post-2020 Global Biodiversity Framework. https://www. iucn. org/sites/dev/files/iucn_key_messages_and_detailed_views_first_draft_post-2020_gbf_0. pdf[2021-07-12].

[61] UNEP. Human Rights and Biodiversity. https://wedocs. unep. org/bitstream/handle/20. 500. 11822/35407/KMBio. pdf? sequence=1&isAllowed=y[2021-02-25].

[62] Legislation. gov. uk. Environment Act 2021: A PART 6 Nature and biodiversity. https://www. legislation. gov. uk/ukpga/2021/30/part/6/enacted[2021-11-10].

[63] Department of Agriculture, Water and the Environment. Threatened Species Strategy 2021-2031. https://www. agriculture. gov. au/sites/default/files/documents/threatened-species-strategy-2021-2031. pdf[2021-05-21].

[64] European Commission Horizon Europe Work Programme 2021-2022: 9. Food, Bioeconomy, Natural Resources, Agriculture and Environment. https://ec. europa. eu/info/funding-tenders/opportunities/docs/2021-2027/horizon/wp-call/2021-2022/wp-9-food-bioeconomy-natural-resources-agriculture-and-environment_horizon-2021-2022_en. pdf[2021-06-15].

[65] European Commission. More Than 290 Million for Nature, Environment and Climate Action Projects. https://cinea. ec. europa. eu/news-events/news/more-eu290-million-nature-environment-and-climate-action-projects-2021-11-25_en[2021-11-25].

[66] Department for Environment, Food & Rural Affairs. Government unveils plans to restore 300,000 hectares of habitat across England. https://www. gov. uk/government/news/government-unveils-plans-to-restore-300000-hectares-of-habitat-across-england[2022-01-06].

[67] UNEP. $43 million boost for developing countries' efforts to reverse species loss. https://www. unep. org/news-and-stories/press-release/43-million-boost-developing-countries-efforts-reverse-species-loss[2022-05-20].

[68] Convention on Biological Diversity. Nations Adopt Four Goals, 23 Targets for 2030 In Landmark UN Biodiversity Agreement. https://prod. drupal. www. infra. cbd. int/sites/default/files/2022-12/221219-CBD-PressRelease-COP15-Final_ 0. pdf [2022-12-19]. [69] WMO. WMO endorses Water Declaration, including the Water and Climate Coalition. https://public. wmo. int/en/media/news/wmo-endorses-water-declaration-including-water-and-climate-coalition[2021-10-18].

[70] U. S. Geological Survey(USGS). Water Resources Research Act Program-Current Status, Development Opportunities, and Priorities for 2020-30. https://www. usgs. gov/publications/water-resources-

research-act-program-current-status-development-opportunities-and[2021-11-24].
[71] UNEP. Freshwater Strategic Priorities 2022-2025. https://wedocs.unep.org/bitstream/handle/20.500.11822/39607/Freshwater_Strategic_Priorities.pdf[2022-03-22].
[72] UNESCO. IHP-IX: Strategic Plan of the Intergovernmental Hydrological Programme: Science for a Water Secure World in a Changing Environment, ninth phase 2022-2029. https://unesdoc.unesco.org/ark:/48223/pf0000381318.
[73] Department of Agriculture, Water and Environment. National Plastics Plan 2021. https://www.environment.gov.au/protection/waste/plastics-and-packaging[2021-03-31].
[74] Arctic Monitoring and Assessment Programme. AMAP Litter and Microplastics Monitoring Plan. https://www.amap.no/documents/doc/amap-litter-and-microplastics-monitoring-plan/3522[2021-05-20].
[75] Ministry for the Environment. National Plastics Action Plan. https://environment.govt.nz/news/acting-on-plastic-waste-the-government-releases-its-national-plastics-action-plan/[2021-09-16].
[76] India Plastics Pact. India Plastics Pact. https://www.indiaplasticspact.org/[2021-09-03].
[77] UNEP. End Plastic Pollution: Towards an International Legally Binding Instrument. https://wedocs.unep.org/bitstream/handle/20.500.11822/38522/k2200647_-_unep-ea-5-l-23-rev-1_-_advance.pdf[2022-03-02].
[78] WHO. WHO global air quality guidelines: Particulate matter ($PM_{2.5}$ and PM_{10}), ozone, nitrogen dioxide, sulfur dioxide and carbon monoxide. https://apps.who.int/iris/handle/10665/345329[2021-09-22].
[79] European Commission. European Green Deal: Commission Proposes Rules for Cleaner Air and Water. https://ec.europa.eu/commission/presscorner/detail/en/ip_22_6278[2022-10-26].
[80] 曲建升,陈伟,曾静静,等. 国际碳中和战略行动与科技布局分析及对我国的启示建议. 中国科学院院刊,2022,37(4):444-458.
[81] Wang L W, Wu W M, Bolan N S, et al. Environmental fate, toxicity and risk management strategies of nanoplastics in the environment: Current status and future perspectives. Journal of Hazardous Materials, 2021, 401: 123415.

Ecology and Environmental Science

Liao Qin, Qu Jiansheng, Zeng Jingjing, Pei Huijuan, Dong Liping, Liu Yanfei, Liu Lina

During 2021～2022, terrestrial ecosystem carbon sinks, biodiversity, groundwater reserves, terrestrial and atmospheric microplastics, and coordinated governance of climate change and air pollution are the core topics of concern in the field of global ecological and environmental sciences. A series of important research progress has been made in the scientific community around these areas. At the same time, under the target of carbon neutrality, major countries and international organizations have actively formulated strategic actions related to carbon neutrality, promoted scientific and technological innovation research and development, and deployed relevant strategies and action plans in soil protection and carbon sinks, biodiversity restoration, water resources management and research, and plastic pollution control.

4.6 地球科学领域发展观察

郑军卫[1] 刘文浩[1] 张树良[1] 刘 学[1] 王立伟[1] 翟明国[2]

(1. 中国科学院西北生态环境资源研究院文献情报中心;
2. 中国科学院地质与地球物理研究所)

近年来，在地球系统科学的理论框架下，开展地球各圈层物质循环及与之相关的资源能源富集、生态环境变迁、地球宜居性和社会可持续发展等问题研究成为新时期地球科学重要的发展方向。2021年以来，地球科学领域①取得了多项重要发现和突破：地球深部地质结构及物质和能量循环研究取得重要进展；科学和技术的进步拓展了探矿和采矿空间；行星地质学在揭示火星内部结构和月球演化方面取得突破；深层地热能潜力评价与高效开发技术研究得到重视；机器学习促进地震与火山预报系统研究不断深入；地球系统观测及模拟研究取得进展；等等。一些国家和国际组织亦围绕上述相关领域进行了研究部署。

一、重要研究进展

1. 地球深部地质结构及物质和能量循环研究取得重要进展

地球深部地质结构及物质和能量循环研究在地壳演化与构成、地幔组分与变化、地核结构与动力过程、板块构造机理与影响等多个领域取得新的突破和认识，对固体地球科学的研究方向和发展目标产生了深远的影响。挪威卑尔根大学联合德国明斯特大学等机构研究人员发现[1]，大陆地壳首次发生风化作用的时间约始于37亿年前，比之前的认识要早5亿年，这对于重新审视早期海洋化学、生命进化和板块构造具有重要意义。法国国家科学研究中心[2]的一项研究也发现，早在37亿年前，地球上就存在跟现今一样的长英质大陆，大陆地壳以5亿～7亿年为周期，发生了6次幕式增生。美国亚利桑那州立大学和芝加哥大学等机构[3]研究指出，温度足够高的深部下地幔具有不同于浅层下地幔的矿物学特征。美国加州大学伯克利分校联合法国国家科学

① 本文所指的地球科学领域主要涉及地质学、地球物理学、地球化学、大气科学、行星科学等学科。

研究中心等机构研究提出了一个在地球内核中不平衡固体对流的模型[4]，描述了在地震各向异性约束下的地球内核动力历史，肯定了地球内核具有相对年轻的年龄（5 亿～15 亿年），该研究成果对深入理解地球内部的演化过程至关重要。美国宝石学院和普渡大学等机构[5]研究发现，地球上地幔和下地幔之间的过渡带含有大量的水，揭示地球内部存在水循环。澳大利亚科廷大学联合奥地利格拉茨大学等[6]的研究也表明，水在早期地球中的运移深度要比此前认为的更深，含水的超镁铁质岩石是早期地球地壳深部和上地幔中水的主要来源。世界主要国家提出“双碳”目标，推动了科学家们对地球深部碳的研究。英国剑桥大学联合德国明斯特大学等机构研究指出，构造板块的缓慢碰撞将更多的碳带入地球内部，地球内部固定的碳比先前理论认为的要多[7]。美国佛罗里达州立大学和莱斯大学的联合研究认为，地球外核可能是地球最大的碳库[8]。澳大利亚悉尼大学联合英国格拉斯哥大学等机构[9]通过重建海洋板块碳储层和利用热力学建模跟踪板块俯冲带碳的去向，重现了地球深部碳循环带的演化历史，为未来的碳循环模型提供了边界条件。

2. 科学和技术的进步拓展了探矿和采矿空间

欧盟[10]、加拿大[11]、美国[12]、英国[13]、澳大利亚[14]等国家和组织不断推出或更新关键矿产清单，推动着全球矿产的研发重点与勘查对象不断调整，从传统大宗矿产向紧缺的关键矿产转移，从地球浅表层向地球深部和偏远地区转移。德国埃尔朗根-纽伦堡大学联合加拿大渥太华大学等机构研究指出，岩浆作用过程中分馏的铁橄榄石可能会含有大量的重稀土元素，并建议将铁橄榄石堆晶作为未来重稀土勘探的潜在目标[15]。2021 年 5 月，法国埃赫曼（Eramet）公司[16]宣布，借助于创新材料和突破性的直接提锂工艺，该公司在法国东部一座试验厂从地热卤水中成功提取出了锂，这是锂矿领域的世界首创，该技术有望在保障欧洲能源转型所需的关键材料方面发挥重要作用。2021 年 10 月，澳大利亚地球科学局（Geoscience Australia）[17]宣布其“探索未来”项目（EFTF）“澳大利亚岩石圈结构大地电磁工程”（AusLAMP）研发出一种新的电导率模型，可用于缩小在勘探不足地区寻找大型成矿系统的范围，帮助行业和政府填补现有矿产和关键矿产知识的空缺，为深部地质勘探研究提供支撑。2022 年 2 月，美国能源部（DOE）[18]宣布投资 1.4 亿美元用于建造美国首个从化石燃料废料中提取稀土的大型设施。

3. 行星地质学在揭示火星内部结构和月球演化方面取得突破

2018 年美国“洞察号”火星探测器在火星着陆以来，一直在尝试获取火星内部的信息，但都被火星外部的黏性沉积物所阻挡，直到 2022 年 5 月 11 日“洞察号”才探测到火星地震的信息。科学家们对探测到的约733次火星地震中的35次数据进行分析，揭示了火星内部结构，估计了火星地核大小、地幔结构和地壳厚度，这对进一步了解火星形成和热演化具有重要意义。2021年7月，《科学》（第6553期）同时发表多篇科研人员根据NASA“洞察号”火星探测器数据，分别从地震调查、大地测量和热传输方面对火星内部进行探索的论文[19-21]及评述文章[22]，报道了这一重大发现。该项研究成果被《科学》评为2021年的世界十大科学突破之一[23]。我国在火星探测方面也取得了重要成就，2021年6月11日，国家航天局公布了由“祝融号”火星车拍摄的着陆点全景、火星地形地貌、“中国印迹”和“着巡合影”等影像图。首批科学影像图的发布，标志着我国首次火星探测任务取得圆满成功，该项成果被列为科学技术部发布的2021年度中国科学十大进展[24]之一。中国科学院地质与地球物理研究所行星与月球内部结构研究团队，联合国家空间科学中心和北京大学等机构的科研人员[25]，在对“祝融号”火星车前113个火星日采集的长度约1171 m的低频雷达数据展开了深入分析后，揭秘了火星乌托邦平原浅表的分层结构。该项成果被列为2022年度中国科学十大进展[26]之一。另外一项被评为2021年度中国科学十大进展之一的成果是由中国科学院地质与地球物理研究所和国家天文台主导、多家研究机构联合，利用“嫦娥五号”样品开展的有关月球演化重要科学问题的研究取得的突破性进展。系列成果发表在《自然》[27-29]和《国家科学评论》[30]上。该系列成果提供了迄今月球上确定的最年轻的2030 Ma玄武岩的证据，改变了科学界以往对月球的热历史和岩浆历史的认识，对人类认识月球的起源和演化具有重要的意义[31]。中国地质科学院地质研究所联合澳大利亚科廷大学等机构对“嫦娥五号”月球玄武岩开展了年代学、元素、同位素分析，证明了月球在19.6亿年前仍存在岩浆活动[32]，为完善月球演化历史供了关键科学证据。

4. 深层地热能潜力评价与高效开发技术研究得到高度重视

地热作为一种清洁的可再生能源，被认为是未来化石能源的可靠替代品之一。随着“双碳”目标的提出，美国、英国等多个国家越来越重视地热能资源的勘探和高效开发。2021年6月，英国地质调查局（British Geological Survey，BGS）宣布开启地热能潜力研究，着眼于发现深层石炭系灰岩地热能资源并评估其潜力，旨在利用石炭纪灰岩中蕴藏的地热作为新的清洁型可再生能源[33]，帮助英国在2050年实现碳的净零排放。2021年9月，美国DOE宣布为7个研究项目提供1200万美元的资助，用于推进增强型地热系统（enhanced geothermal system，EGS）的商业化，以及帮助部署和开发高效和低成本的地热能生产技术[34]；2022年8月，DOE宣布继续投资4400万美元用于推进EGS相关工作，聚焦于地震监测协议、新型油藏增产技术、EGS热抽采效率实验、维持 EGS 石油流动通道的材料以及能够承受高温同时隔离井筒内区域的工具研发[35]；EGS 的开发可以提升美国地热能的利用率。2022 年 12 月，欧盟宣布将通过欧盟创新基金（Innovation Fund）支持 Eavor-Loop 项目，并为创新闭环地热技术的首次商业规模实施提供资金[36]。

5. 机器学习促进地震与火山预报系统研究不断深入

近年来，机器学习在地震与火山预报方面得到普遍应用和快速发展。2021 年 4 月 23 日，在美国地震学会 2021 年年会上，美国斯坦福大学研究人员[37]报告称，其开发的 DeepShake 系统，可基于经过 3.6 万次地震训练的深度时空神经网络对地震信号进行实时分析，并根据地震中最早探测到的地震波的特征，发布强震预警。美国斯坦福大学和英国地质调查局[38]合作，基于监督式机器学习方法开发出新一代地震目录，以前所未有的研究细节揭示了地震活动，再利用无监督机器学习方法分析这些包含完整地震信息的目录，通过追踪最终控制地震前震和余震的精细尺度的应力场演化，来提升对地震的预报能力。新西兰惠灵顿维多利亚大学联合英国卡迪夫大学等机构[39]在对过去 200 万年中全球发生的 13 次超级火山喷发的野外地质、地球化学和岩石学证据进行深入核查的基础上，指出目前还没有一个统一的模型可以刻画超大规模的火山喷发事件及其岩浆体，但使用机器学习算法可以帮助解译，在火山喷发前数小时或数天内岩浆房中的岩浆向地表移动的信号。美国洛斯阿拉莫斯国家实验室的研究人员基于深度学习转换模型，实现了对实验室地震模拟过程中断层发生滑动及时间的预测[40]。

二、重大战略行动

1. 美国提出下一代地球系统科学研究愿景

随着技术进步和生产活动的扩大，人类已经成为与自然过程相当的、影响地球系统运行的重要营力。但目前人类对地球系统的了解还存在许多知识空白，难以跟上地球系统本身的快速变化，人类对地球系统影响的程度、地球系统对人类和生态系统可持续性与恢复力的影响以及应对这些挑战的有效途径等仍不明晰。针对上述问题，应美国国家科学基金会（National Science Foundation，NSF）要求，美国国家科学院、国家工程院和国家医学院（National Academies of Sciences，Engineering，and Medicine）成立了专门委员会从事地球系统科学发展愿景研究。2021 年 9 月 22 日，美国国家科学院、国家工程院和国家医学院发布报告《国家科学基金会的下一代地球系统科学》（Next Generation Earth Systems Science at the National Science Foundation），建议 NSF 创建下一代地球系统科学计划，来探索自然世界和社会之间复杂的相互作用，并增进人类对包括大气圈、水圈、岩石圈、冰冻圈、生物圈在内的复杂地球系统的理解[41]。该报告指出下一代地球系统综合研究方法需要体现以下 6 个关键特征：①在空间、时间和社会组织尺度上推动好奇心和现实需求驱动的地球系统基础研究；②促进社会科学、自然科学、计算科学和工程学的融合，为与地球系统相关的问题提供解决方案；③确保地球系统科学的多样性、包容性、公平性和公正性；④优先考虑与不同利益攸关方的接触和伙伴关系，以造福社会并解决社区、州、国家和国际范围内与地球系统相关的问题；⑤协同观测、计算和建模能力，以加速地球系统科学模型的发现和融合；⑥教育和支持拥有技能和知识的员工，以有效地识别、开展和传播地球系统科学研究。2022 年 11 月，美国 NSF 发布《“俯冲带四维”（SZ4D）实施计划 2022》（SZ4D Implementation Plan 2022）[42]，提出必须解决的 5 个关键科学问题：①破坏性大地震何时何地发生？②地壳过程如何引发岛弧火山喷发？③地球大气、水圈和固体地球内部的事件如何在俯冲带陆-海景观间产生和运输沉积物？④预测一个俯冲带有多少能量用于建造和塑造俯冲带的陆-海景观？⑤如何转变社区的思维方式，将教育、外部网络关系、国际伙伴关系、多样性、公平性、包容性和社会科学作为关键要素？

2.《科学》发布新的地球科学前沿问题

2021 年 4 月 8 日，《科学》期刊联合中国上海交通大学发布了“125 个问题：探索和发现”（125 Questions：Exploration and Discovery），提出了“全世界最前沿的 125 个科学问题”，其中涉及地球科学领域的问题包括：什么可以帮助保护海洋？有机体是如何进化的？为什么恐龙长得如此之大？远古人类是否曾与其他类人祖先杂交？能否复活已灭绝的生物？为什么会发生物种大爆发和大灭绝？水是宇宙中所有生命所必需的吗，还是仅对地球生命必要？是什么阻止了人类进行深空探测？深层生物圈的规模、组成和意义是什么？人类有一天会不得不离开地球吗（还是会在尝试中死去）？我们可以阻止全球气候变化吗？我们能把过量的二氧化碳存到何处？是什么创造了地球的磁场（为什么它会移动）？我们是否能够更准确地预测灾害性事件（海啸、飓风、地震）？如果地球上所有的冰融化会怎样？我们可以生活在一个去化石燃料的世界中吗？氢能的未来是怎样的？[43]

3. 欧盟启动“目标地球”计划

2021 年 3 月 5 日，欧盟委员会（European Commission）启动为期 7～10 年的“目标地球”（Destination Earth，DestinE）计划，将创建一个高度精确的地球数字孪生（digital twin）模型，尽可能准确地进行全球自然资源与现象的数字建模，为应对重大环境退化与灾难提供决策支持[44]。DestinE 计划的目标是利用超高精度的地球数字模型，监测与模拟自然和人类活动，建立能够实现更可持续发展及支持欧洲环境政策的情景，并对这些情景进行测试。DestinE 计划将激发在全球尺度对地球自然资源和相关现象进行数字建模的潜力，如气候变化、水/海洋环境、极地地区和冰冻圈等，以加快绿色转型并帮助制定应对重大环境退化和灾难的规划。

4. 多国密集布局关键矿产资源研发

美国通过成立新机构、支持基础研发等措施，提升其国内关键矿产资源供应能力。2021 年 1 月，DOE 宣布成立一个新的机构——矿产可持续发展司（Division of Minerals Sustainability），以促进建立从上游到中游、下游以及众多相关行业，在环境、经济和地缘政治上可持续的关键矿产供应链。2021 年 3 月，DOE 宣布投入 3000 万美元开展研究，以确保关键矿产的国内供应安全[45]，并在 9 月宣布对 13 项由国家实验室和大学牵头的项目进行资助[46]。资金主要用于研究稀土和铂族元素的基本特性，及发现替代品所需的基础化学、材料科学和地球科学基础研究。为提升澳大利亚的矿产品附加值，2021 年 3 月，澳大利亚工业、科学、能源与资源部（Department of Industry，Science and Resources，DISER）发布面向未来 10 年的《资源技术和关键矿产加工国家制造优先路线图》（Resources Technology and Critical Minerals Processing National Manufacturing Priority Road Map），旨在建立起澳大利亚本土的关键矿产品加工能力，帮助制造商加快生产，将产品商业化并加入全球供应链[47]。5 月，澳大利亚联邦科学与工业研究组织（Commonwealth Scientific and Industrial Research Organisation，CSIRO）发布《关键能源矿产路线图》(Critical Energy Minerals Roadmap)，提出了在未来几十年将可能加速增长的可再生能源技术（太阳能光伏、风能、聚光太阳能发电、电池和氢能），并评估了澳大利亚从供应这些技术所需的关键矿产资源（锂、铝、硅等）中获得价值的潜力，确定了实现这些机遇所需的商业、监管和研发相关的优先投资事项[48]。2021 年 6 月，加拿大众议院自然资源常务委员会发布报告《从矿产勘探到先进制造业：发展加拿大关键矿产的价值链》（From Mineral Exploration to Advanced Manufacturing：Developing Value Chains for Critical Minerals in Canada）[49]，梳理了关键矿产在众多尖端技术应用和支持能源转型方面发挥的作用，明确了与加拿大关键矿产有关的经济、环境和供应问题，确立了在加拿大发展与关键矿产相关的价值链，并提出了当前充分发挥该行业潜力的挑战与相关建议。2022 年 3 月，澳大利亚宣布拨款 2.43 亿澳元用于资助资源技术和关键矿产加工领域的 4 个新项目[50]。6 月，美国内政部（DOI）宣布拨款 7460 万美元资助开展关键矿产潜力研究和填图[51]。

5. 主要国家和组织加强地球系统数值模拟器建设

地球系统数值模拟器是当前进行地球系统研究的重要手段，2021 年，一些国家和组织制定了新的地球模拟器发展战略计划和项目。欧盟委员会除了在 2021 年 3 月 DestinE 计划中提出将创建一个高度精确的地球数字模型外，5 月再次宣布将通过“地平线 2020”计划（Horizon 2020），资助两项新的地球系统模式（earth system model，ESM）开发项目[52]，以进一步改善地球系统模式对关键过程和耦合系统的表达，提升气候模式的模拟、预测和预估能力。2021 年 6 月，由中国科学院大气物理研究所牵头的国家重大科技基础设施地球系统数值模拟装置“寰”（EarthLab）落成启用。这是我国的又一项大国重器，通过超级计算机进行大规模的数值计算，能够重现地球的过去、模拟地球的现在、预测地球的未来，从而进行有针对性的“地球试验”。2021 年 7 月，波茨坦气候影响研究所（The Potsdam Institute for Climate Impact Research，PIK）将其最先进的全球植被动态模式（dynamic global vegetation model）LPJmL5 与美国地球物理流体动力学实验室（Geophysical Fluid Dynamics Laboratory，GFDL）开发的全球耦合气候模式 CM2Mc 进行耦合[53]，得到 CM2Mc-LPJmL v. 1. 0。2021 年 8 月，德国亥姆霍兹波茨坦地球科学研究中心（Helmholtz-Centre Potsdam-German Research Centre for Geosciences，GFZ）基于对地球系统模型（ESMs）和人工智能的系统分析，创造性地提出了“神经地球系统建模”（neural earth system modelling）这一全新概念[54]，为地球系统数值模拟器的发展提供了新的方向。2022 年 8 月，美国 DOE[55]投资 7000 万美元，用于改进地球气候系统超级计算机模型、加速地球系统模型 E^3SM（Energy Exascale Earth System Model）的开发，以帮助科学家增进对气候变化的理解，并帮助美国政府落实其关于解决国内外气候危机的承诺。

三、启示与建议

1. 重视地球系统科学理论在研究中的指导作用

地球系统科学强调将地球作为一个整体来研究，在了解过去的地球、认识现在的地球和预测未来的地球方面发挥着越来越重要的作用。依靠国际地球科学界数十年的不懈努力和现代地球科学研究方法手段的不断进步，科学家们对地球系统科学的理解不断深入，对地球系统当前现状、演化过程和发展规律等的认识有了显著提升，地球系统科学理论不断完善。目前，地球系统科学已成为“未来地球”计划、国际大洋发现计划、“第三极环境”计划等大型国际计划的指导思想。未来，我国应多开展以地球系统科学理论为指导的多圈层、多尺度、多因素、多学科、多手段的地球科学集成研究，深入阐明地球资源生态环境格局、演变规律和机理、发展趋势和控制因素等，为我国地球科学学科的创新发展和经济社会的可持续发展提供支撑。

2. 加强地球深部科学问题研究

目前，地球深部科学研究处于快速发展的机遇期，同时也面临理论原始创新、技术创新、科研组织方式创新等多重挑战，包括地球内部精细的圈层结构、地核物质组成、地球内部磁场、陆壳起源、大陆深俯冲等多个领域仍存在许多未解之谜。为此，建议我国：继续加大对地球科学基础研究的重视和支持，改进科研组织机制，提升地球深部科学领域的原始创新能力；加强顶层设计，加大对地球深部科学的大科学计划、大型设施平台、国家实验室等基础设施投资力度；重视领军人才培养和引进，基于优秀的平台、资源、制度和保障措施吸引国际知名科学家加盟我国深地科学研究事业，产出一批具国际一流水平的地球科学基础研究成果。

3. 强化类地行星地质学研究

开展行星地质学研究是了解行星演化、寻找地外资源、深入认识地球宜居性的重要手段。随着美国、俄罗斯、中国等国家火星探测计划和登月过程的部署和实施，科学家借助美国“洞察号”火星探测器和我国“祝融号”火星车采集到的数据，对火星内部结构和表面地形地貌等有了更深入的认识。同时，“嫦娥五号”月球样品为我国科学家提供了机会，在揭示月球演化方面取得了创新的研究成果。因此，建议我国利用当前在火星探测和探月方面的优势，部署和支持更多的行星地质学研究，以期在对行星地球的理解和认识方面走在世界前面。

致谢：中国地质大学（武汉）马昌前教授、中国科学院地球化学研究所阳杰华研究员、中国地质科学院矿产资源研究所侯可军副研究员等审阅了本文并提出了宝贵的修改意见，中国科学院西北生态环境资源研究院王晓晨、裴惠娟、刘燕飞、李小燕等对本文也做出了贡献，参与了本文的部分资料收集与翻译，在此一并表示感谢。

参考文献

[1] Roerdink D, Ronen Y, Strauss H, et al. The emergence of subaerial crust and onset of weathering 3.7 billion years ago. https://doi.org/10.5194/egusphere-egu21-4701[2021-05-10].

[2] Garçon M. Episodic growth of felsic continents in the past 3.7 Ga. Science Advances, 2021, 7(39): eabj1807.

[3] Ko B, Greenberg E, Prakapenka V, et al. Calcium dissolution in bridgmanite in the Earth's deep mantle. Nature, 2022, 611(7934): 88-92.

[4] Frost D A, Lasbleis M, Chandler B, et al. Dynamic history of the inner core constrained by seismic anisotropy. Nature Geoscience, 2021, 14: 531-535.

[5] Gu T, Pamato M G, Novella D, et al. Hydrous peridotitic fragments of Earth's mantle 660 km discontinuity sampled by a diamond. Nature Geoscience, 2022, 15(11): 950-954.

[6] Hartnady M I H, Johnson T E, Schorn S, et al. Fluid processes in the early Earth and the growth of continents. Earth and Planetary Science Letters, 2022, 594: 117695.

[7] Farsang S, Louvel M, Zhao C, et al. Deep carbon cycle constrained by carbonate solubility. Nature Communications, 2021, 12: 4311.

[8] Bajgain S K, Mookherjee M, Dasgupta R. Earth's core could be the largest terrestrial carbon reservoir. Communications Earth & Environment, 2021, 2: 165.

[9] Müller R D, Mather B, Dutkiewicz A, et al. Evolution of Earth's tectonic carbon conveyor belt. Nature, 2022, 605(7561): 629-639.

[10] European Commission. Critical Raw Materials Resilience: Charting a Path towards greater Security and Sustainability. https://ec.europa.eu/docsroom/documents/42849/attachments/2/translations/en/renditions/native[2020-03-09].

[11] Natural Resources Canada. Canada Announces Critical Minerals List. https://www.canada.ca/en/natural-resources-canada/news/2021/03/canada-announces-critical-minerals-list.html[2021-03-11].

[12] Nassar N T, Fortier S M. Methodology and technical input for the 2021 review and revision of the U.S. Critical Minerals List. https://doi.org/10.3133/ofr20211045[2021-05-07].

[13] Department for Business, Energy & Industrial Strategy. Resilience for the future: The UK's critical minerals strategy. https://www.gov.uk/government/publications/uk-critical-mineral-strategy/resilience-for-the-future-the-uks-critical-minerals-strategy[2022-07-22].

[14] Department of Industry, Science and Resources. 2022 Critical Minerals Strategy. https://www.industry.gov.au/sites/default/files/March%202022/document/2022-critical-minerals-strategy.pdf[2022-03-16].

[15] Brandt S, Fassbender M L, Klemd R, et al. Cumulate olivine: A novel host for heavy rare earth element mineralization. Geology, 2021, 49: 457-462.

[16] Richter A. Success in lithium extraction from geothermal brine, Alsace. https://www.thinkgeoenergy.com/success-in-lithium-extraction-from-geothermal-brine-alsace/[2021-05-11].

[17] Geoscience Australia. Exploring for the Future takes a deeper look at northern Australia. https://www.ga.gov.au/news-events/news/latest-news/exploring-for-the-future-takes-a-deeper-look-at-northern-australia[2021-10-28].

[18] DOE. DOE Launches ＄140 Million Program to Develop America's First-of-a-Kind Critical Minerals Refinery. https://www.energy.gov/articles/doe-launches-140-million-program-develop-americas-first-kind-critical-minerals-refinery[2022-02-14]
[19] Khan A, Ceylan S, Van Driel M, et al. Upper mantle structure of Mars from InSight seismic data. Science, 2021, 373(6553): 434-438.
[20] Knapmeyer-Endrun B, Panning M P, Bissig F, et al. Thickness and structure of the martian crust from InSight seismic data. Science, 2021, 373(6553): 438-443.
[21] Stähler S C, Khan A, Banerdt W B, et al. Seismic detection of the martian core. Science, 2021, 373(6553): 443-448.
[22] Cottaar S, Koelemeijer P. The interior of Mars revealed. Science, 2021, 373(6553): 388-389.
[23] Voosen P. 2021 Breakthrough of the Year: NASA lander uncovers the Red Planet's core. https://www.science.org/content/article/breakthrough 2021#[2021 12 16].
[24] 中华人民共和国科学技术部．2021 年度中国科学十大进展发布．https://www.most.gov.cn/gnwkjdt/202203/t20220301_179578.html [2022-03-01]
[25] Li C, Zheng Y, Wang X, et al. Layered subsurface in Utopia Basin of Mars revealed by Zhurong rover radar. Nature, 2022, 610: 308 -312.
[26] 中华人民共和国科学技术部．2022 年度中国科学十大进展发布．https://www.most.gov.cn/kjbgz/202303/t20230320_185168.html [2023-03-20].
[27] Hu S, He H C, Ji J L, et al. A dry lunar mantle reservoir for young mare basalts of Chang'e-5. Nature, 2021, 600: 49-53.
[28] Li Q L, Zhou Q, Liu Y, et al. Two-billion-year-old volcanism on the Moon from Chang'e-5 basalts. Nature, 2021, 600: 54-58.
[29] Tian H C, Wang H, Chen Y. Non-KREEP origin for Chang'e-5 basalts in the Procellarum KREEP Terrane. Nature, 2021, 600: 59-63.
[30] Li C L, Hu H, Yang M F, et al. Characteristics of the lunar samples returned by the Chang'E-5 mission. National Science Review, 2022, 9(2): nwab188.
[31] 科技处．喜报！中国科学院召开新闻发布会——我所嫦娥五号月球样品研究成果的三篇论文同期发表于《自然》. http://www.igg.cas.cn/xwzx/zhxw/202110/t20211019_6225644.html[2021-10-19].
[32] Che X, Nemchin A, Liu D, et al. Age and composition of young basalts on the Moon, measured from samples returned by Chang'e-5. Science, 2021, 374(6569): 887-890.
[33] BGS. Unlocking the deep geothermal energy potential of the Carboniferous Limestone Supergroup. https://www.bgs.ac.uk/news/unlocking-the-deep-geothermal-energy-potential-of-the-carboniferous-limestone-supergroup/[2021-06-28].
[34] DOE. DOE Announces ＄12 Million to Boost Geothermal Energy Research. https://www.energy.gov/articles/doe-announces-12-million-boost-geothermal-energy-research[2021-09-22].

[35] DOE. DOE Announces up to $44 Million to Advance Enhanced Geothermal Systems. https://www.energy.gov/articles/doe-announces-44-million-advance-enhanced-geothermal-systems[2022-08-15].

[36] European Commission. Innovation Fund: additional large-scale geothermal project invited to prepare grant agreement. https://climate.ec.europa.eu/news-your-voice/news/innovation-fund-additional-large-scale-geothermal-project-invited-prepare-grant-agreement-2022-12-19_en [2022-12-29].

[37] Seismological Society of America. DeepShake uses machine learning to rapidly estimate earthquake shaking intensity. https://phys.org/news/2021-04-deepshake-machine-rapidly-earthquake-intensity.html[2021-04-23].

[38] Beroza G C, Segou M, Mousavi S M. Machine learning and earthquake forecasting—next steps. Nature Communications, 2021, 12: 4761.

[39] Wilson C J N, Cooper G F, Chamberlain K J, et al. No single model for supersized eruptions and their magma bodies. Nature Reviews Earth & Environment, 2021, 2: 610-627.

[40] Wang K, Johnson C W, Bennett K C, et al. Predicting future laboratory fault friction through deep learning transformer models. Geophysical Research Letters, 2022, 49(19): e2022GL098233.

[41] National Academies of Sciences, Engineering, and Medicine. Next Generation Earth Systems Science at the National Science Foundation. https://www.nationalacademies.org/news/2021/09/national-science-foundation-should-create-next-generation-earth-systems-science-initiative-new-report-says [2021-09-22].

[42] NSF. SZ4D Implementation Plan 2022. https://stacks.stanford.edu/file/druid:hy589fc7561/SZ4D%20Implementation%20Plan%202022.pdf [2022-11-08].

[43] Shanghai Jiaotong University. 125 questions: Exploration and discovery. https://www.sciencemag.org/collections/125-questions-exploration-and-discovery[2021-04-08].

[44] European Commission. Destination Earth (DestinE). https://ec.europa.eu/digital-single-market/en/destination-earth-destine[2021-03-05].

[45] DOE. DOE Announces $30 Million for Research to Secure Domestic Supply Chain of Critical Elements and Minerals. https://www.energy.gov/articles/doe-announces-30-million-research-secure-domestic-supply-chain-critical-elements-and[2021-03-18].

[46] DOE. DOE Awards $30M to Secure Domestic Supply Chain of Critical Materials. https://www.energy.gov/articles/doe-awards-30m-secure-domestic-supply-chain-critical-materials [2021-09-02].

[47] DISER. Resources Technology and Critical Minerals Processing National Manufacturing Priority road map. https://www.industry.gov.au/data-and-publications/resources-technology-and-critical-minerals-processing-national-manufacturing-priority-road-map[2021-03-04].

[48] CSIRO. Critical Energy Minerals Roadmap. The global energy transition: Opportunities for Australia's mining and manufacturing sectors. https://www.csiro.au/-/media/Do-Business/

Files/Futures/Critical-energy-minerals-roadmap/21-00041 _ MR _ REPORT _ CriticalEnergy MineralsRoadmap_WEB_210420. pdf[2021-05-19].
[49] Maloney J. From Mineral Exploration to Advanced Manufacturing Developing Value Chains for Critical Minerals in Canada. Report of the Standing Committee on Natural Resources. https://www. ourcommons. ca/Content/Committee/432/RNNR/Reports/RP11412677/rnnrrp06/rnnrrp06-e. pdf[2021-06-17].
[50] Department of Industry, Science and Resources. Over $243 million in grants awarded to resources technology and critical minerals processing manufacturers. https://www. industry. gov. au/news/over-243-million-in-grants-awarded-to-resources-technology-and-critical-minerals-processing-manufacturers[2022-03-16].
[51] USGS. Biden-Harris Administration Invests Over $74 Million in Federal-State Partnership for Critical Minerals Mapping. https://www. usgs. gov/news/national-news-release/biden-harris-administration-invests-over-74-million-federal-state[2022-06-21].
[52] Max Planck Institute. ESM2025: Earth System Models for the future. https://mpimet. mpg. de/en/communication/news/single-news/esm2025-erdsystemmodelle-fuer-die-zukunft[2021-09-06].
[53] Potsdam Institute for Climate Impact Research. Fast & comprehensive: First version of Potsdam Earth Model POEM ready for use. https://www. pik-potsdam. de/en/news/latest-news/fast-comprehensive-potsdam-earth-model-poem-ready-for-use[2021-07-01].
[54] Irrgang C, Boers N, Sonnewald M, et al. Towards neural Earth system modelling by integrating artificial intelligence in Earth system science. Nature Machine Intelligence, 2021, 3: 667-674.
[55] DOE. DOE Announces $70 Million to Improve Supercomputer Model of Earth's Climate System. https://www. energy. gov/articles/doe-announces-70-million-improve-supercomputer-model-earths-climate-system[2022-08-30].

Earth Science

Zheng Junwei, Liu Wenhao, Zhang Shuliang, Liu Xue, Wang Liwei, Zhai Mingguo

In recent years, under the theoretical framework of Earth System Science (ESS), it has become an important direction of earth science development in the new era to carry out research on the material cycles of all spheres of the earth and related issues, such as the enrichment of resources and energy, the change of ecological environment, the livability of the earth and the sustainable development of society, etc. Since 2021, the world has made many important discoveries and breakthroughs in the field of earth science: Important progress has been made in the study of the deep geological structure and material and energy cycle of the earth, advances in science and technology have expanded the exploration and mining space, breakthroughs have been made in planetary geology in revealing the internal structure of Mars and the evolution of the lunar sphere, and attention has been paid to the study of deep geothermal energy potential evaluation and efficient development technology machine learning promotes the continuous deepening of the research on earthquake and volcano prediction system, and the progress of earth system observation and simulation research. Some countries and international organizations have also carried out research and deployment around the above related fields.

4.7 海洋科学领域发展观察

高 峰[1] 王金平[1] 魏艳红[1] 冯志纲[2] 王 凡[2] 牛艺博[1]

(1. 中国科学院西北生态环境资源研究院；2. 中国科学院海洋研究所)

2021～2022年，全球海洋科学研究持续推进，成果显著：物理海洋、海洋地质地貌、海洋生物、海洋环境污染及海洋观测技术等领域取得重要进展；海洋热浪、南北极海冰变化、深海资源勘探、海洋生物多样性和海洋塑料污染研究等方面取得诸多突破。国际组织和主要海洋国家围绕相关海洋研究方向进行了部署。

一、海洋科学领域重要研究进展

1. 物理海洋研究取得新进展

近年来极端海洋热浪事件频发，未来全球多地将陷入永久性海洋热浪状态[1]。美国科罗拉多大学博德分校的研究指出，海洋混合层厚度的变浅是造成近年来极端海洋热浪事件频发的主要原因[2]。加拿大不列颠哥伦比亚大学等的研究表明，Blob海洋热浪事件①可能在短期内抑制海洋生物泵的功能[3]，而海洋生物泵对缓解人类活动对气候的影响具有重要作用。荷兰格罗宁根大学等的研究指出，预计到21世纪末南极西部的热浪事件次数将翻一番，南极东部的热浪事件次数将增加两倍[4]。

英国雷丁大学等的研究指出，南极冰架总面积的34%将面临破裂和塌陷的风险[5]。美国地球物理学会（AGU）秋季会议指出，南极洲的思韦茨冰川（Thwaites）正在逐渐融化，其加速融化趋势可能导致该冰架最快在未来3年内完全崩塌[6]。阿联酋哈利法科学技术大学等的研究表明，南极爆发性气旋会导致冰架崩解，最终造成海平面上升[7]。英国气象局哈德莱中心（MOHC）等的研究指出，冰盖后退时陆地的暴露会增加南极的降雨量，进而导致海冰流失速度加快[8]。英国南极调查局（BAS）研

① 2013年，一股名为“the Blob”的巨大海洋热浪在阿拉斯加海岸外形成，并很快沿着北美太平洋沿岸向南延伸至墨西哥。它的持续时间很长，摧毁了渔业，引发了有毒藻类的大量繁殖，扰乱了海带森林，并导致海鸟饿死。

究指出，热带风暴带来的暖湿气流在南极半岛上空移动时导致极端高温事件和冰融化，是导致南极半岛东部和拉森C冰架出现创纪录高温和冰层融化的新因素。

美国哥伦比亚大学等研究指出，北极海冰可能在21世纪消失，到2050年北极夏季海冰将会急剧变薄[9]。德国阿尔弗雷德·魏格纳极地与海洋研究所（AWI）2021年9月16日发文称，北冰洋海冰面积于2021年9月12日降至年度最低值[10]。英国伦敦大学学院牵头的研究指出，北极沿海地区的海冰变薄速度可能是预期变薄速度的两倍[11]。芬兰气象局（Finnish Meteorological Institute）的研究表明，在过去43年中北极的升温速度接近全球平均升温速度的4倍，这一速度显著高于之前文献中的研究数据[12]。

综上可知，著名的Blob海洋热浪事件破坏了浮游植物生长、海洋鱼类种群数量暴跌、海鸟死亡，还引发了全球范围内的珊瑚大面积白化。南极爆发性气旋的增加、北极“热量炸弹”事件的发生，均加剧了冰架坍塌和海冰融化的趋势。

2. 海洋地质地貌研究开启新方向

欧洲海洋局（EMB）发布《海洋地质灾害：保护社会和蓝色经济免受隐性威胁》报告，概述了欧洲沿海地区海洋地质灾害的类型和分布，以及其对蓝色经济相关行业的影响[13]。国际海底管理局（ISA）等召开线上会议，评估了当前和未来深海海底矿物资源勘探和开发方面的技术以及关键行动[14]。英国南极调查局（BAS）的研究指出，借助3D地震反射波技术首次揭示了数千年至数百万年前覆盖英国和西欧大部分地区的巨大冰盖下形成的冰下湖景观[15]。英国利兹大学的新研究表明，海洋深处的火山喷发强度非常高，释放的能量足以为整个美国提供动力[16]。澳大利亚北领地政府宣布永久禁止该地区的海底采矿活动，该永久禁令将涉及澳大利亚沿北方海岸线17.5%的海域[17]。美国斯克利普斯海洋研究所的新研究，揭示了雨水和海浪在海岸悬崖中起到的不同作用[18]。

3. 海洋生物学研究取得新发现

美国罗格斯大学完成的研究，提出通过衡量珊瑚产生的代谢物识别受热应激影响珊瑚的新方法[19]。日本筑波大学的新研究，证实了海洋酸化对珊瑚藻类钙化过程的负面影响[20]。英国南极调查局等的研究表明，南极冰架下方的生物数量比预期的更多[21]。美国斯坦福大学的研究，揭示了大型海洋哺乳动物的急剧减少对海洋生态系统的破坏[22]。美国南卡罗来纳大学等的研究指出，极度濒危的北大西洋露脊鲸数量将在未来几十年内继续下降并有可能灭绝[23]。

美国罗德岛大学的研究指出，未来全球海洋群落的结构将随气候变化而变化[24]。

美国俄勒冈州立大学的研究，通过追踪浮游植物揭示了微生物（microbes）在海洋碳循环中发挥的重要作用[25]。美国哈佛大学等的研究指出，海洋中的微生物对地球温度起到了重要的调节作用[26]。

4. 海洋环境污染及其影响研究获得新认识

据估计，每年有 800 万 t 塑料废弃物进入海洋，危害海洋生物和海洋生态系统，并且很难追踪和清理[27]。英国普利茅斯大学的研究指出，海上船只的绳索在拖曳过程中每年可产生数十亿微塑料碎片，成为海洋微塑料污染的一大来源[28]。希腊海洋研究中心（HCMR）等的研究指出，每年进入地中海的塑料废弃物总量约为 17 600 t[29]。世界自然基金会（WWF）等通过对 2592 项研究进行评估，总结了塑料污染对海洋物种及生态系统影响的现状，指出全球面临着形势严峻的塑料污染危机[30]。

英国南极调查局等的研究指出，在塑料污染和海洋酸化的叠加作用下，南大洋南极磷虾的生长受限[31]。美国国家海洋与大气管理局（NOAA）宣布为美国沿海和五大湖水域的有害藻华研究计划提供 1520 万美元的资助[32]，以加强对有害藻华的早期预警、监测及其控制和影响的研究。澳大利亚塔斯马尼亚大学等针对有害藻华趋势的首个全球统计分析表明，全球范围内的有害藻华事件并没有显著增加[33]。美国杜克大学的研究发现，2019 年 12 月至 2020 年 3 月的澳大利亚野火导致南大洋出现大面积浮游生物赤潮，不仅影响当地的陆地生态系统，而且通过大气影响数千英里①外的海洋生态系统[34]。

5. 海洋观测技术发展步入新阶段

联合国发布消息称，正在使用卫星成像技术绘制全球“珊瑚礁地图集”[35]。美国伍兹霍尔海洋研究所（WHOI）牵头的机器人研发计划，将开辟利用机器人研究珊瑚礁的新途径[36]。意大利国家环境保护研究所（ISPRA）部署了新的海平面观测网络，以加强对地中海地区海啸和其他海平面危害的观测和监测[37]。斐济南太平洋大学等签署的“太平洋岛屿海洋生物入侵警报网络”（PACMAN）计划协议，将极大地提升斐济对高风险海洋外来入侵物种的监测能力[38]。美国国家海洋与大气管理局等投放了约 100 个全新 Argo 浮标，以支持海洋、天气和气候研究与预测[39]。英国南极调查局牵头开发的 IceNet 人工智能工具，可准确预测北极海冰的变化并支持新的预警系统[40]。WHOI 开发的新观测网络将覆盖大西洋西北部 25 万 km^2 的区域，并收集有关暮光区的全天候数据[41]。

---

① 1英里（mi）≈1609.344 m。

6. 全球海洋综合评估取得丰硕成果

2021年2月，联合国教科文组织（UNESCO）发布《UNESCO海洋世界遗产地：全球蓝碳资产的守护者》报告，首次评估了UNESCO海洋世界遗产地中的“蓝碳”资产[42]。4月，联合国发布《第二次全球海洋综合评估报告》[43]，阐明来自人类活动的压力持续对海洋造成危害。10月，联合国环境规划署（UNEP）发布《从污染到解决方案：海洋垃圾与海洋污染的全球评估》报告[44]，指出大幅减少不必要、可避免和有问题的塑料对于应对全球污染危机至关重要。联合国政府间气候变化专门委员会（IPCC）第六次评估报告（AR6）指出：海洋在发生着明显的变化，现在海洋变暖的速度比至少1.1万年以来的任何时候都快[45]。自1900年以来，全球海平面上升了约20 cm；其上升速度比至少3000年来的任何时候都快，而且还在加速；海洋从大气中吸收CO_2，海洋酸化也在加剧。

二、海洋科学领域重要研究部署

1. 国际组织

以联合国为代表的国际组织持续关注“蓝碳”经济、海洋生态环境、海洋生物多样性以及海洋空间规划等方向。2021年1月，“联合国海洋科学促进可持续发展十年（2021—2030）”（以下简称“海洋十年”）计划正式启动，国际组织及多国积极响应，纷纷发布其行动计划。以国际海底管理局、世界气象组织（WMO）为代表的国际组织发布其战略优先事项；联合国相关机构公布“海洋十年”获批行动。美国、加拿大正式启动“海洋十年”计划；英国、日本、挪威、巴西等国发布本国“海洋十年”计划具体举措与优先事项。10月，UNESCO宣布启动eDNA计划[46]，通过提取环境样品遗传物质（eDNA）① 监测海洋物种并加强对海洋生物多样性的保护。10月，UNESCO等发布《海洋空间规划全球指南》[47]，帮助各国政府、合作伙伴和从业人员开展海洋空间规划活动。12月，可持续海洋经济高级别小组（HLP）启动新的全球联盟“海洋行动2030”[48]，通过制定和实施可持续海洋计划，解决与可持续海洋经济过渡相关的问题。

---

① 环境样品遗传物质（eDNA）是一种基因监测技术，研究人员可以通过测序获得环境样本（如水、土壤或空气）中的物种信息。eDNA方法速度更快，灵敏度更高，对环境的破坏性更小，收集大量样本不需要分类专业知识，有助于直接解决传统测量方法的局限性。

2. 美国

美国持续推进海洋科学领域的发展与创新。2021 年 1 月，美国国家海洋与大气管理局发布《蓝色经济战略计划（2021—2025 年）》[49]，以加强美国海上运输、海洋勘探、海产品竞争力、旅游休闲业以及沿海韧性等的能力建设。3 月，美国国家海洋与大气管理局等发布《美国海洋和淡水中有害藻华的社会经济影响报告》[50]，对美国淡水和海水中有害藻华的社会经济影响进行全面评估。10 月，美国海洋保护协会（Ocean Conservancy）等发布《海洋数据在推动保护和管理方面的挑战与机遇》报告[51]，评估了在推动海洋保护、认识蓝色经济和可持续管理海洋资源潜力方面存在的挑战。12 月，美国国家科学院、国家工程院与国家医学院（NASEM）发布《反思美国在全球海洋塑料垃圾中的角色》报告[52]，指出美国是全球塑料垃圾的最大贡献者。

3. 欧洲

欧洲持续关注海洋健康、海洋观测及海洋安全、海洋交通等方向。2021 年 3 月，英国国家海洋学中心（NOC）等创建英国大西洋地区生物地球化学 Argo 网络[53]，用于监测海洋健康。4 月，英格兰环境与野生动物联盟（Link）发布题为“2021 年会成为海洋超级年吗？英国政府海洋健康进展打分表”的报告[54]，对海洋健康的 5 项主要指标开展了评估。6 月，欧洲海洋局发布《在数字海洋时代持续开展海洋原位观测活动》的政策简报[55]，重点关注海洋原位观测，并指出原位观测的优势、资助情况和管理挑战。9 月，欧洲环境署（EEA）等发布《欧洲海洋运输报告》[56]，这是对海洋交通行业的首次全面健康检查。2022 年 8 月，英国政府发布《国家海洋安全战略》[57]，制定了英国今后 5 年的海洋安全方针。

4. 澳大利亚

澳大利亚致力于珊瑚礁状况、渔业管理以及蓝色经济发展等方面。2021 年 4 月，大堡礁海洋公园管理局（GBRMPA）等发布《年度珊瑚礁概要：2020—2021 年夏季》[58]，总结了大堡礁的夏季状况以及不同地区珊瑚的健康状况。4 月，澳大利亚莫里森政府宣布提供 1 亿澳元的投资[59]，用于引领全球和澳大利亚的海洋栖息地与沿海环境管理。7 月，澳大利亚海洋科学研究所（AIMS）发布《2020/2021 年珊瑚礁状况年度总结报告》[60]，指出大堡礁在经历了大范围干扰事件之后，目前正处于恢复期。8 月，AIMS 等启动“澳大利亚珊瑚礁恢复力倡议”[61]，提出了全球首创的对珊瑚的生境选择和生长产生积极影响的生物多样性手段，即创新性地结合了模仿健康珊瑚礁的声音特征吸引鱼类的技术和珊瑚礁种植技术，有望彻底变革全球珊瑚礁恢复行动。9

月，澳大利亚联邦科学与工业研究组织（CSIRO）等发布《针对气候变化的渔业管理适应力手册》[62]，旨在帮助渔业管理人员和从业者利用循证方法确定应对气候变化的有效措施。11月，澳大利亚国家海洋科学委员会（NMSC）发布了《2015—2025年国家海洋科学计划：中期评估与展望》报告[63]，呼吁进一步采取行动以推动蓝色经济发展并释放其潜力。

5. 其他国家/地区

2021年2月，拉丁美洲和加勒比地区通过了《生态系统恢复十年行动计划》[64]，通过10项行动来促进未来十年陆地、海洋和沿海生态系统的恢复。12月，巴西正式发布《巴西"海洋十年"国家计划》[65]，提出推动以海洋科学为基础的变革性解决方案的实现路径。12月，韩国海洋水产部（MOF）发布《2050年海洋与渔业碳中和路线图》[66]，为绿色航运、港口、"蓝碳"经济和海洋能源4个领域指明全面的政策方向。2022年6月，芬兰发布《联合国"海洋十年"国家实施计划》[67]，旨在助力保护和可持续利用海洋及其资源，应对海洋环境的变化，保护自身免受与海洋有关的危害。

三、启示与建议

1. 持续推进海洋生态环境保护，助力实现一个清洁、健康的海洋

2021年1月，世界经济论坛"达沃斯议程"（Davos Agenda）提出警告：到2050年海洋中塑料的总重量将超过鱼类的总重量，甚至只剩下塑料；90%的珊瑚礁将会死亡，从而触发大面积海洋生物死亡，海洋面临温度过高、酸化和缺氧等问题[68]。联合国环境规划署提出，加快从化石燃料向可再生能源过渡、取消化石燃料补贴和向可循环方案转型，将有助于减少塑料废弃物[69]。未来，我国应采取有效措施查明海洋污染源，推进实现更清洁、更健康的海洋。

2. 继续深化南北极研究，重点关注海洋生态系统的连锁反应

在全球气候变暖加剧的背景下，海洋热浪的强度和频率不断增加，南极冰盖快速融化，北极海冰面积急剧减少，导致全球海平面上升，极端天气事件频发，破坏了海洋生态系统稳定性。鉴于极地的战略重要性与应对气候变化的需要，各国纷纷加快了南北极地区的研究部署，未来我国应进一步深化对海洋-热浪-海冰-大气相互作用的认识。

3. 借助新兴数字化技术，提升海洋观测的数字化、网络化、智能化水平

大数据、人工智能、云计算、物联网等新兴数字化技术越来越多地用于海洋观测活动。通过卫星成像技术绘制全球“珊瑚礁地图集”，IceNet 人工智能工具可准确预测北极海冰的变化，自主无人潜水器（AUV）开辟了利用机器人研究珊瑚礁的新途径，无人水面艇（USV）捕捉到大西洋内部的大型飓风，机器人浮标首次实现了全球海洋碳循环的定量测量。这些新兴技术将为探索海洋未知环境、海洋变化和极端天气的形成与发生等提供支撑。未来，我国应持续推进海洋观测的数字化、网络化、智能化水平，利用新兴数字化技术来识别、理解和预测海洋系统的发展变化。

4. 提升海洋生态系统多样性、稳定性、持续性，推动以国家海洋公园为主体的生物多样性和自然保护地建设

受海洋资源过度利用等人为因素的影响，海洋生态系统和生物多样性指标正在迅速恶化，已波及全球近 2/3 的海域。海洋为人类的生存和发展提供了坚实的保障，我国是世界上海洋生物多样性最为丰富的国家之一，但自 2019 年以来海洋生物多样性衰退现象出现，今后亟须采取一些有效的措施。党的二十大提出，提升生态系统多样性、稳定性、持续性；加快实施重要生态系统保护和修复重大工程；推进以国家公园为主体的自然保护地体系建设[70]。未来，我国应持续推动以国家海洋公园为主体的生物多样性和自然保护地建设。

致谢：中国科学院海洋研究所的李超伦研究员、中国海洋大学的高会旺教授、自然资源部第一海洋研究所的王宗灵研究员、中国海洋大学的于华明教授对本文初稿进行了审阅并提出了宝贵修改意见，在此表示感谢！

参考文献

[1] Viglione G. Fevers are plaguing the oceans and climate change is making them worse. Nature,2021,593(7857):26-28.

[2] Amaya D J, Alexander M A, Capotondi A, et al. Are long-term changes in mixed layer depth influencing North Pacific marine heatwaves? Bulletin of the American Meteorological Society,2021,102(1):S59.

[3] Traving S J,Kellogg C T E,Ross T,et al. Prokaryotic responses to a warm temperature anomaly in northeast subarctic Pacific waters. Communications Biology,2021,4(1):1217.

[4] Feron S,Cordero R R,Damiani A,et al. Warming events projected to become more frequent and last longer across Antarctica. Scientific Reports,2021,11(1):19564.

[5] Gilbert E, Kittel C. Surface melt and runoff on Antarctic ice shelves at 1.5℃, 2℃, and 4℃ of future warming. Geophysical Research Letters, 2021, 48(8): e2020GL091733.

[6] Science News. Antarctica's Thwaites Glacier ice shelf could collapse within five years. https://www.sciencenews.org/article/antarctica-thwaites-glacier-ice-shelf-collapse-climate-5-years[2021-12-13].

[7] Francis D, Mattingly K S, Lhermitte S, et al. Atmospheric extremes caused high oceanward sea surface slope triggering the biggest calving event in more than 50 years at the Amery Ice Shelf. The Cryosphere, 2021, 15(5): 2147-2165.

[8] Bradshaw C D, Langebroek P M, Lear C H, et al. Hydrological impact of Middle Miocene Antarctic ice-free areas coupled to deep ocean temperatures. Nature Geoscience, 2021, (14): 429-436.

[9] BAS. Tropical storms trigger Antarctic ice melt. https://www.bas.ac.uk/media-post/tropical-thunderstorms-contribute-to-antarctic-ice-melt/[2022-07-13].

[10] AWI. Annual sea ice minimum in the Arctic. https://www.awi.de/en/about-us/service/press/single-view/jaehrliches-meereisminimum-in-der-arktis.html[2021-09-16].

[11] Mallett R D C, Stroeve J C, Tsamados M, et al. Faster decline and higher variability in the sea ice thickness of the marginal Arctic seas when accounting for dynamic snow cover. The Cryosphere, 2021, 15(5): 2429.

[12] Finnish Meteorological Institute. The Arctic has warmed nearly four times faster than the globe since 1979. https://www.nature.com/articles/s43247-022-00498-3[2022-08-11].

[13] EMB. Marine geohazards: Safeguarding society and the Blue Economy from a hidden threat. https://marineboard.eu/sites/marineboard.eu/files/public/publication/EMB_PP26_Marine_Geo_Hazards_v5_web.pdf[2021-11-05].

[14] ISA. ISA-NOC expert meeting defines pathways to advance innovation and technology development for sustainable exploitation of deep-sea minerals in the Area. https://www.isa.org.jm/index.php/news/isa-noc-expert-meeting-defines-pathways-advance-innovation-and-technology-development[2021-11-05].

[15] Kirkham J D, Hogan K A, Larter R D, et al. Tunnel valley infill and genesis revealed by high-resolution 3-D seismic data. Geology, 2021, 49(12): 1516-1520.

[16] Pegler S S, Ferguson D J. Rapid heat discharge during deep-sea eruptions generates megaplumes and disperses tephra. Nature Communications, 2021, 12(1): 2292.

[17] Mining Weekly. Northern Territory bans seabed mining. https://www.miningweekly.com/article/northern-territory-bans-seabed-mining-2021-02-05[2021-02-05].

[18] Young A P, Guza R T, Matsumoto H, et al. Three years of weekly observations of coastal cliff erosion by waves and rainfall. Geomorphology, 2021, (375): 107545.

[19] Williams A, Chiles E N, Conetta D, et al. Metabolomic shifts associated with heat stress in coral holobionts. Science Advances, 2021, 7(1): eabd4210.

[20] Cornwall C E, Harvey B P, Comeau S, et al. Understanding coralline algal responses to ocean

acidification: Meta-analysis and synthesis. Global Change Biology, 2022, 28(2): 362-374.

[21] Griffiths H J, Anker P, Linse K, et al. Breaking all the rules: the first recorded hard substrate sessile benthic community far beneath an Antarctic Ice Shelf. Frontiers in Marine Science, 2021, 8: 642040.

[22] Savoca M S, Czapanskiy M F, Kahane-Rapport S R, et al. Baleen whale prey consumption based on high-resolution foraging measurements. Nature, 2021, 599: 85-90.

[23] Meyer-Gutbrod E L, Greene C H, Davies K T A, et al. Ocean regime shift is driving collapse of the North Atlantic right whale population. Oceanography, 2021, 34(3): 22-31.

[24] Anderson S I, Barton A D, Clayton S, et al. Marine phytoplankton functional types exhibit diverse responses to thermal change. Nature Communications, 2021, 12(1): 6413.

[25] Brandon K, Zhou L, Samuel B, et al. Phytoplankton exudates and lysates support distinct microbial consortia with specialized metabolic and ecophysiological traits. Proceedings of the National Academy of Sciences of the United States of America, 2021, 118(41): e21011781.

[26] Marlow J J, Hoer D, Jungbluth S P, et al. Carbonate-hosted microbial communities are prolific and pervasive methane oxidizers at geologically diverse marine methane seep sites. Proceedings of the National Academy of Sciences, 2021, 118(25): e2006857118.

[27] Evans M C, Ruf C S. Toward the Detection and Imaging of Ocean Microplastics with a Spaceborne Radar. IEEE Transactions on Geoscience and Remote Sensing, 2022, (60): 1-9.

[28] Napper I E, Wright L S, Barrett A C, et al. Potential microplastic release from the maritime industry: abrasion of rope. Science of the Total Environment, 2021, 804(347): 150155.

[29] Tsiaras K, Hatzonikolakis Y, Kalaroni S, et al. Modeling the pathways and accumulation patterns of micro-and macro-plastics in the mediterranean. Frontiers in Marine Science, 2021, 8: 743117.

[30] WWF. Impacts of Plastic Pollution in the Oceans on Marine Species, Biodiversity and Ecosystems. https://www.wwf.de/fileadmin/fm-wwf/Publikationen-PDF/Plastik/WWF-Impacts_of_plastic_pollution_in_the_ocean_on_marine_species__biodiversity_and_ecosystems.pdf[2022-02-08].

[31] Rowlands E, Manno C, Galloway T, et al. The effects of combined ocean acidification and nanoplastic exposures on the embryonic development of Antarctic Krill. Frontiers in Marine Science, 2021, (8): 709763.

[32] NOAA. NOAA Awards $15.2M for Harmful Algal Bloom Research. https://oceanservice.noaa.gov/news/oct21/2021-hab-awards.html[2021-07-05].

[33] Hallegraeff G M, Anderson D M, Belin C, et al. Perceived global increase in algal blooms is attributable to intensified monitoring and emerging bloom impacts. Communications Earth & Environment, 2021, (2): 117.

[34] Tang W, Llort J, Weis J, et al. Widespread phytoplankton blooms triggered by 2019-2020 Australian wildfires. Nature, 2021, 597: 370-375.

[35] 联合国. 保护和恢复珊瑚礁:科学家利用卫星成像技术绘制全球珊瑚礁地图集. https://news.un.org/zh/story/2021/01/1075292[2021-01-08].

[36] WHOI. Development of a curious robot to study coral reef ecosystems awarded $1.5 million by the National Science Foundation. https://www.whoi.edu/press-room/news-release/development-of-a-curious-robot-to-study-coral-reef-ecosystems-awarded-1-5-mil[2021-11-10].

[37] UNESCO. Italy launches new sea level observation network to help enhance the detection, monitoring of tsunamis, and other sea level hazards in the Mediterranean Region. https://ioc.unesco.org/news/italy-launches-new-sea-level-observation-network-help-enhance-detection-monitoring-tsunamis[2021-09-23].

[38] UNESCO. BAF and USP partner to strengthen detection of marine invasive species thanks to the PacMAN Project. https://ioc.unesco.org/news/baf-and-usp-partner-strengthen-detection-marine-invasive-species-thanks-pacman-project[2021-11-10].

[39] NOAA. New ocean floats to boost global network essential for weather, climate research. https://www.noaa.gov/news-release/new-ocean-floats-to-boost-global-network-essential-for-weather-climate-research[2021-12-15].

[40] Andersson T R, Hosking J S, Pérez-Ortiz M, et al. Seasonal Arctic sea ice forecasting with probabilistic deep learning. Nature Communications, 2021, (12): 5124.

[41] WHOI. New observation network will provide unprecedented, long-term view of life in the ocean twilight zone. https://www.whoi.edu/press-room/news-release/new-observation-network-will-provide-unprecedented-long-term-view-of-life-in-the-ocean-twilight-zone/[2021-02-08].

[42] UNESCO. Custodians of the globe's blue carbon assets. https://unesdoc.unesco.org/ark:/48223/pf0000375565[2021-02-25].

[43] UN. The Second World Ocean Assessment(WOA II). https://www.un.org/regularprocess/woa2launch[2021-04-21].

[44] UNEP. Comprehensive assessment on marine litter and plastic pollution confirms need for urgent global action. https://www.unep.org/resources/pollution-solution-global-assessment-marine-litter-and-plastic-pollution[2021-10-21].

[45] IPCC. Climate Change 2021: The Physical Science Basis. https://report.ipcc.ch/ar6/wg1/IPCC_AR6_WGI_FullReport.pdf[2021-08-09].

[46] UN. UNESCO 'eDNA' initiative to 'unlock' knowledge for biodiversity protection. https://news.un.org/en/story/2021/10/1103352[2021-10-18].

[47] UNESCO. UNESCO and European Commission launch new flagship guide on Marine/Maritime Spatial Planning. https://en.unesco.org/news/unesco-and-european-commission-launch-new-flagship-guide-marinemaritime-spatial-planning[2021-10-07].

[48] UNESCO. New global coalition will support the development and implementation of Sustainable Ocean Plans. https://ioc.unesco.org/news/new-global-coalition-ocean-action-2030-forms-support-

development-and-implementation[2021-12-07].
[49] NOAA. Blue Economy Strategic Plan 2021—2025. https://aamboceanservice. blob. core. windows. net/oceanservice-prod/economy/Blue-Economy% 20Strategic-Plan. pdf[2021-01-19].
[50] WHOI. WHOI and NOAA Release Report on U. S. Socio-economic Effects of Harmful Algal Blooms. https://www. whoi. edu/press-room/news-release/habs-socio-economic-impacts/[2021-03-23].
[51] Ocean Conservancy. Challenges and Opportunities for Ocean Data to Advance Conservation and Management. https://oceanconservancy. org/wp-content/uploads/2021/05/Ocean-Data-Report-FINAL. pdf[2021-10-07].
[52] NATIONAL ACADEMIES. Reckoning with the U. S. Role in Global Ocean Plastic Waste. https://www. nap. edu/catalog/26132/reckoning-with-the-us-role-in-global-ocean-plastic-waste[2021-12-01].
[53] NOC. New fleet of advanced robotic floats to understand ocean health. https://noc. ac. uk/news/new-fleet-advanced-robotic-floats-understand-ocean-health[2021-03-23].
[54] WCL. 2021: The Marine Super Year? A scorecard for Government progress on ocean health. https://www. wcl. org. uk/docs/assets/uploads/WCL_Marine_Scorecard_Report_April_2021. pdf[2021-04-27].
[55] EMB. Sustaining in Situ Ocean Observations in the Age of the Digitalocean. https://www. marineboard. eu/sites/marineboard. eu/files/public/publication/EMB_PB9_Sustaining_OO_web_HQ. pdf[2021-06-09].
[56] EEA. European Maritime Transport Environmental Report. https://www. eea. europa. eu/publications/maritime-transport/[2021-09-01].
[57] UK Government. New maritime security strategy to target latest physical and cyber threats. https://www. gov. uk/government/news/new-maritime-security-strategy-to-target-latest-physical-and-cyber-threats[2022-08-15].
[58] Australian Government. Reef Snapshot Summer 2020-21. https://elibrary. gbrmpa. gov. au/jspui/bitstream/11017/3813/2/Reef-Summer-Snapshot-2020-21. pdf[2021-04-27].
[59] Prime Minister. Australia announces $100 million initiative to protect our oceans. https://www. pm. gov. au/media/australia-announces-100-million-initiative-protect-our-oceans[2021-04-23].
[60] AIMS. Long-Term Monitoring Program Annual Summary Report of Coral Reef Condition 2020/2021. https://www. aims. gov. au/news-and-media/reef-recovery-window-after-decade-disturbances[2021-07-19].
[61] AIMS. Innovative marine science approach is a "game-changer" for global reef recovery efforts. https://www. aims. gov. au/news-and-media/innovative-marine-science-approach-game-changer-global-reef-recovery-efforts[2021-08-10].
[62] CSIRO. Climate Adaptation Handbook. https://research. csiro. au/cor/home/climate-impacts-adaptation/climate-adaptation-handbook/[2021-09-21].

[63] NMSC. National Marine Science Plan 2015-2025: The Midway Point. file:///C:/Users/lenovo/Desktop/NMSC_Midway_Point_Report_Card_FINAL_WEB_July27_2021. pdf[2021-11-05].
[64] UNEP. Action Plan for the Decade on Ecosystem Restoration. https://wedocs. unep. org/handle/20. 500. 11822/34950;jsessionid=D61C6FC35E74B3A129471E0405D247F5[2021-02-02].
[65] UNESCO. Brazil launches the National Plan for the Ocean Science Decade. https://en. unesco. org/news/brazil-launches-national-plan-ocean-science-decade[2021-12-07].
[66] MOF. 2050 GHG emissions target for the marine and fisheries sector confirmed as 3. 24 million tons. https://www. mof. go. kr/en/board. do? menuIdx=1491&bbsIdx=78533[2021-12-17].
[67] Finnish Government. Finland's National Implementation Plan for the UN Decade of Ocean Science aims to achieve the Sustainable Development Goals. https://valtioneuvosto. fi/en/-/10616/finland-s-national-implementation-plan-for-the-un-decade-of-ocean-science-aims-to-achieve-the-sustainable-development-goalsehityksen-tavoitteiden-saavuttamiseen[2022-06-08].
[68] World Economic Forum. The Davos Agenda 2021:Global Engagement. https://cn. weforum. org/reports/davos-agenda-2021[2021-01-25].
[69] UNEP. From Pollution to Solution: a global assessment of marine litter and plastic pollution. https://www. unep. org/resources/pollution-solution-global-assessment-marine-litter-and-plastic-pollution[2021-10-21].
[70] 习近平. 高举中国特色社会主义伟大旗帜为全面建设社会主义现代化国家而团结奋斗——在中国共产党第二十次全国代表大会上的报告. https://www. gov. cn/xinwen/2022-10/25/content_5721685. htm[2022-10-16].

Marine Science

Gao Feng, Wang Jinping, Wei Yanhong,
Feng Zhigang, Wang Fan, Niu Yibo

In 2021~2022, significant achievements have been made in the field of global oceanography. Important progress has been made in the fields of physical oceans, marine geology and geomorphology, marine biology, marine environmental pollution, and marine observation technology, and many breakthroughs have been made in marine heat waves, Arctic and Antarctic sea ice changes, deep-sea resource exploration, marine biodiversity, and marine plastic pollution research. International organizations and major maritime countries have conducted numerous deployments around relevant marine research directions.

4.8 空间科学领域发展观察

杨 帆 韩 淋 王海名 范唯唯

（中国科学院科技战略咨询研究院）

2021～2022年，“帕克”太阳探测器（Parker Solar Probe）首次飞越日冕，“洞察号”（InSight）火星探测器揭示火星内部结构，“贝皮科伦坡号”（BepiColombo）水星探测器初尝水星科学美妙滋味，阿联酋、中国、美国探测器齐聚火星，美国关注地月空间、宇宙奥秘和地外生命，绘制天文学和天体物理学、行星科学和天体生物学发展路线图，欧洲确定2035～2050年大型空间科学任务科学主题，中国和俄罗斯联合发布建设国际月球科研站路线图，詹姆斯·韦布空间望远镜（JWST）、中国科学院空间科学战略性先导科技专项卫星“夸父一号”成功发射，美欧批准新的金星探测任务，中国开始建造长期有人照料的空间站……空间科学探索聚焦宇宙、太阳系、地球、物质运动规律等重大前沿科学问题，向纵深加速发展，高潮迭起。

一、重要研究进展

1.“尼尔·格雷尔斯雨燕天文台”等任务首次观测到潮汐破坏事件中发射的高能中微子

利用美国NASA“尼尔·格雷尔斯雨燕（Swift）天文台”等天基和地基望远镜的观测数据，研究人员首次观测到潮汐破坏事件AT2019dsg中发射的高能中微子[1]；但事件中中微子产生的时间和方式与理论预期的有所不同。未来需要对这些现象究竟是如何发生的进行更加深入的研究[2,3]。

2021年9月7日，中国科学院空间科学战略性先导科技专项首发星暗物质粒子探测卫星（DAMPE）“悟空号”正式发布2016年1月1日至2018年12月31日的伽马光子科学数据[4]，有望为宇宙起源、暗物质探测等科学前沿问题研究提供重要数据支持。2022年12月，发射两周年的中国科学院空间科学战略性先导科技专项卫星“怀柔一号”发布首批科学数据，包括首批75个伽马射线暴的详细观测数据，有助于国内外天文学家开展伽马射线暴的多波段、多信使联合观测研究[5]。

2. 研究人员绘制出全新银河系最外层全天星图

利用欧洲空间局（ESA）“盖亚”（Gaia）空间望远镜和NASA“近地天体宽视场红外巡天探索者”（NEOWISE）任务的长期观测数据，研究人员绘制出银河系最外层区域的新版全天星图，并揭示出大麦哲伦星系在穿过银晕时，其引力在身后产生的距离银河系中心20万～32.5万光年的恒星尾迹[6]；研究还分析了大麦哲伦星系在新版星图中留下的尾迹特性，结果发现根据冷暗物质理论推算的结果与实际观察到的星图相对而言更加吻合。2022年6月，“盖亚”发布第三次完整观测数据集（DR3），在这一迄今最详细的银河系巡天结果中，提供了对星震、恒星DNA、不对称运动等一系列现象的最新见解[7]。

3. “哈勃空间望远镜”给出迄今最精确哈勃常数局部值的综合测量结果

“基于超新星、哈勃常数研究暗能量状态方程”（SH0ES）科学合作项目基于哈勃空间望远镜（HST）近30年的观测结果，完成了对哈勃常数的最大规模（也可能是最后一次）重大更新；通过测量42颗超新星标记物，确定哈勃常数值为（73±1）(km/s)/Mpc，其不确定度已达1%水平，出错几率仅为百万分之一[8]。

4. NASA确认已累计发现超5000颗系外行星

从1992年发现首颗系外行星至今，经过持续30年的观测，截止2022年6月8日，已确认发现了分布在3775个恒星系统中的5035颗系外行星，另有9017颗待确认[9]。哈勃空间望远镜和斯皮策（Spitzer）空间望远镜在Kepler-138恒星系统中发现围绕红矮星运转的两颗充满水的系外行星存在的证据[10]。

5. “帕克”太阳探测器首次飞越日冕

NASA宣布“帕克”太阳探测器在人类航天史上首次穿越太阳高层大气——日冕，并进行了粒子采样和磁场原位测量，这标志着太阳物理学研究实现重大飞跃[11]。在第八次飞掠太阳期间，“帕克”太阳探测器在太阳表面上方18.8个太阳半径处观测到特定的磁性和粒子条件，证实探测器首次越过太阳大气层终结和太阳风开始的临界面（阿尔文临界面），进入了太阳的大气层[12]。

我国首颗大型X射线天文卫星“慧眼”完整探测到了第24太阳活动周最大耀斑的高能辐射过程，获得了耀斑过程中非热电子的谱指数演化，为理解太阳高能辐射随时间演化提供了新的观测结果[13]。我国综合性太阳探测专用卫星“夸父一号”发布首批太阳观测科学图像，实现了多项国内外首次，在轨验证了有效载荷的观测能力和先进性[14]。

6. “嫦娥五号”揭示月球演化奥秘

“嫦娥五号”于2020年12月17日携带1731 g月球样品返回地球。截至2022年9月，国家航天局探月中心已经完成了4批152份共计53 625.7 mg的月球样品发放[15]。“嫦娥五号”科学探测及月球样品研究取得多项重要发现。例如，证明“嫦娥五号”月球样品为一类新的月海玄武岩，对着陆区岩浆年龄、源区性质给出全新的认识[16]；在月表原位靠近采集样品的位置探测到水并估算出水的具体含量[17]；首次在月球上发现新矿物，并命名为“嫦娥石”，这是人类在月球上发现的第六种新矿物；等等。

7. 火星探测掀起新高潮

“洞察号”火星探测器揭示火星内部结构。基于“洞察号”火星探测器地震仪的探测数据对火星内部结构研究的首批成果登上《科学》封面。研究发现，火星壳厚度为24～72 km，岩石圈深约500 km，岩石圈下方可能存在与地球相似的低速层[18]；火星壳可能富含起到辅助加热作用的放射性元素[19]；巨大的火星核呈液态，直径约1830 km，这意味着火星幔只有一个岩石层，与地幔截然不同[20]。此次探测是对地球以外的另一颗岩质行星的壳、幔、核结构的首次直接测量，可用于与地球的相关特性进行比较研究。相关成果入选《科学》2021年度十大科学突破[21]。“洞察号”探测到一次强度达到5级的火震，这是在地球之外的行星上探测到的强度最高的地震[22]。2022年12月，“洞察号”任务结束使命[23]。

阿联酋“希望号”[24]、中国“天问一号”[25]、美国“毅力号”（Perseverance）[26]火星探测器先后抵达火星，成为国际火星探测新的里程碑。2022年9月，“天问一号”发布系列科学成果[27]，揭示了火星风沙与水活动对地质演化和环境变化的影响，为火星乌托邦平原曾经存在海洋的猜想提供了有力支撑，丰富了人类对火星地质演化和环境变化的科学认知。“毅力号”成功在火星制氧[28]，并陆续采集火星岩芯样本，将通过未来的NASA、ESA联合任务送返地球。此外，ESA与俄罗斯联邦航天局（RKA）联合部署的“火星生命探测计划2016任务”（ExoMars-2016）取得在火星低纬度地区发现大量地下水冰等重要发现[29]。

8. “朱诺号”加深人类对木星极光和大气的理解

研究人员利用“朱诺号”（Juno）木星探测器搭载的极光相机首次拍摄木星夜侧极光过程，揭示了木星最重要的极光现象——木星晨暴的起源和发展阶段[30]。研究人员利用“XMM牛顿望远镜”和“朱诺号”任务的观测数据，揭示了极光产生的完整链条，解决了困扰科学家长达40年之久的木星X射线极光成因之谜[31]。“朱诺号”搭载的微波辐射计对木星云顶下方及众多涡旋风暴结构的探测结果，更全面地展示了木星独特多彩的大气特征，为了解云层下的未知过程提供了线索[32]。

9. 水星、金星、小行星探测任务取得重要成果

欧洲与日本合作开展的“贝皮科伦坡号”水星探测器以最近199 km的距离飞越水星，探测了水星周围的磁场和粒子环境，并利用加速度计记录了水星引力[33]。日本“拂晓号”（Akatsuki）金星探测器首次实现了在日间和夜间分别测量金星不同云层高度处的风速，并发现在夜间会产生一股与白天相反的、吹向赤道的南北风[34]，该项研究增进了人类对金星超高速大气环流机制的理解。研究人员利用“黎明号”（Dawn）小行星探测器在2011～2012年对灶神星的观测数据开展数值模拟，发现灶神星多次遭受大型岩质天体的撞击，且时间比此前认为的要早得多。该项研究描绘了关于早期太阳系碰撞历史的新图景[35]。日本宇宙航空研究开发机构（JAXA）对“隼鸟2号”（Hayabusa 2）送返地球的龙宫小行星样本分析，确定了龙宫小行星从形成到碰撞破坏的历史，包括其形成及在太阳系中的位置、来源材料、冰的类型、小行星表面和内部与水相互作用的化学演变、碰撞的影响等[36]。

10. 国际空间站持续产出重要成果，中国空间站进入科研应用新阶段

国际空间站进入充分科研应用阶段，新成果持续涌现。例如，气体燃料球形火焰的烟尘产生与控制研究增进了我们对火灾行为的理解；用于夜间工作的多用途紫外望远镜“极端宇宙空间天文台多波长成像新仪器”（Mini-EUSO）观测到地球大气辉光和紫外辐射的变化，并跟踪了空间碎片和超高能宇宙线；在NASA为实现太空建造开展的空间水泥研究中，研究人员发现微重力下铝酸三钙和石膏混合物能够形成独特的微观结构，并可能影响材料强度；对空间站微生物分布开展的监测实验确定了数百种微生物；中子星内部构成探测器（NICER）对新发现天体Swift J1555.2-5402的观测确定其是一颗磁星；利用在轨的冷原子实验室设施生成冷原子云，研究原子的基本行为和量子特性；等等[37,38]。

中国空间站天和核心舱、问天实验舱、梦天实验舱相继就位，于2022年底正式完成在轨建造，随即进入10年以上的应用与发展阶段[39]。中国空间站开展了首次太空授课活动，航天员演示并讲解了微重力环境下细胞学实验、人体运动、液体表面张力等现象及其科学原理，在全社会引起热烈反响[40]。

11. 中国、美国等国家成功发射一批新的空间科学任务平台

中国科学院空间科学战略性先导科技专项卫星“先进天基太阳天文台”（“夸父一号”）、中国首颗太阳探测科学技术试验卫星“羲和号”和“可持续发展科学卫星1号”，美国NASA旗舰级任务“詹姆斯·韦布空间望远镜”、首个X射线偏振任务X射线偏振成像探测器”、太阳高能粒子起源实验“紫外光谱日冕探路者”、小行星探测任务“露西号”和“双小行星重定向测试”，以及韩国“探路者”月球轨道器（KPLO）、美法联合“地表水和海洋地形”（SWOT）和德国“环境测绘与分析计划”（EnMAP）卫星等成功发射，为未来空间科学取得更多重大突破奠定基础。

二、重大战略行动

1. 美国关注地月空间、宇宙奥秘和地外生命，绘制未来10年天文学和天物理学、行星科学和天体生物学发展蓝图，明确载人月球探索科学目标

白宫发布首份《国家地月空间科技战略》，提出最大化地利用地月空间开展科学研究[41]；《NASA战略规划2022》明确通过新的科学发现拓展人类知识，以及拓展人类在月球和火星的存在以便进行可持续的长期探索、开发和利用的战略目标[42]；NASA发布《科学2020—2024：2021年卓越科学愿景》报告[43]；未来NASA科学任务部将聚焦发现宇宙奥秘、搜寻地外生命、保护和改善地面与空间生活；NASA增设“天体物理学探测器计划”，旨在为未来的大型空间望远镜奠定基础，同时填补大型旗舰任务和小型探测器计划之间的空白[44]。

美国国家科学院发布《21世纪20年代天文学和天体物理学的发现之路》，提出未来10年（2022～2032年）美国应重点围绕系外行星系统、新信使和新物理学、宇宙系统三个宽泛的科学主题开展研究，对开辟通往宜居世界的道路、打开动力学宇宙研究的新窗口、揭示星系增长的驱动力等三个科学领域予以优先资助；同时建议NASA设立一项大型天文台任务和技术成熟计划，启动前沿的大型（约6 m口径）红外/可见光/紫外空间望远镜等大型项目，继续发展和平衡资助中小型科学项目，夯实、培育和维持天文学和天体物理学研究的专业基础和研究基础[45,46]。

美国国家科学院发布《起源、世界和生命：2023—2032年行星科学和天体生物学十年战略》报告[47]，确定未来十年行星科学、天体生物学、行星防御领域的科学主题（包括太阳系和地球是如何起源的，这样的系统在宇宙中是常见的还是罕见的？行星体是如何从原始状态演变成今天所见的各类天体的？是什么条件造就了地球上的宜居

环境和生命的诞生，其他地方是否也曾出现生命?），以及优先科学问题、优先任务及资助建议。

NASA 陆续披露“阿尔忒弥斯”载人月球探索计划（Artemis Program）的科学研究与应用目标体系[48]，七大顶层科学目标包括：①了解行星过程；②了解挥发物周期；③解释地-月系统撞击史；④揭示太阳远古历史；⑤利用独特的月基位置观测宇宙；⑥在月球环境下开展科学实验；⑦研究深空探索对人体的风险及减缓措施。

2. ESA 选定“旅程 2050”空间科学规划大型任务科学主题

ESA 成员国部长级中期会议再次强调，未来将通过开展冰月采样返回任务和载人空间探索等重大空间活动，加强欧洲在科学发现、技术发展和教育激励方面的领导地位[49]。ESA 科学计划委员会宣布“旅程 2050”空间科学规划中计划在 2035～2050 年时间框架内开展的 3 项大型任务的优先科学主题：巨行星卫星、系外行星或银河系、早期宇宙[50]。同时，ESA 正在太阳系科学、天体测量学、天文学、天体物理学、基础物理等各领域中识别中型任务主题，包括可能参与 NASA 下一代旗舰天文望远镜以及外太阳系任务等大型国际合作任务。ESA 还提出对用于制造原子钟的冷原子干涉测量等前沿技术进行资助，以期在 21 世纪后半叶取得突破性的科学回报。

3. 俄罗斯确定建设俄罗斯空间站

俄罗斯明确提出要建设俄罗斯本国的空间站[51]。俄罗斯国家航天集团公司（Roscosmos）科学技术委员会和俄罗斯科学院空间理事会共同向俄罗斯内阁提交关于组建俄罗斯轨道服务站的建议[52]，由俄罗斯“能源”火箭公司负责开发[53]，科学动力舱将是未来俄罗斯轨道服务站的首个舱段[54]，但用于空间科学实验的设备将置于空间站外部，而不像目前国际空间站上那样位于内部[55,56]。

4. 中俄联合发布国际月球科研站路线图和合作伙伴指南

2021 年 6 月，中俄联合发布《国际月球科研站路线图（V1.0）》和《国际月球科研站合作伙伴指南（V1.0）》[57]，介绍了国际月球科研站项目的概念、科学领域、实施途径和合作机会建议等，欢迎国际伙伴参与。项目已确定八大科学目标：①月球地形地貌与地质构造；②月球物理与内部结构；③月球化学（物质成分与年代学）；④地月空间环境；⑤月基天文观测；⑥月基对地观测；⑦月基生物医学实验；⑧月球资源原位利用。项目工程建设分为勘（2021～2025 年）、建（2026～2035 年）、用（2035 年后）三个阶段。项目顶层设计包含地月运输设施、月球表面长期支持设施、月球运输和运行设施、科学设施、地面支持和应用设施等，目前已初步确定 5 项任务。

5. 美欧双边、金砖国家多边气候变化战略合作愈加活跃

2021年7月，NASA和ESA确立地球科学和气候变化研究战略伙伴关系，双方将在地球科学观测、研究和应用方面加强合作[58]。2021年8月，中国、巴西、俄罗斯、印度、南非共同签署《关于金砖国家遥感卫星星座合作的协定》[59]，标志着金砖国家在和平利用空间领域开展互利合作方面取得重要成果，同时实现从双边协议向多边合作的转变[60]。

6. 美欧加速空间科学新任务部署

NASA遴选“康普顿光谱仪和成像仪”小型任务以及四个“先驱者”小型天体物理学任务概念。NASA“探索者”计划日球层物理学概念研究项目选定“多缝隙太阳探测器”（MUSE）和HelioSwarm。NASA正式批准“达·芬奇+”（DaVinci+）和“真理号”（VERITAS）两项金星探测任务，ESA批准“想象号”（EnVision）金星探测任务，三项任务将开展合作研究。美国、意大利、加拿大、日本将共同开发“火星冰测绘”轨道器任务。阿联酋计划启动一项小行星带探索任务。NASA为地球风险计划遴选了“对流上升气流调查”（INCUS）任务，计划实施“地球系统天文台”星座。NASA“地球空间动力学星座”（GDC）遴选出综合极光降水实验（CAPE）、用于热等离子体的大气电动力学探测器（AETHER）、用于大气和电离层表征的模块化光谱仪（MoSAIC）三项地球高层大气研究新任务。ESA为“地球探索者”计划遴选了新任务“和谐”（Harmony）和多项未来任务概念，并确定了第二项“侦查”任务——HydroGNSS卫星。

三、启示与建议

空间科学领域继续取得重要进展。系外行星探秘、新物理学探测、宇宙系统研究领域奠定了未来10年乃至更远期空间科学发展基调；月球科研持续升温，气候变化相关科学研究成为合作重点议题；未来空间站形态和组织模式在探索与竞争合作中仍存变数。詹姆斯·韦布空间望远镜、“夸父一号”等新平台成功发射，金星探测、火星冰测绘等新任务拉开帷幕。未来空间科学领域将呈现更加激动人心的光明前景。

我国空间科学研究影响逐渐提升，月球和火星探测、暗物质研究等重大成果世界瞩目，卫星发射活动与美国各领风骚，国际月球科研站、中国空间站等未来战略部署

充分体现合作共赢的务实态度、开放情怀和大国担当。未来，我国应进一步聚焦宇宙的起源与演化、暗物质与暗能量的本质、高能宇宙线的起源与加速机制、主导极端致密天体的物理规律、太阳爆发活动与日地空间多圈层的耦合关系等科学问题，加强任务系列部署，鼓励提出更多创新概念，充分挖掘和拓展领域发展对技术、产业、经济发展的带动性，提高国际影响力和吸引力，在引领世界空间科学发展、推动全人类和平探索利用太空、构建人类命运共同体方面发挥更大作用。

致谢：中国科学院国家空间科学中心王赤院士、中国科学院科技战略咨询研究院张凤研究员对本文的撰写提出许多宝贵的修改意见，特此致谢。

参考文献

[1] NASA. NASA's Swift helps tie neutrino to star-shredding black hole. https://www. nasa. gov/feature/goddard/2021/nasa-s-swift-helps-tie-neutrino-to-star-shredding-black-hole[2021-02-23].

[2] Stein R, van Velzen S, Kowalski M, et al. A tidal disruption event coincident with a high-energy neutrino. Nature Astronomy, 2021, (5): 510-518.

[3] Winter W, Lunardini C. A concordance scenario for the observed neutrino from a tidal disruption event. Nature Astronomy, 2021, (5): 472-477.

[4] 中国科学院国家空间科学中心 ."悟空"号暗物质粒子探测卫星伽马光子科学数据正式公开发布. http://www. nssc. ac. cn/xwdt2015/xwsd2015/202109/t20210907_6194609. html[2021-09-07].

[5] 中国科学院 ."怀柔一号"极目卫星发布首批科学数据 . https://www. cas. cn/sygz/202212/t20221210_4857722. shtml[2022-12-10].

[6] Conroy C, Naidu R P, Garavito-Camargo N, et al. All-sky dynamical response of the Galactic halo to the Large Magellanic Cloud. Nature, 2021, 592(7855): 534-536.

[7] ESA. Gaia sees strange stars in most detailed Milky Way survey to date. https://www. esa. int/Science_Exploration/Space_Science/Gaia/Gaia_sees_strange_stars_in_most_detailed_Milky_Way_survey_to_date [2022-06-13].

[8] NASA. Hubble Reaches New Milestone in Mystery of Universe's Expansion Rate. https://www. nasa. gov/feature/goddard/2022/hubble-reaches-new-milestone-in-mystery-of-universes-expansion-rate[2022-05-19].

[9] NASA. Cosmic Milestone: NASA Confirms 5,000 Exoplanets. https://www. nasa. gov/feature/jpl/cosmic-milestone-nasa-confirms-5000-exoplanets[2022-03-21].

[10] NASA. Two Exoplanets May Be Mostly Water, NASA's Hubble and Spitzer Find. https://www. nasa. gov/feature/goddard/2022/two-exoplanets-may-be-mostly-water-nasas-hubble-and-spitzer-find[2022-12-16].

[11] NASA. NASA's TESS discovers stellar siblings host "teenage" exoplanets. https://www. nasa. gov/

feature/goddard/2021/nasa-enters-the-solar-atmosphere-for-the-first-time-bringing-new-discoveries [2021-07-12].

[12] Kasper J C, Klein K G, Lichko E, et al. Parker Solar Probe enters the magnetically dominated solar Corona. Physical Review Letters, 2021, 127(25): 255101.

[13] 中国科学院国家空间科学中心. 中国科学院空间科学先导专项集中发布一批重大科学成果. http://www.nssc.ac.cn/xwdt2015/xwsd2015/202107/t20210721_6143648.html[2021-07-21].

[14] 中国科学院."夸父一号"首批太阳观测科学图像发布 实现多项国内外首次. https://www.cas.cn/yw/202212/t20221213_4857909.shtml[2022-12-13].

[15] 国家航天局. 嫦娥五号科学成果发布. http://www.cnsa.gov.cn/n6758967/n6758969/n6760416/index.html[2022-09-09].

[16] 中国科学院. 20 亿年前玄武岩进一步揭示月球演化奥秘. https://www.cas.cn/zt/kjzt/cewhcg/ky/202110/t20211020_4810376.shtml[2021-10-19].

[17] 中国科学院. 嫦娥五号或找到月球有"本地水"证据. https://www.cas.cn/cm/202201/t20220118_4822562. shtml[2022-01-18].

[18] Knapmeyer-Endrun B, Panning M P, Bissig F, et al. Thickness and structure of the Martian crust from InSight seismic data. Science, 2021, 373(6553): 438-443.

[19] Khan A, Ceylan S, van Driel M, et al. Upper mantle structure of Mars from InSight seismic data. Science, 2021, 373(6553): 434-438.

[20] Stähler S C, Khan A, Banerdt W B, et al. Seismic detection of the martian core. Science, 2021, 373(6553): 443-448.

[21] Science. 2021 breakthrough of the year. https://www.science.org/content/article/breakthrough-2021[2021-12-16].

[22] NASA. NASA's InSight Records Monster Quake on Mars. https://www.nasa.gov/feature/jpl/nasa-s-insight-records-monster-quake-on-mars[2022-05-10].

[23] NASA. NASA Retires InSight Mars Lander Mission After Years of Science. https://www.nasa.gov/press-release/nasa-retires-insight-mars-lander-mission-after-years-of-science[2022-12-21].

[24] CNN. The UAE's Hope Probe has successfully entered orbit around Mars. https://edition.cnn.com/2021/02/09/world/uae-hope-probe-mars-mission-orbit-scn-trnd/index.html[2021-02-09].

[25] 国家航天局. 国家航天局举办新闻发布会 介绍我国首次火星探测任务情况. http://www.gov.cn/xinwen/2021-06/12/content_5617394.htm[2021-06-12].

[26] NASA. Touchdown! NASA's Mars Perseverance rover safely lands on Red Planet. https://www.nasa.gov/press-release/touchdown-nasas-mars-perseverance-rover-safely-lands-on-red-planet[2021-02-19].

[27] 国家航天局. 首次火星探测任务科学研究成果研讨会召开. http://www.cnsa.gov.cn/n6758823/n6758838/c6840903/content.html[2022-09-19].

[28] NASA. NASA's Perseverance Mars Rover Extracts First Oxygen from Red Planet. https://www.nasa.gov/press-release/nasa-s-perseverance-mars-rover-extracts-first-oxygen-from-red-planet

[2022-04-22].
[29] ESA. ExoMars discovers hidden water in Mars' Grand Canyon https://www.esa.int/Science_Exploration/Human_and_Robotic_Exploration/Exploration/ExoMars/ExoMars_discovers_hidden_water_in_Mars_Grand_Canyon[2021-12-15].
[30] NASA. NASA's Juno reveals dark origins of one of Jupiter's grand light shows. https://www.nasa.gov/feature/jpl/nasa-s-juno-reveals-dark-origins-of-one-of-jupiter-s-grand-light-shows[2021-03-16].
[31] ESA. The mystery of what causes Jupiter's X-ray auroras is solved. https://www.esa.int/Science_Exploration/Space_Science/The_mystery_of_what_causes_Jupiter_s_X-ray_auroras_is_solved[2021-07-09].
[32] NASA. NASA's Juno:Science results offer first 3D view of Jupiter atmosphere. https://www.nasa.gov/press-release/nasa-s-juno-science-results-offer-first-3d-view-of-jupiter-atmosphere[2021-10-29].
[33] ESA. BepiColombo's first tastes of Mercury science. https://www.esa.int/Science_Exploration/Space_Science/BepiColombo/BepiColombo_s_first_tastes_of_Mercury_science[2021-12-31].
[34] Fukuya K, Imamura T, Taguchi M, et al. The nightside cloud-top circulation of the atmosphere of Venus. Nature, 2021, 595(7868): 511-515.
[35] Zhu M H, Morbidelli A, Neumann W, et al. Common feedstocks of late accretion for the terrestrial planets. Nature Astronomy, 2021, 5: 1286-1296.
[36] JAXA. Asteroid Explorer Hayabusa2 Initial Analysis Stone Team reveals the formation and evolution of carbonaceous asteroid Ryugu. https://global.jaxa.jp/press/2022/09/20220923-1_e.html[2022-09-23].
[37] NASA. What we learned from the space station this past year. https://www.nasa.gov/mission_pages/station/research/news/what-we-learned-from-iss-2021[2021-12-22].
[38] NASA. What We Learned from Scientific Investigations on the Space Station in 2022. https://www.nasa.gov/mission_pages/station/research/what-we-learned-from-scientific-investigations-on-iss-2022/[2023-01-28].
[39] 新华社．中国空间站未来有望获系列重大科学发现和大批创新科技成果．https://baijiahao.baidu.com/s?id=1774577068954719247&wfr=spider&for=pc [2023-08-18].
[40] 中国载人航天工程网．中国空间站首次太空授课活动取得圆满成功．http://www.cmse.gov.cn/xwzx/zhxw/202112/t20211209_49179.html[2021-12-09].
[41] National Sciences & Technology Council. National Cislunar Science & Technology Strategy. https://www.whitehouse.gov/wp-content/uploads/2022/11/11-2022-NSTC-National-Cislunar-ST-Strategy.pdf/[2022-11-17].
[42] NASA. NASA Strategy Plan 2022. https://www.nasa.gov/sites/default/files/atoms/files/fy_22_strategic_plan.pdf /[2022-3-28].
[43] NASA. Science 2020-2024: A Vision for Scientific Excellence in 2021. https://science.nasa.gov/

science-pink/s3fs-public/atoms/files/2020-2024 _ NASA _ Science _ Plan _ YR _ 21-22 _ Update _ FINAL. pdf/[2022-1-25].

[44] SpaceNews. NASA to start astrophysics probe program. https://spacenews. com/nasa-to-start-astrophysics-probe-program//[2022-1-12].

[45] The National Academies of Sciences, Engineering, and Medicine. New report charts path for next decade of astronomy and astrophysics; recommends future ground and space telescopes, scientific priorities, investments in scientific community. https://www. nationalacademies. org/news/2021/11/new-report-charts-path-for-next-decade-of-astronomy-and-astrophysics-recommends-future-ground-and-space-telescopes-scientific-priorities-investments-in-scientific-community[2020-11-04].

[46] The National Academies of Sciences, Engineering, and Medicine. Pathways to discovery in astronomy and astrophysics for the 2020s. https://www. nap. edu/catalog/26141/pathways-to-discovery-in-astronomy-and-astrophysics-for-the-2020s[2021-12-31].

[47] The National Acadamies of Sciences, Engineering, and Medicine. Origins, Worlds, and Life: A Decadal Strategy for Planetary Science and Astrobiology 2023-2032. https://nap. nationalacademies. org/download/26522＃/[2022-4-19].

[48] NASA. Artemis III science definition team report. https://www. nasa. gov/sites/default/files/atoms/files/artemis-iii-science-definition-report-12042020c. pdf[2020-12-04].

[49] ESA. N° 39-2021: Decisions from the Intermediate Ministerial Meeting 2021. https://www. esa. int/Newsroom/Press_ Releases/Decisions _ from _ the _ Intermediate _ Ministerial _ Meeting _ 2021 [2021-11-19].

[50] ESA. Voyage 2050 sets sail: ESA chooses future science mission themes. https://www. esa. int/Science_Exploration/Space_Science/Voyage_2050_sets_sail_ESA_chooses_future_science_mission_themes[2021-06-11].

[51] Смотрим. Россия создаст свою национальную космическую станцию. https://smotrim. ru/article/2552166[2021-04-18].

[52] Роскосмос. НТС Роскосмоса и Бюро Совета РАН рекомендовали представить в Кабмин предложения о создании РОСС. https://www. roscosmos. ru/32672/[2021-09-21].

[53] Роскосмос. Россия запустит собственный орбитальный космопорт. https://www. roscosmos. ru/32343/[2021-08-27].

[54] Роскосмос. Роскосмос показал первый модуль новой орбитальной станции. https://www. roscosmos. ru/30858/[2021-04-23].

[55] Роскосмос. Выступление Дмитрия Рогозина на общем собрании РАН. https://www. roscosmos. ru/30832/[2021-04-21].

[56] 俄罗斯卫星通讯社．俄罗斯新轨道站的科学设备将放置在外部．http://sputniknews. cn/science/202104211033535456/[2021-04-21].

[57] 国家航天局．国际月球科研站合作伙伴指南．http://www. cnsa. gov. cn/n6758823/n6758839/

c6812148/content. html[2021-06-16].
[58] ESA. ESA and NASA join forces to understand climate change. https://www. esa. int/Applications/Observing_the_Earth/ESA_and_NASA_join_forces_to_understand_climate_change[2021-07-13].
[59] 国家航天局．金砖国家航天机构签署遥感卫星数据共享合作协定．http://www. cnsa. gov. cn/n6758823/n6758838/c6812394/content. html[2021-08-19].
[60] Роскосмос. Руководители космических агентств БРИКС подписали соглашение о сотрудничестве в области обмена данными ДЗЗ. https://www. roscosmos. ru/32228/[2021-08-18].

Space Science

Yang Fan, Han Lin, Wang Haiming, Fan Weiwei

In 2021-2022, the scientific research of space science focuses on major frontier scientific issues such as the Universe, the Solar System, the Earth, and the laws of matter movement, and accelerates its development in depth and culminates. The Parker Solar Probe entered the Sun's corona for the first time, Insight revealed the deep interior of Mars, BepiColombo first tasted the beauty of Mercury science. The United Arab Emirates, China, and the United States all launched projects to Mars. The United States focused on cislunar space, the mysteries of the universe and extraterrestrial life, and developed an astrophysics roadmap and a decadal strategy for planetary science and astrobiology. Europe determined the large-class science missions for the 2035-2050 timeframe. China and Russia announced plans to build the International Lunar Research Station. The Kuafu-1, a satellite for the Strategic Priority Research Programme of Chinese Academy of Sciences, was successfully launched. JWST was successfully launched, NASA and ESA announced a new Venus exploration mission, and China began construction of a permanent manned space station.

4.9 信息科技领域发展观察

唐 川 杨况骏瑜 张 娟 谢 黎 黄 茹
（中国科学院成都文献情报中心）

全球各国持续推进信息科技创新和产业应用，聚焦新一代信息技术，深化布局关键技术研发，推动产业生态建设，旨在在全球科技竞赛中掌握主动权。本文以 2021～2022 年全球人工智能（AI）、半导体、量子信息、高性能计算和下一代移动通信五个关键领域为对象，重点剖析领域重要研究进展与各国战略规划，以揭示信息科技领域的战略新动向、技术新趋势、未来关键挑战与机遇。

一、重要研究进展

1. 人工智能加速破解重大科学前沿问题和关键产业技术瓶颈，多学科迎来科研范式变革

（1）人工智能模型巨量化和多模态趋势明显。谷歌、微软、英伟达、北京智源人工智能研究院和阿里巴巴达摩院等新研发的模型参数达万亿级别。此外，面对越来越复杂的场景，深度学习正从语音、文字、图像等单模态向多模态智能发展。Facebook 公司①、哥伦比亚大学等机构共同推出 AI 新框架——Vx2Text 模型[1]，可从多元数据中提取信息，再生成字幕或回答问题。中国科学院自动化研究所推出的“紫东太初”平台以视觉、文本、语音三模态预训练模型为核心，可支撑全场景的人工智能应用[2]。

（2）人工智能加速破解重大前沿科学问题，有望促进多个学科研究范式变革。DeepMind 公司和华盛顿大学先后研制出蛋白质结构预测能力与人类科学家不相上下的人工智能工具 AlphaFold2[3] 和 RoseTTAFold[4]，2022 年 DeepMind 利用深度强化学习成功控制核聚变反应堆内过热的等离子体[5]。此外，谷歌公司深度学习模型 ProtCNN 可以利用氨基酸序列准确预测蛋白质功能，将蛋白质家族数据库 Pfam 中注释的蛋白质

① 2021 年 10 月 28 日，Facebook 公司宣布更名为 Meta 公司。

的覆盖范围至少扩大了 9.5%[6]；斯坦福大学开发出可准确预测 RNA 三维结构的人工智能工具 ARES[7]。DeepMind 公司还在纯数学领域的拓扑学[8]和表征理论、物理和化学领域的电子相互作用[9]等方面取得重大突破，加速了前沿科学问题的解答。

（3）利用量子计算与人工智能联合攻关科技难题正成为新趋势。美国密歇根大学通过量子计算和深度学习来研究黑洞引力，求解出可以描述这一引力的量子矩阵模型[10]。日本昭和电工与富士通公司利用量子人工智能方法预测半导体材料的特性，比传统人工智能方法快约 7.2 万倍[11]。Allosteric Bioscience 公司和药物设计公司 Polaris Quantum Biotech 共同利用量子计算和人工智能来开发新型药物，开发一种涉及衰老的关键蛋白抑制剂[12]。

2. 新一代光刻技术成为研究热点，第三代半导体产业正快速发展

（1）晶体管将在数年内挺进埃（Å）① 尺度，定向自组装（directed self-assembly，DSA）成为下一代光刻技术的主要候选方案。比利时微电子研究中心（IMEC）预测 2025 年后晶体管将进入埃尺度，同时高数值孔径（high-NA）EUV 光刻机将在 2023 年底实现样机，2025 年实现量产[13]。电气和电子工程师协会（IEEE）《国际器件与系统路线图》[14]（2020 年）将 EUV 光刻、DSA 光刻和纳米压印光刻（NIL）列为下一代光刻技术的主要候选方案。

（2）碳化硅产业临近爆发拐点。在国际碳中和背景下，以碳化硅为代表的第三代半导体产业正在迎来拐点。据多方预测，到 2030 年，碳化硅将在充电基础设施、电动汽车领域得到广泛应用，碳化硅器件市场规模将达 100 亿美元（2020 年为 6 亿美元）[15]。此外，第三代半导体芯片持续向实用化演进，关键技术逐渐完善。意法半导体公司与美国 MACOM 公司共同宣布取得一项里程碑式突破，生产出面向射频应用的硅基氮化镓晶圆及器件原型，实现了规模经济效益的成本和性能目标[16]。

3. 全球量子计算研究热潮持续升温，多机构开始利用量子技术研究重大科学问题

（1）2022 年量子计算各项性能获得持续性突破。在量子计算容错研究方面，耶鲁大学联合团队提出了容错量子计算新方法，成功提高了中性原子量子计算容错阈值[17]；在量子控制研究方面，日本国立自然科学研究所在冷原子体系下实现了世界最快双量子比特门操作，操作耗时仅 6.5 ns[18]；在量子比特构建研究方面，IBM 公司研制出 433 qbit 的超导量子处理器“鱼鹰”（Osprey），打破世界纪录[19]；芝加哥大学

① 1 Å=0.1 nm。

在实验室中创纪录地实现了 512 qbit 的中性原子体系[20]；除此之外，硅基量子计算芯片在制造能力上也达到关键节点。伦敦大学和牛津大学研究团队取得了对硅芯片上量子器件测量的世界纪录，为使用现有工艺大规模生产硅基量子芯片奠定了基础[21]；英特尔公司制造的硅基量子计算芯片在刷新硅自旋量子比特数量纪录的同时，良品率达到 95％并接近量产[22]。

（2）多机构利用量子技术研究重大科学问题。欧洲核子研究中心（CERN）《量子技术倡议战略和路线图》提出利用量子计算、量子传感等技术帮助解决高能物理问题[23]。英国研究与创新署斥资 3100 万英镑支持 7 个项目，包括利用尖端量子技术进行早期宇宙、黑洞和暗物质等基础物理学研究[24]。美国能源部也资助了多个面向核物理、聚变能源问题的量子技术应用项目[25]。

4. 高性能计算技术自主性成为全球共同追求，并与人工智能、大数据分析日趋融合

（1）日本和欧盟借力 ARM 或 RISC-V 处理器追求 E 级超算①的关键核心技术自主。日本采用 ARM 处理器打造的超级计算机“富岳”（Fugaku）在运算速度、模拟计算方法、人工智能学习、大数据处理等方面一直稳居世界第一。此外，欧洲处理器计划（European Processor Initiative）拟开发的“瑞亚”（Rhea）通用处理器已从 ARM 转向 RISC-V 架构，并计划用于 2023 年推出的 E 级超算。

（2）高性能计算、人工智能和大数据分析日趋融合，经典超算仍具巨大潜力。“珀尔马特”（Perlmutter）超级计算机被用于拼接有史以来最大的可见宇宙 3D 地图，以及处理“宇宙摄像机”暗能量光谱仪的海量数据。中国超算应用“超大规模量子随机电路实时模拟”斩获 2021 年戈登·贝尔奖[26]，打破了 2019 年谷歌公司宣称的量子计算机在特定问题上对传统超算具有“量子霸权”② 的神话。

5. 6G 研发稳步推进，太赫兹相关研究是重点

6G 太赫兹频段无线信号传输测试持续突破。三星电子顺利通过了全球第一个 6G 原型系统测试[27]，实现了在 15 m 之内的数据传输，速率达到 6.2 Gbps，并将在实验环境测试 6G 网络下的手机运行情况。随后 LG 电子的 6G 太赫兹频段的无线信号传输测试打破了三星电子的纪录[28]，将 6G 传输距离提升到 100 m 以上。我国清华大学在硅基太赫兹功放芯片设计[29]、中国科学院上海高等研究院在 6G 新型多址接入技术[30]

---

① E 级超算指每秒可进行百亿亿次数学运算的超级计算机。

② 量子霸权，也称作“量子优势”，是指量子计算机拥有的超越所有经典计算机的计算能力。

上均取得进展，中国科学技术大学还提出拍赫兹（PHz）通信新框架[31]。美国联邦通信委员会（FCC）向美国是德科技公司（Keysight）发放了首个亚太赫兹频段（95 HZ以上）6G技术开发许可[32]。韩国三星集团发布6G频谱白皮书[33]，呼吁大规模扩展毫米波频段，在亚太赫兹频段范围内研究和部署6G技术。

二、重要战略规划

1. 欧美全面推进人工智能战略计划，人工智能治理成为关注焦点

美国、欧盟、英国等国家和组织全面推进人工智能计划，更新人工智能战略，全球人工智能治理成为关注焦点，国际人工智能竞争进一步加剧。

（1）美国落实“人工智能计划法案”，多方面推进国家人工智能计划。美国《2020年国家人工智能计划法案》正式生效[34]，国家人工智能计划办公室、国家人工智能研究资源工作组、国家人工智能计划咨询委员会也相继成立。美国国家科学基金会（NSF）投入2.2亿美元支持11个国家人工智能研究所的建设[35]，从机构组织、资源经费、咨询建议等方面为持续推进美国人工智能的研发和应用提供保障。白宫科技政策办公室发布《人工智能权利法案蓝图》，要求在人工智能和自动化系统的设计、开发和部署中确保落实系统的安全有效性、数据隐私保护和算法歧视保护[36]。

（2）欧盟注重人工智能协同发展、安全、可信。欧盟更新了《人工智能协调计划》[37]，提出通过一系列联合行动促进人工智能协同发展。此外，欧盟还发布了《制定统一的人工智能规则（人工智能法案）并修正某些联合立法》提案[38]，将根据安全、可信等标准对在欧盟市场上投放和使用的人工智能系统进行监管。欧盟委员会通过《人工智能责任指令》提案，计划建立一套统一的信息获取规则并减轻人工智能系统造成损害时的举证负担，旨在通过更广泛的保护来促进人工智能的发展[39]。

（3）英国制定详细路线图，确立国家人工智能战略。英国发布《人工智能路线图》[40]和《国家人工智能战略》[41]，从研发与创新、数据、基础设施和公共信任、人工智能标准等方面提出发展建议和计划。

（4）韩国、日本、澳大利亚相继出台国家战略，加剧国际人工智能竞争。韩国出台《人工智能半导体产业发展战略》[42]，聚焦人工智能和半导体的综合发展。日本发布《人工智能战略2021》[43]，从研究开发、人才培养等方面提出具体的实施方案。澳大利亚发布《人工智能行动计划》[44]，提出开发人工智能未来科学平台等目标。

2. 半导体供应链安全和自主可控成为各国竞争核心

美国、欧盟、韩国、日本等世界主要国家和组织分别采取举措保障半导体供应链

的安全和自主可控，包括制定战略计划和法案、建立联盟和推动技术创新等。

（1）美国主要通过自身立法和向他人施压的方式来强化其半导体供应链，例如美国国会出台《2021 创新与竞争法案》，拟投超千亿美元强化国内半导体供应链[45]；美国商务部以提高芯片"供应链透明度"为由，要求台积电、三星等 150 余家全球知名半导体企业交出被视为商业机密的库存量、订单、销售记录等数据[46]。美国政府推出 500 亿美元规模的"美国芯片基金"发展计划和实施战略[47]，大力发展先进逻辑和内存芯片制造业，构建充足稳定的成熟节点半导体供应链，加强下一代半导体等技术的研发领导力。

（2）欧盟通过"欧洲处理器和半导体技术倡议"，计划投资 1450 亿欧元推动数字化转型和半导体研究[48]。欧盟及其成员国拟投入 43 亿欧元开展关键数字技术伙伴关系，加强处理器和半导体等技术创新[49]。欧盟《2030 数字指南针：欧洲数字十年之路》还提出到 2030 年使欧洲的尖端半导体（5 nm 及以下节点）产值占全球 20% 以上[50]。

（3）韩国《K-半导体战略》计划 10 年内投资 510 兆韩元（约 4500 亿美元）建立全球最大规模的半导体制造基地。同时，韩国发布"下一代功率半导体技术开发及产能扩充方案"，并启动第三代半导体国产化项目，加速发展第三代半导体[51]。

（4）日本"半导体和数字产业战略"希望通过国际合作强化半导体供应链，并提出半导体产业提升计划：升级和强化现有的半导体生产基地；与美国合作研发下一代半导体技术；推动技术创新，开发颠覆性技术[52]。

3. 全球大力部署量子信息技术研究，并开展广泛合作以加快技术应用进程

美国、欧盟、俄罗斯、英国等国家和组织大力部署量子信息技术研究，并广泛开展合作加快量子信息技术的军事化和商业化应用进程。

（1）美国深入推进"国家量子行动计划"，强化量子技术军事化应用，且超计划倍增经费。美国根据"国家量子计划法案""国家量子行动计划"等已有战略深入推进量子技术发展，能源部拨款 7300 万美元资助 29 个量子信息研发项目[53]。同时，美国空军研究实验室、美国国防部高级研究计划局（DARPA）等多个军事部门资助了 20 余项量子军事化应用的研究，美国国会研究服务部（CRS）等重要智库也呼吁开发量子技术的军事应用潜力[54]。此外，2019～2022 财年，美国政府为"国家量子行动计划"安排的研发预算已超 27.9 亿美元，远远超出最初 5 年 12 亿美元的计划。美国启动"量子技术制造路线图"项目[55]，评估未来量子产业所面临的需求、挑战和制造障碍。

（2）欧盟、法国、德国、俄罗斯、英国各自确立量子研发目标与发展战略。欧盟深入推进“量子旗舰计划”，并计划在2025年研制出具有量子加速功能的计算机；法国启动“量子技术国家战略”，计划5年内投入18亿欧元促进多类量子信息技术研究[56]；德国出资3亿欧元启动“慕尼黑量子谷计划”，并计划投入20亿欧元研制量子计算机[57]；俄罗斯布局建设量子通信网络，计划在10～15年后投入使用[58]；英国进一步加强量子技术商业化布局，计划拨款5000万英镑（约4.3亿元人民币）推进12个商业化导向的量子技术项目，涉及量子时钟、智能量子比特控制系统、量子气体传感器、量子计算药物研发、卫星通信网络量子安全加密等方面[59]。

（3）国际合作和国家级产业联盟成为发展量子技术的重要方式。美国、英国、日本等7国宣布斥资7000万美元联合研发“联合量子系统”[60]。法国和荷兰签署量子技术合作谅解备忘录[61]。美国分别与澳大利亚[62]和英国[63]签署了量子科技合作协议，将加强量子技术创新和商业化合作，并通过共享知识和市场来壮大量子产业。另外，欧洲成立了欧洲量子产业联盟，以构建欧洲的量子技术生态系统。日本成立了量子革命战略产业联盟，以推动量子技术的研发和产业化应用[64]。

4. 多国出台高性能计算研发计划，并与建设安全、高效的数据资源和软件生态系统协同进行

美国在《引领未来先进计算生态系统：战略计划》[65]中强调，要打造未来先进计算生态系统，确保软件和数据生态系统的稳定性和可持续性，并发布了加强网络安全的行政令和关键软件备忘录。欧盟“数字欧洲”计划拟在2021～2027年拨付22亿欧元研发超算技术，重点打造E级超算及后E级超算设施，并促进超算在公共部门的应用。韩国发布《国家超高性能计算创新战略》[66]，提出开发基于自主处理器的E级超级计算机，建立超高性能计算数据实验室和软件生态系统。中国提出构建全国一体化大数据中心协同创新体系及相应的新型算力网络体系，通过“东数西算”工程加以落实，以促进数据要素流通应用，解决东西部算力供需失衡问题[67]。

5. 全球部署6G顶层战略，推进科研攻关与标准制定

全球主要国家基于6G顶层战略部署，进一步推进科研攻关与标准制定，6G有望在2030年实现商用，未来3～5年将是其技术研发的关键窗口期。

（1）美国抓紧布局6G，希望谋求跨越式发展，改变当前美国在5G技术领域的不利局面。美国NSF征集RINGS计划[68]，其目标是从不同的角度探讨下一代无线和移动通信、网络、传感和计算系统的设计。美国参议院通过《下一代电信技术法案》[69]，计划设立“下一代电信委员会”，就开发和应用6G技术、评估内部职责、

制定6G标准等方面向国会提供建议，并制定以6G为核心的国家电信战略。Next G联盟发布《6G路线图：构建北美6G领导力基础》[70]，规划了6G生命周期和时间表，并将组件技术、无线电技术、系统和网络架构、可信度等确立为6G技术的关键研究领域。

（2）欧盟计划在2030年实现6G商用，启动了6G研究旗舰项目Hexa-X，以制订6G发展路线图和研发基础技术。欧盟举办全球6G峰会，以争取国际话语权。Hexa-X项目[71]预算总额达1200万欧元，旨在建立“智能连接、多网聚合、可持续、全球覆盖、极致体验和可信赖”的X使能架构。

（3）韩国希望于2028年在全球率先实现6G商用，该国已召开官民联合6G战略会议并制定“6G研发实行计划”，将在低轨道通信卫星、超精密网络技术等领域投资2200亿韩元[72,73]。日本提出在2030年实现6G商用，已出台国家级发展战略和路线图，重点措施包括发挥日本在通信材料领域的优势、推动日本技术纳入国际标准等。

三、启示与建议

1. 聚焦重大科技问题攻关，重视信息技术带来的科研范式变革

随着人工智能、量子技术等新一代信息技术的发展，利用新技术改变科研范式的新趋势已日益凸显。在国外相继取得重要突破之际，我国应对重大科技问题开展深入的分析研判，并结合我国自身发展需求与基础，遴选出有望借助新技术取得突破的重大科技问题，并在求解这些科技问题的过程中加强对人工智能、量子技术、超算的研究和利用，力争取得重大突破并引领科研范式变革。

2. 立足双循环，持续加强关键技术和设备的自主可控

目前我国许多关键技术和设备在很大程度上仍依赖进口，“卡脖子”问题较突出。对此，一方面，我国需通过国家整体布局，强壮“内循环”。例如，通过实施税收优惠、奖励科技成果转化等方式促进科研机构、投资者和企业的协同合作，突破核心元器件、关键材料、配套软件以及先进工艺的工程化瓶颈。另一方面，也需激活“外循环”，通过民用企业以持股、并购以及合作研发等方式有效掌控国际先进技术和产品，以共同支撑我国关键技术和设备自主可控。

3. 推动建设产业生态体系，成立国家级产业联盟

新一代信息技术正处于从基础研究向实际应用迈进的阶段，关系到国民经济和国

家安全，我国必须推进相关产业生态体系的建设，引领国际信息技术产业发展。我国应将政府机构、科研单位、产业界、投资方和民间组织联合起来，从顶层设计上规划、部署新一代信息技术的研发与应用，在国际标准的建设方面抢夺更多话语权，成立“政产学研投用”一体化的国家级产业联盟，共同开辟并壮大产业生态体系，并促使人工智能、高性能计算和量子计算等新一代信息技术协同发展。

致谢：中国科学院计算技术研究所谭光明研究员、中国科学院半导体研究所魏钟鸣研究员、中国科学技术大学郭国平教授、电子科技大学许渤教授、电子科技大学鲁力教授等专家审阅了本文并提出了宝贵意见，在此一并感谢！

参考文献

[1] Lin X D, Bertasius G, Wang J, et al. VX2TEXT: End-to-End Learning of Video-Based Text Generation from Multimodal Inputs. https://doi.org/10.48550/arXiv.2101.12059[2021-01-29].

[2] 中国科学院. 自动化所研发出跨模态通用人工智能平台“紫东太初”. https://www.cas.cn/syky/202107/t20210712_4798152.shtml[2021-07-14].

[3] DeepMind. AlphaFold: a solution to a 50-year-old grand challenge in biology. https://deepmind.com/blog/article/alphafold-a-solution-to-a-50-year-old-grand-challenge-in-biology[2020-11-03].

[4] Baek M, DiMaio F, Anishchenko I, et al. Accurate prediction of protein structures and interactions using a three-track neural network. Science, 2021, 373(6557): 871-876.

[5] Jonas D, Federico F, Jonas B, et al. Magnetic control of tokamak plasmas through deep reinforcement learning. Nature, 2022, 602: 414-419.

[6] Maxwell L, David B, Drew H, et al. Using deep learning to annotate the protein universe. Nature Biotechnology, 2022, 40: 932-937.

[7] Townshend R, Eismann S, Watkins A, et al. Geometric deep learning of RNA structure. Science, 2021, 373: 1047-1051.

[8] Nature. DeepMind's AI helps untangle the mathematics of knots. https://www.nature.com/articles/d41586-021-03593-1[2021-12-01].

[9] Kirkpatrick J, McMorrow B, Turban D P, et al. Pushing the frontiers of density functionals by solving the fractional electron problem. Science, 374: 1385-1389.

[10] University of Michigan. What's inside a black hole? Physicist uses quantum computing, machine learning to find out. https://phys.org/news/2022-02-black-hole-physicist-quantum-machine.html[2022-02-14].

[11] SDK. Showa Denko Drastically Accelerates Exploration of Optimal Formulation of Semiconductor Materialswith Quantum Computing Technology. https://www.sdk.co.jp/english/news/2022/41713.html[2022-02-10].

[12] PRNewswire. Quantum Computing Targets Improved Human Aging and Longevity in new Agreement between Allosteric Bioscience and Polaris Quantum Biotech. https://www. marketscreener. com/news/latest/Quantum-Computing-Targets-Improved-Human-Aging-and-Longevity-in-new-Agreement-between-Allosteric-Bio--37735096/[2022-02-03].

[13] IMEC. High-NA EUVL: the next major step in lithography. https://www. imec-int. com/en/articles/high-na-euvl-next-major-step-lithography[2022-05-06].

[14] IEEE. International Roadmap for Devices and Systems (IRDS™) 2020 Edition. https://irds. ieee. org/editions/2020[2022-05-06].

[15] 华为技术有限公司．数字能源 2030. https://www-file. huawei. com/-/media/CORP2020/pdf/giv/industry-reports/Digital_Power_2030_cn. pdf

[16] STMicroelectronics. STMicroelectronics and MACOM RF Gallium-Nitride-on-Silicon prototypes achieve technology and performance milestones. https://newsroom. st. com/media-center/press-item. html/t4449. html[2022-05-13].

[17] Wu Y, Kolkowitz S, Puri S, et al. Erasure conversion for fault-tolerant quantum computing in alkaline earth Rydberg atom arrays. NatureCommunications, 2022, 13: N4657.

[18] Chew Y, Tomita T, Mahesh T P, et al. Ultrafast energy exchange between two single Rydberg atoms on a nanosecond timescale. Nature Photonics, 2022, 16: 724-729.

[19] IBM. IBM Quantum's mission is to bring useful quantum computing to the world. https://research. ibm. com/blog/next-wave-quantum-centric-supercomputing[2022-11-09].

[20] Singh K, Anand S, Pocklington A, et al. Dual-element, two-dimensional atom array with continuous-mode operation. Physical Review X, 2022, 12: 011040.

[21] Quantum Motion. Quantum industry milestone brings mass production of quantum chips closer. https://quantummotion. tech/quantum-industry-milestone-brings-mass-production-of-quantum-chips-Scloser/[2022-10-26].

[22] Intel. Intel Hits Key Milestone in Quantum Chip Production Research. https://www. intel. com/content/www/us/en/newsroom/news/intel-hits-key-milestone-quantum-chip-research. html#gs. euc37d[2022-10-05].

[23] CERN. CERN Quantum Technology Initiative unveils strategic roadmap shaping CERN's role in next quantum revolution. https://home. cern/news/press-release/knowledge-sharing/cern-quantum-technology-initiative-unveils-strategic-roadmap[2021-10-14].

[24] UK Research and Innovation. Quantum projects launched to solve the universe's mysteries. https://www. ukri. org/news/quantum-projects-launched-to-solve-the-universes-mysteries/[2021-01-13].

[25] Office of Science. Department of Energy Announces $11 Million for Research on Quantum Information Science for Fusion Energy Sciences. https://www. energy. gov/science/articles/department-energy-announces-11-million-research-quantum-information-science-fusion[2021-04-26].

[26] Fu H H, Team SWQSIM. Closing the "Quantum Supremacy" Gap: Achieving Real-Time Simulation of

a Random Quantum Circuit Using a New Sunway Supercomputer. https://sc21. supercomputing. org/proceedings/tech_paper/tech_paper_pages/gb103. html

[27] SamSury Newsroom. Samsung Electronics and University of California Santa Barbara Demonstrate 6G Terahertz Wireless Communication Prototype. https://news. samsung. com/global/samsung-electronics-and-university-of-california-santa-barbara-demonstrate-6g-terahertz-wireless-communication-prototype [2021-06-16].

[28] LG Mewsroom LG Records 6G THz Band Milestone. https://www. lgnewsroom. com/2021/08/lg-records-6g-thz-band-milestone/[2021-09-19].

[29] 微波与天线研究所．清华大学微波所在硅基太赫兹功放芯片设计上取得进展．https://mp. weixin. qq. com/s/wNrlrej0Ny05iLaRDVr8KA[2021-09-02].

[30] 中国科学院．上海高研院等在6G新型多址接入技术研究方面取得进展．https://www. cas. cn/syky/202109/t20210914_4805780. shtml[2021-09-15].

[31] 吴长锋．我学者提出拍赫兹通信新框架 助力未来6G发展．http://digitalpaper. stdaily. com/http_www. kjrb. com/kjrb/html/2021-12/27/content_528119. htm? div=-1[2021-12-27].

[32] Keysigt. Keysight Technologies First to Receive FCC Spectrum Horizons License for Developing 6G Technology in Sub-Terahertz. https://www. keysight. com/us/en/about/newsroom/news-releases/2022/0308-nr22031-keysight-technologies-first-to-receive-fcc-spectrum. html[2022-03-08].

[33] Samsung. Company's vision and plans to prepare for the next-generation communication system. https://news. samsung. com/global/samsung-unveils-6g-spectrum-white-paper-and-6g-research-findings[2022-05-08].

[34] The National Artificial Intelligence Initiative (NAII) Aboat artificial intelligence. https://www. ai. gov/about/.

[35] NSF. NSF partnerships expand National AI Research Institutes to 40 states. https://www. nsf. gov/news/news_summ. jsp? cntn_id=303176 [2021-07-29].

[36] The White House. Blueprint for an AI Bill of Rights. https://www. whitehouse. gov/ostp/ai-bill-of-rights/[2022-10-04].

[37] European Commission. Communication on Fostering a European approach to Artificial Intelligence. https://digital-strategy. ec. europa. eu/en/library/communication-fostering-european-approach-artificial-intelligence[2021-04-21].

[38] European Commission. Laying down harmonised rules on artificial intelligence (artificial intelligence act) and amending certain union legislative acts. https://eur-lex. europa. eu/legal-content/EN/TXT/?qid=1623335154975&uri=CELEX%3A52021PC0206[2021-04-21].

[39] European Commission. Liability Rules for Artificial Intelligence. https://commission. europa. eu/business-economy-euro/doing-business-eu/contract-rules/digital-contracts/liability-rules-artificial-intelligence_en[2022-9-28].

[40] GOV. UK. AI Roadmap. https://www. gov. uk/government/publications/ai-roadmap[2021-01-06].

[41] GOV. UK. National AI Strategy. https://www. gov. uk/government/publications/national-ai-strategy [2021-09-22].

[42] 관계부처 합동.「인공지능 강국」실현을 위한인공지능 반도체 산업 발전전략-시스템반도체 비전과 전략 2.0. https://motie. go. kr/common/download. do? fid=bbs&bbs_cd_n=42&bbs_seq_n=240&file_seq_n=5 [2020-10-12].

[43] 内閣府 . AI戦略 2021～人・産業・地域・政府全てにAI～. https://www8. cao. go. jp/cstp/ai/aistrategy2021_honbun. pdf[2021-06-11].

[44] Department of Industry, Science and Resources. Australia's Artificial Intelligence (AI) Action Plan. https://www. industry. gov. au/data-and-publications/australias-artificial-intelligence-action-plan [2021-06-18].

[45] THE WHITE HOUSE. Statement of President Joe Biden on Senate Passage of the U. S. Innovation and Competition Act. https://www. whitehouse. gov/briefing-room/statements-releases/2021/06/08/statement-of-president-joe-biden-on-senate-passage-of-the-u-s-innovation-and-competition-act/ [2021-06-06].

[46] Shin-young Park. US pressures Samsung, chipmakers to disclose key internal data. https://www. kedglobal. com/semiconductor-shortages/newsView/ked202109260001[2021-09-26].

[47] U. S. Department of Commerce. Biden Administration Releases Implementation Strategy for $50 Billion CHIPS for America program. https://www. commerce. gov/news/press-releases/2022/09/biden-administration-releases-implementation-strategy-50-billion-chips[2022-09-06].

[48] Alexander M, Kirschstein T. How partnerships with non-European providers and the EU's new initiative can forge a global powerhouse. https://www. rolandberger. com/en/Insights/Publications/A-path-to-success-for-the-EU-semiconductor-industry. html #:～:text = In% 20December% 202020% 2C% 2019% 20European% 20Union% 20member% 20states, as% 20a% 20global% 20powerhouse%20in%20the%20semiconductor%20industry[2021-02-21].

[49] European Commission. Key Digital Technologies: new partnership to help speed up transition to green and digital Europe. https://ec. europa. eu/digital-single-market/en/news/key-digital-technologies-new-partnership-help-speed-transition-green-and-digital-europe[2021-03-10].

[50] Europran Commission. 2030 Digital Compass: the European way for the Digital Decade. https://eufordigital. eu/wp-content/uploads/2021/03/2030-Digital-Compass-the-European-way-for-the-Digital-Decade. pdf[2021-03-09].

[51] Korea JoongAng Daily. All in for chipmaking. https://koreajoongangdaily. joins. com/2021/05/16/opinion/editorials/semiconductor-industry-semiconductor-strategy-industry-collaboration/20210516202400377. html[2021-05-16].

[52] 日本経済産業省 .「半導体戦略の進捗と今後」. https://www. meti. go. jp/policy/mono_info_service/joho/conference/semicon_digital/0004/03. pdf[2021-11-15].

[53] DOE. Department of Energy Announces $73 Million for Materials and Chemical Sciences Research for Quantum Information Science. https://science. osti. gov/-/media/bes/pdf/Funding/2021/2449_

BES_QIS_Awards_List. pdf? la=en&hash=0A30746455D320C1817CA8049451C4916A1CF87C [2022-05-06].

[54] CRS. Defense Primer: Quantum Technology. https://sgp. fas. org/crs/natsec/IF11836. pdf[2022-05-06].

[55] SRI International. SRI International Developing First-Ever Quantum Manufacturing Technology Roadmap. https://www. prnewswire. com/news-releases/sri-international-developing-first-ever-quantum-manufacturing-technology-roadmap-301566497. html[2022-06-13].

[56] EE Times. French President Details 1. 8b Quantum Plan. https://www. eetimes. eu/french-president-details-e1-8b-quantum-plan/[2021-01-22].

[57] Wiesmayer P. Munich Quantum Valley to accelerate quantum research. https://innovationorigins. com/en/the-munich-quantum-valley-set-to-accelerate-quantum-research/[2021-01-12].

[58] Muhammad Irfan. Quantum Communication Network to Be Fully Operational in russia In 10-15 Years-Peskov. https://www. urdupoint. com/en/world/quantum-communication-network-to-be-fully-ope-1282844. html[2021-06-21].

[59] UK Research and Innovation. £50 million in funding for UK quantum industrial projects. https://www. ukri. org/news/50-million-in-funding-for-uk-quantum-industrial-projects/[2021-11-04].

[60] 全球技术地图．美英等 7 国正联合开发基于卫星的量子加密网络．https://mp. weixin. qq. com/s/EFa0mPMtDjoEJxze7l2Yvg[2021-06-22].

[61] Quantum Delta. A joint initiative by France and the Netherlands. https://jobs. quantumdelta. nl/jobs[2021-08-31].

[62] Australian Government. Australia signs quantum technology cooperation agreement with United States. https://www. industry. gov. au/news/australia-signs-quantum-technology-cooperation-agreement-with-united-states[2021-11-19].

[63] Department for Business, Energy & Industrial Strategy of UK. New joint statement between UK and US to strengthen quantum collaboration. https://www. gov. uk/government/news/new-joint-statement-between-uk-and-us-to-strengthen-quantum-collaboration[2021-11-04].

[64] NTT. Establishment of Quantum Strategic Industry Alliance for Revolution (Q-STAR). https://group. ntt/en/newsrelease/2021/09/01/210901a. html[2021-09-01].

[65] The White House. Pioneering the Future Advanced Computing Ecosystem: A Strategic Plan. https://www. whitehouse. gov/wp-content/uploads/2020/11/Future-Advanced-Computing-Ecosystem-Strategic-Plan-Nov-2020. pdf[2020-11-18].

[66] 전석남 사무관. 국가초고성능컴퓨팅 혁신전략 발표. https://www. msit. go. kr/bbs/view. do? sCode=user&mId=113&mPid=112&pageIndex=4&bbsSeqNo=94&nttSeqNo=3180299&searchOpt=ALL&searchTxt=.

[67] 中华人民共和国国家发展和改革委员会．关于印发《全国一体化大数据中心协同创新体系算力枢纽实施方案》的通知．https://www. ndrc. gov. cn/xxgk/zcfb/tz/202105/t20210526_1280838.

html? code=&state=123[2021-05-24].

[68] Resilient & Intelligent NextG Systems (RINGS). https://www. nsf. gov/pubs/2021/nsf21581/nsf21581. htm? org=NSF[2021-08-12].

[69] US Congress. Next Generation Telecommunications Act. https://www. congress. gov/bill/117th-congress/senate-bill/3014/text? r=4&s=1[2022-03-22].

[70] Next G Alliance. NGA Report: Roadmap to 6G. https://nextgalliance. org/wp-content/uploads/2022/01/NextG_FMG_Roadmap_Report_Summary_27Jan22. pdf[2022-01-27].

[71] European Commission. A flagship for B5G/6G vision and intelligent fabric of technology enablers connecting human, physical, and digital worlds. https://cordis. europa. eu/project/id/101015956[2021-01-01].

[72] 韩联社 . 韩国未来 5 年将投 12 亿元抢占 6G 主导权 . https://cn. yna. co. kr/view/ACK20210623001900881? section=search[2021-06-23].

[73] Korea JoongAng Daily. Korea to spend ₩ 220 billion on 6G network development. https://koreajoongangdaily. joins. com/2021/06/23/business/tech/6G-KoreaUS-alliance-China/20210623170700361. html[2021-01-23].

Information Science and Technology

Tang Chuan, Yang-Kuang Junyu, Zhang Juan, Xie Li, Huang Ru

Since 2021, countries around the world will continue to promote IT innovation and industrial applications, focus on the new generation of information technology, deepen the layout of key technology research and development, and promote the construction of industrial ecology, in order to grasp the initiative in the global technology race. This paper focuses on the analysis of important research progress and national strategic plans in the leading edge of five key areas in IT, i. e. artificial intelligence, semiconductor, quantum information, high-performance computing and next generation mobile technology, to reveal new strategic trends, technological trends, key challenges and opportunities in the future.

4.10 能源科技领域发展观察

陈 伟[1] 岳 芳[1] 汤 匀[1] 李岚春[1] 李娜娜[1]
孙玉玲[2] 秦阿宁[2] 滕 飞[2] 彭 皓[2]
（1. 中国科学院武汉文献情报中心；2. 中国科学院文献情报中心）

2021年以来，新冠疫情延宕反复和俄乌冲突不断升级对国际能源格局造成巨大冲击，能源安全成为全球优先议题。各国重新审视激进脱碳政策，近中期务实发挥化石能源的兜底保障作用，加强能源产业链供应链自主化、多样化的战略布局；而着眼长远实施碳中和行动，强化布局科技创新，提高应对国家发展和安全风险的能力，着力攻关和示范应用高效可再生能源、规模储能、绿色氢能、清洁电能深度替代及智慧系统等关键技术，加快规划建设新型能源体系。

一、重要研究进展

1. 化石能源利用向高效低碳方向转型

（1）新型热力循环与高效热功转换系统聚焦灵活多源发电、新型工质热工转换以实现燃烧过程高效低排放。2022年7月，日本JERA公司宣布将于2023年启动燃煤机组高比例混氨（20%热值）共燃项目，希望到2028年实现50%混氨比例[1]。中国科学院工程热物理研究所“2000吨/年气化细灰流化-熔融燃烧技术”[2]和“循环流化床高温后燃超低 NO_x 燃烧技术”[3]通过科技成果鉴定，处于国际领先水平。日本三菱电力公司成功投运全球最大规模整体煤气化联合循环发电机组（543 MW），发电效率达到48%[4]。英国启动全球首个Allam-Fetvedt循环燃气发电项目，以推进先进超临界 CO_2 布雷顿循环技术[5]。

（2）集成储能、余热回收等技术提升燃煤发电调峰潜力成为重要发展方向。韩国科学技术研究院采用压缩 CO_2 储能与火力发电蒸汽循环耦合系统来提升负荷跟随能力，实现了46%的最大往返效率和36 kWh/m^3 的储能最大能量密度[6]。东南大学开发新型高效余热回收热电联产系统，集成了超临界再压缩再生布雷顿循环、跨临界再生布雷顿循环、CO_2 制冷循环、除湿系统和碳捕集系统，有效提升燃煤发电总功率和热效率[7]。

（3）碳基能源高效催化转化不断创新。碳基材料低成本转化取得突破。例如，美国国家能源技术实验室与莱斯大学合作利用低成本焦耳热闪蒸技术，成功将煤炭等含碳矿石转化成石墨烯并实现工艺量产[8]。甲烷催化转化实现低温路线创新。例如，美国斯坦福大学和比利时鲁汶大学利用笼蔽效应，通过沸石分子筛催化剂实现了甲烷向甲醇的室温稳定转化[9]；中国科学院上海高等研究院开发出富含氮缺陷的石墨相氮化碳（g-C_3N_4）固定铜双原子催化剂，低温催化甲烷转化效率比单原子催化剂高 300%，甚至高于已报道的贵金属催化剂[10]。

2. 新一代可再生能源技术逐步实现高效率、低成本转化利用，装备器件绿色循环利用受到关注

（1）太阳能高效低成本转化利用新技术不断取得突破。钙钛矿太阳电池解决材料稳定性与规模制造工艺挑战；太阳能燃料从实验室走向示范，逐步解决运行与降解机理、高性能长寿命催化剂制备、系统集成耦合等问题。沙特阿卜杜拉国王科技大学钙钛矿/晶硅叠层电池认证效率创造 33.2%的纪录[11]。韩国蔚山国立科学技术研究院利用挥发性烷基氯化铵实现单结钙钛矿电池认证效率创造 25.73%的纪录[12]。中国科学院上海微系统与信息技术研究所研发出高柔韧性单晶硅太阳电池，可以像纸一样弯曲[13]。中国科学院大连化学物理研究所千吨级液态太阳燃料合成（太阳能电解水制氢+CO_2 催化合成绿色甲醇）示范项目运行成功，具有完全自主知识产权，整体技术国际领先[14]。该研究所还在国际上首次拍到光生电荷转移演化全时空图像，为突破太阳能光催化反应瓶颈提供了新的认识和研究策略[15]。英国剑桥大学开发首个漂浮式钙钛矿-$BiVO_4$人造树叶装置，助力太阳能燃料规模化生产[16]。日本东京大学实现光催化全分解水制氢表观量子效率大于 50%新突破[17]。

（2）风电向远海高空、大型化、智能化、高可靠性发展，重点研究大容量风电机组、新型叶片材料研制以及全生命周期回收利用等关键问题。美国桑迪亚国家实验室提出不依赖稀土磁体的低成本新型风力涡轮机设计，在无维护情况下可运行超 30 年[18]。中船海装风电有限公司自主研制的最大功率 18 MW、叶轮直径最大 260 m 海上风电机组样机下线[19]，明阳智慧能源集团研发的最大功率 18 MW、叶轮直径超 280 m 的风电机组获设计认证[20]。西门子歌美飒利用新型树脂材料开发全球首款陆上风机可回收叶片，并实现应用[21]。明阳智慧能源集团研制的可回收热固性树脂叶片成功下线，回收率可达 95%以上[22]。

（3）生物质高效转化与高值规模化利用新路线和新工艺取得新进展，重点聚焦先进低成本双功能催化剂开发、催化过程的原子/分子级精确调控、无溶剂催化环境构

筑等。比利时鲁汶大学开发木质纤维素一体化精炼工艺，通过结合催化还原分馏、脱烷基处理和糖化发酵等多步过程，实现了木质纤维素全组分高效降解利用制化学品，整体转化率达 78%（基于投料质量计算）[23]。中国科学院大连化学物理研究所设计并构筑具有金属-酸“限域毗邻”结构的分子筛双功能催化剂，实现了无溶剂体系下“一锅法”高效制备戊酸酯类生物柴油新路线[24]。

3. 四代裂变堆研发及聚变堆实验突破推进核能安全高效可持续发展

（1）核电强国积极布局四代核电技术研发应用，在下一代新型反应堆、新型核燃料材料及循环技术、反应堆理论研究等方面取得了一系列突破进展。中国 20 万 kW 球床模块式高温气冷堆成功实现并网发电，成为全球首座正式投运的第四代核电机组。俄罗斯成功实现全球唯一浮动式核电站商业运行[25]，其全球首座全堆芯装载铀-钚混合氧化物（MOX）燃料的钠冷快堆 BN-800 实现满功率运行[26]，并开建全球首台 BREST-OD-300 铅冷快堆[27]。美国爱达荷国家实验室与南方公司、泰拉能源等合作，将设计和建设一座熔盐实验快堆，计划在 21 世纪 20 年代末投入运行[28]。美国泰拉能源公司热功率 1 MW 的氯化物熔盐堆综合效应试验装置业已建成[29]，三结构同向性型（TRISO）颗粒燃料制造设施建设正式启动[30]，预计在 2025 年进行调试和运行。

（2）可控核聚变等离子体理论研究、材料开发和运行试验方面不断涌现突破性成果。中国科学院等离子体物理研究所全超导托卡马克核聚变实验装置（EAST）成功实现 1 MA 等离子体电流、1 亿℃等离子体电子温度、1000 s 连续运行时间三大科学目标，标志着我国在稳态高参数磁约束聚变研究领域引领国际前沿[31]。EAST 还成功实现了 403 s 稳态长脉冲高约束模式等离子体运行，创造了托卡马克装置高约束模式运行的新世界纪录[32]。中国科学技术大学在新一代“神威”超级计算机上首次实现聚变堆全装置动力学等离子体演化模拟，研究成果入围有“超算领域诺贝尔奖”之称的美国计算机学会戈登·贝尔奖（ACM Gordon Bell Prize）[33]。美国国家点火装置惯性约束聚变实验实现了聚变能量增益，实验输出能量（3.15 MJ）大于激光输入能量（2.05 MJ），首次实验验证了聚变自持燃烧产生增益的可行性，这是世界首次成功的激光“核聚变点火”（即核聚变反应所产生的能量等于或超过输入能量）[34]。英国“兆安球形托卡马克”（MAST）装置成功完成升级并首次获得等离子体[35]。日本国立聚变科学研究所和美国 TAE 技术公司首次在磁约束聚变等离子体中实现了氢-硼聚变实验[36]。欧洲联合环聚变实验装置启动了等量氘-氚聚变实验，5 s 内产生了 59 MJ 的稳定等离子体，创造新的聚变能量输出纪录[37]。

4. 低碳氢/氨技术不断创新突破并推动多元场景应用

（1）绿色氢/氨制备技术不断突破，向高效、温和条件发展。由深圳大学/四川大学谢和平院士团队与东方电气集团联合研制的全球首套海水无淡化原位直接电解制氢技术装备海试成功[38]。德国乌尔姆大学开发出一种可在-20℃环境运行的太阳能热耦合电解水制氢系统[39]。澳大利亚莫纳什大学通过电极-电解质界面区域中的致密离子分层实现高效合成氨，法拉第效率接近100%[40]。荷兰乌特勒支大学开发出温和条件下无过渡金属的氮还原制氨催化剂，性能可与经典贵金属催化剂媲美[41]。丹麦技术大学连续流氮还原制氨电解槽法拉第效率达到61%[42]。

（2）氢气储运技术聚焦开发更高效、易运输的载体材料。美国劳伦斯伯克利国家实验室评估发现，金属有机框架（MOF）储氢有望成为10 MW级固定电源的首选[43]。韩国科学技术院开发出用于室温储氢的单原子铂非晶态缺陷富勒烯，可实现高质量储氢密度（6.8 wt%）和体积储氢密度（64.9 kg/m^3），接近DOE储氢目标[44]。

（3）氢/氨向灵活、高效的多元场景应用。钢铁巨头安赛乐米塔尔公司将建造全球首个融合太阳能发电、电解水制氢的绿氢直接还原炼钢工厂[45]。欧盟资助全球首个氨燃料电池船舶项目，开发船用2 MW级氨燃料电池[46]。法国阿尔斯通创纪录实现氢燃料电池火车连续行驶1175 km[47]。日本东京燃气公司和住友电气公司合作开发出日本首个氢燃气轮机热电联产系统[48]。瑞典钢铁集团（SSAB）投运用于无碳炼钢的大型储氢试点设施[49]。英国混凝土公司（Tarmac）示范氢气替代天然气用于商业规模生石灰生产[50]。挪威船级社启动开发工业规模新型浮动式绿氨生产装置[51]。美国GE公司与日本IHI公司合作研发100%氨燃料燃气轮机[52]。

5. 新型低成本规模化储能技术正处在重要突破关口

（1）新型压缩空气储能向大容量、长寿命、高效率和集成化发展。清华大学开发出转换效率在60%以上的世界首个非补燃压缩空气储能电站[53]。中国科学院工程热物理研究所国际首套百兆瓦［100 MW/400（MW·h）］先进压缩空气储能国家示范项目顺利并网发电，系统设计效率在70%以上，是目前世界单机规模最大、效率最高的新型压缩空气储能电站[54]，300 MW级项目也已开工建设[55]。英国50 MW/250（MW·h）液态压缩空气储能设施投入建设，并计划于2024年建设4座液态压缩空气储能设施[56]。

（2）全固态电池作为下一代高能量密度储能技术受到广泛关注。韩国首尔大学和三星公司成功开发出满足商业要求的世界最大单体容量全固态锂电池，在3 mA/cm^{2}条件下具有超过4000 mAh/cm^{2}的累积容量[57]。南京大学制备出能量密度高达541 Wh/kg全固态锂硫电池[58]，创造了新纪录。

（3）金属-空气电池仍需不断探索，以实现实际应用。德国明斯特大学与电子科技大学合作，利用非碱性电解质开发出稳定运行超过600 h的实用性锌-空气电池[59]。德国德累斯顿工业大学利用原子分散的五配位Zr催化剂开发出功率密度创纪录的锌-空气电池[60]。美国南卡罗莱纳大学开发新型低成本固体氧化物铁-空气电池首次实现12.5小时的长时储能，且能量密度高达625 Wh/kg[61]。美国伊利诺伊理工大学设计了基于氧化锂的锂-空气电池，能量密度突破685 Wh/kg[62]。

(4) 开发低成本、高性能钠离子电池是非锂电池路线的重要方向。中科海钠科技责任有限公司正式投运全球首套1 MW·h钠离子电池光储充智能微网系统，综合能量效率达到了86.8%[63]；全球首条吉瓦时（GW·h）级钠离子电池生产线产品也已下线[64]。日本发布世界首个氧化物全固态钠电池原型，其中正负极均采用无机氧化物微晶玻璃材料，无需锂、钴等稀缺资源[65,66]。

6. 多能融合智慧能源系统成为新的战略竞争焦点

(1) 发达国家和组织开始探索一体化、智能化多能融合体系的架构设计。美国2020年启动首个氢能-核能复合能源系统示范项目，2021年发布《综合能源系统：协同研究机遇》报告提出构建多能流综合能源系统[67]。欧盟发布一系列近中长期研发规划[68-70]，推动发展高度融合可再生能源、深度电气化、广泛数字化、完全碳中和的泛欧综合能源系统。丹麦、德国等地正在部署融合可再生电力多元转化（Power-to-X）和氢能的多能融合系统。日本“福岛系能源社会”示范项目计划探索到2040年构建100%可再生能源供应、基于氢能、发展智慧社区的未来多能融合能源系统，已建成全球最大的可再生能源电力制氢示范厂[71]。

(2) 数字技术正加速与多能融合系统的深度融合。美国能源部人工智能与技术办公室发布2021～2022年工作计划，加快人工智能的研发和应用[72]。壳牌与微软等公司联合推出“开放式人工智能能源计划”，推动能源产业的数字化转型[73]。我国华为公司推出数字能源解决方案[74]，并提出能源T^3战略①，建设零碳智慧能源体系。英国电力系统运营商启动全行业计划，将开发涵盖全英能源系统的“数字孪生”[75]。德国意昂电力公司推出数字化整体解决方案助力电网智能化发展[76]。C40城市集团与谷歌公司启动24/7② 城市零碳能源计划，利用人工智能等创新工具推进城市能源完全脱碳[77]。

7. 碳捕集、利用与封存（CCUS）作为碳中和的兜底技术受到各国广泛重视

(1) CO_2 捕集技术在研发与应用示范方面取得进展。中国科学院过程工程研究所开发出离子液体 CO_2 捕集分离的吸收剂，CO_2 捕集率高于90%，投资及捕集成本较

---

① 华为能源T^3战略提出，建设零碳智慧能源体系应构建“零碳转型、能源转型、数字化转型”三大核心能力。

② 24/7指一周7天，1天24小时。

传统乙醇胺（MEA）溶剂化学吸收工艺降低 30%[78]。日本川崎汽船株式会社宣布在散货轮船上安装全球首个船载 CO_2 捕集装置[79]。瑞士 Climeworks 公司建造的世界上最大的直接空气碳捕集设施在冰岛上线[80]。我国山西大唐国际云冈热电有限责任公司建设的世界首个煤电 CO_2 捕集及资源化利用全产业链生产线进入试生产[81]。

（2）CO_2 资源化利用探索向高价值产品的多元转化。瑞士苏黎世联邦理工学院开发包含直接空气碳捕集、太阳能热化学驱动 CO_2 还原、合成气合成液态烃/甲醇的一体化装置[82]。中国科学院天津工业生物技术研究所实现全球首次 CO_2 到淀粉的从头合成[83]。

8. 重点工业行业减排技术获得高度关注

（1）加快低碳燃料替代，提高工业领域非化石能源比重是低碳转型的重点。我国安徽海螺水泥股份有限公司建成国内首套生物质替代燃料系统，替代率超过 40%，实现 20 万 t/a 的减排量[84]。阿联酋环球铝业集团打造出世界上第一个采用太阳能电力的铝冶炼厂[85]。墨西哥水泥公司（CEMEX）与瑞士企业 Synhelion 合作建成 1 MW 太阳能供热水泥煅烧示范项目，并拟建设全球首个通过高温太阳能供热和燃料生产（捕集 CO_2）实现零排放的水泥厂[86]。该公司还将引进并试验氢供热技术[87]。

（2）低碳冶金流程再造工艺路线取得进展，氢基直接还原铁技术取得商业化突破。美国波士顿金属公司围绕熔融氧化物电解（molten oxide electrolysis，MOE）冶炼工艺进行 1000 倍试验放大，将在 2023 年进行半工业化验证[88]。美国犹他大学将开展氢闪速熔炼铁反应器中试[89]。瑞典钢铁公司交付全球首批氢直接还原技术冶炼的钢材[90]。安赛乐米塔尔集团启动年产 250 万 t 的直接还原铁工厂建设，预计 2026 年完工[90]，拟实现全绿氢生产模式。无碳氢基流化床直接还原炼铁工艺示范取得实质性进展。英国普锐特冶金技术公司与韩国浦项制铁公司[92]、鞍钢集团与中国科学院工程过程研究所分别开展了工艺合作研发，鞍钢集团将于 2023 年投产万吨级绿氢零碳流化床氢气炼铁工程示范[93]。

（3）水泥工业原料替代技术研究取得突破性进展。德国马丁路德·哈勒维腾贝格大学采用 Belterra 黏土取代 50%～60% 的石灰石，新工艺可以减少 2/3 以上的 CO_2 排放量[94]。美国斯坦福大学发明了采用火山岩替代石灰石的低碳水泥，可将材料制造过程中的碳排放量减少近 2/3[95]。

二、重大战略行动

1. 保障能源安全成为能源转型的优先议题，构建弹性、安全自主和可持续供应链

美国首次发布保护清洁能源供应链的全面战略[96]，旨在建立本土化、多样化清洁

能源产业基础，摆脱对他国的过度依赖；投入近 30 亿美元重塑本土化电池关键供应链体系[97]；从国家安全高度启用冷战时期《国防生产法》，加速国内清洁能源制造业发展及锂、镍等矿产资源投资[98]；同时联合澳大利亚、加拿大、芬兰等国及欧盟签订“矿产安全伙伴关系”计划[99]，旨在构筑西方主导的关键矿产供应链。欧盟委员会成立欧盟能源平台工作组[100]以确保油气资源供应安全，发布“REPowerEU”能源计划[101]，从节能、供应多样化、加速推进可再生能源三方面推动清洁能源转型，构建更具弹性的能源系统。英国重组能源安全与净零排放部[102]，发布能源安全战略[103]、促进能源安全和净零增长的一揽子计划[104]，提出加强可再生能源、核能产业投资，实现能源生产多样化、脱碳化和本土化，巩固并提升其在全球绿色能源技术领域的领导地位。

2. 碳中和成为后疫情时代全球最为关注的议题，各国碳中和行动正加速绿色低碳转型大潮

主要国家均从制定法律、制定长期战略、实施产业行动、着力科技支撑等多方面下好碳中和时代推动经济高质量发展的先手棋。例如，美国制定《迈向 2050 净零排放长期战略》[105]，并出台了长时储能、氢能、负碳等系列攻关计划；欧盟出台《欧洲绿色协议》，并将“地平线欧洲”科研框架计划 35% 的预算用于支持气候目标[106]；日本公布《绿色增长战略》[107]，设立了 2 万亿日元“绿色创新基金”，已启动低成本海上风电、下一代电池技术、氢/氨供应链等 19 个为期十年的大型研发项目。

当前，各国/组织碳中和战略布局具有以下四个特点：①构建零碳能源体系是各国战略布局的核心。重点是大力发展可再生能源，逐步减少化石燃料使用，推动能源终端消费电气化。②促进产业低碳转型是各国建立绿色经济的着力点。重点发展循环经济模式，建立低碳工业示范集群，加速建筑节能改造和绿色转型，推进交通电气化、低碳化。③保护并增强陆地和海洋生态系统固碳能力是各国提高气候治理水平的重要途径。重点布局增强农、林、渔及海洋自然碳汇，并构建 CCUS 等多元负排放技术体系。④交叉技术融合创新成为各国日益关注的焦点。

2. 实现碳中和是一项艰巨挑战，要求各国清洁能源技术创新发生质的飞跃

根据国际能源署（IEA）的统计和预测[108]，当前各国的研究部署与实现碳中和目标还存在巨大的差距，全球电气化、氢能、生物能源以及 CCUS 等关键技术创新投入仅为成熟技术研发投入的 1/3，到 2050 年将有一半的减排量来自目前还处于示范或原型开发阶段的技术，必须聚焦碳中和重大科学问题，通过加强跨领域综合交叉研究

来引导相关技术突破，形成全面支撑碳中和愿景实现的核心技术体系。

三、启示与建议

1. 发挥新型举国体制优势，强化碳中和科技支撑体系建设

强化面向碳中和重大科技需求的国家战略科技力量布局，改组、新建一批高水平全国重点实验室、国家工程研究中心、国家技术创新中心等，开展碳中和重大科技问题系统性研究与应用示范。有序推进碳中和国家重大科技专项和重点研发计划专项研发布局，建立前沿探索与基础研究、应用研究与技术示范有机联动的碳中和科技研发与示范模式，打造自主可控、国际领先的碳中和核心技术体系。

2. 重点推动能源革命，以能源技术革命推动建立新能源为主体的零碳能源结构

在保障国家能源安全和产业链安全的基础上，支撑清洁高效煤电由主体电源向电力保障和调峰的基础性调节性电源转变，加强煤炭低碳高效转化利用，存量化石能源转化利用重心由“碳燃料”向“碳材料”转变。中长期保障新能源安全可靠替代，优先发展新一代高效低成本可再生能源、安全先进核能系统、新型电化学能源等颠覆性零碳能源技术，大力发展高比例消纳、先进电网、多能互补与供需互动、大规模储能等关键技术构建新型电力系统，发展氢/氨燃料、生物能源、低品位余热利用等低碳零碳技术满足高品位热能、高密度燃料等非电用能需求。最终显著提高非化石能源在能源结构中的比重，构建清洁低碳、安全高效的现代能源体系。

3. 加快构建低碳产业，打造变革性绿色智能工业技术体系

重点通过电气化应用、燃料/原料替代、高效节能技术，大幅削减工业过程原料反应和化石能源使用造成的碳排放。发展物质能量循环与再利用技术，包括原生资源高效加工转化、废弃物再资源化技术，加强资源的全生命周期管理与利用，以及在重点/难减排领域利用氢/氨、可再生能源等技术开展颠覆性零碳/低碳工业流程再造，如氢还原炼铁、绿色化工、生物冶金等。

4. 前瞻部署 CCUS 等负排放技术研发与示范

以 CO_2 规模化减排和资源化利用为重点，加快 CO_2 转化高价值化学品等转化利用技术的研发和推广，实现化学吸附、膜分离等新型碳捕集技术突破。中远期有序推

进 CCUS 技术在火电、化工、钢铁等产业的全流程融合示范，加强跨行业、跨领域的技术集成，着眼长远前瞻部署 CO_2 生物转化、直接空气碳捕集（DAC）、生物能源碳捕集与封存（BECCS）等前沿负排放技术的研发与示范。

5. 促进集成耦合与协同优化，加强系统解决方案部署

推动碳中和技术系统集成耦合与产业、区域发展协同优化，加快打造全产业链、跨产业链多能融合智慧系统，协同解决能源转化和工业生产过程的高能耗、高排放难题。系统评估关键技术跨系统、跨区域大规模应用的经济-社会-环境-气候-健康综合影响，统筹推进分区域和分部门的低碳、零碳和负碳技术发展。推动人工智能、数字化等新一代信息技术与能源、工业和生态领域的融合创新。

致谢：中国科学院大连化学物理研究所刘中民院士、蔡睿研究员，中国科学院上海高等研究院魏伟研究员，中国科学院广州能源研究所赵黛青研究员，中国科学院科技战略咨询研究院郭剑锋研究员等审阅了本文并提出了宝贵的修改意见，特致谢忱。

参考文献

[1] Patel S. Power Digest[July 2022]. https://www.powermag.com/power-digest-july-2022/.

[2] 梁晨. 研究所“2000 吨/年气化细灰流化-熔融燃烧技术”通过科技成果鉴定. https://www.iet.cas.cn/news/kyjz/202301/t20230112_6598890.html[2023-01-12].

[3] 宋维健. 研究所“循环流化床高温后燃超低 NO_x 燃烧技术”通过科技成果鉴定. https://www.iet.cas.cn/news/zh/202304/t20230410_6729795.html[2023-04-10].

[4] Mitsubishi Power. Nakoso IGCC Plant Completed, the World's Largest Plant began Operations on April 16 in Japan. https://power.mhi.com/news/20210419.html[2021-04-19].

[5] Sonal Patel. UK's First Gas-Fired Allam Cycle Power Plant Taking Shape. https://www.powermag.com/uks-first-gas-fired-allam-cycle-power-plant-taking-shape/[2021-07-15].

[6] Chae Y J, Lee J I. Thermodynamic analysis of compressed and liquid carbon dioxide energy storage system integrated with steam cycle for flexible operation of thermal power plant. 2022, 256: 115374.

[7] Su Z X, Yang L. A novel and efficient cogeneration system of waste heat recovery integrated carbon capture and dehumidification for coal-fired power plants. Energy Conversion and Management, 2022, 255: 115358

[8] Luong D X, Bets K V, Algozeeb W A, et al. Gram-scale bottom-up flash graphene synthesis. Nature, 2020, 577(7792): 647-651.

[9] Snyder B E R, Bols M L, Rhoda H M, et al. Cage effects control the mechanism of methane hydroxylation in zeolites. Science, 2021, 373(6552): 327-331.

[10] Yang N T, Zhao Y H, Wu P, et al. Defective C_3N_4 frameworks coordinated diatomic copper catalyst: towards mild oxidation of methane to C1 oxygenates. Applied Catalysis B: Environmental, 2021, 299: 120682.

[11] King Abdullah University of Science and Technology. KAUST team sets world record for tandem solar cell efficiency. https://www.kaust.edu.sa/en/news/kaust-team-sets-world-record-for-tandem-solar-cell-efficiency[2023-04-16].

[12] Park J, Kim J, Yun H S, et al. Controlled growth of perovskite layers with volatile alkylammonium chlorides. *Nature*, 2023, DOI: 10.1038/s41586-023-05825-y.

[13] Liu W Z, Liu Y J, Yang Z Q, et al. Flexible solar cells based on foldable silicon wafers with blunted edges. Nature, 2023, 617: 717-723.

[14] 中国科学院大连化学物理研究所. 由大连化物所研发千吨级"液态太阳燃料合成示范项目"通过科技成果鉴定. http://www.dicp.ac.cn/xwdt/mtcf/202010/t20201020_5719609.html[2020-10-20].

[15] Ruotian Chen, Zefeng Ren, Yu Liang, et al., Spatiotemporal imaging of charge transfer in photocatalyst particles. Nature, 2022, DOI: 10.1038/s41586-022-05183-1.

[16] Virgil A, Ucoski G M, Pornrungroj C, et al. Floating perovskite-BiVO4 devices for scalable solar fuel production. Nature, 2022, DOI: 10.1038/s41586-022-04978-6.

[17] Suguro T, Kishimoto F, Kariya N, et al. A hygroscopic nano-membrane coating achieves efficient vapor-fed photocatalytic water splitting. Nature Communications, 2022, DOI: 10.1038/s41467-022-33439-x.

[18] Sandia National Laboratories. Propelling wind energy innovation. https://energy.sandia.gov/propelling-wind-energy-innovation/[2022-08-03].

[19] 中国船舶集团有限公司. 全球单机功率最大、风轮直径最大的 H260-18 兆瓦海上风电机组研制成功. http://www.sasac.gov.cn/n2588025/n2588124/c26980781/content.html[2023-01-16].

[20] 明阳智能. 张传卫:助力以海洋能源为支撑的新型能源体系建设. http://www.myse.com.cn/jtxw/info.aspx? itemid=945[2023-01-12].

[21] 风能专委会 CWEA. 西门子歌美飒推出全球首款陆上风机可回收叶片. https://mp.weixin.qq.com/s/FFnfxdLqshjXNLzMnKM93w[2022-09-25].

[22] 王晶晶. 风机叶片再利用,时代新材、明阳智能下线 95%可回收叶片. https://www.thepaper.cn/newsDetail_forward_22401230[2023-03-22].

[23] Liao Y H, Koelewijn S F, van den Bossche G, et al. A sustainable wood biorefinery for low-carbon footprint chemicals production. Science, 2020, 367(6484): 1385-1390.

[24] He J, Wu Z J, Gu Q Q, et al. Zeolite-tailored active site proximity for the efficient production of pentanoic biofuels. Angewandte Chemie(International Ed in English), 2021, 60(44): 23713-23721.

[25] Rosatom. ROSATOM: world's only floating nuclear power plant enters full commercial exploitation. https://rosatom.ru/en/press-centre/news/rosatom-world-s-only-floating-nuclear-power-plant-enters-full-commercial-exploitation/[2020-05-20].

[26] World Nuclear News. Beloyarsk BN-800 fast reactor running on MOX. https://www. world-nuclear-news. org/Articles/Beloyarsk-BN-800-fast-reactor-running-on-MOX[2022-09-13].

[27] TVEL Fuel Company of Rosatom. Rosatom starts construction of unique power unit with BREST-OD-300 fast neutron reactor. https://www. tvel. ru/en/press-center/news/? ELEMENT_ID=8787[2021-06-08].

[28] Idaho National Laboratory. Southern Company Signs Agreement with U. S. Department of Energy to Demonstrate World's First Fast-Spectrum Salt Reactor in Collaboration with Terrapower, Idaho National Laboratory. https://inl. gov/article/southern-company-signs-agreement-with-u-s-department-of-energy-to-demonstrate-worlds-first-fast-spectrum-salt-reactor-in-collaboration-with-terrapower-idaho-national-laboratory/[2021-11-18].

[29] Office of Nuclear Energy. Southern Company Services and TerraPower Build World's Largest Chloride Salt System. https://www. energy. gov/ne/articles/southern-company-services-and-terrapower-build-worlds-largest-chloride-salt-system[2022-10-18].

[30] X-energy. TRISO-X Breaks Ground on North America's First Commercial Advanced Nuclear Fuel Facility. https://x-energy. com/media/news-releases/triso-x-breaks-ground-on-north-americas-first-commercial-advanced-nuclear-fuel-facility[2022-10-13].

[31] 中国科学院等离子体物理研究所 . EAST 装置物理实验创造 1.2 亿度 101 秒等离子体运行的世界纪录 . http://www. ipp. ac. cn/xwdt/ttxw/202105/t20210528_641312. html[2021-05-28].

[32] 中国科学院合肥大科学中心 . EAST 实现世界上最长时间可重复的高约束模等离子体运行 . http://hfsc. ustc. edu. cn/2023/0414/c18928a598735/pagem. htm[2023-04-14].

[33] 中国科学技术大学 . 中国科大在新一代神威超级计算机上首次实现 EAST 和 CFETR 聚变堆全装置动理学等离子体演化模拟 . https://kyb. ustc. edu. cn/2021/1203/c6076a537181/page. htm[2021-12-03].

[34] National lgnition Facility & Photon Science. Star Power: Blazing the Path to Fusion Ignition. https://lasers. llnl. gov/news/star-power-blazing-the-path-to-ignition[2023-02-23].

[35] UK Atomic Energy Authority, Department for Business, Energy & Industrial Strategy. All systems go for UK's £55M fusion energy experiment. https://www. gov. uk/government/news/all-systems-go-for-uks-55m-fusion-energy-experiment[2020-10-29].

[36] Magee R M, Ogawa K, Tajima T, et al. First measurements of $p^{11}B$ fusion in a magnetically confined plasma. Nature Communications, 2023, 14:955.

[37] Gibney E. Nuclear-fusion reactor smashes energy record. Nature, 2022, DOI:602:371.

[38] 张朝登 . 全球首个海上风电无淡化海水直接电解制氢在福建海试成功 . http://jjdf. chinadevelopment. com. cn/xw/2023/06/1842197. shtml[2023-06-02].

[39] Kölbach M, Rehfeld K, May M M. Efficiency gains for thermally coupled solar hydrogen production in extreme cold. Energy & Environmental Science, 2021, 14(8):4410-4417.

[40] Du H L, Chatti M, Hodgetts R Y, et al. Electroreduction of nitrogen with almost 100% current-to-

ammonia efficiency. Nature,2022,609:722-727.

[41] Chang F, Tezsevin I, de Rijk J W, et al. Potassium hydride-intercalated graphite as an efficient heterogeneous catalyst for ammonia synthesis. Nature Catalysis,2022,5:222-230.

[42] Fu X, Pedersen J B, Zhou Y, et al. Continuous-flow electrosynthesis of ammonia by nitrogen reduction and hydrogen oxidation. Science,2023,379:707-712.

[43] Peng P, Anastasopoulou A, Brooks K, et al. Cost and potential of metal-organic frameworks for hydrogen back-up power supply. Nature Energy,2022,7:448-458.

[44] Lee H, Park D G, Park J, et al. Amorphized Defective Fullerene with a Single-Atom Platinum for Room-Temperature Hydrogen Storage. Advanced Energy Materials,2023,13:2300041.

[45] World-Energy Media. ArcelorMittal to Develop the World's First Full-Scale Zero Carbon-Emissions Steel Plant Using Hydrogen. https://www.world-energy.org/article/18953.html[2021-07-14].

[46] Fraunhofer Institute for Microengineering and Microsystems. The world's first high-temperature ammonia-powered fuel cell for shipping. https://www.fraunhofer.de/en/press/research-news/2021/march-2021/worlds-first-hightemperature-ammonia-powered-fuel-cell-for-shipping.html[2021-03-01].

[47] Alstom. Alstom's Coradia iLint successfully travels 1,175 km without refueling its hydrogen tank. https://www.alstom.com/press-releases-news/2022/9/alstoms-coradia-ilint-successfully-travels-1175-km-without-refueling-its[2022-09-16].

[48] 東京ガス株式会社、サンレー冷熱株式会社．日本初！水素専焼、ガスタービンコージェネ用追焚きバーナの開発．https://www.tokyo-gas.co.jp/news/press/20230228-01.html[2023-02-28].

[49] HYBRIT. HYBRIT: Milestone reached -pilot facility for hydrogen storage up and running. https://www.hybritdevelopment.se/en/hybrit-milestone-reached-pilot-facility-for-hydrogen-storage-up-and-running[2022-09-23].

[50] TARMAC. UK lime kiln in world first net zero hydrogen trial. https://tarmac.com/news/uk-lime-kiln-in-world-first-net-zero-hydrogen-trial[2022-07-05].

[51] DNV. DNV awards AiP for a floating ammonia production unit developed by SWITCH2 and BW Offshore. https://www.dnv.com/news/dnv-awards-aip-for-a-floating-ammonia-production-unit-developed-by-switch2-and-bw-offshore--240876[2023-03-10].

[52] GE. GE and IHI Sign Memorandum of Understanding to Develop Gas Turbines that Can Operate on 100% Ammonia. https://www.ge.com/news/press-releases/ge-and-ihi-sign-memorandum-of-understanding-to-develop-gas-turbines-that-can-operate[2023-01-18].

[53] 清华大学．清华电机系合作研发的金坛盐穴压缩空气储能国家试验示范项目并网试验成功. https://www.tsinghua.edu.cn/info/1181/87521.htm[2021-10-02].

[54] 新华网．国际首套百兆瓦先进压缩空气储能国家示范项目并网发电．http://www.news.cn/science/2022-09/30/c_1310667448.htm[2022-10-01].

[55] 新华社．我国首台(套)300 兆瓦级压缩空气储能示范工程开工．http://www.news.cn/photo/2022-07/26/c_1128865942.htm[2022-07-26].

[56] Highview Power. Financial Times: UK group plans first large-scale liquid air energy storage plant. https://highviewpower.com/press _ coverage/financial-times-uk-group-plans-first-large-scale-liquid-air-energy-storage-plant[2022-11-16].

[57] Kim S, Kim J S, Miara L, et al. High-energy and durable lithium metal batteries using garnet-type solid electrolytes with tailored lithium-metal compatibility. Nature Communications, 2022, 13: 1883.

[58] Pan H, Zhang M, Cheng Z, et al. Carbon-free and binder-free Li-Al alloy anode enabling an all-solid-state Li-S battery with high energy and stability. Science Advances, 2022, 8: eabn4372.

[59] Sun W, Küpers V, Wang F, et al. A non-alkaline electrolyte for electrically rechargeable zinc-air batteries with long-term operation stability in ambient air. Angewandte Chemie, 2022, 134: e202207353.

[60] Wang X, An Y, Liu L, et al. Atomically dispersed pentacoordinated-zirconium catalyst with axial oxygen ligand for oxygen reduction reaction. Angewandte Chemie, 2022, 134: e202209746.

[61] Tang Q, Zhang Y, Xu N, et al. Demonstration of 10+ hour energy storage with ϕ″laboratory size solid oxide iron-air batteries. Energy & Environmental Science, 2022, 15: 4659-4671.

[62] Kondori A, Esmaeilirad M, Harzandi A M, et al. A room temperature rechargeable Li_2O-based lithium-air battery enabled by a solid electrolyte. Science, 2023, 379: 499-505.

[63] 中国能源网．全球首套 1MWh 钠离子电池储能系统正式投运，助力“双碳”目标达成．http://www.cnenergynews.cn/chuneng/2021/07/01/detail_20210701100059.html[2021-07-01].

[64]澎湃新闻．中科海钠·阜阳全球首条 GWh 级钠离子电池生产线产品下线仪式举行．https://www.thepaper.cn/newsDetail_forward_20957839[2022-11-29].

[65] Nippon Electric Glass. Development of the World's First All-Oxide All-Solid-State Sodium (Na) Ion Secondary Battery. https://www.neg.co.jp/en/news/20211118-4417.html[2021-11-18].

[66] Yamauchi H, Ikejiri J, Tsunoda K, et al. Enhanced rate capabilities in a glass-ceramic-derived sodium all-solid-state battery. Scientific Reports, 2020, 10(1): 9453.

[67] U. S. Department of Energy. Hybrid Energy Systems: Opportunities for Coordinated Research. https://www.nrel.gov/docs/fy21osti/77503.pdf[2021-04-29].

[68] ETIP SNET. ETIP SNET R&I Implementation Plan 2021-2024. https://www.etip-snet.eu/wp-content/uploads/2020/05/Implementation-Plan-2021-2024_WEB_Single-Page.pdf[2020-05-13].

[69] ETIP SNET. ETIP SNET R&I Roadmap 2020-2030. https://www.etip-snet.eu/wp-content/uploads/2020/02/Roadmap-2020-2030_June-UPDT.pdf[2020-02-27].

[70] ETIP SNET. ETIP SNET VISION 2050-Integrating Smart Networks for the Energy Transition: Serving Society and Protecting the Environment. https://www.etip-snet.eu/wp-content/uploads/2018/06/VISION2050-DIGITALupdated.pdf[2018-06-27].

[71] The New Energy and Industrial Technology Development Organization. The world's largest-class

hydrogen production, Fukushima Hydrogen Energy Research Field (FH2R) now is completed at Namie town in Fukushima. https://www.nedo.go.jp/english/news/AA5en_100422.html[2020-03-07].

[72] U. S. Department of Energy. Artificial Intelligence & Technology Office FY21/22 Program Plan and FY23 Forecast. https://www.energy.gov/sites/default/files/2021-09/AITO%20Program%20Plan%2009-16-2021.pdf[2021-09-01].

[73] Business Wire. Shell, C3 AI, Baker Hughes, and Microsoft Launch the Open AI Energy Initiative, an Ecosystem of AI Solutions to Help Transform the Energy Industry. https://www.businesswire.com/news/home/20210201005936/en/[2021-02-01].

[74] 华为．数字能源产品与解决方案．https://digitalpower.huawei.com/[2022-08-26].

[75] National Grid. Introducing the Virtual Energy System. https://www.nationalgrideso.com/news/introducing-virtual-energy-system[2021-11-05].

[76] E. ON. E. ON sets all course for digitalization and establishes new subsidiary E. ON One. https://www.eon.com/en/about-us/media/press-release/2022/e.on-sets-all-course-for-digitalization.html[2022-03-11].

[77] C40 CITIES. C40 and Google launch 24/7 Carbon-Free Energy for Cities programme. https://www.c40.org/news/c40-and-google-launch-24-7-carbon-free-energy[2022-10-20].

[78] 中国化工报．离子液体：让二氧化碳接受“再教育”．http://www.ipe.cas.cn/xwdt_/kyjz/202105/t20210519_6030600.html[2019-12-06].

[79] Kawasaki Kisen Kaisha, Ltd. World's first CO_2 Capture Plant on Vessel Installed on Coal Carrier “CORONA UTILITY” Launch of the “CC-OCEAN” Project Demonstration. https://www.kline.co.jp/en/news/csr/csr7601431474845700352/main/0/link/210805EN.pdf[2021-08-05].

[80] Climeworks. Climeworks begins operations of Orca, the world's largest direct air capture and CO storage plant. https://climeworks.com/news/climeworks-launches-orca[2021-09-08].

[81] 山西经济日报．云冈热电：世界首个煤电 CO_2 捕集及资源化利用全产业链生产线进入试生产. http://epaper.sxrb.com/shtml/sxjjrb/20210123/590991.shtml[2021-01-22].

[82] Schäppi R, Rutz D, Dähler F, et al. Drop-in fuels from sunlight and air. Nature, 2022, 601(7891): 63-68.

[83] Cai T, Sun H B, Qiao J, et al. Cell-free chemoenzymatic starch synthesis from carbon dioxide. Science, 2021, 373(6562): 1523-1527.

[84] 人民网．枞阳海螺国内首条生物质替代燃料系统试生产．http://ah.people.com.cn/n2/2020/1110/c227767-34406396.html[2020-11-10].

[85] Dubai Electricity & Water Authority. EGA and DEWA join hands to make the UAE the world's first country to produce aluminium using solar power. https://www.dewa.gov.ae/en/about-us/media-publications/latest-news/2021/01/ega-and-dewa-join-hands[2021-01-18].

[86] Synhelion. CEMEX, Sandia Labs, and Synhelion to scale solar energy technology to produce

cement. https://synhelion.com/news/cemex-sandia-labs-and-synhelion-to-scale-solar-energy-technology-to-produce-cement[2023-02-16].

[87] CEMEX. CEMEX to introduce hydrogen technology to reduce CO_2 emissions in four cement plants in Mexico. https://www.cemex.com/w/cemex-to-introduce-hydrogen-technology-to-reduce-co2-emissions-in-four-cement-plants-in-mexico[2022-12-06].

[88] Boston Metal. Transforming Metal Production. https://www.bostonmetal.com/transforming-metal- production/[2022-08-26].

[89] Sohn H Y, Fan D Q, Abdelghany A. Design of novel flash ironmaking reactors for greatly reduced energy consumption and CO_2 emissions. Metals, 2021, 11(2): 332.

[90] SSAB. The World's First Fossil-free Steel Ready for Delivery. https://www.ssab.com/news/2021/08/the-worlds-first-fossilfree-steel-ready-for-delivery[2021-08-18].

[91] ArcelorMittal. ArcelorMittal breaks ground on first transformational low-carbon emissions steelmaking project. https://corporate.arcelormittal.com/media/press-releases/arcelormittal-breaks-ground-on-first-transformational-low-carbon-emissions-steelmaking-project[2022-10-13].

[92] Primetals Technologies. Primetals Technologies and POSCO to develop new green steel demonstration plant. https://www.primetals.com/fileadmin/user_upload/press-releases/2022/2022082601/PR2022082601en.pdf[2022-08-31].

[93] 鞍钢集团．全球首套绿氢零碳流化床高效炼铁新技术示范项目在鞍钢开工．http://www.heic.org.cn/newshow.asp?id=1005[2022-09-29].

[94] Negrão L B A, Pöllmann H, da Costa M L. Production of low-CO_2 cements using abundant bauxite overburden"Belterra Clay". Sustainable Materials and Technologies, 2021, 29: e00299.

[95] MacFarlane J, Vanorio T, Monteiro P J M. Multi-scale imaging, strength and permeability measurements: understanding the durability of Roman marine concrete. Construction and Building Materials, 2021, 272: 121812.

[96] Department of Energy. DOE Releases First-Ever Comprehensive Strategy to Secure America's Clean Energy Supply Chain. https://www.energy.gov/articles/doe-releases-first-ever-comprehensive-strategy-secure-americas-clean-energy-supply-chain[2022-02-24].

[97] Office of Energy Efficiency & Renewable Energy. Biden Administration, U.S. Department of Energy to Invest $3 Billion to Strengthen U.S. Supply Chain for Advanced Batteries for Vehicles and Energy Storage. https://www.energy.gov/eere/articles/biden-administration-us-department-energy-invest-3-billion-strengthen-us-supply-chain[2022-02-11].

[98] Department of Energy. President Biden Invokes Defense Production Act to Accelerate Domestic Manufacturing of Clean Energy. https://www.energy.gov/articles/president-biden-invokes-defense-production-act-accelerate-domestic-manufacturing-clean?utm_medium=email&utm_source=govdelivery[2022-06-06].

[99] U.S. Department of State. Minerals Security Partnership. https://www.state.gov/minerals-

security-partnership-june-14-2022[2022-06-14].
[100] European Commission. REPowerEU: Commission establishes the EU Energy Platform Task Force to secure alternative supplies. https://ec. europa. eu/commission/presscorner/detail/en/IP_22_3299[2022-05-25].
[101] European Commission. REPowerEU: A plan to rapidly reduce dependence on Russian fossil fuels and fast forward the green transition. https://ec. europa. eu/commission/presscorner/detail/en/IP_22_3131[2022-05-18].
[102] Cabinet Office and Prime Minister's Office, 10 Downing Street. Making Government Deliver for the British People (HTML). https://www. gov. uk/government/publications/making-government-deliver-for-the-british-people/making-government-deliver-for-the-british-people-html[2023-02-07].
[103] Department for Energy Security & Net Zero, Prime Minister's Office, 10 Downing Street, and Department for Business, Energy & Industrial Strategy. British energy security strategy. https://www. gov. uk/government/publications/british-energy-security-strategy[2022-04-07].
[104] Department for Energy Security and Net Zero. Powering up Britain. https://www. gov. uk/government/publications/powering-up-britain[2023-03-30].
[105] The White House. The Long-Term Strategy of the United States: Pathways to Net-Zero Greenhouse Gas Emissions by 2050. https://www. whitehouse. gov/wp-content/uploads/2021/10/US-Long-Term-Strategy. pdf[2021-11-01].
[106] European Commission. Horizon Europe Work Programme 2021-2022: 8. Climate, Energy and Mobility. https://ec. europa. eu/info/funding-tenders/opportunities/docs/2021-2027/horizon/wp-call/2021-2022/wp-8-climate-energy-and-mobility_horizon-2021-2022_en. pdf[2021-06-16].
[107] Ministry of Economy, Trade and Industry. Green Growth Strategy Through Achieving Carbon Neutrality in 2050. https://www. meti. go. jp/english/policy/energy_environment/global_warming/ggs2050/pdf/ggs_full_en1013. pdf[2021-06-18].
[108] International Energy Agency. Net Zero by 2050—A Roadmap for the Global Energy Sector. https://iea. blob. core. windows. net/assets/deebef5d-0c34-4539-9d0c-10b13d840027/NetZeroby2050-ARoadmapfortheGlobalEnergySector_CORR. pdf[2021-05-18].

Energy Science and Technology

Chen Wei, Yue Fang, Tang Yun, Li Lanchun, Li Nana
Sun Yuling, Qin Aning, Teng Fei, Peng Hao

Energy is a strategic contested field related to national development and security. Low-carbon technology innovation and disruptive breakthroughs in energy technology are the most critical factors in promoting the energy revolution and industrial revolution and implementing the national strategy of "carbon peaking and carbon neutrality". Under the guidance of the national demand-driven and carbon neutrality strategy, the global energy system is changing from the dominance of fossil energy to the integration of multi-energy and low-carbon energy. The development and deep integration of new energy technologies and a series of emerging technologies have promoted profound changes in the entire chain of energy production, conversion, transmission, storage, and consumption. The major strategic plans for energy science and technology developed by major developed countries and regions, as well as the progress and important achievements of energy technology since 2021, are systematically sorted out and analyzed in this paper. Finally, it puts forward some recommendations for the development of energy science and technology in China.

4.11 材料制造领域发展观察

万 勇 黄 健 冯瑞华 姜 山 董金鑫
（中国科学院武汉文献情报中心）

2021 年以来，世界主要国家持续重视材料制造领域对科技与经济发展的基础性带动作用，加强顶层设计，出台创新型发展战略，通过技术路线图、项目计划等推动实施，在先进材料制造技术突破、材料与制造关键技术集群建设、战略性关键原材料供应链安全、制造业绿色低碳转型等方面取得重要进展与成效。

一、重要研究进展

1. 关键战略材料工艺与工程化取得新进展

（1）稀土及其功能材料高效利用持续受到重视。美国国防部高级研究计划局（DARPA）启动稀土生物开采研究，利用微生物和生物分子工程相关技术，开发稀土资源分离与提纯方法，以有效利用稀土资源，填补供应链缺口[1]。美国 DOE 埃姆斯国家实验室（Ames Laboratory）基于稀土数据库开发出机器学习模型用于评估新发现的稀土化合物的稳定性，并开展了磁性预测、制造工艺过程控制和力学行为优化等研究[2]。中国科学院赣江创新研究院、中国科学院长春应用化学研究所等创新性地提出增强对钇的选择性及改善萃取过程的分子设计思路，合成并筛选出新型含磷羧酸萃取剂，并完成中试[3]。

（2）碳纤维工艺及产业化获得突破。日本新能源产业技术综合开发机构（NEDO）开发出一种可快速固化的碳纤维增强预浸料片材，可在 30 s 内固化，创当前最短固化时间，并可在室温下储存[4]。由中国建材集团有限公司投资建设的我国首个万吨碳纤维生产基地在青海西宁投产，首次实现了单线年产 3000 t 高性能碳纤维生产线设计和高端成套技术自主可控[5]。

（3）金属材料关键研制技术实现重大进展。中国科学院金属研究所牵头研制成功直径 8 m 主轴承，这是目前我国制造的首台套直径最大、单重最大的盾构机用主轴承，标志着我国已掌握盾构机主轴承的自主设计、材料制备、精密加工、安装调试和检测评价等集成技术[6]。中国钢研科技集团有限公司首次试制出目前我国最大规格的高温合金涡轮盘整体模锻件，重 13.5 t，直径为 2380 mm，强度等级为 1200 MPa，可用于 650℃温度条件，打破国外垄断[7]。

2. 前沿新材料制备、物性探索等基础研究成果显著

（1）新型材料和新物质结构不断涌现。德国马尔堡大学和芬兰阿尔托大学在金表面制备出石墨烯的新同素异形体——联苯烯网络，其独特的金属性有望提升电池存储容量[8]。中国科学院化学研究所利用聚合-剥离两步法，创制出新型碳材料——单层聚合 C_{60}，其具有较高的结晶度、良好的热力学稳定性和适度的禁带宽度[9]。

（2）物性探索与材料应用研究取得新的进展，理论变为现实。德国于利希研究中心首次证实，在二维材料中存在一种奇异的电子态——费米弧，该研究为新型量子材料及其在新一代自旋电子学和量子计算中的潜在应用奠定了基础[10]。中国科学院高能物理研究所首次验证了大尺寸铁基超导线圈在高磁场领域应用的可行性，助力我国铁基超导材料向高磁场应用迈进[11]。上海交通大学与美国麻省理工学院采用分子束外延技术，通过低温强磁场扫描隧道显微镜在由拓扑绝缘体 Bi_2Te_3/超导体 $NbSe_2$ 组成的异质结体系中，首次实验观察到了 50 多年前理论预言的“分段费米面”[12]。

3. 材料引发器件形态及机理革新，助力高性能应用

材料结构设计与性质调控不断深入，推动电子器件多功能化、轻薄化、柔性化。以色列特拉维夫大学利用六方氮化硼研制出当前最薄的技术装置——仅有两个原子厚度的可储存电子信息的微型设备，有望使电子设备速度更快、密度更低、能效更高[13]。澳大利亚蒙纳士大学使用硫化锡纳米片制造了迄今最薄的 X 射线探测器，厚度不到 10 nm，具有灵敏度高、响应速度快的特点，有助于实现细胞生物学的实时成像[14]。美国杜克大学利用三种碳基墨水打印制成的新型晶体管，成为全球首个完全可回收的印刷电子产品，能将电子器件便捷地打印在纸张或其他柔性环保材料表面[15]。

4. 微纳精密制造技术呈现交叉融合发展

（1）制造工艺融合发展，注重高效。美国 IBM 公司开发出全球首款 2 nm 芯片制造工艺，利用底部介电隔离技术、内部空间干燥工艺和 2 nm 极紫外光刻技术等，实现约 3.33 亿个晶体管/mm^2，与 7 nm 芯片相比，同功率下性能提升 45%，同性能下能耗降低 75%[16]。瑞士洛桑联邦理工学院开发出用于超导电路光学机械系统的新型纳米制造技术，建造了首个大型可配置的超导电路光机晶格，可克服量子光学机械系统的尺度挑战[17]。

（2）激光无掩模光刻技术突破光学衍射极限限制。中国科学院理化技术研究所与暨南大学利用超衍射纳米光刻技术高效制备了数百微米尺度与纳米尺度并存的跨尺度微纳结构，实现了多种跨尺度图案的批量化制备[18]。

（3）信息技术与制造系统加速融合。武汉华中数控股份有限公司研发出世界首台具备自主学习、自主优化补偿能力的 iNC 智能数控系统，加工精度达 6 μm，进入批量市场推广阶段[19]。

5. 增材制造技术创新与应用领域不断扩展

（1）工艺打破传统模式，成果质量创新纪录。南京航空航天大学提出“材料-结构-工艺-性能”一体化并行模式，在复杂整体构件内部同步实现多材料设计与布局、多层级结构创新与打印制造[20]。南方科技大学、德国马普学会钢铁研究所和新加坡先进制造技术研究院通过选择性激光熔化（SLM）和热处理技术，制备出的 AlZnMgCuScZr 合金的最高强度达 647MPa，这是采用 SLM 技术生产的铝合金当今国际报道的最高值[21]。

（2）在航天军工、新能源汽车和医疗等行业加速落地。中国航天科工集团第六研究院突破航天液体动力领域 3D 打印全流程技术，实现了 230 余种复杂精密构件 3D 打印成型[22]。德国黑石技术公司利用 3D 打印“厚层技术”工艺生产电池电极和锂离子电池隔膜，能量密度达 220 Wh/kg，比传统锂电池提高 20%，体积减小 15%，电池材料成本下降 20 欧元/kW · h[23]。加拿大首个本地制造的下颌骨板 3D 打印医疗植入物获加拿大卫生部批准[24]。

（3）我国国际标准化工作实现零的突破。《信息技术 3D 打印和扫描增材制造服务平台架构》（ISO/IEC 23510：2021）正式发布，成为我国在增材制造领域牵头制定的第一项国际标准[25]。

二、重要战略规划

1. 美国：更新材料战略规划，启动“制造业美国”技术路线图研究，国防增材制造受重视

2021 年，指导美国材料领域科技发展的两大计划——“材料基因组计划”（MGI）和“国家纳米技术计划”（NNI）均发布了新版战略规划，提出了未来五年的新发展目标。美国国家科技委员会发布的 2021 年版《材料基因组计划战略规划》对 2014 年版进行了更新。新版规划压缩为材料创新基础设施、材料数据和人员培养三个目标，

未明确提及具体的材料研究方向，而是更加强调材料基因组计划对推动材料创新，尤其是推动新材料走向应用方面具有的潜力[26]。美国白宫科技政策办公室和国家纳米技术协调办公室联合发布的2021年版《国家纳米技术计划战略规划》对2016年版做出较大调整。在发展愿景上，围绕技术与产业变革，将“引发”调整为“已经发生”，反映出美国科技界对科技革命态势的最新判断。在发展目标上，继续关注开展技术研发、推进商业化、加强基础设施建设和负责任发展等方向，并将公众参与和劳动力培养的相关内容单独列出，体现了对人才培养的重视[27]。

2022年10月，美国白宫发布《国家先进制造业战略》，对2018年《美国先进制造业领导力战略》进行了更新，继续聚焦先进制造技术、劳动力和供应链三大方向；在先进制造技术方面，持续重点关注智能制造未来趋势，电子制造更加聚焦半导体领域，并着重强调清洁能源与制造工艺脱碳技术，以及生物制造和生物质加工[28]。

美国制造业创新网络截至2022年末已建立16家研究所，成员机构达2000余家，构建了覆盖全美的创新生态系统。2021年6月，美国国家标准与技术研究院（NIST）启动“制造业美国”技术路线图项目，支持行业联盟制定相关技术路线图，以应对具有高优先级的研究挑战，推动美国先进制造业发展。除了关注未来新建研究所的技术方向外，当前有望带来制造业变革的前沿技术也将成为关注的主题[29]。

增材制造技术发展持续受到重视。2021年1月和6月，美国先后发布首份《国防部增材制造战略》[30]和《增材制造在国防部的使用》[31]等顶层战略文件，进一步明确了增材制造战略目标及具体政策措施，旨在加速推动增材制造从新型技术向规模化、普适性应用阶段迈进，使其成为类似机械加工、铸造等在国防领域广泛应用的制造技术。2022年5月，美国联邦政府推出“增材制造推进计划”，旨在提升中小企业供应链韧性与创新性，克服增材制造面临的应用挑战，更好地建设美国本土区域制造业生态系统[32]。

2. 欧洲：注重可持续韧性发展

2021年1月，欧盟委员会研究科研与创新总司发布首份“工业5.0”白皮书。白皮书指出，工业4.0考虑更多的是经济或技术因素，对于社会公平价值观以及可能的供应中断（尤其是在疫情环境下）考虑较少，需对当前工业4.0模式进行补充和扩展，要将以人为中心、可持续发展等理念融入欧洲工业的发展过程当中，并称之为“工业5.0”[33]。从当前文本内容来看，其研究主要集中在概念阐述、特征分析和人机协同等方面，研讨范围多局限在欧洲。

2022年12月，欧洲材料联盟组织发布《材料2030路线图》，提出推动材料开发数字化，加速材料设计与开发；加强新材料加工和规模化的支撑活动等行动建议。该路线图围绕健康医疗、建筑、新能源、运输、家庭与个人护理、包装、农业、纺织品和电子电器等领域九大类材料创新市场，阐述了面临的研发挑战与优先事项，以及预期的社会经济效益等[34]。

3. 英国：建设先进材料与制造关键技术集群

2021 年 7 月，英国商业、能源和产业战略部发布《英国创新战略：创新未来引领未来》报告，希冀通过充分利用研发和创新体系来支持私营部门创新，并提出了英国政府的创新愿景：到 2035 年，使英国成为全球创新中心。该报告将“先进材料与制造”确定为未来助推英国经济的七项关键技术集群之一，实现先进材料的批量化制造，并将安全性评估与可持续发展融入材料的设计与创新。该报告将超材料、二维材料、智能仿生自修复材料、复合材料结构与涂层技术、增材制造等列为有发展潜力的机遇方向[35]。

4. 日本：构建以数据为基础的材料创新体系

2021 年 4 月，日本发布由文部科学省和经济产业省高层战略会议形成的《材料创新力强化战略》[36]。报告认为，面向未来科学技术和社会经济发展，材料将发挥重要的推动作用；通过创新发展战略能够快速高效解决当前日本材料行业中存在的发展瓶颈与问题。报告提出，建立以数据为基础的材料创新体系，推动数据驱动型材料研究，以强化日本材料创新能力。围绕材料开发与应用、数据驱动研发、国际竞争力三个维度，报告提出了行动计划方案。其中，“数据驱动研发”是该战略的主要举措布局，将整合以数据为基础的材料研发平台，构建数据驱动型创新体系。

5. 英美及欧盟：战略性关键原材料供应安全依旧备受关注

战略性关键原材料作为高技术产业链中的重要一环，具有支撑性、引领性和颠覆性作用，其供应的安全性日益成为世界主要经济体重点关切的战略布局方向之一。2021 年 4 月，英国伯明翰大学战略元素与关键材料中心发布的《保卫对英国高科技行业至关重要的金属》将稀土（钕/镨/钐/镝/铽）、电池材料（镍/钴/锂/天然石墨）、铂族（钯/铑/铂）、航空航天金属（铼/钽）等列为高技术行业所必需的关键金属，建议成立专门机构开展跨部门合作[37]。6 月，美国国防部发布《关键矿物与材料百日审查报告》，建议采取“全政府参与”的方式使供应链多样化，推动全球市场获得可持续、可靠的关键矿物和材料生产来源，同时采取“逐个矿物”的战略，探索和扩大可持续的本土生产、加工，以及本土关键矿物和材料的回收利用[38]。9 月，欧洲原材料联盟发布的《稀土磁体和电机：欧洲行动呼吁》报告分析了欧洲当前及未来对稀土元素的需求，提出了打造安全的稀土供应链、畅通稀土项目的融资渠道、制定可持续稀土磁体和电机的标准和认证方案等行动建议方案[39]。11月，欧盟委员会第三版《原材料记分牌》报告强调指出，欧洲在某些原材料生产方面高度依赖其他地区，应循环利用原材料实现脱碳，确保就业机会和附加值[40]。2022年7月，英国关键矿物情报中心正式启动，这是英国首个收集和分析关键矿产供应信息的机构，由英国地质调查局负责运营，将通过为政策制定者提供有关供应、需求和市场动态的最新数据与分析，提高英国关键矿产供应链的韧性[41]。

6. 日欧：碳中和/零碳理念逐步凸显，助力绿色化转型升级

节能环保、可循环、低碳排放等绿色化发展成为共识，呈现出持续向绿色低碳转型升级发展的态势。日本经济产业省设立2万亿日元的绿色创新基金，旨在推动低碳绿色技术从研发到示范，再到成果的社会化推广。"氢气炼钢"是首批关注的技术，将开发高炉氢气还原技术和氢气直接还原铁矿石的炼钢技术[42]。"地平线欧洲"2021～2022年工作计划在"数字、工业与空间"领域明确了与制造相关的研发目标及内容。其中，"气候中立、循环化和数字化生产"主题聚焦绿色柔性先进制造、先进数字制造技术、流程工业可再生能源与电气化的整合等方向[43]。英国启动"国家跨学科循环经济研究"计划，围绕纺织品、化学工业、建筑材料和金属等，通过利用更少的资源、重复使用和回收产品及材料，实现环境与经济效益双赢[44]。

三、启示与建议

1. 加强问题导向的材料科学基础研究

材料科学是典型的以应用为目标的基础学科。我国在材料领域原创性、突破性的基础前沿成果较少，与产业需求、国家重大发展需求脱节。未来我国材料科学研究需要以原创性思想、变革性实践、突破性进展、标志性成果为导向，注重从国家重大战略需求、国民经济发展需求中提炼核心科学问题，加强以应用目标为导向的材料科学应用基础研究，在应用基础研究组织模式、人才引进等方面积极探索，解决更多应用场景中凝练的基础科学问题和"卡脖子"问题背后的深层次科学问题。建议重点关注的方向包括高温合金、轻质金属、宽禁带半导体、量子材料、5G/6G信息功能材料、能源存储与转换材料、生物医用材料等；涉及问题包括极端环境下材料与结构力学、后摩尔时代半导体能耗边界与速度极限、矿物-微生物相互作用机理等。

2. 进一步强化材料领域数据积累

美国、日本等均重视材料领域数据的积累与应用，并从国家层面出台了指导性的

战略规划。与此相比，我国相关工作尽管备受关注，中国科学院物理研究所与怀柔科学城共建的材料基因组研究平台、中国科学院上海硅酸盐研究所无机材料基因科学创新中心、上海大学材料基因组工程研究院等机构形成了一定的研究基础，但尚未达到国家或行业层面。可将材料制造领域海量数据积累列入“东数西算”工程等新基建投资重点，从更高层面夯实材料制造创新体系数据基础。

3. 推动我国制造业绿色低碳转型升级

我国是制造业大国，需加快构筑制造业新型绿色技术创新体系，打造绿色低碳产业集群。利用工业互联网、大数据、人工智能、5G 等新一代信息技术，积极赋能绿色制造，提升能源、资源、环境管理水平，深化生产制造过程的数字化应用，推进重点行业和领域低碳工艺革新及数字化转型。

4. 保障战略性关键原材料供应链安全

加强国家战略性关键原材料各类计划的有效衔接，统筹完善关键原材料领域政策体系。针对我国当前及未来对关键原材料的需求，开展原材料供应链、竞争力和贸易流动等方面的深入分析，开展跨部门合作，打造安全的战略资源供应链，并强化对全球矿产资源供应链的主导力。

5. 加速推动增材制造高质量发展

增材制造作为制造业极具代表性的颠覆性技术，已成为世界各国抢占未来产业制高点的焦点领域。我国应抓住“换道超车”的历史性发展机遇，面向国家战略性产品和战略性领域的重大需求，加强增材制造顶层设计，发展自主创新的增材制造技术，解决“卡脖子”问题，增强增材制造基础研究能力，提升增材制造实用化程度，完善增材制造产业链，从而推动我国增材制造高质量发展。

致谢：中国兵器工业集团北方科技信息研究所高彬彬研究员、中国航空工业发展研究中心胡燕萍高级工程师、中国科学院长春应用化学研究所王鑫岩研究员、中国科学院宁波材料技术与工程研究所应华根研究员、中南大学李明君副教授对本文初稿进行了审阅并提出了宝贵的修改意见，在此表示感谢！

参考文献

[1] DARPA. Developing Cohesive, Domestic Rare Earth Element (REE) Technologies. https://www.darpa.mil/news-events/2021-07-13[2021-07-13].

[2] Singh P, Del Rose T, Vazquez G, et al. Machine-learning enabled thermodynamic model for the design of new rare-earth compounds. Acta Materialia, 2022, 229: 117759.

[3] 中国科学院赣江创新研究院．南方离子吸附型稀土矿中高丰度钇元素萃取分离工艺研究获进展. http://www.gia.cas.cn/gzdt_166111/202107/t20210714_6132504.html[2021-07-14].

[4] Aalto University. A new form of carbon. https://www.aalto.fi/en/news/a-new-form-of-carbon [2021-05-20].

[5] 国务院国有资产监督管理委员会．我国首个万吨碳纤维生产基地投产．http://www.sasac.gov.cn/n2588025/n2588124/c20634858/content.html[2021-09-10].

[6] 中国科学院金属研究所．我国首台套盾构机用超大直径主轴承研制成功．http://www.imr.cas.cn/xwzx/zhxw/202212/t20221215_6584906.html[2022-12-15].

[7] 航空产业网．中国钢研攻克超大尺寸高温合金涡轮盘核心制造技术．https://www.chinaerospace.com/article/show/c46e4391b23514d55ff43c784125cdd6[2021-12-27].

[8] Fan Q T, Yan L H, Tripp M W, et al. Biphenylene network: a nonbenzenoid carbon allotrope. Science, 2021, 372(6544): 852-856.

[9] Hou L, Cui X, Guan B, et al. Synthesis of a monolayer fullerene network. Nature, 2022, 606: 507-510.

[10] Chen Y, Hanke J, Hoffmann M, et al. Spanning fermi arcs in a two-dimensional magnet. Nature Communications, 2022, 13: 5309.

[11] Zhang Z, Wang D L, Wei S Q, et al. First performance test of the iron-based superconducting racetrack coils at 10 T. Superconductor Science and Technology, 2021, 34: 035021.

[12] Zhu Z, Papaj M, Nie X A, et al. Discovery of segmented Fermi surface induced by Cooper pair momentum. Science, 2021, 374(6573): 1381-1385.

[13] Tel Aviv University. The World's Thinnest Technology—Only Two Atoms Thick. https://english.tau.ac.il/news/worlds-thinnest-tech[2021-07-04].

[14] Shabbir B, Liu J, Krishnamurthi V, et al. Soft X-ray detectors based on SnS nanosheets for the water window region. Advanced Functional Materials, 2022, 32(3): 2105038.

[15] Williams N X, Bullard G, Brooke N, et al. Printable and recyclable carbon electronics using crystalline nanocellulose. Nature Electronics, 2021, 4: 261-268.

[16] IBM. IBM unveils world's first 2 nanometer chip technology, opening a new frontier for semiconductors. https://newsroom.ibm.com/2021-05-06-IBM-Unveils-Worlds-First-2-Nanometer-Chip-Technology, Opening-a-New-Frontier-for-Semiconductors[2021-05-06].

[17] Youssefi A, Kono S, Bancora A, et al. Topological lattices realized in superconducting circuit optomechanics. Nature, 2022, 612: 666-672.

[18] Liu Y-H, Zhao Y-Y, Jin F, et al. λ/12 super resolution achieved in maskless optical projection nanolithography for efficient cross-scale patterning. Nano Letters, 2021, 21 (9): 3915-3921.

[19] 付丽丽．能自主学习决策 全球首台智能数控系统发布．科技日报, 2021-04-13(3).

[20] Gu D, Shi X, Poprawe R, et al. Material-structure-performance integrated laser-metal additive manufacturing. Science, 2021, 372 (6545): eabg1487.

[21] Zhu Z, Ng F L, Seet H L, et al. Superior mechanical properties of a selective-laser-melted AlZnMgCuScZr alloy enabled by a tunable hierarchical microstructure and dual-nanoprecipitation. Materials Today, 2022, 52: 90-101.

[22] 张梅．航天六院突破航天液体动力领域 3D 打印全流程技术．陕西日报，2021-08-16(1).

[23] Blackstone Resources AG. Launch of 3D-printed lithium batteries: High attendance and exciting content. https://www.blackstoneresources.ch/en/press/2021/12/09/launch-of-3d-printed-lithium-batteries-high-attendance-and-exciting-content/[2021-12-09].

[24] 3D Printing Industry. Health Canada has approved its first Canadian-made 3D printed medical implant. https://3dprintingindustry.com/news/health-canada-has-approved-its-first-canadian-made-3d-printed-medical-implant-201111/[2021-12-09].

[25] 中机生产力促进中心．新突破、新征程——我国牵头制定的首项增材制造国际标准正式发布. https://www.cifmt.cn/JCY/contents/1257/10934.html[2021-10-14].

[26] MGI. Materials Genome Initiative Strategic Plan. https://www.mgi.gov/sites/default/files/documents/MGI-2021-Strategic-Plan.pdf[2021-11-01].

[27] NNI. 2021 National Nanotechnology Initiative Strategic Plan. https://www.nano.gov/2021strategicplan[2021-10-08].

[28] U. S. Department of Commerce. Announcing the National Strategy for U. S. Leadership in Advanced Manufacturing. https://www.commerce.gov/news/blog/2022/10/announcing-national-strategy-us-leadership-advanced-manufacturing[2022-10-07].

[29] NIST. NIST Launches New Manufacturing USA Technology Roadmap Grant Competition. https://www.nist.gov/news-events/news/2021/06/nist-launches-new-manufacturing-usa-technology-roadmap-grant-competition[2021-06-17].

[30] Chief Technology Officer. DoD Additive Manufacturing Strategy. https://www.cto.mil/dod-additive-manufacturing-strategy/[2021-01-01].

[31] Washington Headquarters Services. Use of Additive Manufacturing in the DoD. https://www.esd.whs.mil/Portals/54/Documents/DD/issuances/dodi/500093p.PDF? ver=JM7vpZGnbXAFX5uv91rXOQ%3D%3D[2021-06-10].

[32] White House. Fact Sheet: Biden Administration Celebrates Launch of AM Forward and Calls on Congress to Pass Bipartisan Innovation Act. https://www.whitehouse.gov/briefing-room/statements-releases/2022/05/06/fact-sheet-biden-administration-celebrates-launch-of-am-forward-and-calls-on-congress-to-pass-bipartisan-innovation-act/[2022-05-06].

[33] European Commission. Industry 5.0: towards more sustainable, resilient and human-centric industry. https://ec.europa.eu/info/news/industry-50-towards-more-sustainable-resilient-and-human-centric-industry-2021-jan-07_en[2021-01-07].

[34] The Materials 2030 Manifesto. The Materials 2030 roadmap. https://www.ami2030.eu/wp-content/uploads/2022/12/2022-12-09_Materials_2030_RoadMap_VF4.pdf[2022-12-01].
[35] UK Government. UK Innovation Strategy: leading the future by creating it. https://www.gov.uk/government/publications/uk-innovation-strategy-leading-the-future-by-creating-it[2021-07-22].
[36] CAO. Materials Innovation Strategy. https://www8.cao.go.jp/cstp/material/material_honbun_en.pdf[2021-04-27].
[37] University of Birmingham. Securing technology-critical metals for Britain. https://www.birmingham.ac.uk/documents/college-eps/energy/policy/policy-comission-securing-technology-critical-metals-for-britain.pdf[2021-04-01].
[38] DoD. The defense department's strategic and critical materials review. https://www.defense.gov/Newsroom/Releases/Release/Article/2649649/the-defense-departments-strategic-and-critical-materials-review/[2021-06-08].
[39] Eit. Rare Earth Magnets and Motors: A European Call for Action. https://eit.europa.eu/library/rare-earth-magnets-and-motors-european-call-action [2021-11-03].
[40] European Commission. Raw Materials Scoreboard highlights urgency to strengthen the resilience and sustainability of raw materials supply in the EU. https://ec.europa.eu/jrc/en/news/raw-materials-scoreboard-highlights-urgency-strengthen-resilience-and-sustainability-raw-materials [2021-11-17].
[41] Department for Business, Energy & Industrial Strategy. UK's first Critical Minerals Intelligence Centre to help build a more resilient economy. https://www.gov.uk/government/news/uks-first-critical-minerals-intelligence-centre-to-help-build-a-more-resilient-economy[2022-07-04].
[42] METI. An R&D and Social Implementation Plan for "Hydrogen Use in Steelmaking Processes" Projects Formulated. https://www.meti.go.jp/english/press/2021/0914_001.html[2021-09-14].
[43] European Commission. Horizon Europe Work Programme 2021-2022: 7. Digital, Industry and Space. https://ec.europa.eu/info/funding-tenders/opportunities/docs/2021-2027/horizon/wp-call/2021-2022/wp-7-digital-industry-and-space_horizon-2021-2022_en.pdf[2021-03-10].
[44] UKRI. Hub to provide national circular economy leadership. https://www.ukri.org/news/hub-to-provide-national-circular-economy-leadership/[2021-01-20].

Materials and Manufacturing

Wan Yong, Huang Jian, Feng Ruihua, Jiang Shan, Dong Jinxin

Since 2021, significant breakthroughs have been made in the preparation methodology and engineerizationof critical materials. A variety of novel materials and physical structures have been developed, as well as the theoretical predictions of physical properties have been validated and justified. Moreover, these novel materials are contributing to the transformation of device morphology and their applications are being extended to key areas. In the manufacturing sector, the development of micro/nano manufacturing technologies shows a cross integration tendency and focuses on the efficiency. Additive manufacturing technologies have undergone considerable development and have been implemented in various industries. Major developed countries are strengthening the top-level design, introducing innovative strategies, and attaching great importance to the influence of materials and manufacturing sectors on the development of science, technology and economy. Furthermore, they are also paying attention to the security of the supply chain of critical raw materials and the green transformation of the manufacturing sector.

4.12 重大科技基础设施发展观察

董 璐 李宜展 王志强 郭世杰 魏 韧 李泽霞

（中国科学院文献情报中心）

新冠疫情仍然影响全球的方方面面，对重大科技基础设施（本文以下简称“重大设施”）发展的负面影响也仍在继续，重大设施的现场访问和合作交流大幅减少，许多设施的建设和升级进度都受到影响。但总体上，重大设施的发展仍稳中有进，大量的新技术推动了重大设施的发展，能源领域的重大设施建设和实验进展取得重大进步。随着科研范式的转变，美国重新定义重大设施的构成要素，以数据链对重大设施进行治理。

一、领域重要进展

1. 国际重大科技基础设施建设稳步推进

2121年1月，西班牙科技部批准了西班牙同步辐射光源（ALBA）二期项目，将更换部分加速器部件、建设新的光束线、升级现有光束线组件，将ALBA升级改造为第四代同步辐射光源[1]。4月，欧洲科学家开始在西西里岛海岸外的地中海海底投放探测器阵列，开始着手建造立方千米中微子望远镜（Kilometer Cube Neutrino Telescope，KM3NeT①）。5月，欧盟委员会决定建立欧洲极端光研究基础设施联合体(ELI ERIC②)，意味着位于不同地点的几个ELI实验室可以作为一个组织合法地联合运营[2]。6月，平方公里阵列天文台（Square Kilometre Array Observatory，SKAO）理事会批准启动平方公里阵列（Square Kilometre Array，SKA）射电望远镜的建设工作[3]，这是有史以来规模最大、最复杂的射电望远镜网络。7月，美国电子离子对撞机关键决策点1（CD-1）获DOE批准，标志着该项目完成概念设计阶段[4]；中东同

① KM3NeT. Exciting deployment on the ARCA site! https：//www. km3net. org/exciting-deployment-on-the-arca-site/ [2021-04-15].

② ERIC是欧盟管理组织欧盟层面研究基础设施的一种特定法律形式，以非经济的方式运营基础设施。

步辐射光源（SESAME）的材料科学光束线启用[5]。2022 年 2 月，美国布鲁克海文国家实验室 sPHENIX 探测器外部强子热量计装配完成，升级工作取得里程碑进展。该探测器是 PHENIX 实验的升级版，有助于科学家进一步了解夸克-胶子等离子体（QGP）[6]。4 月，美国能源部正式批准授权质子改进计划（PIP-Ⅱ）项目全面开工建设，开展对费米国家加速器实验室加速器设施的必要升级，升级完成后该加速器将成为世界上能量最高、功率最高的线性粒子加速器之一[7]。同月，CERN 加速器综合体结束 3 年多的升级改造工作，大型强子对撞机（LHC）的质子束能量从 6.5 TeV 增加到 6.8 TeV，创下新的世界纪录[8]。8 月，德国仿星器 Wendelstein 7-X 核聚变装置完成升级，开启新一轮实验预期产生最长可达 30 min 的等离子体脉冲[9]。9 月，美国劳伦斯伯克利实验室完成激光加速器 BELLA 的升级，拍瓦激光器的第二条光束线建设完成，这为下一代粒子加速器的开发奠定基础[10]。

2. 新技术、新材料的应用推动重大设施新发展

2021 年 2 月，美国能源部三个实验室（阿贡国家实验室、劳伦斯伯克利国家实验室和费米国家实验室）经过 15 年合作，研制成功一种由铌三锡合金构成的先进、强大的超导磁体[11]，当磁场强度达到 23 T 时仍可保持超导性，将应用于欧洲核子研究中心（CERN）的大型强子对撞机和费米国家加速器实验室未来的 PIP-II 直线加速器。这种磁体非常适合于制成波荡器，可应用于下一代同步辐射光源和自由电子激光设施上，并有望降低加速器成本。6 月，美国科学家开发出一种类似于将数据筛选和图像重建技术相结合从而降低宇宙噪声的工具[12]，使中微子信号在宇宙射线背景中以 5∶1 的信噪比突显出来。2022 年 1 月，美国洛斯阿拉莫斯国家实验室研究人员开发出可用于下一代电子束加速器的仅原子厚度的石墨烯涂层[13]。3 月，美国哈佛大学研发出用于高功率激光器的金刚石镜，研究人员表示，采用的单材料反射镜方法消除了传统反射镜在受到较大光功率照射时产生的热应力问题，有可能改善或创造高功率激光器的新应用[14]。机器学习方法是下一代加速器和激光性能的关键推动力，能够为新的光子源和未来的粒子对撞机提供动力。7 月，美国劳伦斯伯克利国家实验室研究人员开发了一种新的机器学习平台，能够自动补偿加速器光束和其他组件（如磁铁）的实时变化[15]。

3. 聚变设施的建设升级和实验应用获重大进展

2021 年 1 月，美国能源部宣布启动多个由企业和国家实验室联合开展的聚变能源创新网络项目，以推动聚变能的应用[16]。5 月，英国原子能管理局（UKAEA）公布其升级版兆安培球形托卡马克（MAST-U）装置的初步实验结果[17]，成功测试了一

种能承受核聚变高温的排气系统概念设备，为实现聚变能发电消除了一个重要障碍。8月，美国劳伦斯利弗莫尔国家实验室的国家点火装置实验取得重要进展，实现了超过1.3 MJ的能量输出[18]。

4. 提出重大设施的新概念设计

2021年1月，日本大阪大学研究人员提出一个下一代超强激光概念[22]，采用广角非共线光学参量啁啾脉冲放大（WNOPCPA）技术和优化相位匹配，完全避免了泵浦干扰，实现具有两个宽光谱的超宽带宽，从而实现了小于10 fs的高能激光放大，结合后压缩技术有望将当前的激光脉冲功率纪录从当前的10 PW提高到500 PW。1月，英国研究与创新署（UKRI）宣布资助720万英镑研发一种全新的仪器——10 m原子干涉仪[23]，这将是英国的第一个大型原子干涉仪，建成后将大大提升探测暗物质和引力波的能力。2022年6月，荷兰阿姆斯特丹大学制造出可持续的原子激光器，成功解决了创造连续玻色-爱因斯坦凝聚体的难题，未来有望用于测试基本物理常数和工程精密技术[24]。

5. 依托重大设施实现多领域重大科技突破

（1）支撑天文和粒子物理重大进展

2021年2月，CERN大型强子对撞机（LHC）上的ATLAS实验组发现希格斯玻色子衰变为两个轻子（带相反电荷的电子或缪子对）和一个光子的首个证据[25]。6月，美国激光干涉引力波天文台（LIGO）、欧洲“处女座”（Virgo）引力波探测器和日本神冈引力波探测器（KAGRA）的天文学家在LIGO和Virgo的第三轮运行中，连续发现两例来自黑洞-中子星合并的引力波事件——GW200105和GW200115[26]。7月，天文学家使用与欧洲南方天文台（ESO）合作的Atacama大型毫米波/亚毫米阵列波（ALMA），首次明确观测到太阳系外行星周围存在原初行星盘。这一观测结果将为年轻恒星系统中卫星和行星的形成方式提供新的线索[27]。2022年1月，科学家利用NOIRLab观测站及全球其他望远镜的观测数据和档案数据，在银河系420 ly外创纪录地发现至少70颗“自由漂浮”行星[28]。5月，CERN的LHC的ALICE合作组首次直接观察到粒子物理学中的死角效应，或将提供测量夸克质量的新方法[29]。

（2）支撑材料科学的研究和发展

2021年1月，加拿大科学家团队和欧洲同步辐射光源（ESRF）的研究人员利用ID11光束线，使用3D同步辐射X射线衍射技术，了解材料在原子层级发生的变化，并将其与材料的工作原理联系起来[30]。10月，美国石溪大学研究人员利用多模同步辐射技术探索打印参数与材料缺陷状态的关联，揭示了激光增材制造 316L 不锈钢（一种广泛用于海军设施的耐腐蚀金属）的腐蚀行为与底层材料结构之间的联系，为开展更耐腐蚀的打印合金的工程设计提供路径[25]。2022 年 3 月，美国艾姆斯实验室、艾奥瓦州立大学研究团队及其合作者发现了新型费米弧，可通过施加磁脉冲非常快速地打开或关闭电弧，为电子产品提供新的发展路径[32]。7 月，由德国锡根大学和欧洲 X 射线自由电子激光（XFEL）的国际科学家团队合作在 XFEL 上首次以纳米和皮秒分辨率观察超快表面过程，有望为激光材料加工和高能量密度科学开辟新的前景[33]。

（3）支撑新能源的探索和发展

2021 年 2 月，美国佐治亚理工学院、普渡大学、韩国蔚山国立科学技术研究所的研究人员使用美国先进光子源（APS）的 2-BM 束线，利用超亮 X 射线计算机断层扫描技术，捕获了充放电循环期间电池内部发生的结构变化的实时 3D 图像[34]。10 月，剑桥大学领导的国际研究团队发现了有机太阳能电池中的一条损耗途径，它影响了有机太阳能电池的光电转化效率，使其低于硅基电池；研究团队提出了一种操纵太阳能电池内部的分子抑制这条损耗途径的方法，防止电流因不良状态（称为三重态激子）引起损失[35]。2022 年 4 月，英国欧洲联合环（JET）实验在 5 秒钟内产生 59 兆焦耳的能量，打破在受控、持续的核聚变反应产生的能量记录[36]。5 月，日本国立聚变科学研究所与美国威斯康星大学的研究团队合作首次发现了大型螺旋装置（LHD）中等离子体在热量逸出时湍流的移动速度比热量快。这种湍流的特征使预测等离子体温度的变化成为可能，对其观测或将有助于未来开发一种实时控制等离子体温度的方法[37]。

（4）支撑解决生命健康的重大科学问题

2021 年 6 月，CERN 的紧凑型直线对撞机（Compact Linear Collider，CLIC）在临床环境中通过高梯度加速产生极高能量电子（very high energy electron，VHEE）光束，相较于其他粒子束疗法具有更高性价比，可以为医生提供高度可靠的诊疗技术手段；此外，VHEE 光束能与 FLASH 放射疗法相互兼容，FLASH 放射疗法可以在几乎瞬间（不到一秒）将高能粒子传送到病理组织中[30]。6 月，国际研究团队使用英国钻石（Diamond）光源和美国劳伦斯伯克利国家实验室的先进光源（Advanced Light Source，ALS）进行脑组织成像，首次在人脑淀粉样斑块中识别出金属元素铜和磁性元素铁[31]。这些新发现为理解金属促成神经退行性变性疾病的脑细胞死亡模式提供了思路，也可能有助于开发恢复患病大脑金属平衡的新疗法，减缓或防止疾病的发展。11 月，美国辉瑞制药公司借助美国能源部的先进光子源（Advanced Photon Source，APS）完成 COVID-19 抗病毒药物的研发[40]。2022 年 3 月，德国科学家使用激光-等离子体加速器产生的质子束，以超高峰值剂量率照射模式动物的肿瘤，论证了其遏制肿瘤、改善癌症放射治疗的潜力[41]。11 月，国际研究小组利用欧洲同步辐射光源揭示 COVID-19 患者肺部纤维化机制[42]。

二、重要战略计划与部署

1. 多国发布重大设施发展规划

2021 年 8 月，德国亥姆霍兹联合会发布新的设施路线图[32]，共规划了 40 个设施。其中，新规划 31 个设施，主要集中在物质科学（8 个）、地球与环境（6 个）和能源（6 个）领域，总投资较高的为国际聚变材料放射测试中子源示范设施及重力回溯和气候实验设施；持续建设 9 项设施，分别是地球能源科学地下实验室（GeoLaB）、高功率电网实验室（HPGL）、城市环境变化观测设施（UrbENO）、全球大气观测卫星（AtmoSat）、神经退行性疾病临床研究平台（KFNE）、Tandem-L 地表观测卫星、衍射极限同步辐射光源升级（PETRA Ⅳ）、柏林同步辐射电子存储环Ⅲ（BESSY Ⅲ）、全球宇宙射线观测设施（GCOS）。

2021 年 11 月，美国国家科学院、工程院和医学院发布关于天体学和天体物理学未来十年规划报告[33]，提出 2023～2032 年天文学和天体物理学在空间和地面的综合规划。在科学目标驱动下，该报告列举了建议优先投资的项目，包括开拓前沿的大型项目、维持和平衡科学发展的中型项目和基础活动。在大型项目方面，首要建议建设大型（约 6 m 口径）红外/光学/紫外（IR/O/UV）空间望远镜，5 年后实施远红外和 X 射线探测任务，两者成本约 30 亿～50 亿美元，持续推进美国极大望远镜计划（US-ELTP）、地基宇宙微波背景（CMB）研究、下一代超大阵列（ngVLA）和冰立方第 2 代天文台建设。中型项目方面，建议开展时域和多信使天文观测计划，拓展中等规模地基项目，大力支持对先进引力波干涉仪技术的开发。

2021 年 12 月，欧洲研究基础设施战略论坛（ESFRI）发布最新版本的《欧洲研究基础设施战略论坛路线图 2021》。这是继 2006 年 ESFRI 发布第一版研究基础设施路线图和 2016 年、2018 年发布其更新版本之后，对未来泛欧洲研究基础设施的建设和发展进行战略规划和部署的又一重大举措。《欧洲研究基础设施战略论坛路线图 2021》中包括了 22 个未来十年重点支持建设的基础设施项目（即 ESFRI 项目，ESFRI Projects），以及 41 个未来十年重点支持运行/升级的基础设施（即 ESFRI 地标，ESFRI Landmarks），涉及数字、能源、环境、健康和食品、物理科学和工程、社会和文化创新六大领域；新布局 11 个设施，其中多数为数据（数字）类型。

2022 年 3 月，法国高等教育、研究与创新部（MESRI）发布最新版本《国家研究基础设施战略 2021》暨《国家研究基础设施路线图 2021》[46]。该路线图是继 2008 年、2012 年、2016 年和 2018 年之后发布的第 5 版路线图，保留了路线图 2018 中的 91 个设施并新增 17 个设施。该路线图可看出法国重视物质前沿科学、生命健康与环境生态、科学数据设施的建设。

2022 年 3 月，英国研究与创新署（UKRI）发布《一起变革未来：UKRI 战略 2022—2027》报告[47]，提出将确保英国在重要国际研究基础设施方面的合作伙伴地位，并利用其对国际设施投资的杠杆效应，从跨境的知识、数据和能力共享中受益。同时，将发展和维护最前沿的研究与创新基础设施，包括数据基础设施，并与研究人员和企业一起确定必要的战略性基础设施以推动研究与创新集群的增长。

2022 年 4 月，澳大利亚发布《国家研究设施路线图 2021》[48]，概述了未来十年为保持研究的卓越性、提高创新和转移转化、应对新的研究挑战所需的国家研究基础设施。该路线图明确提出制定《国家数字研究基础设施战略》，通过提供充分利用数据所需的计算资源、数字工具、数据管理框架和专业知识，支持所有领域的研究人员计算、存储、分析需求和大规模合作，协调和整合国家数字研究基础设施生态系统。

2022 年 6 月，欧洲加速器光子源联盟（LEAPS）发布《2022 欧洲加速器光子源战略》，旨在巩固欧洲加速器光子源处于全球领导地位[49]。该战略提出将保持以发现为驱动的服务，支持和开发颠覆性新想法；基于加速器光源为其合作伙伴设施、用户和工业供应商开辟创新路线，以保持欧洲在创新技术发展的竞争中处于全球前沿。6 月，英国经济和社会研究委员会（ESRC）发布《数据基础设施五年战略》[50]，将引导 ESRC 在数据收集与服务方面的投资，为世界领先的社会科学研究提供支持。

2022 年 8 月，美国以《2022 芯片和科学法案》和《通货膨胀削减法案》为牵引，全面推进重大设施的建设和升级，加大对中型研究基础设施的支持。《2022 芯片和科学法案》概述了研究基础设施雄心勃勃的愿景，涉及大型科学设施建设项目以及对量子计算、高强度激光器和研究反应堆的支持[51]。美国能源部科学办公室在《通货膨胀削减法案》中获得了 15.5 亿美元资助，用于加速科研基础设施升级和国家实验室基础设施项目，包括先进光源（ALS）升级、国家同步加速器Ⅱ（NSLS-Ⅱ）升级等[52]。

2. 重大设施建设支撑“双碳”目标的实现

2021 年 9 月，英国科学与技术设施理事会（STFC）宣布启动“净零计划”[53]，旨在探索将研究成果转化为解决现实问题的方案和技术，以实现温室气体净零排放。10 月，美国能源部向太平洋西北国家实验室投入 9000 万美元，集成实验室现有和新的科学仪器建设新的能源科学中心，以加速在化学、材料科学和计算科学方面的科学发现[54]。

3. 重新定义重大设施构成要素，重塑治理链条

2021年10月，美国国家科学技术委员会发布《国家研究与发展设施战略概述》[55]，指出研发基础设施的三要素为实验和观测、知识基础、研究网络基础。以科学数据为核心重新定义重大设施构成要素，重塑治理链条。同时将科学馆藏、文献数据库、知识产权和人力资源信息作为重大设施构成要素的重要组成部分，发挥科学数据汇聚、存储、管理、共享等功能，并提供相应的标准、协议等机制保障；提升设施及其研发生态的整体水平，展现出强化科学数据基础设施建设、推动数据密集型科研范式转变的努力。

4. 不断深化基于重大设施的国际合作

2021年2月，平方公里阵列天文台（SKAO）正式成立[56]，它是第二个致力于天文学研究的政府间合作组织，包括我国在内的多个国家深度参与其中，标志着全球射电天文学进入了一个联合观测和研究的新时代。5月，英国科学和技术设施委员会（STFC）与美国费米国家实验室就建造PIP-Ⅱ加速器签署合作协议，STFC将为粒子加速器设计、制造和测试重要组件[57]。10月，英国STFC宣布将参与EIC设施的新探测器研发工作[58]。

2022年1月，俄罗斯的西伯利亚环形光子源加入欧洲加速器光源联盟[59]。2月，加拿大和巴西同步辐射光源签署谅解备忘录，在同步辐射和加速器研究方面开展合作共同推进农业研究[60]。7月，美国国家科学基金会（NSF）和捷克科学基金会（GACR）宣布，由美国加利福尼亚大学圣迭戈分校和捷克ELI Beamlines合作新项目，旨在展示密集伽马射线光束的高效产生[61]。10月，英国UKRI发布科学与技术设施理事会战略交付计划（2022—2025）[62]，提出未来3年通过参与日本超级神冈中微子实验、美国西蒙斯天文台、CERN、SKAO等大型设施深化国际科技合作，在大型国际研究基础设施战略、技术、管理中发挥领导作用，实现科学外交利益最大化。

三、启示与建议

1. 持续规划重大设施发展，重视设施升级换代

前沿科学的重大发现和突破越来越依赖于重大科技手段的提升，这需要始终保持

观测和实验技术手段的先进性。欧盟从 2002 年开始对重大设施的发展战略进行持续性研究，并根据相关结果每两年调整路线图规划的内容。我国重大设施 5 年规划在一定程度上对设施发展起到了极大的推动作用，但柔性和灵活性稍显不足，建议在整个规划周期，根据设施实施的实际情况动态调整规划的内容。

2. 重新调整重大设施的管理模式，以应对科研范式转变

大量创新成果的应用，推动了科学研究变得更加网络化、灵活和可重构，科研范式逐步发生转变，数据要素已经成为科研的主要驱动力。重大科技基础设施大量观测和实验会产生大量的数据，已经形成“大数据”＋“大计算”的大科学模式。因此，在布局、规划和管理重大设施的过程中也应当考虑科研范式对设施的影响，研究新的符合新范式发展规律的重大设施管理模式，优化对重大设施后端的科研数据的治理。

3. 加强对未来新设施概念和新技术的研发支持力度

欧美在重大设施概念预研方面已经形成很好的研究传统和氛围，以及良好的探讨机制，因此不断有新的概念设施提出，保持了其在设施研发方面的先进水平。但是我国在重大设施新概念和原创性新技术的提出方面仍较为匮乏，建议设立稳定的资助计划对新概念和新技术研究给予一定的支持。

4. 围绕能源领域重大设施布局先进核能技术方向

欧美在聚变和裂变能源重大设施方面均加强了战略引导和布局。我国核能技术在过去几十年获得极大进步，在重大设施战略发展方向进行了重点布局，“十二五”和“十三五”期间布局了加速器嬗变研究装置、高效低碳燃气轮机和聚变堆主机关键系统综合研究设施，建成后将为我国未来先进能源技术发展提供实验和研究环境。建议前瞻考虑围绕这些能源重大设施的先进核能研发问题，进行先期的准备和布局。特别针对聚变设施，建议在聚变设施研发早期就加强与企业和产业界的合作，提高企业相关技术的承接能力，加快实现人类利用核聚变能源的速度。

5. 强化重大设施研究对产业的辐射、带动和提升作用

依托重大设施产生的重大成果，以及重大设施研建过程中克服的技术难题都有可能促进国家的产业发展。过去一年依托平台型重大设施，在新材料、新能源和生命健康等方面取得一系列重大进展，例如提高有机太阳能电池光电转化效率、研制可作为制造屏幕原材料的复合玻璃等将促进新兴产业的发展，因此建议从国家层面顶层设

计、引导重大设施对产业的支撑。

致谢：中国科学院高能物理研究所张闯研究员、彭良强研究员和中国科学院物理研究所金铎研究员审阅了全文并提出宝贵的修改意见和建议，谨致谢忱！

参考文献

[1] LEAPS. ALBA to become 4th generation synchrotron——LEAPS. https://leaps-initiative.eu/alba-to-become-4th-generation-synchrotron/[2021-01-13].

[2] ESFRI. ELI ERIC officially established by the European Commission. https://www.esfri.eu/latest-esfri-news-project-landmarks-news/eli-eric[2021-05-06].

[3] UKRI. Construction starts on world's largest radio telescope network. https://www.ukri.org/news/construction-starts-on-worlds-largest-radio-telescope-network/[2021-06-29].

[4] BNL. Electron-Ion Collider Achieves Critical Decision 1 Approval. https://www.bnl.gov/newsroom/news.php?a=118765[2021-07-06].

[5] SESAME. Inauguration of SESAME's Materials Science (MS) Beamline. https://www.sesame.org.jo/news/inauguration-sesames-materials-science-ms-beamline[2021-07-12].

[6] BNL. sPHENIX Detector Upgrade Clears Assembly Milestone. https://www.bnl.gov/newsroom/news.php?a=219443[2022-03-18].

[7] FNAL. New accelerator at Fermilab approved for construction start. https://news.fnal.gov/2022/04/new-accelerator-at-fermilab-approved-for-construction-start/[2022-04-20].

[8] CERN. Accelerator Report: Crescendo at the LHC following the first stable beams at 6.8 TeV. https://home.cern/news/news/accelerators/accelerator-report-crescendo-lhc-following-first-stable-beams-68-tev[2022-04-27].

[9] Max Planck Insitute. Federal Research Minister Stark-Watzinger and Minister Martin visit IPP Greifswald. https://www.ipp.mpg.de/5265532/02_22[2022-08-09].

[10] IBL. Upgraded Laser Facility Paves the Way for Next-Generation Particle Accelerators. https://newscenter.lbl.gov/2022/09/07/next-generation-particle-accelerators/[2022-09-07].

[11] Argonne National Laboratory. Scientists Create Advanced, More Powerful Superconducting Magnet for Next Generation Light Sources. https://scitechdaily.com/scientists-create-advanced-more-powerful-superconducting-magnet-for-next-generation-light-sources/[2021-02-01].

[12] BNL. Physicists Achieve Significant Improvement in Spotting Accelerator-produced Neutrinos in a Cosmic Haystack. https://www.bnl.gov/newsroom/news.php?a=117324[2021-06-09].

[13] SciTechDaily. Atomic Armor for Next-Generation, Electron-Beam Accelerators. https://scitechdaily.com/atomic-armor-for-next-generation-electron-beam-accelerators/[2022-01-027].

[14] Harvard. Diamond mirrors for high-powered lasers. https://www.seas.harvard.edu/news/2022/05/diamond-mirrors-high-powered-lasers[2022-05-027].

[15] IBL. Machine Learning Paves Way for Smarter Particle Accelerators. https://newscenter. lbl. gov/2022/07/19/ml-particle-accelerators/[2022-07-19].

[16] Princeton Plasma Physics Laboratory. Three new public-private INFUSE projects to speed development of fusion energy selected for PPPL. https://www. pppl. gov/news/2021/01/three-new-public-private-infuse-projects-speed-development-fusion-energy-selected-pppl[2021-02-01].

[17] UKAEA. First experiment results point to fusion energy solution. https://www. ukri. org/news/first-experiment-results-point-to-fusion-energy-solution/[2021-05-26].

[18] LLNL. National Ignition Facility experiment puts researchers at threshold of fusion ignition. https://www. llnl. gov/news/national-ignition-facility-experiment-puts-researchers-threshold-fusion-ignition[2021-08-08].

[19] Kyoto Fusioneering. World-first Integrated Testing Facility for Fusion Power Plant Equipment to be Constructed in Japan. https://kyotofusioneering. com/en/news/2022/07/06/768[2022-07-06].

[20] Nature. A sustained high-temperature fusion plasma regime facilitated by fast ions. https://www. nature. com/articles/s41586-022-05008-1[2022-09-07].

[21] LLNL. Lawrence Livermore National Laboratory achieves fusion ignitionhttps://www. llnl. gov/archive/news/lawrence-livermore-national-laboratory-achieves-fusion-ignition[2022-12-14].

[22] Science Daily. Towards Exawatt-Class Lasers: New Concept for Next-Generation Ultra-Intense Lasers. https://scitechdaily. com/towards-exawatt-class-lasers-new-concept-for-next-generation-ultra-intense-lasers/[2021-01-15].

[23] UKRI. STFC scientists design instrument to understand dark matter. https://www. ukri. org/news/stfc-scientists-design-instrument-to-understand-dark-matter/[2021-01-15].

[24] SciTechDaily. Eternal Matter Waves: Physicists Build Atom Laser That Can Stay On Forever. https://scitechdaily. com/eternal-matter-waves-physicists-build-atom-laser-that-can-stay-on-forever/[2022-06-16].

[25] CERN. ATLAS finds evidence of a rare Higgs boson decay. https://home. cern/news/news/physics/atlas-finds-evidence-rare-higgs-boson-decay[2021-02-08].

[26] SINA. 引力波再立大功！人类首次确认黑洞-中子星合并事件. https://finance. sina. com. cn/tech/2021-06-30/doc-ikqciyzk2718483. shtml[2021-06-30].

[27] ESO. Astronomers make first clear detection of a moon-forming disc around an exoplanet. https://www. eso. org/public/news/eso2111/[2021-07-22].

[28] NSF. Astronomers identify record number of free-floating planets. https://new. nsf. gov/news/astronomers-identify-record-number-free-floating[2022-01-25].

[29] Tech Explorist. First direct observation of the dead-cone effect in particle physics. https://www. techexplorist. com/first-direct-observation-dead-cone-effect-particle-physics/47444/[2022-05-19].

[30] ESRF. Tracking how engineering materials deform to extend their lifetime. http://www. esrf. eu/home/news/general/content-news/general/tracking-how-engineering-materials-deform-to-extend-

their-lifetime. html[2021-01-19].

[31] BNL. Additively Manufacturing a Better Steel: The Key Could be in Synchrotron X-ray Techniques. https://www. bnl. gov/newsroom/news. php? a=219202[2021-10-26].

[32] AMES. New Fermi arcs could provide a new path for electronics. https://www. ameslab. gov/news/new-fermi-arcs-could-provide-a-new-path-for-electronics[2022-03-24].

[33] XFEL. Ultrafast surface processes observed. https://www. xfel. eu/news_and_events/news/index_eng. html? openDirectAnchor=1974&two_columns=0[2022-07-13].

[34] Anl. Inside the battery in 3D: Powerful X-rays watch solid state batteries charging and discharging. https://www. anl. gov/article/inside-the-battery-in-3d-powerful-xrays-watch-solid-state-batteries-charging-and-discharging[2021-02-03].

[35] The Cavendish Laboratory. Researchers identify and clear efficiency hurdle for organic solar cells. https://www. phy. cam. ac. uk/news/researchers-identify-and-clear-efficiency-hurdle-organic-solar-cells[2021-01-10].

[36] SciTechDaily. Scientists Shatter Record for the Amount of Energy Produced During a Controlled, Sustained Fusion Reaction. https://scitechdaily. com/scientists-shatter-record-for-the-amount-of-energy-produced-during-a-controlled-sustained-fusion-reaction/[2022-04-12].

[37] NIFS. 高速で移動するプラズマ乱流を世界で初めて発見. https://www. nifs. ac. jp/news/researches/220519. html[2022-05-19].

[38] CERN. CLEAR study paves the way for novel electron-based cancer therapy. https://home. cern/news/news/knowledge-sharing/clear-study-paves-way-novel-electron-based-cancer-therapy [2021-06-24].

[39] DIAMOND. Diamond helps discover microscopic metallic particles in the brain. https://www. diamond. ac. uk/Home/News/LatestNews/2021/10-06-21. html[2021-06-10].

[40] ANL. Advanced Photon Source helps Pfizer create COVID-19 antiviral treatment. https://www. anl. gov/article/advanced-photon-source-helps-pfizer-create-covid19-antiviral-treatment[2021-12-22].

[41] Nature. Laser-Plasma Accelerators Ready for translational research. https://www. natureasia. com/en/research/highlight/14009[2022-03-14].

[42] ESRF. Long COVID and pulmonary fibrosis better understood thanks to innovative technique. https://www. esrf. fr/home/news/general/content-news/general/long-covid-and-pulmonary-fibrosis-better-understood-thanks-to-innovative-techniques. html[2022-11-01].

[43] Helmholtz. Roadmap Forschungsinfrastrukturen 2021. https://www. helmholtz. de/fileadmin/medien_upload/21_Helmholtz_FIS_Roadmap_Deutsch. pdf[2021-8-15].

[44] National Academies of Sciences, Engineering, and Medicine. Pathways to Discovery in Astronomy and Astrophysics for the 2020s. Washington, DC: The National Academies Press, 2021.

[45] ESFRI. Roadmap2021. https://roadmap2021. esfri. eu/[2021-12-08].

[46] Ministère de l'Enseignement supérieur et de la. La Feuille de route nationale des Infrastructures de

recherche 2021. https://enseignementsup-recherche. gouv. fr/fr/la-feuille-de-route-nationale-des-infrastructures-de-recherche-2021-84056[2022-03-08].

[47] UKRI. UKRI strategy 2022 to 2027: transforming tomorrow together. https://www. ukri. org/publications/ukri-strategy-2022-to-2027/[2022-03-16].

[48] Australian Department of Education. 2021 National Research Infrastructure Roadmap. https://www. dese. gov. au/national-research-infrastructure/2021-national-research-infrastructure-roadmap [2022-04-29].

[49] LEAPS. The European Strategy for Accelerator-based Photon Science. https://leaps-initiative. eu/wp-content/uploads/2022/05/LEAPS-ESAPS-Broschure_final-20052022-3. pdf[2022-05-23].

[50] UKRI. ESRC launches five-year strategy for data infrastructure. https://www. ukri. org/news/esrc-launches-five-year-strategy-for-data-infrastructure/[2022-07-09].

[51] AIP. Research Infrastructure Initiatives in the CHIPS and Science Act. https://ww2. aip. org/fyi/2022/research-infrastructure-initiatives-chips-and-science-act[2022-09-08].

[52] DOE. Fact Sheet: Inflation Reduction Act Supporting the Future of DOE Science. https://www. energy. gov/science/articles/fact-sheet-inflation-reduction-act-supporting-future-doe-science [2022-11-04]. [53] STFC. Experts in innovation and research launch net zero schemes. https://www. ukri. org/news/experts-in-innovation-and-research-launch-net-zero-schemes/[2021-09-24].

[54] PNNL. New 90M PNNL Center to Focus on Solving Clean Energy's Biggest Puzzles. https://www. pnnl. gov/news-media/new-90m-pnnl-center-focus-solving-clean-energys-biggest-puzzles[2021-10-25].

[55] National Science and Technology Council. National strategic overview for research and development infrastructure. https://www. whitehouse. gov/wp-content/uploads/2021/10/NSTC-NSO-RDI-_REV_FINAL-10-2021[2021-08-15].

[56] UKRI. "Ambitious" global radio astronomy organisation launched. https://www. ukri. org/news/ambitious-global-radio-astronomy-organisation-launched/[2021-02-04].

[57] FERMI. UK to play vital role in creating the world's most powerful neutrino beam. https://news. fnal. gov/2021/05/uk-to-play-vital-role-in-creating-the-worlds-most-powerful-neutrino-beam/[2021-05-12].

[58] UKRI. UK to lead detector development for powerful particle collider. https://www. ukri. org/news/uk-to-lead-detector-development-for-powerful-particle-collider/[2021-10-11].

[59] LEAPS. LEAPS welcomes new twinning partner SKIF. https://leaps-initiative. eu/leaps-welcomes-new-twinning-partner-skif/[2022-1-17].

[60] Canadian Light Source. Canadian and Brazilian synchrotrons sign MOU to advance agricultural research. https://www. lightsource. ca/public/news/2021-22-q4-jan-march/canadian-and-brazilian-synchrotrons-sign-mou-to-advance-agricultural-research. php[2022-2-14].

[61] NSF. U. S. and Czech scientists collaborate to explore gamma-ray production with high-power lasers. https://beta. nsf. gov/news/us-and-czech-scientists-collaborate-explore-gamma-ray-production-

high-power-lasers[2022-07-01].

[62] UKRI. STFC strategic delivery plan. https://www.ukri.org/publications/stfc-strategic-delivery-plan/[2022-09-02].

Major Research Infrastructure

Dong Lu, Li Yizhan, Wang Zhiqiang,
Guo Shijie, Wei Ren, Li Zexia

The COVID-19 epidemic still affected the entire world. And the negative impact on the development of major research infrastructure was continued. Significant reduction was in site visits and collaborative exchanges at major research infrastructure. The progress of construction and upgrading of many facilities was affected.

Overall, however, the development of major research infrastructure is still progressing steadily. A large number of new technologies have promoted the development of major research infrastructure. Significant progress has been made in the construction and experimental progress of major research infrastructure in the energy field. With the transformation of scientific research paradigms, the United States is redefining the elements of major research infrastructure and taking advantage of data chain to manage them.

4.13 世界主要国家和组织科技创新战略与体制机制发展观察

叶 京[1] 李 宏[1,2,3] 张秋菊[1] 王建芳[1] 惠仲阳[1]
（1. 中国科学院科技战略咨询研究院；2. 中国科学院大学；
3. 粤港澳大湾区战略研究院）

随着新一轮科技革命和产业变革深入推进，全球科技创新进一步加快。数字技术、生物医药等领域的创新活动空前活跃，极大地推动了全球信息通信和医疗卫生产业的发展。面临机遇的同时，全球科技竞争加剧，全球供应链动荡等新形势给全球科技创新带来更多挑战。尤其是，以美国出台《2021 年美国创新与竞争法案》[1]和《2022 年芯片和科学法案》[2]为代表的对华遏制打压行动，极大恶化了全球科技创新发展的国际环境。因此，新形势下提升国家的关键核心技术创新能力、牢牢掌握竞争和发展主动权就显得至关重要。为此，世界主要国家和组织加快调整新一轮科技创新战略布局与方向。

一、综合性科技战略规划的总体目标

为适应全球科技经济发展变化、提升自身的创新能力和竞争能力，2021 年以来，英国、日本、西班牙等陆续出台国家层面的创新战略和总体规划，全局性谋划未来 3～5 年科技创新发展方向和目标。各国重视科技创新工作的顶层设计和宏观指导，提高各项科技创新政策的协同性和有效性，以实现以下三个方向的科技创新战略与规划目标。

1. 以科技创新赋能社会、经济和环境的可持续发展

日本于 2021 年 3 月发布《第六期科学技术创新基本计划》[3]，作为指导 2021～2025 年科学技术创新发展的纲领性规划，提出建设韧性社会、强化研究能力、重视培养人才的战略方向，旨在实现“Society 5.0”（超智能社会）的总目标；并于 2022 年 6 月发布《统合创新战略 2022》[4]，作为 2022 年日本科技创新工作的年度指导战略。2021 年 3 月，欧盟委员会发布《“地平线欧洲”2021—2024 年战略计划》[5]，主要对

标“全球性挑战与产业竞争力”部分，明确了“地平线欧洲”第一阶段的战略投资优先方向，围绕技术、环境、经济和社会“四大关键战略目标”开展有针对性的研发与创新。

2. 通过释放创新要素活力，促进国家创新生态系统完善

2022 年 3 月，英国发布《2022—2027 年战略：共同改变未来》[6]，围绕世界一流的人才和事业、创新基地、创新思想、创新行动、影响和组织等 6 个方面的世界级战略目标构建卓越的科研体系。2021 年 7 月，巴西发布《国家创新战略》[7]，提出了促进创新和刺激私人投资的计划和行动、促进打造适于创新的技术知识基础、传播创业创新文化、刺激创新产品和服务市场的发展、发展创新教育体系等 5 个战略方向，力求提高国家有关支持创新各项政策的凝聚力、协同性和有效性。

3. 营造良好研发环境，促进企业创新能力提升

2021 年 2 月，西班牙发布《2021—2027 年西班牙科技创新战略》[8]，提出优先解决环境挑战、促进科学研发转移转化、培养吸引和留住人才、激励企业创新等 4 个战略目标。7 月，英国发布《英国创新战略：通过创造引领未来》[9]，希望利用英国的研发和创新系统来支持企业的创新，使英国成为全球创新中心之一。2022 年 8 月，韩国发布《第一次研究产业振兴基本计划（2022—2026 年）》[10]，提出“由研究产业主导国家研发生产力创新”的愿景，以及扩大研究产业市场规模、增加专门企业数量、提高国产研究装备比重等三个目标。

二、综合性创新战略的部署重点

为推动整个国家的创新能力和生产力水平提升，世界主要国家和组织的创新战略聚焦人才、技术、企业等创新关键要素，旨在通过优化创新要素配置的模式和结构，充分释放技术、人才和企业的创新潜能，进而促进国家经济高质量增长。

1. 技术研发支持与创新产业培育

一是选定一批能够带来变革性经济影响、对国家安全至关重要的关键技术领域，作为参与全球科技竞争的研发目标，以谋求技术主权。美国计划投入 1000 亿美元，推动人工智能、高性能计算、量子信息科学和技术、机器人技术等 10 个关键技术领域的研究进步[1]。英国[9]对影响未来经济发展的关键技术选定了人工智能、数字和高级计算；生物信息学和基因组学；工程生物学；电子、光子学和量子科学；能源与环

境技术；机器人和智能机器等领域。德国[11]希望在多个领域实现技术主权，包括下一代电子部件、通信技术、软件和人工智能、数据技术、量子计算机、价值创造系统、循环经济、材料创新、电池研究、绿色氢能、疫苗研发等。日本则选定了基础领域的人工智能技术、生物技术、量子技术、材料技术[3]。韩国[12]选定了半导体与显示器、二次电池、先进移动通信、下一代核能、前沿生物、航空航天与海洋、氢能、网络安全、人工智能、下一代通信、前沿机器人与制造、量子科技等 12 项国家战略技术。

二是面向促进社会经济发展的重点应用领域，提出培育创新产业的重点领域方向。德国选定产业现代化技术，包括 5 个核心领域：①气候影响、生物多样性、可持续发展；②地球系统、可持续农业和粮食系统；③现代卫生体系；④技术主权和数字化潜力；⑤空间和海洋研究[13]。日本选定了环境能源、安全舒适生活、健康医疗、空间、海洋、食品和农林水产业等应用领域[3]。西班牙确定了 6 个重点发展领域，包括：①健康；②促进包容性社会发展；③国家安全；④数字化发展、制造业和空间防御；⑤气候、能源和交通，食品；⑥环境资源。

2. 加大创新型人才吸引与培养力度

一是调整移民政策，提升国际创新人才吸引能力。美国宣布将 22 个新增研究领域的学生和交流访问学者计划（SEVP）纳入 STEM 可选实践培训计划[15]。英国拟定和实施最优奖学金项目，通过扩大签证种类和数量，吸引和留住具备高技能、高潜力的全球创新人才[6]。法国为欧洲技术人才提供标准合同，允许跨境享受社会权利；为非欧洲人才提供“欧盟科技签证”快速通道以及税收优惠政策[16]等。

二是增加基础教育阶段中的科技创新内容。巴西提出发展创新导向的教育体系，从基础教育阶段起，针对有望对国家生产部门产生影响的技术开发增设奖学金；在本科和研究生课程中加入发展创业和创新精神的实践和跨学科课程[7]。日本提出培养中小学生对数理化课程的兴趣，在大学设立个性化课程满足多样化学习需求[17]。德国提出 STEM（科学、技术、工程和数学）行动计划 2.0 并资助 4500 万欧元[18]。

三是扩大创新人才能力培训和跨机构流动。英国通过“辅助成长和管理计划”提出扩大跨部门的人员培训和流动，提高工作成效；由英国研究与创新署提供技能培训，确保研究人员具备相关的知识储备和专业的技术技能[9]。西班牙提出促进国际和跨部门的人才流动，将其作为研发和创新人员职业生涯的一个组成部分[14]。

3. 坚持发挥以企业为创新主体的功能

一是资助支持企业开展研发活动。韩国[19]提出在国家战略技术发展过程中全面支持企业的创新活动和国际合作互助、技术保护，加大提高企业参与度力度。西班牙提出增加对研发创新企业的资助和激励，鼓励成立研发创新企业，提升企业的科研竞争力[8]。英国计划对创新英国核心项目的资助到2024～2025财年增加66%，达到11亿英镑，帮助企业获得创新和发展所需的资本、技能和产学研合作[20]。

二是促进产学研合作创新与成果转化。英国[6]提出促进产学研间的合作和共同投资，启动新的“繁荣经济合作伙伴关系”，产业界、大学和政府共同投资5900万英镑，建立以企业为主导的研究项目以开发变革性新技术，并支持加快成果转化。西班牙[8]计划资助面向社会挑战的前沿领域科研项目，将基础科研知识应用于企业新技术的开发，促进科研知识和成果向企业和社会转移转化，并促进公私合作，增强公共机构、企业与社会协同发展的能力。

三是加强对中小企业的创新资助与合作。日本[17]、巴西[7]等国提出加强中小企业研发补助制度，促进中小企业获得创新补贴和资金来源。西班牙提出加强科研机构与企业间的双向联系，尤其是与中小企业的合作交流，进一步增强对研发需求的相互理解，确保给予研发创新企业尤其是中小企业足够的税收优惠[8]。挪威提出要强化商业界与企业之间，特别是与中小企业之间的合作，动员参与“地平线欧洲”计划[21]。德国从融资、人才、数据、监管等方面，全面加强和促进德国初创企业生态系统[22]。

三、完善科技创新体制机制的主要举措

1. 强化创新研发活动安全措施，应对全球技术竞争

随着全球科技竞争加剧，各国将研发安全提升至国家安全的层面，通过出台规定和防范指南予以保障研发安全。

美国国家科学技术委员会（NSTC）研究安全小组委员会和研究环境联合委员会共同发布《关于在美国政府资助下研发的国家安全战略的国家安全总统备忘录（NSPM-33）实施指南》，要求联邦机构统一披露利用外国研究资金的规则，保护美国联邦政府资助的研究免受外国政府干预和盗用[23]。欧盟委员会发布关于如何减轻外国对研究和创新干预的工具包，提出支持欧盟高等教育机构和研究执行组织维护其基本价值观，并保护其员工、学生、研究成果和资产[24]。

2. 设置专门科技创新管理服务机构，提升政策实施效果

设立综合管理机构等新的管理机制能够进一步提升创新政策实施效果。各国通过专门机构加强综合领导职能，保障科技创新政策的实施效果。

美国提出在国家科学基金会（NSF）设立一个新的“技术与创新学部”（DTI），

加速技术商业化，加强美国在关键技术方面的领导地位[1]。英国宣布将政府科学办公室扩编为科学技术战略办公室，有效协调政府跨部门工作，首要任务是审查英国应该支持的技术投入，并规划英国科技创新的优先领域和战略优势[25]。日本设立“内阁府科技创新推进事务局”，确保综合科学技术创新会议的核心领导职能，协调与知识产权战略本部、健康医疗战略推进本部等其他肩负领域领导职能机构的关系[17]。俄罗斯新设立了政府科学技术发展委员会，在制定重大国家创新项目时，协调联邦权力执行机构和各组织的活动并保障项目实施[26]。

3. 建立资助跨领域研发的创新资助体系

为促进交叉学科和融合技术发展，一些国家采取去中心化模式，纷纷设立“横向”资助部门等跨领域资助机制。

美国国家科学基金会新设“技术、创新与合作学部”专注于支持“以应用为基础”的研发，包括小企业研发计划、创新团队、创业教育计划，以及作为推进多学科研究十大创意而建立的“融合加速器”[27]。德国成立转移与创新局，支持中小型大学和应用学科大学以创新为导向的合作，促进面向应用的研究和知识转移，可持续地建立并加强区域和跨区域创新生态系统[28]。欧盟委员会正式启动新的资助机构——欧洲创新理事会（EIC），集合新兴技术研发、加速器计划和专门的股权基金等多类型资助措施，以开发和扩展突破性创新[29]。

4. 加大研发活动减税与金融制度支持

一些国家采取扩大减税范围、简化资助复杂流程等立法保障和制度支持，以便为企业提供创新动力。

俄罗斯[30]将未来产业先进技术、节能高效环保技术、俄罗斯空天防御系统技术纳入可减税的研发类型清单，为实现技术创新而创造有利的税收条件。日本将根据《税制修改大纲》完善针对研发活动的税额减免、税收抵扣等扶持政策，吸引企业投资科研活动[17]。英国计划通过在“创新英国”战略在英国商业银行设立在线金融和创新中心，降低创新公司获得金融支持程序的复杂程度[9]。俄罗斯简化创新产品研发补贴规则，考虑到企业因受到制裁等客观原因，延长违规行为界定期限，并降低对优先创新产品最低销售额的要求[31]。

5. 加强创新研发数据平台设施建设和使用

一些国家开始重视对研发数据的利用，促进开展数据驱动型高附加值创新活动。德国计划借助《研究数据法》全面改进和简化公共和私人研究对研究数据的获取，将开放获取确立为通用标准，进一步发展国家研究数据基础设施，推进欧洲研究数据空间[13]。韩国提出应建立国家必备战略技术的相关数据基础设施[19]。巴西提出构建一个整合巴西研发与创新数据、研究、立法和指南的平台[7]。英国计划集成研究与创新署的数据和系统，进行项目进度监控和信息共享[6]。日本提出将完善“日本研究数据基础系统”建设，在2023年实现基于公共资金开展研究的科研数据格式统一[3]。

四、启示与建议

2021年以来的全球科技创新环境形势与以往相比更为复杂，世界主要国家和组织纷纷认识到实现本国科技独立自主的重要性，期望通过系统性推进和实施创新战略和计划实现关键技术自主可控。

综合来看，全球科技创新战略与体制机制发展体现出四个主要特征。一是突出技术创新引领，追求制定领先型的创新战略以确保掌握关键核心技术的优势和主导权；二是支持建立多学科、跨学科合作网络，促进交叉学科和融合技术发展来应对全球挑战，提倡跨界多元合作式研发；三是鼓励民间投资及私营企业资助创新，保障科技研发资助的稳定性和连续性；四是重视科技创新人才在基础教育阶段的培养与锻炼，提升人才储备质量和未来发展潜能。

因此，结合我国实际需求和发展情况，提出以下三点建议：一是建议开展有针对性的研发与创新资助，建立资助跨领域研发的创新资助体系；二是建议充分发挥企业创新主体的功能，提高企业在国家关键核心技术研发过程中的参与度，全面支持企业的创新活动和国际合作、技术保护等；三是建议加强科技创新人才成长全周期的支持体系建设，发展科技创新导向的教育体系。

致谢：中国科学院武汉文献情报中心刘清研究员、中国科学院成都文献情报中心张志强研究员、中国科学院文献情报中心胡智慧研究员审阅了全文并提出宝贵的修改意见和建议，谨致谢忱！

参考文献

[1] United States Senate. S. 1260-United States Innovation and Competition Act of 2021. https://www.congress.gov/bill/117th-congress/senate-bill/1260/all-actions[2021-06-08].

[2] United States Senate. The CHIPS and Science Act of 2022s. https://www.commerce.senate.gov/2022/8/view-the-chips-legislation[2022-08-09].

[3] 内閣府．第6期科学技術・イノベーション基本計画．https://www8.cao.go.jp/cstp/kihonkeikaku/6honbun.pdf[2021-03-06].

[4] 内閣府．統合イノベーション戦略 2022. https://www8. cao. go. jp/cstp/siryo/haihui061/siryo1-2-1. pdf[2022-06-02].

[5] European Commission. Horizon Europe's first strategic plan 2021-2024: Commission sets research and innovation priorities for a sustainable future. https://ec. europa. eu/commission/presscorner/detail/en/ip_21_1122[2021-03-15].

[6] UK Research and Innovation. UKRI Strategy 2022 to 2027. https://www. ukri. org/publications/ukri-strategy-2022-to-2027/ukri-strategy-2022-to-2027/[2022-03-17].

[7] MCTI. Publicada a Estratégia Nacional de Inovação. https://www. gov. br/mcti/pt-br/acompanhe-o-mcti/noticias/2021/07/publicada-a-estrategia-nacional-de-inovacao[2021-07-26].

[8] Gobierno de España. Estrategia Española de Ciencia, Tecnología e Innovación 2021-2027. https://www. ciencia. gob. es/stfls/MICINN/Ministerio/FICHEROS/EECTI-2021-2027. pdf[2021-02-01].

[9] Department for Business, Energy and Industrial Strategy. UK Innovation Strategy: leading the future by creating it. https://www. gov. uk/government/publications/uk-innovation-strategy-leading-the-future-by-creating-it/uk-innovation-strategy-leading-the-future-by-creating-it-accessible-webpage [2021-07-22].

[10] 과학기술정보통신부. 제 1 차 연구산업 진흥 기본계획('22~'26) 발표. https://www. msit. go. kr/bbs/view. do?sCode=user&mId=113&mPid=112&bbsSeqNo=94&nttSeqNo=3182062[2022-08-26].

[11] Bundesministerium für Bildung und Forschung. Rat für technologische Souveränität nimmt Arbeit auf. https://www. bmbf. de/bmbf/shareddocs/pressemitteilungen/de/2021/08/020921-Rat-technologische- Souveraenitaet. html[2021-09-02].

[12] 과학기술정보통신부.국가전략기술육성방안 발표. https://www. msit. go. kr/bbs/view. do? sCode=user&mId=113&mPid=112&pageIndex=2&bbsSeqNo=94&nttSeqNo=3182291[2022-10-28].

[13] Bundesregierung. Mehr Fortschritt wagen. https://www. spd. de/fileadmin/Dokumente/Koalitionsvertrag/Koalitionsvertrag_2021-2025. pdf[2021-12-07].

[14] Gobierno de España. Plan Estatal de Investigación Científica, Técnica y de Innovación 2021-2023. https://www. ciencia. gob. es/site-web/Estrategias-y-Planes/Planes-y-programas/Plan-Estatal-de-Investigacion-Cientifica-y-Tecnica-y-de-Innovacion-PEICTI-2021-2023. html[2021-08-01].

[15] The White House. FACT SHEET: Biden-Harris Administration Actions to Attract STEM Talent and Strengthen our Economy and Competitiveness. https://www. whitehouse. gov/briefing-room/statements-releases/2022/01/21/fact-sheet-biden-harris-administration-actions-to-attract-stem-talent-and-strengthen-our-economy-and-competitiveness/[2022-01-21].

[16] Sifed. Scale-Up Europe: How to build global tech leaders in Europe. https://content. sifted. eu/wp-content/uploads/2021/06/15162949/Scale-Up-Europe-Report. pdf[2021-06-15].

[17] 内閣府．統合イノベーション戦略 2021. https://www8. cao. go. jp/cstp/tougosenryaku/togo2021_honbun. pdf[2021-06-18].

[18] Bundesministerium für Bildung und Forschung. Stark-Watzinger: Mit einem MINT-Aktionsplan 2.0 gegen die Fachkräfterlücke. https://www. bmbf. de/bmbf/shareddocs/pressemitteilungen/de/2022/05/010622-MINT. html[2022-06-01].

[19] 과학기술정보통신부. 제 20 회 과학기술관계장관회의（2021. 12. 22. ）. https://www. msit. go. kr/bbs/view. do? sCode=user&mId=206&mPid=89&pageIndex=&bbsSeqNo=122&nttSeqNo=19&searchOpt=ALL&searchTxt=[2021-12-22].

[20] Department for Business, Energy & Industrial Strategy, UK Space Agency. Government announces plans for largest ever R&D budget. https://www. gov. uk/government/news/government-announces-plans-for-largest-ever-rd-budget[2022-03-14].

[20] Ministry of Education and Research. Strategy for Norway's participation in Erasmus+and the European Education Area. https://www. regjeringen. no/en/dokumenter/strategy-for-norways-participation-in-erasmus-and-the-european-education-area/id2863355/[2021-12-03].

[22] BUNDESMINISTERIUM FÜR WIRTSCHAFT UND KLIMASCHUTZ. Start-up Fahrplan steht: Kabinett beschlieBt erste umfassende Start-up-Strategie. https://www. bmwk. de/Redaktion/DE/Pressemitteilungen/2022/07/20220726-start-up-fahrplan-steht-kabinett-beschliesst-erste-umfassende-startup-strategie. html[2022-07-27].

[23] National Science and Technology Council. Guidance for Implementing National Security Presidential Memorandum 33（NSPM-33）On National Security Strategy for United States Government-Supported Research and Development. https://www. whitehouse. gov/wp-content/uploads/2022/01/010422-NSPM-33-Implementation-Guidance. pdf? utm_medium=email&utm_source=FYI&dm_i=1ZJN,7OKSB,E29EKO,VB5UF,1[2022-01-01].

[24] European Commission. Commission publishes a toolkit to help mitigate foreign interference in research and innovation. https://ec. europa. eu/info/news/commission-publishes-toolkit-help-mitigate-foreign-interference-research-and-innovation-2022-jan-18_en[2022-01-18].

[25] Prime Minister's Office. Prime Minister sets out plans to realise and maximise the opportunities of scientific and technological breakthroughs. https://www. gov. uk/government/news/prime-minister-sets-out-plans-to-realise-and-maximise-the-opportunities-of-scientific-and-technological-breakthroughs [2021-06-21].

[26] Президент резиде. Указидентww. gov. uk/government/news/prime-minister-sets-out-. http://www. kremlin. ru/acts/bank/46506[2021-03-15].

[27] U. S. National Science Foundation. NSF establishes new Directorate for Technology, Innovation and Partnerships. https://www. nsf. gov/news/news_summ. jsp? cntn_id=304624[2022-03-16].

[28] Bundesministerium für Bildung und Forschung. Deutsche Agentur für Transfer und Innovation（DATI）. https://www. bmbf. de/bmbf/shareddocs/downloads/_ pressestelle/pressemitteilung/2022/04/Eckpunktepapier. pdf? __blob=publicationFile&v=7[2022-04-11].

[29] European Commission. Commission launches European Innovation Council to help turn scientific ideas into breakthrough innovations. https://ec. europa. eu/commission/presscorner/detail/en/ip_21_1185[2021-03-22].

[30] Правительствоesscorner/detail/en/ip_21_1185/detail/en/ip_21_1185uncil to help turn scientific

ideas into breakthrough innovations. log. http://government. ru/docs/44671/[2022-02-18].
[31] Правительствоment. ru/docs/44671//ip_21_1185/detail/en/ip_21_1185uncil to help turn scientific ideas into breakthrough innovat. http://government. ru/docs/44667/[2022-02-26].

Strategy and Planning of S&T Innovation

Ye Jing, Li Hong, Zhang Qiuju,
Wang Jianfang, Xi Zhongyang

The further spread of the COVID-19 pandemic has significantly slowed down productivity and innovation, and further impaired global socio-economic development. In particular, the impact of the pandemic has disrupted the normal functioning of innovation systems and jeopardized key production and innovation capabilities. In order to adapt to the development and changes in the global science, technology and economy, and to enhance their own innovation capacity and competitiveness, major countries and organizations have established national-level S&T innovation strategy and planning. It aims to achieve the planning objectives in the following 3 directions through the top-level design: promoting the improvement of enterprise innovation capability, promoting the improvement of national innovation ecosystem, and promoting the sustainable development of social economy. By sufficiently launching the innovation potential of talents, technologies and enterprises, they promote high-quality economic development.

4.14 世界主要国家和组织国际科技合作与竞争态势发展观察

王文君[1] 李 宏[1,2,3] 刘 栋[1] 葛春雷[1] 贾晓琪[1]

(1. 中国科学院科技战略咨询研究院；2. 中国科学院大学；
3. 粤港澳大湾区战略研究院)

2021～2022年，在新一轮科技革命和产业变革的加速演进以及新冠疫情、中美摩擦和俄乌冲突的深远影响下，国际发展格局发生深刻变化：全球化遭遇逆流，单边主义、保护主义上升；重大传染性疾病、气候变化、网络安全等非传统安全威胁持续蔓延；大国间科技竞争更加激烈，国际关系发生显著变化。在全球发展形势不稳定性和不确定性明显增加的背景下，科技创新成为国际战略博弈的主要战场，各国间科技合作与竞争交织并存，高端人才、技术标准及供应链成为竞合焦点，同时数字化和绿色化转型发展等领域竞争与合作也在不断凸显。

一、国际科技合作与竞争格局复杂交织

冷战结束以来，国际科技合作与竞争是推动知识经济加速发展的重要因素。在新冠疫情起伏反复造成全球经济增长疲软的背景下，各国经济发展亟须寻找新的增长动力，因此，大国间竞争博弈更加突出，实用主义的“联盟式”科技合作倾向突显，科技领域呈现出国际竞争与合作交织并存的复杂态势。

1. 制定国际合作战略政策，提升全球竞争力

在战略顶层设计方面，相关国家及地区组织通过制定国际科技合作相关战略及政策，实施全球合作差异化布局，构建利益共同体，以此压制、约束其他竞争对手发展，国际科技合作显现出“阵营化”态势。

2021年3月，美国参议院两党议员联合提出《民主技术合作法案》[1]，号召通过由美国主导的国际技术合作对抗中国在新兴及关键技术领域日益增长的影响力，并撬动“全球民主国家”的力量，在针对新兴及关键技术领域的国际标准与规范制订、联

合研究、出口管控、投资审查等方面联手。7 月，美国国务卿布林肯在全球新兴技术峰会上的讲话[2]，概述了拜登政府在所谓民主国家间建立技术伙伴关系的愿景，并阐明落实《2021 年美国创新和竞争法案》、建立技术合作伙伴关系办公室等具体行动重点。

2021 年 5 月，欧盟委员会发布《全球研究与创新方法：变化世界中的欧洲国际合作战略》[3]，强调坚持多边主义、开放和互惠原则，同时也指出“在地缘政治局势紧张、一些国家通过歧视性措施寻求技术领先背景下，欧洲需要协调努力，以最大限度发挥其影响并加强其领导作用”。12 月，欧盟委员会宣布正式启动“全球门户”（The Global Gateway）基础设施投资战略[4]，确定了数字化、气候与能源、交通运输、卫生健康、教育和研究等五大优先投资领域，构建利益共同体，增强欧盟与全球合作伙伴国的产业链韧性。

2021 年 7 月，俄罗斯总统普京正式批准发布新版《国家安全战略》[5]，该文件首次明确提出要发展与中国的全面伙伴关系和战略协作，以及发展与印度的特惠战略伙伴关系，强调要应对以美国为首的北约步步紧逼等威胁。2022 年 6 月，北约发布《北约战略概念 2022》，该战略提出加强北约-欧盟战略伙伴关系，在气候变化对安全的影响、新兴与颠覆性技术、应对网络和混合威胁等共同关注的问题上加强合作[6]。

2. 推进联盟式合作，“阵营化”“集团化”趋势初显

有关国家通过发表联合宣言及倡议、设立双边技术理事会、建立多边联盟等方式，加强数字经济、绿色能源等关键及新兴技术领域的联合，并期望在合作中实施对可能的竞争对手的打压。2021 年 3 月，美国、日本、澳大利亚、印度 4 国领导人首次举行“四方安全对话”（QUAD），并发表联合声明[7]，将在新冠疫苗、气候变化、新兴技术三大领域加强合作。7 月，欧盟委员会启动“处理器和半导体技术联盟”及“欧洲工业数据和边缘/云计算联盟”，旨在发展下一代微芯片和工业云/边缘计算技术。9 月，美国、英国和澳大利亚宣布成立名为“澳英美”（AUKUS）的新三方安保联盟，加强在先进海军武器技术（包括网络、人工智能、量子科技等）以及核军事技术等领域合作。9 月，美欧贸易与技术委员会（TTC）召开首次会议[8]，成立了技术标准、投资审查、出口管制、数据治理与技术平台、供应链安全等 10 个技术领域的工作组，以加强美欧在技术和工业方面的世界领导地位。2022 年 2 月，俄乌冲突发生后，德国政府宣布停止与俄罗斯在科学、研究、职业教育和培训方面的长期合作，随后美国、欧盟和日本相继发起技术制裁，以削弱俄罗斯的经济能力，停止对其的关键技术出口[9]。

二、高端人才、技术标准、供应链成为竞合焦点

世界主要国家及组织在国际合作和竞争中，将争夺高端科技创新人才、扩大自身技术标准输出，以及保障关键科技产品和原材料的供应链安全、韧性发展等列为优先事项。

1. 高端科技人才成为争夺重点，人才合作交流受阻

在各国加大对国际高端科技人才吸引的同时，某些大国间博弈导致科技国界性增强，阵营间科技人才合作交流受到严重阻碍。2021年1月，英国人工智能委员会发布《人工智能路线图》[10]，提出不断吸引并留住来自世界各地的顶尖人才，寻找能将研究人员、学科和部门汇集在一起的新方法等政策措施。5月，美国国会参议院商务、科学与交通委员会审议通过了《无尽前沿法案》(Endless Frontier Act，EFA)[11]，该提案强调关注科学、技术、工程和数学（STEM）教育人才，在“强化科研安全”方面，指出禁止联邦科研人员参与以中国为首的外国政府人才招募计划，禁止中国参加本法案资助的项目，禁止有中国政府背景的实体参与基站建设，禁止国家科学基金会向与孔子学院合作的高校提供资金。3月，澳大利亚安全情报组织（ASIO）负责人称，中国的人才计划是中国成为世界科技领导者并保持经济和军事优势的战略计划的自然延伸，未来ASIO将与美国、加拿大、英国、新西兰4国的情报部门定期交流外国干预行动及相关风险。2022年10月，美国商务部宣布从事高端芯片制造业的“美国人”，在华工作未获许可证或受民事和刑事处罚，导致中国16家半导体企业中的43名美籍人才在芯片禁令后离开中国[12]。

2. 在国际合作中追求标准话语权及自身技术标准输出

国际技术标准的制定已成为技术领域竞赛的激烈战场，部分大国及地区组织将输出技术标准作为与他国开展国际合作的条件，以扩大自身技术的控制力和影响力。例如，2021年3月美国参议院两党议员联合提出《民主技术合作法案》[13]，强调推动技术政策与标准协调；采用数据隐私、数据共享和数据存档的共同标准等，旨在技术合作中形成对特定国家的“排他”效应。

其中，信息通信领域国际技术标准的竞合最为突出。2021年3月，美国、日本、印度、澳大利亚在“四方安全对话”峰会上设立了4国“关键技术和新兴技术工作组”[14]，促进技术标准制定协调，包括技术标准机构之间的协调以及与广泛合作伙伴的合作。5月，日本政府宣布将制定新的技术标准和规则，以考虑对本国信息通信及

电力等14个行业的重要基础设施在IT设备和云服务采用方面限制海外技术。[15]6月，欧盟委员会发布《控制者和处理者数据使用标准合同条款》[16]和《国际数据转移标准合同条款》[14]，提供标准化和批准模板帮助企业符合数据安全传输要求。

3. 供应链政策愈加强调安全、弹性、可控和本国利益

在疫情影响下，发达国家和地区组织认识到供应链在经济发展中的不可替代作用，将半导体、关键材料等方面的供应链政策作为提升产业竞争力的关键手段，以及推进多边合作的战略工具。2021年6月，美国白宫发布100天供应链审查报告《建立弹性供应链、振兴美国制造业和促进广泛增长》[15]，为加强美国供应链弹性，将重新布局半导体、关键矿物和材料、药品供应链国际合作，与盟国建立伙伴关系，发展安全、弹性的供应链，避免过度依赖非共同利益国家，利用政策解决市场失衡等问题。5月，欧盟更新了其于2020年3月提出的《欧洲产业战略》[19]，基于对原材料、电池、活性药、氢、半导体及云技术等6个领域欧洲的战略依存关系的深入审查结果，提出将提高开放战略自主权，创建多元化的国际伙伴关系的更新措施。

三、强化绿色化及数字化双轨转型

面对当前脆弱复苏的世界经济，绿色化和数字化对经济社会复苏与转型发展起到推动作用。各国在这两个领域积极开展相关科技合作，推动绿色化及数字化转型联动增效。

1. 加强绿色科技领域合作

面对气候变化、能源短缺等严峻挑战，相关国家及组织积极开展绿色科技领域国际合作。2021年11月，欧盟理事会通过《单一基本法案》[20]，推动欧盟、成员国及产业界联合启动9个新的欧洲伙伴关系，其中“可循环的生物欧洲计划”“清洁氢能”“清洁航空”等3个联合计划旨在加快欧洲向绿色、气候中性的过渡。7月，美国总统拜登和德国时任总理默克尔在美国白宫启动了“美德气候与能源伙伴关系”[21]，深化两国在加速全球净零碳排放未来所需的政策和可持续技术方面的合作。在7月发布的《法国经济复苏计划》[22]中，法国和德国将牵头关于氢能技术与应用的欧洲共同利益重要项目，共同打造欧洲氢能产业链并减少碳排放。2022年底，在加拿大蒙特利尔闭幕的《生物多样性公约》第十五次缔约方大会（COP15）上，各国就制止并最终扭转物种衰退和生态系统退化趋势达成协议[23]。

2. 通过国际合作加快数字化转型

以互联网、大数据、人工智能、云计算等为代表的数字技术是数字经济的实现基础和动力来源，为加速数字化转型，各国积极发展数字技术领域合作伙伴。2021 年 7 月，法国发布《国家经济复苏计划》[24]，在数字化转型方面，法国、德国、西班牙等将合作关于云的欧洲共同利益重要项目，目的是保障欧洲数字主权、弥补欧洲云平台基础设施不足。9 月，美欧贸易与技术委员会设立数据治理和技术平台工作组，就美欧各自的数据治理与技术平台治理方法交换信息，在可行的情况下寻求一致性和互操作性。11 月，欧盟通过《单一基本法案》，除绿色伙伴关系构建外，还提出“单一欧盟空间 ATM 研究计划”[25]，旨在加速空中交通管理的数字化转型，使欧洲空域成为世界上最高效、最安全和最环保的飞行天空，同时支持航空业的竞争力和疫情危机后行业的复苏。此外，“智能网络与服务计划”也将支撑欧洲数字服务的发展。2022 年 2 月，俄罗斯政府分别批准与塞尔维亚和阿塞拜疆签署信息安全领域的国际合作协议[26]。

四、启示与建议

当今世界百年未有之大变局加速演进，疫情肆虐导致全球经济增长态势出现疲软，各国都在寻找经济增长和社会发展的新动能，科技创新被视为主要的突破口。2021～2022 年世界主要国家和组织在国际科技合作及竞争态势方面给我国以下几点启示建议。

1. 塑造国际竞合新优势，以全球视野谋划科技创新

一是聚焦关键核心技术，在半导体、人工智能、材料科学、清洁能源、生物技术、先进制造等领域有针对性地部署一批重大科研项目，精准发力培育“撒手锏”技术，以破解美西方等国的封锁遏制，达到局部技术平衡、牵制与非对称反制的效果；二是构建人才、技术、设施平台、供应链等全方位、深层次的国际科技合作新格局，在保证安全可控的前提下，加深我国与其他国家高技术产业领域的相互依赖关系，提高其他国家“去中国化”的成本。

2. 传播合作共赢理念，拓宽国际科技创新合作之路

一是瞄准主要科技强国大国和创新型小国，结合我国的各类创新需求，精准选择合作领域，鼓励民间科技交流，开辟多元化科技合作渠道；二是结合共建“一带一路”倡议，推动中国科技创新“走出去”，面向共建“一带一路”合作国家中的发展

中国家推广转移具有“适用性”或“包容性”的技术；三是完善多边合作机制，增强我国在创新合作议题设置和规则制定中的主动权和话语权，加强与国际标准化机构和联合国全球技术法规机构间的合作，参与全球科技治理；四是提升产业链、供应链的韧性和安全水平，重新布局合作伙伴，构建互利共赢的产业链、供应链国际合作体系。

3. 激发科研要素流动活力，提升人才服务保障能力，建设开放创新生态

一是提升科技管理服务水平，为科技人才出入境提供便利，推动外事服务、外国人才服务、法律咨询服务等科技保障类工作更加专业化、国际化，在特殊情况下采取“一事一议”机制，为集聚国际人才和国际创新资源营造富有吸引力的创新生态；二是依托我国的大科学装置平台、科研网络平台，在气候变化、公共卫生、数字科技等全球共同关注合作领域主动发起国际创新议题，加速汇聚全球创新资源，助力全球科学家扩大科研合作的范围；三是联合各国的相关科技管理机构，共同打造系列国际科技合作传播产品，在全球范围讲好“中国科技故事”，提升中国科技创新的国际影响力。

致谢：中国科学院武汉文献情报中心刘清研究员、中国科学院成都文献情报中心张志强研究员、中国科学院文献情报中心胡智慧研究员对本文进行了审阅，并提出了宝贵的修改建议，特致谢忱。

参考文献

[1] U. S. Senate. Bipartisan Senators Introduce Legislation to Reassert Democratic Leadership in Technology Strategy & Development. https://www.warner.senate.gov/public/_cache/files/8/9/895e0a40-65ee-43cc-8629-450555faefe7/AC6A0E54DB992E1612161C48BB34FC57.democracy-technology-partnership-act-two-pager-explainer.pdf[2022-12-01].

[2] American Institute of Physics. Secretary of State Tony Blinken Sets Out Vision for Global Technology Diplomacy. https://www.aip.org/fyi/2021/secretary-state-tony-blinken-sets-out-vision-global-technology-diplomacy[2022-12-01].

[3] European Union. Global Approach to Research and Innovation Europe's strategy for international cooperation in a changing world. https://ec.europa.eu/info/sites/default/files/research_and_innovation/strategy_on_research_and_innovation/documents/ec_rtd_com2021-252.pdf[2022-12-01].

[4] European Union. Global Gateway. https://commission.europa.eu/strategy-and-policy/priorities-2019-2024/stronger-europe-world/global-gateway_en[2022-12-01].

[5] Президент России. Стратегия национальной безопасности Российской Федерации до 2020 года. http://www.kremlin.ru/acts/bank/47046[2022-12-01].

[6] Xinhua. NATO 2022 Strategic Concept. http://www.chinadaily.com.cn/a/202207/03/WS62c0e

074a310fd2b29e6a00d. html[2022-12-30].

[7] The White House. Quad Joint Leaders'Statement. https://www. whitehouse. gov/briefing-room/statements-releases/2022/05/24/quad-joint-leaders-statement/[2022-12-01].

[8] European Union. EU-US Trade and Technology Council Inaugural Joint Statement. https://ec. europa. eu/commission/presscorner/detail/en/STATEMENT_21_4951[2022-12-01].

[9] Jacobson R. Russia's Invasion of Ukraine Has Forever Changed How the World Does Science. https://www. thedailybeast. com/russias-invasion-of-ukraine-has-forever-how-the-world-does-science[2022-12-30].

[10] UK AI Council. UK AI Council AI Roadmap. https://assets. publishing. service. gov. uk/government/uploads/system/uploads/attachment_data/file/949539/AI_Council_AI_Roadmap. pdf[2022-12-01].

[11] U. S. Industry and Security Bureau. Implementation of Additional Export Controls: Certain Advanced Computing and Semiconductor Manufacturing Items; Supercomputer and Semiconductor End Use; Entity List Modification, https://www. federalregister. gov/documents/2022/10/13/2022-21658/implementation-of-additional-export-controls-certain-advanced-computing-and-semiconductor[2022-12-30].

[12] U. S. Senate. Endless Frontier Act. https://www. democrats. senate. gov/imo/media/doc/DAV21A48. pdf [2022-12-30].

[13] U. S. Senate. Democracy Technology Partnership Act. https://www. warner. senate. gov/public/_cache/files/8/9/895e0a40-65ee-43cc-8629-450555faefe7/AC6A0E54DB992E1612161C48BB34FC57. democracy-technology-partnership-act-two-pager-explainer. pdf[2022-12-30].

[14] The White House. Quad Leaders' Joint Statement: "The Spirit of the Quad". https://www. whitehouse. gov/briefing-room/statements-releases/2021/03/12/quad-leaders-joint-statement-the-spirit-of-the-quad/[2022-12-30].

[15] 日经中文网. 日本将限制 14 个领域基础设施运用海外 IT. https://cn. nikkei. com/politicsaeconomy/politicsasociety/44766-2021-05-18-09-37-57. html[2022-12-30].

[16] The European Commission. Standard Contractual Clauses (SCC). https://commission. europa. eu/law/law-topic/data-protection/international-dimension-data-protection/standard-contractual-clauses-scc_en [2022-12-30].

[17] The European Commission. Standard contractual clauses for international transfers. https://commission. europa. eu/publications/standard-contractual-clauses-international-transfers_en[2022-12-30].

[18] The White House. Building Resilient Supply Chains, Revitalizing American Manufacturing, and Fostering Broad-Based Growth: 100-Day Reviews Under Executive Order 14017. https://www. whitehouse. gov/wp-content/uploads/2021/06/100-day-supply-chain-review-report. pdf[2022-12-30].

[19] The European Commission. Updating the 2020 Industrial Strategy: towards a stronger Single Market for Europe's recovery. https://ec. europa. eu/commission/presscorner/detail/en/IP_21_1884[2022-12-30].

[20] The European Commission. Commission welcomes approval of 10 European Partnerships to accelerate

the green and digital transition. https://ec. europa. eu/info/news/commission-welcomes-approval-10-european-partnerships-accelerate-green-and-digital-transition-2021-nov-19_en[2022-12-01].

[21] The White House. FACT SHEET: U. S. -Germany Climate and Energy Partnership. https://www. whitehouse. gov/briefing-room/statements-releases/2021/07/15/fact-sheet-u-s-germany-climate-and-energy-partnership/[2022-12-01].

[22] 新华社."昆明-蒙特利尔全球生物多样性框架"成功通过. https://www. gov. cn/xinwen/2022-12/19/content_5732712. htm[2022-12-30].

[23] Ministère de l'Économie, des Finances et de la Souveraineté industrielle et numérique. L'Union européenne valide le plan national de relance et de résilience de la France. https://www. economie. gouv. fr/union-europeenne-valide-plan-national-relance-resilience-france[2022-12-01].

[24] European Union. Proposal for a COUNCIL REGULATION establishing the Joint Undertakings under Horizon Europe. https://eur-lex. europa. eu/legal-content/EN/TXT/? uri=COM: 2021: 87: FIN[2022-12-30].

[25] Правительства Российской. О подписании Соглашения между Правительством Российской Федерации и Правительством Республики Сербии о сотрудничестве в области обеспечения информ ационной безопасности. http://publication. pravo. gov. ru/Document/View/0001202203020011?index=3&rangeSize=1[2022-12-30].

International S&T Cooperation and Competition

Wang Wenjun, Li Hong, Liu Dong, Ge Chunlei, Jia Xiaoqi

In 2021 and 2022, under the accelerated evolution of a new round of scientific and technological revolution and industrial reform, as well as the far-reaching impact of the COVID-19, the China-USA friction and the Russia — Ukraine conflict, the international development has undergone profound changes: the undercurrent against international cooperation is gaining momentum, unilateralism and bullying are forcing their way in the world; Non-traditional security threats such as major infectious diseases, climate change and Internet security are more possible than ever; Scientific and technological competition among major countries has become more prominent, and the form and content of alliances have changed significantly. In the new era full of uncertainty, scientific and technological innovation has become the main battleground of the international strategic game.

S &T cooperation and competition among countries are intertwined. Talents, technical standards and supply chains have become the focus. Meanwhile, international cooperation related to digital and green transformation has been constantly strengthened.

第五章

重要科学奖项巡礼

Introduction to Important Scientific Awards

5.1　呼唤复杂系统研究，应对自然与社会挑战
——2021 年度诺贝尔物理学奖评述

陈晓松

（北京师范大学）

2021 年的诺贝尔物理学奖授予三位科学家真锅淑郎（Syukuro Manabe）、克劳斯·哈塞尔曼（Klaus Hasselmann）和乔治·帕里西（Giorgio Parisi）（图 1），以表彰他们“对我们理解复杂物理系统的开创性贡献”。真锅淑郎，1931 年出生于日本爱媛县新宫，1958 年获得东京大学理学博士学位；毕业后到美国地球物理流体动力学实验室（GFDL）工作，目前在普林斯顿大学担任高级气象学家。克劳斯·哈塞尔曼，1931 年出生于德国汉堡，1955 年获汉堡大学物理学与数学学士学位，1957 年获哥廷根大学和马克斯·普朗克流体动力学研究所①物理学博士学位；先后在汉堡大学、马克斯·普朗克气象研究所、德国气候计算中心等机构工作。乔治·帕里西，1948 年出生于意大利罗马，1970 年获罗马大学物理学博士学位；先后在意大利弗拉斯卡蒂国家实验室、罗马大学等机构工作。真锅淑郎和哈塞尔曼因为“地球气候的物理建模，量化可变性并可靠地预测全球变暖”的研究共享诺贝尔物理学奖的一半奖金，帕里西因为“发现了从原子尺度到行星尺度物理系统中的无序和涨落的相互作用”获得诺贝尔物理学奖的另一半奖金。

纵观历史，诺贝尔物理学奖早期奖励实验重大发现，后来逐步授予实验验证的理论及重要实验方法和技术。2021 年实现了一个突破，即奖励了还在发展中的理论，但保持了诺贝尔物理学奖的一贯宗旨和特色——推动科学技术发展、促进人类进步。

一、什么是复杂系统

复杂系统是由大量个体组成的系统，个体间相互作用导致其性质不是个体性质的简单加和，而是呈现超越个体性质的关联、合作和涌现等集体行为。正如 1977 年诺

① 2004 年更名为马克斯·普朗克动力学与自组织研究所。

真锅淑郎　　　　克劳斯·哈塞尔曼　　　　乔治·帕里西

图1　2021年度诺贝尔物理学奖获得者

贝尔物理学奖获得者菲利普·沃伦·安德森（Philip W. Anderson）1972年在《科学》期刊上发表的文章[1]中所述，不能依据少数个体的性质简单地外推出多个体复杂系统的行为，不同层次系统会呈现全新的性质。

综观自然与社会，存在着大量不同尺度、不同类型的复杂系统。例如，人类用于感知世界并思考的大脑是由海量神经元组成的[2]，认知、学习及思考与神经元的协同及涌现行为等相关；语言和知识系统可看作是由文字、名词及知识点等组成的复杂系统，教育过程与复杂性密切相关[3]；生命系统基本单元的生物分子由原子构成，大量生物分子聚集形成具有生命功能的细胞，再进一步细胞形成生命组织以及系统；社会的基本单元是人，人与人之间的相互关系与影响导致不同层次的社会结构；凝聚态物理中的高温超导相变、非晶材料的玻璃化转变是还没解决的重大科学问题，都涉及复杂物理系统的集体及涌现行为。可以说，当前迫切需要推进复杂系统研究。

研究复杂系统的复杂性科学（complexity science）兴起于20世纪80年代，主要研究系统结构与功能之间的关系，以及演化和调控规律，是一门新兴的交叉性、综合性学科。在还原论思想的指引下，近代科学取得了巨大发展，建立了越来越细的学科及领域，科学家个体在越来越窄的领域知道得越来越多。而需要研究的系统涉及多学科，要从个体的性质得到系统整体的性质，必须跳出还原论科学范式，时代呼唤复杂系统研究。

二、获奖成果简介

地球气候系统是一个典型的复杂系统，主要由大气圈、水圈、冰冻圈、岩石圈和生物圈五大圈层组成。各个圈层内部和圈层之间均具有高度交叉的物理、化学和生物过程以及反馈回路。1967 年，真锅淑郎和韦瑟拉尔德（Wetherald）第一次开发出一维辐射对流模式[4]，将大气圈简化为一个铅直的大气柱，详细考虑大气内的辐射过程和不同辐射过程之间的相互作用，以及一维的温度垂直分布。此模式建立在以下两个原理之上：①任何高度上的太阳辐射和长波辐射通量与对流热通量保持平衡；②因辐射差异引起的温度垂直分布的不稳定由对流调整而使其达到平衡。为理解 CO_2 浓度增加如何导致气温升高，真锅淑郎把空气团因对流而产生的垂直输送以及水蒸气的潜热纳入其中。采用该模型，他们发现地球表面和对流层的温度随着 CO_2 浓度升高而升高，而平流层的温度则随着 CO_2 浓度升高而降低。进一步还发现，氧和氮对地表温度的影响可忽略不计，而 CO_2 的影响则很明显，即大气中 CO_2 浓度若增加一倍，全球平均温度将上升 2.36℃。1969 年，真锅淑郎和布莱恩（Bryan）开发了第一个海洋-大气耦合环流模式，研究了十年到百年时间尺度的气候演化。20 世纪 90 年代，真锅淑郎研究小组还将海洋-大气耦合环流模式应用于过去气候变化的研究，包括在古气候记录中揭示北大西洋的淡水如何影响气候突变。

哈塞尔曼开发了描述气候变率的哈塞尔曼模型[5]。将宏观气候因素（如海洋）作为慢变量、介观天气变化（如大气）作为快变量，得到一种时间尺度可分离的耦合系统；通过投影算符和泰勒展开方法进行简化，将快变量看作是随机白噪声，得到慢变量的变化满足朗之万方程；由此可推导出福克尔-普朗克（Fokker-Planck）方程，从而研究一个气候系统在给定时间处于给定状态的概率。此外，朗之万方程的涨落耗散关系意味着可通过天气的噪声强度计算出宏观气候的涨落。

哈塞尔曼还创建了随机气候模式，可用来分离自然噪声和人类活动噪声的影响，从而可识别人类对气候系统的影响。例如，太阳辐射、火山爆发或温室气体水平的变化会留下独特的信号和印记，可以被分离出来，进一步从科学上证实由人为原因导致的 CO_2 排放导致了全球变暖的结论。

帕里西提出了能够揭示无序与涨落之间相互影响的方法，能够适用从原子到行星尺度的物理系统，且对数学、生物学、神经科学研究等产生影响。为描述无序磁性系统，1975 年爱德华兹（Edwards）和安德森（Anderson）提出了自旋玻璃模型（EA 模型），模型的自旋之间存在铁磁和反铁磁两种相互作用，分别对应自旋相同和相反，具有更低的能量。这种相互矛盾的相互作用引起阻挫，从而使系统存在大量的亚稳态。当自旋相互作用是无限长程，即每个自旋与系统剩余所有自旋都相互作用，EA 模型成为平均场自旋玻璃模型（SK 模型）。帕里西的重要贡献之一是基于复本理论（replica theory）精确求解了 SK 模型[6]。

帕里西严格求解自旋玻璃模型的理论和思维现已被广泛应用于很多无序体系，包括玻璃化转变、物种的进化、人脑的建模、机器学习模型等，极大地帮助我们整体理解复杂系统中无序和涨落的作用。

三、复杂系统研究回顾

20 世纪 80 年代复杂系统研究兴起之后，世界各地陆续成立了许多复杂系统研究机构。国外目前有 40 多个专门从事复杂性研究的机构，在复杂系统基础理论、社会系统复杂性、生命系统复杂性、地球系统、大数据科学等方面开展研究。复杂系统基础理论涉及涌现行为、自组织、动力学、远离平衡态统计物理、复杂网络及动力学等。社会复杂系统研究涉及人群行为、社会稳定性与冲突、经济系统与气候系统、疾病传播、教育等。生命复杂系统方面开展了认知与脑科学、计算生物学、生物群体运动、人类疾病等研究。为应对全球气候变暖的挑战，人们开始将地球作为复杂系统进行研究。国内复杂系统研究与系统科学学科建设密切相关，相关的研究机构现有 20 多个，但其中从事系统工程研究的较多、开展复杂系统研究的较少。下面，我们具体对比两个著名的研究机构，以及简介经济物理学的发展状况。

1984 年成立的美国圣塔菲研究所是国际最负盛名的复杂系统研究机构，聚集了一批物理、经济、理论生物、计算机科学等领域的著名学者开展跨学科复杂系统研究。该研究所虽不断提出新概念、新思想，引领复杂系统研究的潮流，但由于研究方向不断变化、研究人员不断流动，并没有产出对某个领域有实质性、根本性影响的科研成果。1992 年成立的德国波茨坦气候影响研究所，建所以后一直坚持气候影响研究，现已成为全球气候变化方面最具影响力的研究机构，产出了许多非常有影响的研究成果。现有研究队伍规模已达 400 多人，分布在地球气候分析、气候韧性、发展路径转变、复杂性科学四个研究部门。针对社会经济系统，美国著名物理学家尤金·斯坦利（H. Eugene Stanley）于 20 世纪 90 年代初提出了经济物理学的概念，试图将物理学思想和方法应用于经济学，打破主流经济学的僵化理论。尽管努力了二十多年，但并没有产出真正对经济学有影响的研究成果，经济学家们不满意物理学家们对市场的观点。2006 年《自然》期刊顾问编辑菲利普·鲍尔（Philip Ball）发表题为“文化崩溃”的文章[7]，详细介绍和评述了相关情况。

总体来说，虽然复杂系统研究取得了不小的进展，但离实质性解决复杂系统重大问题还存在较大差距，复杂系统研究方兴未艾。

四、展望未来发展

人类在还原论指引下取得了科学技术的长足进步与发展，也深刻改变了人类社会及人们赖以生存的地球。当前，科学研究、社会治理及应对全球气候变暖都迫切需要复杂系统科学的支持与帮助，亟待进一步加强复杂系统研究。

展望未来的复杂系统研究，一定要聚焦复杂物理系统、社会系统和地球系统中的重大问题，加强基于大数据的唯象学研究，将数学逻辑与实验融合。正如爱因斯坦所说“仅通过逻辑思维，我们无法获得任何关于经验的知识；所有关于现实的知识都来自经验并终于经验”。复杂物理系统研究中，很有必要将复杂性科学与凝聚态物理及材料科学进行融合。社会系统研究要结合国家战略需求。教育、科技、人才是全面建设社会主义现代化国家的基础性、战略性支撑，为实现我国教育、科技、人才的科学发展，需要将其作为复杂系统进行科学、定量的研究。地球系统的南极、北极、亚马孙雨林、大西洋环流等临界要素的研究尤为重要，需要采用复杂系统的思想和方法研究这些临界要素是否发生“相变”以及它们之间的相互影响。选定重大科学问题后，须将相关学科与领域的专家聚集在一起，建立各类不同的跨学科复杂系统研究机构，采用适合其发展的体制与机制进行管理。推进科学范式的变革[8]，发挥各学科优势及汇聚各科学家智慧，实现新思想和方法的涌现，携手解决重大科学问题。

参考文献

[1] Anderson P W. More is different：broken symmetry and nature of hierarchical structure of science. Science，1972，177：393.

[2] Koch C，Laurent G. Complexity and the nervous system. Science，1999，284：96.

[3] Jacobson M J，Levin J A，Kapur M. Education as a complex system：conceptual and methodological implications. Educational Researcher，2019，48，112.

[4] Manabe S，Wetherald R. Thermal equilibrium of the atmosphere with a given distribution of relative humidity. Journal of the Atmospheric Sciences，1967，24：241-259.

[5] Hasselmann K. Stochastic climate models：Part I. Theory. Tellus，1976，28(6)：473.

[6] Mézard M，Parisi G，Virasoro M A. Spin Glass Theory and Beyond：An Introduction to the Replica Method and Its Applications. Singapore：World Scientific Publishing Company，1987.

[7] Ball P. Culture crash. Nature，2006，441：686-688.

[8] Li J H，Huang W L. Paradigm shift in science with tackling global challenges. National Science Review，2019，6：1091-1093.

Call for Complex System Research to Meet Natural and Social Challenge
——Commentary on the 2021 Nobel Prize in Physics

Chen Xiaosong

The 2021 Nobel Prize in Physics has been awarded to three renowned scientists Syukuro Manabe, Klaus Hasselmann and Giorgio Parisi, "for groundbreaking contributions to our understanding of complex physical systems". Syukuro Manabe and Klaus Hasselmann shared half of the prize "for the physical modelling of Earth's climate, quantifying variability and reliably predicting global warming". Giorgio Parisi received the other half of the prize "for the discovery of the interplay of disorder and fluctuations in physical systems from atomic to planetary scales". Although the research on complex systems has made a lot of progress, there is still a big gap from the substantive solution of the major problems of natural and social complex systems. Looking ahead to the future of complex system research, we need to address the challenges in complex physical systems, social systems and Earth systems based on big data and new scientific paradigm.

5.2　不对称有机催化

——2021年度诺贝尔化学奖评述

余金生　周　剑

（华东师范大学化学与分子工程学院）

2021年诺贝尔化学奖授予马克斯·普朗克科学促进学会煤炭研究所本杰明·李斯特（Benjamin List）教授和美国普林斯顿大学戴维·麦克米伦（David MacMillan）教授（图1），以表彰他们推动“不对称有机催化发展”而做出的杰出贡献。至此，不对称催化的三种基本方法——金属催化（2001年）、酶催化（2018年）和有机催化（2021年）均已获授诺贝尔化学奖，彰显了不对称催化在构建手性化合物这一领域的勃勃生机和巨大研究价值。

本杰明·李斯特

戴维·麦克米伦

图1　2021年诺贝尔化学奖获得者

一、获奖成果简介

不对称有机催化，即利用非金属手性有机小分子化合物来催化不对称反应，是继金属催化和酶催化后的第三种发展不对称催化反应的策略。本次获奖的李斯特和麦克米伦两位科学家关于手性胺催化的研究工作，极大地促进并形成了世界范围内的有机催化研究热潮。其中，李斯特教授和勒纳（Lerner）教授以及巴巴斯（Barbas Ⅲ）教授合作，报道了手性脯氨酸催化的高对映选择性的丙酮和醛的分子间羟醛反应，颠覆了以往氨基酸催化局限于分子内反应的刻板印象，奠定了现代不对称烯胺催化的基础[1]。麦克米伦教授利用手性二级胺现场活化共轭烯醛（酮）来形成亲电性更高的不饱和亚胺正离子与共轭双烯发生高立体选择性的［4+2］环加成反应，从而开创了现代不对称亚胺催化的先河[2]。

二、不对称有机催化的研究进展与重大研究成果

随着手性物质在医药、农药、香料、材料和信息等领域应用日益广泛，其经济高效合成变得尤为重要。早期，人们利用天然手性化合物作为原料来制备所需的手性物质，但一个手性原料只能合成一个手性产物。为了实现手性增殖，不对称催化的方法应运而生，并在过去半个世纪得到了蓬勃发展，成功发展了金属催化、酶催化和有机催化三种策略。

如图 2 所示，不对称有机催化的探索可追溯到 20 世纪初利用生物碱类天然产物作为催化剂发展不对称反应，如利用番木鳖碱促进丙二酸脱羧制备手性羧酸[3]，以及利用奎尼丁促进的氰化氢对苯甲醛加成制备手性腈醇[4]。这些早期尝试的结果均不理想，直到 20 世纪 70 年代初，首例高对映选择性的有机催化不对称反应才被报道。多位化学家发现脯氨酸能催化分子内不对称羟醛反应，最高取得 93%的对映选择性[5,6]，但因局限于分子内反应未受到足够重视。此后 20 余年不对称有机催化依然鲜有报道，80 年代的唯一亮点是默克制药公司研究人员利用手性季铵盐实现了不对称甲基化反应[7]。

从 20 世纪 90 年代中后期开始，不对称有机催化的研究逐步迎来小高潮，新的催化模式和催化体系陆续出现。例如，香港大学杨丹教授和美国科罗拉多州立大学史一安教授开拓了手性酮催化[8,9]；德国亚琛工业大学恩德斯（Enders）发展了氮杂环卡宾催化[10]；麻省理工学院傅（G. C. Fu）等发展了手性吡啶衍生物亲核催化[11]；美国宾州州立大学张绪穆教授发现了手性叔膦催化[12]；哈佛大学雅各布森（Jacobsen）教授报道了氢键给体催化[13]；布兰迪斯大学邓力教授拓展了金鸡纳碱手性叔胺双功能催化[14]。这些原创性研究工作展示了不同有机催化模式的功能和特点，多角度展示了不对称有机催化的巨大发展潜力，在当时金属催化主导不对称催化的时代背景下让人耳目一新，如星星之火开始燃起对有机催化的研究热情。

2000 年，李斯特和麦克米兰伦位教授发表的关于不对称胺催化的研究工作，彻底

点燃了不对称有机催化的研究热潮。李斯特教授等利用 L-脯氨酸实现的高对映选择性丙酮和醛的分子间不对称羟醛反应，是手性金属催化剂也难以在温和条件下实现的挑战性反应，打破氨基酸催化局限于分子内反应的历史局限，为现代不对称烯胺催化奠定了基础[1]。几乎同时，麦克米伦教授报道了手性二级胺催化的共轭烯醛与共轭二烯的不对称［4＋2］环加成反应。这项研究的关键是利用手性胺现场活化共轭烯醛（酮）来生成活性更高的不饱和亚胺正离子，开创了现代不对称亚胺催化的先河[2]。

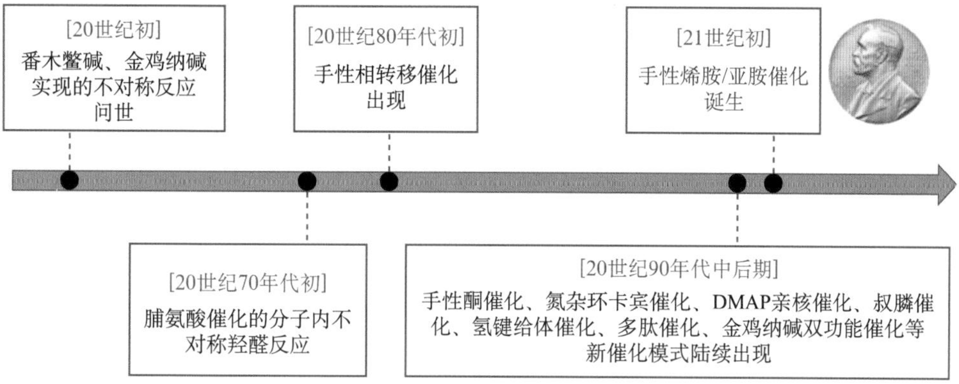

图 2　不对称有机催化发展简史

这两份诺贝尔奖成果展示了不对称胺催化的一些独特优势：具有丰富的活化模式（烯胺催化或亚胺催化）、选择性控制能力强（通过共价键作用等来进行手性诱导）、操作简便（无须严格无水无氧操作）、可避免重金属离子对产物的污染等。更有意思的是，烯胺催化和亚胺催化可在一定条件下相互转化。这一特性（李斯特教授于 2006 年形象地以“不对称胺催化的阴和阳”进行了总结介绍[15]），非常有助于模块化组合几个反应在一个反应瓶内进行，而无需分离提纯中间产物，从而提高合成效率。李斯特[16]、麦克米伦[17]和丹麦奥胡斯大学的约尔根森（Jørgensen）教授[18]在 2005 年几乎同时证明了同一个手性胺催化剂，可先形成亚胺正离子活化共轭烯醛发生亲核加成反应，之后再形成烯胺中间体与亲电试剂发生反应。恩德斯教授还证明通过巧妙设计多次切换手性胺催化剂的角色，可模块化组合更多的反应，从而实现从多个简单原料出发“一锅法”合成复杂多手性中心分子[19]。

不对称胺催化在 2000 年取得的突破进展犹如声声战鼓，促使世界各国有机化学研究人员快马加鞭进入这一前沿领域——利用不同的手性胺催化剂来活化醛（酮）生成亲核性更强的烯胺中间体，与不同亲电试剂进行反应，或活化共轭烯醛（酮）生成亲电性更高的不饱和亚胺正离子来发生反应，进而多样性合成在天然产物、药物和精细化工品合成中具有重要应用价值的手性醛（酮）等手性分子。因此，不对称胺催化的发展极大地促进了新型有机催化剂和不对称催化反应的研发，奠定了有机催化与金属催化和酶催化三足鼎立的地位。

不对称胺催化的进展带动了不对称有机催化研究的迅猛发展，新的有机催化策略和催化体系不断出现。例如，2004 年，日本学习院大学的彰山（Akiyama）教授[20]和东北大学的寺田（Terada）教授[21]几乎同时报道的手性磷酸催化，开启了手性质子酸催化这一新的研究领域；芝加哥大学山本（Yamamoto）教授[22]和李斯特教授[23]进一步设计发展了酸性更强的手性酸催化体系，使不对称有机催化的效率达到了一个新高度。不对称胺手性催化的模式也得到了进一步发展。瑞典斯德哥尔摩大学科尔多瓦（Córdova）教授率先尝试了手性烯胺和光协同催化[24]；约尔根森教授[25]和四川大学陈教授[26]分别发展了手性二烯胺和三烯胺催化；李斯特教授结合手性磷酸催化丰富了烯胺催化和亚胺催化的范畴，发展了新型不对称抗衡阴离子催化策略[27]；麦克米伦教授发展了手性胺的单电子占有轨道（SOMO）活化模式[28,29]，即利用手性烯胺中间体被氧化后产生的自由基物种与烷基化试剂发生反应。

不对称有机小分子催化的发展使得结合金属催化或酶催化来实现单一催化所不能实现的不对称反应。2001 年，中国科学院成都有机化学研究所的龚流柱等率先尝试结合手性相转移催化和钯催化来发展不对称反应[30]，揭开了发展有机小分子和金属协同催化的序幕。随后，结合各种有机小分子催化方式和不同金属来发展不对称反应以及串联合成得到了蓬勃发展，至今方兴未艾。特别是结合手性烯胺催化和手性金属催化来调控反应的非对映选择性，同时获得多手性中心的化合物所有可能光学异构体，开辟了不对称催化的立体发散性合成的新方向[31]。

三、未来展望

不对称催化研究三获诺贝尔奖看似象征着手性合成的研究已经比较成熟。其实不然，这一领域还有诸多问题等待进一步探索，实现工业化的高效不对称催化合成方法还很有限。按照 2001 年诺贝尔奖获得者野依良治教授所指出的高标准：“未来的合成化学必须是经济的、安全的、环境友好的以及节省资源和能源的化学，化学家需要为实现‘完美的反应化学’而努力，即以 100% 的选择性和 100% 的收率只生成需要的产物而没有废物产生。”[32]现有的大部分不对称催化方法从效率来考察都有很大提升空间。不论寻求“手性起源”这一重要科学问题的答案，还是发展“理想”的不对称催化合成方法来提供各种手性功能分子的多样化精细化合成，从而满足人类的生活和发展需求，都还需要上下求索。

值得一提的是我国有机化学家从 20 世纪 80 年代开始发愤图强，在不对称催化领域奋起直追，大大缩小了我国在该领域与国际顶级水平的差距。设计发展了以周氏手性螺环配体、冯氏手性双氮氧配体为代表的一批享誉海内外的优势手性配体和催化剂，实现了以罗斯坎普-冯（Roskamp-Feng）反应为代表的系列高效不对称催化新方法。目前，我国已经形成了以中青年骨干为主的梯度合理的活力研究队伍，致力于发展具有自主知识产权的手性技术服务于国家新药研发、生态安全、高新材料战略需求，努力为解决国家“卡脖子”问题做出应有贡献。中国科研工作者一定会为实现不对称催化“精准化”和“实用化”以及为人类在手性世界的探索做出重要贡献。

参考文献

[1] List B, Lerner R A, Barbas Ⅲ C F. Proline-catalyzed direct asymmetric aldol reactions. Journal of the American Chemical Society, 2000, 122(10): 2395-2396.

[2] Ahrendt K A, Borths C J, MacMillan D W C. New strategies for organic catalysis: the first highly enantioselective organocatalytic Diels-Alder reaction. Journal of the American Chemical Society, 2000, 122(17): 4243-4244.

[3] Marckwald W. Ueber asymmetrische synthese. Berichte Der Deutschen Chemischen Gesellschaft, 1904, 37(1): 349-354.

[4] Bredig G, Fiske P S. Durch Katalysatoren Bewirkte Asymmetrische Synthese. Biochemische Zeitschrift, 1912, 46: 7-23

[5] Eder U, Sauer G, Wiechert R. New type of asymmetric cyclization to optically active steroid CD partial structures. Angewandte Chemie International Edition in English, 1971, 10(7): 496-497.

[6] Hajos Z G, Parrish D R. Asymmetric synthesis of bicyclic intermediates of natural product chemistry. The Journal of Organic Chemistry, 1974, 39(12): 1615-1621.

[7] Dolling U, Davis P, Grabowski E. Efficient catalytic asymmetric alkylations. 1. Enantioselective synthesis of (+)-indacrinone via chiral phase-transfer catalysis. Journal of the American Chemical Society, 1984, 106(2): 446-447.

[8] Yang D, Yip Y C, Tang M W, et al. A C_2 symmetric chiral ketone for catalytic asymmetric epoxidation of unfunctionalized olefins. Journal of the American Chemical Society, 1996, 118(2): 491-492.

[9] Tu Y, Wang Z X, Shi Y A. An efficient asymmetric epoxidation method for *trans*-olefins mediated by a Fructose-derived ketone. Journal of the American Chemical Society, 1996, 118(40): 9806-9807.

[10] Enders D, Teles J H, Breuer K. A novel asymmetric benzoin reaction catalyzed by a chiral triazolium salt-Preliminary communication. Helvetica Chimica Acta, 1996, 79(4): 1217-1221.

[11] Ruble J C, Latham H A, Fu G C. Effective kinetic resolution of secondary alcohols with a planar—chiral analogue of 4-(dimethylamino) pyridine. use of the Fe(C_5Ph_5) group in asymmetric Catalysis. Journal of the American Chemical Society, 1997, 119(6): 1492-1493.

[12] Zhu G X, Chen Z G, Jiang Q Z, et al. Asymmetric [3+2] cycloaddition of 2,3-butadienoates with electron-deficient olefins catalyzed by novel chiral 2,5-dialkyl-7-phenyl-7-phosphabicyclo[2. 2. 1] heptanes. Journal of the American Chemical Society, 1997, 119(16): 3836-3837.

[13] Sigman M S, Jacobsen E N. Schiff base catalysts for the asymmetric strecker reaction identified and optimized from parallel synthetic libraries. Journal of the American Chemical Society, 1998, 120(19): 4901-4902.

[14] Chen Y G, Tian S K, Deng L. A highly enantioselective catalytic desymmetrization of cyclic anhydrides with modified *Cinchona* alkaloids. Journal of the American Chemical Society, 2000, 122 (39): 9542-9543.

[15] List B. The Ying and Yang of asymmetric aminocatalysis. Chemical Communications, 2006, (8): 819-824.

[16] Yang J W, Hechavarria Fonseca M T, List B. Catalytic asymmetric reductive Michael cyclization. Journal of the American Chemical Society, 2005, 127(43): 15036-15037.

[17] Huang Y, Walji A M, Larsen C H, et al. Enantioselective organo-cascade catalysis. Journal of the American Chemical Society, 2005, 127(43): 15051-15053.

[18] Marigo M, Schulte T, Franzén J, et al. Asymmetric multicomponent domino reactions and highly enantioselective conjugated addition of thiols to α, β-unsaturated aldehydes. Journal of the American Chemical Society, 2005, 127(45): 15710-15711.

[19] Enders D, Hüttl M R M, Grondal C, et al. Control offour stereocentres in a triple cascade organocatalytic reaction. Nature, 2006, 441(7095): 861-863.

[20] Akiyama T, Itoh J, Yokota K, et al. Enantioselective Mannich-type reaction catalyzed by a chiral Brønsted acid. Angewandte Chemie, 2004, 43(12): 1566-1568.

[21] Uraguchi D, Terada M. Chiral Brønsted acid-catalyzed direct Mannich reactions via electrophilic activation. Journal of the American Chemical Society, 2004, 126(17): 5356-5357.

[22] Nakashima D, Yamamoto H. Design of chiral *N*-triflyl phosphoramide as a strong chiral Brønsted acid and its application to asymmetric Diels-Alder reaction. Journal of the American Chemical Society, 2006, 128(30): 9626-9627.

[23] Tsuji N, Kennemur J L, Buyck T, et al. Activation of olefins via asymmetric Brønsted acid catalysis. Science, 2018, 359(6383): 1501-1505.

[24] Córdova A, Sundén H, Engqvist M, et al. The direct amino acid-catalyzed asymmetric incorporation of molecular oxygen to organic compounds. Journal of the American Chemical Society, 2004, 126 (29): 8914-8915.

[25] Jensen K L, Dickmeiss G, Jiang H, et al. The diarylprolinol silyl ether system: a general organocatalyst. Accounts of Chemical Research, 2012, 45(2): 248-264.

[26] Jia Z J, Jiang H, Li J L, et al. Trienamines inasymmetric organocatalysis: Diels-Alder and tandem reactions. Journal of the American Chemical Society, 2011, 133(13): 5053-5061.

[27] Mayer S, List B. Asymmetric counteranion-directed catalysis. Angewandte Chemie International Edition, 2006, 45(25): 4193-4195.

[28] Beeson T D, Mastracchio A, Hong J B, et al. Enantioselective organocatalysis using SOMO activation. Science, 2007, 316(5824): 582-585.

[29] Nicewicz D A, MacMillan D W C. Merging photoredox catalysis with organocatalysis: the direct asymmetric alkylation of aldehydes. Science, 2008, 322(5898): 77-80.

[30] Chen G S, Deng Y J, Gong L Z, et al. Palladium-Catalyzed Allylic Alkylation of *tert*-Butyl (diphenylmethylene)-glycinate with Simple Allyl Esters under Chiral Phase Transfer Conditions. Tetrahedron: Asymmetry, 2001, 12(11): 1567-1571.

[31] Krautwald S, Sarlah D, Schafroth M A, et al. Enantio- and diastereodivergent dual catalysis: α-allylation of branched aldehydes. Science, 2013, 340(6136): 1065-1068.

[32] Noyori R. Synthesizing our future. Nature Chemistry, 2009, 1: 5-6.

Asymmetric Organocatalysis
——Commentary on the 2021 Nobel Prize in Chemistry

Yu Jinsheng, Zhou Jian

On October 6, 2021, the Nobel Prize in Chemistry is awarded to two scientists, including Prof. Benjamin List from the Max-Planck-Institut für Kohlenforschung, Germany, Prof. David W. C. MacMillan from Princeton University, USA, for the development of asymmetric organocatalysis. In this article, we review their representative works and their contributions, as well as the development and outlook of asymmetric organocatalysis.

5.3 温度及触觉受体的发现及研究
——2021年度诺贝尔生理学或医学奖评述

闫致强

(深圳湾实验室分子生理学研究所)

人类拥有多种感觉，如视觉、听觉、嗅觉、味觉、触觉，此外还有温觉（包括热觉和冷觉）、痛觉等。人体是如何感知物理世界的问题一直吸引着人类。然而，人类关于这些感觉的基础生物学层面的理解仍十分有限。2021年诺贝尔生理学或医学奖授予感知觉研究领域，以表彰美国加利福尼亚大学旧金山分校（University of California，San Francisco，UCSF）的戴维·朱利叶斯（David Julius）和斯克利普斯研究所（Scripps Research）的雅顿·帕塔普蒂安（Ardem Patapoutian）在感知温度与触觉受体的发现上做出的深远而广泛的贡献（图1）。本文着重对本次诺贝尔生理学或医学奖进行解读，并简介其他感知觉受体的发现及研究历程。

戴维·朱利叶斯

雅顿·帕塔普蒂安

图1　2021年诺贝尔生理学或医学奖获得者

一、温度觉受体的研究

热感觉受体的发现离不开辣椒素（capsaicin）的贡献。辣椒因其独特的辛辣感被世界各地广泛在烹调中采用，辣椒素是辣椒的活性成分。一定量的辣椒素接触人的口腔和嘴唇等区域会引发灼热感，并引起出汗，这一现象被称为味觉出汗。朱利叶斯实验室[1]通过分离啮齿动物背根神经节（dorsal root ganglion，DRG）制备了 cDNA 文库，用该文库转染辣椒素不敏感细胞，最终分离出一个基因，依据同源比对表明它属于瞬时受体电位（transient receptor potential，TRP）阳离子通道超家族；接着通过异源细胞表达 TRPV1 受体，使用电生理技术对其进行功能表征，发现辣椒素诱发的电生理特性与感觉神经元中的类似，同时拮抗剂可以阻断电流。在随后的研究中，朱利叶斯证明了 TRPV1 受体对高温（高于 40°C，这个温度对人体已经起到伤害作用）敏感，发现热能显著激活 TRPV1 并引起细胞 Ca^{2+} 内流[1]。研究人员进一步发现，敲除 *TRPV1* 基因后，小鼠对于热的急性反应出现了部分缺失及不敏感[2]。针对表达 TRPV1 蛋白的神经纤维的研究发现，其不在响应触觉、本体感觉等的感知无害性刺激的 A 类神经纤维上表达，而在表达 P 物质、神经激肽 A 和 CGRP 的多肽能 C 类神经纤维上表达。TRPV1 受体也随后作为一种感知伤害性刺激的神经纤维标志物而被领域内的科学家们广泛认可并采用。后续的研究更发现 TRPV1 受体除了直接参与感知热痛，也在热痛后的痛觉超敏、过敏反应以及在痒觉信息的传递中发挥了重要作用。

TRPV1 受体同时是辣椒素受体和热感觉离子通道的发现打开了温度觉领域研究的大门，代表了人类在探索热传感的分子和神经基础方面的研究取得的里程碑式的成就。未来的研究可能会在这个领域提供更多的见解，发现更多的温度觉的受体。朱利叶斯并不满足于只鉴定受体和通道的功能性研究，他和同事华人学者程亦凡（Yifan Cheng）通过冷冻电子显微镜解析了 TRPV1 受体的结构，将对该通道的理解提升到了一个新的高度[3]。

二、触觉受体的发现及研究

触觉是我们日常生活中最常见的感觉之一。从对手中事物的纹理、形状、重量等的静态感知，再到风拂过脸庞和情感性触碰的动态感觉，都在很大程度上依赖触觉机械力感知。在日常生活中，我们可以通过机械力的介导识别不同的物体，以及感受自身空间位置和肢体运动，这些看似平常的活动受到精细的调节，涉及不同神经元的独特功能。

触觉主要通过皮肤这一人体最大的器官上的机械感觉末端小体以及支配它们的低阈值的机械敏感受体（low threshold mechnoreceptor，LTMR）的组合所编码传递。早期研究发现，皮肤上有4种机械感觉末端小体（迈斯纳小体、梅克尔细胞、环层小体和鲁菲尼小体）以及支配它们的神经末梢，感受范围各异，对不同的刺激频率和持续时间敏感。为了找到哺乳动物的机械敏感受体，加利福尼亚州斯克利普斯研究所的帕塔普蒂安实验室开发了一种新的筛选方法——压力钳结合膜片钳的方法。使用这种方法，研究人员发现Neuro2A细胞系具有机械敏感性，该细胞系受到机械力刺激后会出现机械敏感电流[4]。随后帕塔普蒂安的研究鉴定出72个可能作为机械敏感受体的候选基因，通过RNAi敲低每个候选基因，再利用压力钳结合膜片钳的方法来确定给细胞施加机械力是否会产生电流变化。当敲低候选基因*FAM38A*（后被命名为*PIEZO1*基因）后，机械敏感电流消失，证明该蛋白具有机械敏感性，此后该蛋白从希腊语单词piesi中得名，命名为PIEZO1；随后通过同源序列比对发现了第二个机械敏感通道蛋白，命名为PIEZO2。

以上的研究证明了PIEZO蛋白具有机械敏感性，但是否为触觉受体仍不得而知。为了找到证明PIEZO2蛋白是触觉受体的直接证据，2014年，帕塔普蒂安实验室证明梅克尔细胞在接受机械刺激后出现机械力依赖的电流，并且该电流可以导致神经末梢出现动作电位，从而直接证明了PIEZO2是真正的触觉受体[5]；2014年的另一项研究显示，*PIEZO2*基因敲除小鼠严重缺乏触觉，但其他感觉不受损害[6]，从行为学上再次证明PIEZO2是触觉受体。

PIEZO蛋白是一类全新的脊椎动物机械敏感通道，拥有38个跨膜结构域的最大跨膜离子通道亚单位。帕塔普蒂安和其他实验室的工作揭示了PIEZO1和PIEZO2的高分辨率结构，PIEZO蛋白为三聚体结构，包括中心离子通道和三个外围机械感应螺旋桨形叶片。当对膜施加机械力时，弯曲的叶片变平并导致中心孔的开放。带有弯曲叶片的螺旋桨状结构产生了较大的平面内膜面积膨胀，这可能解释了PIEZO通道的机械敏感性[7]。

三、如何开展具有冒险精神的、“诺贝尔奖”级别的研究

朱利叶斯实验室关于温度受体的研究[1]，利用辣椒素能引起热觉（辣）来克隆可能的温度觉受体，其中最重要的实验是克隆表达背根神经节的分子，利用钙成像分析鉴别辣椒素受体，也就是可能的温度觉受体。这无疑是一个很有创意的实验，但同时也存在着很大的风险。最大的风险就是，体外表达的cDNA无法复制体内的状况，从而无法确定表达克隆的策略的可行性。朱利叶斯实验室面对这样的风险是如何做的

呢？在其发表文章 Figure 1 第二列里，已经基本排除了这种风险，他们通过将整个背根神经节 cDNA 基因文库转染到 293 细胞，发现辣椒素可以使其中一些转染的 293 细胞被激活，证明了实验的可行性（图 2）。笔者认为这是朱利叶斯获得诺贝尔奖最重要的一幅图或一个实验，能在一个体系里表达被激活，接下来就是按部就班进行筛选，其风险在这幅图中已经得到了完美的规避。

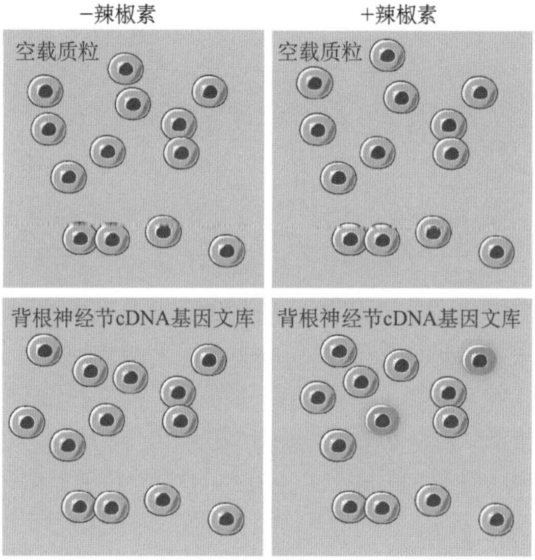

图 2　辣椒素使表达了辣椒素受体的培养细胞产生钙反应

理论上讲，克隆背根神经节神经元的触觉受体，同样可以使用 1997 年钙成像[1]的方法，实际上，朱利叶斯实验室也是这么做的[8]。他们用一个膜拉扯细胞，然后膜复位产生一个钙信号，但机械力敏感通道激活比辣椒素受体激活进行钙成像难度大很多，这使得实验本身很不稳定，实验设计有很大问题。反观最终克隆出 PIEZO 的文章[4]，帕塔普蒂安实验室采用了不同的策略，并没有“路径依赖”，不再坚持使用钙成像方法，而是直接利用电生理记录，找到一个具有机械力敏感电流的细胞系，反过来利用 RNAi 敲低的方法进行筛选。电生理记录比钙成像做机械力敏感通道灵敏得多。触觉是在背根神经节发生的，但如果直接在背根神经节神经元做 RNAi 敲降，其工作量之大单个实验室几乎是不可能完成的，所以整个工作最大的亮点是，其把风险和工作量降到最低的点也是发现了一个具有机械力敏感电流的细胞系 Neuro2A（图 3）。在培养细胞系进行电生理记录，同样的筛选，工作量可能只有在背根神经节的 1/10，可行性强，也极大地规避了工作量巨大从而无法完成的风险。

总结这两个“诺贝尔奖”级别的发现，作者都是通过筛选基因组，得到一个重要

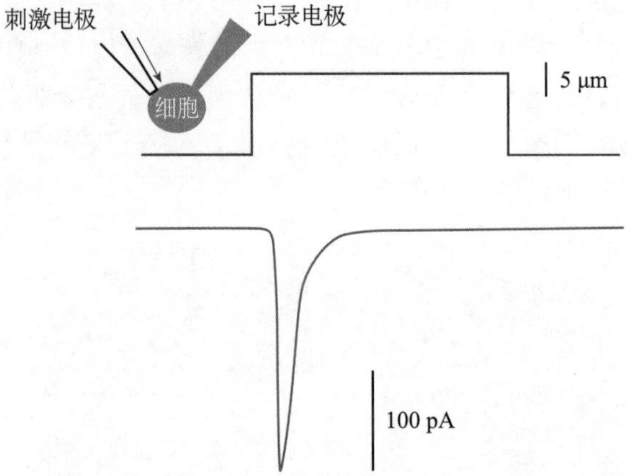

图 3　N2A 培养细胞在机械电极刺激下，产生机械敏感电流

分子，即温度或者触觉的受体；一开始毫无头绪，看上去风险极大，但通过一些工作量不多的探索性试验后，风险骤然降低，后续工作就顺理成章。可见，在优秀的设计和初步的实验后，这些看上去极大风险的重要课题，其实风险很小，两篇文章图 1 中的数据，才是科研的精华所在，分析这些之后更可以体会到，做一些“诺贝尔奖”级别的工作，并不是“那么险”。做好科学，最大的风险不在于这些技术风险，而在于：有动力找到重大的问题；有信心克服障碍付诸实践；有技巧找到合适的方法降低风险。

四、未来诺贝尔奖展望

在感知觉研究的历史中，视觉研究发展较快。1967 年诺贝尔生理学或医学奖颁发给美国科学家乔治・沃尔德（George Wald）、霍尔登・哈特兰（Haldan K. Hartline）及瑞典科学家拉格纳・格拉尼特（Ragnar A. Granit），以表彰三人发现了眼睛的初级生理及化学视觉过程。随后则是嗅觉的研究，1991 年，美国科学家理查德・阿克塞尔（Richard Axel）和博士后琳达・巴克（Linda B. Buck）的研究首次揭示了嗅觉受体分子家族，阿克塞尔和巴克共同于 2004 年获得诺贝尔生理学或医学奖。人类有五种基本味觉，包括酸、甜、苦、咸、鲜。在这一领域，查尔斯・祖克（Charles S. Zuker）和尼古拉斯・里巴（Nicholas J. P. Ryba）做出了重要的贡献，从 2000 年开始，二人实验室就陆续发现了感受五种基本味觉的细胞类型以及除酸味以外的另外四种味觉的分子受体[9]。听觉是人类的主要感官之一，但是听觉受体及机械转导机制却一直未知。内耳科蒂器（organ of Corti）中的毛细胞是感知声音的第一级神经元。2002 年，安德鲁·格里菲斯（Andrew J. Griffith）实验室和卡伦·斯蒂尔（Karen P. Steel）实验室分别在先天性耳聋患者和耳聋小鼠模型中发现同一个致病基因 *TMC1*。2011 年，安德鲁·格里菲斯实验室和杰弗里·霍尔特（Jeffrey R. Holt）实验室合作研究发现，在毛细胞中除 TMC1 蛋白之外，TMC 蛋白家族同源分子 TMC2 在听觉转导过程中同样发挥重要作用，只有在把 *TMC1* 和 *TMC2* 基因同时敲除的情况下，才能完全消除毛细胞的机械传导电流[10]。上述研究初步证明，TMC 蛋白家族可能是参与听觉机械传导的关键蛋白。此后，陆续发现一些分子对于毛细胞的机械传导也很重要，到底哪个分子是机械传导离子通道孔道区核心亚基，仍无定论。2018 年，霍尔特（Holt）和科里（Corey）团队[11]合作证明 TMC1 蛋白可以形成二聚体，并且预测 TMC1 蛋白跨膜区 S4～S7 结构域可能构成了毛细胞机械传导通道的孔道区。直到 2020 年，笔者实验室将绿海龟的 TMC1（CmTMC1）与虎皮鹦鹉的 TMC2（MuTMC2）表达纯化，再重构在脂质体里进行电生理记录，证明了 CmTMC1 和 MuTMC2 蛋白均具备离子通道活性并可以被机械力激活[12]，为 TMC1/2 蛋白是听觉受体提供了重要证据。听觉、味觉受体的研究工作能否获得诺贝尔奖，值得期待与关注。

参考文献

[1] Caterina M J, Schumacher M A, Tominaga M, et al. The capsaicin receptor: a heat-activated ion channel in the pain pathway. Nature, 1997, 389(6653): 816-824.

[2] Caterina M J, Leffler A, Malmberg A B, et al. Impaired nociception and pain sensation in mice lacking the capsaicin receptor. Science, 2000, 288(5464): 306-313.

[3] Zhang K, Julius D, Cheng Y. Structural snapshots of TRPV1 reveal mechanism of polymodal functionality. Cell, 2021, 184(20): 5138-5150. e12.

[4] Coste B, Mathur J, Schmidt M, et al. Piezo1 and Piezo2 are essential components of distinct mechanically activated cation channels. Science, 2010, 330(6000): 55-60.

[5] Ikeda R, Cha M, Ling J, et al. Merkel cells transduce and encode tactile stimuli to drive Aβ-afferent impulses. Cell, 2014, 157(3): 664-675.

[6] Ranade S S, Woo S H, Dubin A E, et al. Piezo2 is the major transducer of mechanical forces for touch sensation in mice. Nature, 2014, 516(7529): 121-125.

[7] Wang L, Zhou H, Zhang M, et al. Structure and mechanogating of the mammalian tactile channel PIEZO2. Nature, 2019, 573(7773): 225-229.

[8] Bhattacharya M R, Bautista D M, Wu K, et al. Radial stretch reveals distinct populations of mechanosensitive mammalian somatosensory neurons. Proceedings of the National Academy of Sciences of the United States of America, 2008, 105(50): 20015-20020.

[9] Yarmolinsky D A, Zuker C S, Ryba N J. Common sense about taste: from mammals to insects. Cell, 2009, 139(2): 234-244.

[10] Kawashima Y, Géléoc G S, Kurima K, et al. Mechanotransduction in mouse inner ear hair cells requires transmembrane channel-like genes. The Journal of Clinical Investigation, 2011, 121(12): 4796-4809.
[11] Pan B, Akyuz N, Liu X P, et al. TMC1 forms the pore of mechanosensory transduction channels in vertebrate inner ear hair cells. Neuron, 2018, 99(4): 736-753. e6.
[12] Jia Y, Zhao Y, Kusakizako T, et al. TMC1 and TMC2 proteins are pore-forming subunits of mechanosensitive ion channels. Neuron, 2020, 105(2): 310-321. e3.

Discovery of Temperature and Touch Receptors —Commentary on the 2021 Nobel Prize in Physiology or Medicine

Yan Zhiqiang

Institute of Molecular Physiology, Shenzhen Bay Laboratory Human has five senses, including vision, touch, hearing, taste, smell and taste, as well as temperature(including heat and cold) and pain sensation. How humans perceive the physical world through somatic sensation has fascinated us for centuries. The 2021 Nobel Prize in Physiology or Medicine was awarded to David Julius from the University of California, San Francisco (UCSF) and Ardem Patapoutian from Scripps Research for their discoveries of thermal and mechanical transducers. The discovery of TRPV1 by David Julius opened the door and represented a landmark achievement to a molecular understanding of thermosensation. Ardem Patapoutian mainly focuses on the research of mechanosensitive ion channels. The discovery of PIEZO proteins provides insights into how humans perceive internal and external environments. This review focuses on the 2021 Nobel Prize in Physiology or Medicine and recent research on sensory perceptions is also briefly introduced.

5.4　2021年度图灵奖获奖者简介

唐　川　黄　茹

（中国科学院成都文献情报中心）

2022年3月30日，国际计算机协会（Association for Computing Machinery，ACM）宣布将2021年度图灵奖（Turing Award）授予杰克・唐加拉（Jack J. Dongarra，图1），以表彰他在数值算法和库方面的开创性贡献。他的工作使高性能计算软件能够在40多年来跟上硬件的指数级发展[1]，并对人工智能、计算机图形学等诸多研究领域产生了深远影响。

杰克・唐加拉

图1　2021年度图灵奖获得者

杰克・唐加拉1950年出生于美国芝加哥，拥有芝加哥州立大学数学学士学位、伊利诺伊理工大学计算机科学硕士学位、新墨西哥大学应用数学博士学位，是美国田纳西大学杰出教授和橡树岭国家实验室的杰出研究人员，同时还是莱斯大学计算机科学系兼职教授。自2007年以来，他还一直担任英国曼彻斯特大学的图灵研究员（Turing Fellow）。唐加拉是美国国家工程院（National Academy of Engineering）院士和英国皇家学会（The Royal Society）外籍院士，还是国际计算机协会（Association for Computing Machinery，ACM）、美国电气与电子工程师协会（Institute of Electrical and Electronics Engineers，IEEE)、美国工业和应用数学学会（Society for Industrial and Applied Mathematics，SIAM)、美国科学促进会（American Association for the Advancement of Science，AAAS)、国际超级计算会议（International Supercomputing Conference，ISC）和国际工程技术协会（International Engineering and Technology Institute，IETI）的会士（Fellow)[1]。他曾被授予IEEE计算机先驱奖（IEEE Computer Pioneer Award)，SIAM/ACM计算科学与工程奖（SIAM/ACM Prize in Computational Science and Engineering)、ACM/IEEE肯尼迪奖（ACM/IEEE（CS）Ken Kennedy Award）等多项荣誉。

近40年来，硬件的性能呈指数级增长，唐加拉对线性代数运算的高效数值算法、并行计算编程机制和性能评估工具的研究工作在很大程度上促使高性能数值软件的性能跟上了硬件进步的步伐。他为单处理器、并行计算机、多核节点和多图形处理器（GPU）创建了一系列开源软件库和相关标准，其采用线性代数作为中间语言，可以被各种应用程序使用。他还在此基础上做出许多重要的创新，包括自动调整、混合精度计算和批处理计算。现在，从笔记本电脑到世界上最快的超级计算机，基于线性代数的软件库被普遍用于处理高性能科学与工程计算，协助科学家和工程师在大数据分析、医疗保健、可再生能源、天气预报、基因组学和经济学等领域取得重要发现和颠覆性创新[2]。

唐加拉于1972年在美国阿贡国家实验室实习，使用矩阵特征系统软件包（EISPACK）运行测试问题并向开发人员报告错误，这段经历激发了他在数值算法上的兴趣。在阿贡国家实验室工作期间，他相继参与创建线性系统软件包（LINPACK）和数学软件库（NETLIB)。1989年，他离开阿贡国家实验室，就职于田纳西大学和橡树岭国家实验室，正式成为一名教授，带领学生开启了在数值算法和库方面的研究工作。唐加拉教授项目中的代码已被整合到MATLAB、Maple、Mathematica和R编程语言等工具中，帮助促进了计算机体系结构的跨越式发展[1]。

国际计算机协会主席加布里埃尔·科奇什（Gabriele Kotsis）称赞唐加拉的开创性工作可以追溯到1979年，他在引领高性能计算领域的成功发展中发挥了核心作用，仍然是高性能计算社区中最重要且积极参与的领导者之一。谷歌高级研究员、谷歌研究兼谷歌健康高级副总裁杰夫·迪恩（Jeff Dean）也表示唐加拉在使用最频繁的数据库的核心领域开展了深入而重要的工作，奠定了科学计算各个领域的基础，他为表征各种计算机性能所做出的努力促使适合数值计算的计算机体系结构取得重大进步[2]。

参考文献

［1］ A. M. Turing Award. DR. Jack Dongarra. https://amturing.acm.org/award_winners/dongarra_3406337.cfm［2022-12-22］.

［2］ A. M. Turing Award. ACM Turing Award Honors Jack J. Dongarra For Pioneering Concepts And Methods Which Have Resulted In World-changing Computations. https://amturing.acm.org/［2022-12-22］.

Introduction to the 2021 A. M. Turing Award

Tang Chuan, Huang Ru

The 2021 Turing Award Winners for Computer Achievement was Jack J. Dongarra who has been a University Distinguished Professor at the University of Tennessee and a Distinguished Research Staff Member at the Oak Ridge National Laboratory since 1989. He received the award for his pioneering contributions to numerical algorithms and libraries. His work has enabled high-performance computing software to keep pace with the rapid development of hardware for more than 40 years, and has had a profound impact on artificial intelligence, computer graphics and many other fields.

5.5 2021年度未来科学大奖获奖者简介

叶 京

（中国科学院科技战略咨询研究院）

2021年9月12日，未来科学大奖科学委员会在北京公布2021年获奖名单[1]。袁国勇、裴伟士凭借“发现了冠状病毒（SARS-CoV-1）为导致2003年全球严重急性呼吸综合征（SARS）病原，以及由动物到人的传染链，为人类应对MERS①和COVID-19冠状病毒引起的传染病产生了重大影响”的贡献摘得“生命科学奖”；张杰因“通过调控激光与物质相互作用产生精确可控的超短脉冲快电子束，并将其应用于实现超高时空分辨高能电子衍射成像和激光核聚变的快点火研究”的贡献获得“物质科学奖”；施敏因“在对金属与半导体间载流子互传的理论认知做出的贡献，促成了过去50年中按‘摩尔定律’速率建造的各代集成电路中如何形成欧姆和肖特基接触的关键技术”取得的成就荣膺“数学与计算机科学奖”。

一、未来科学大奖生命科学奖

未来科学大奖生命科学奖获奖人香港大学教授袁国勇、裴伟士（图1）于2003年治疗了中国香港的第一例严重急性呼吸综合征患者，并从临床标本中分离出冠状病毒，为病症诊断和鉴定提供了必要信息[2]。

袁国勇对野生蝙蝠中SARS类冠状病毒的持续研究，大大扩展了我们对人畜共患病的宿主、跨物种传播障碍、发病机制与疾病诊断的认识。鉴于野生蝙蝠衍生的SARS类冠状病毒的高流行率，袁国勇和裴伟士的研究预测了类似SARS的流行病可能再次出现，并强调了公共卫生防备的重要性。正如研究所料，蝙蝠冠状病毒HKU4/5被认为是引起了流行性中东呼吸综合征的MERS-CoV病毒的前身。

从2003年的全球严重急性呼吸综合征到2019新型冠状病毒感染（COVID-19），袁国勇和裴伟士的研究为我们认识和治疗这种新兴传染疾病作出了重大贡献，为我们

① MERS为Middle East Respiratory Syndrome Coronavirus的简写，中文为“中东呼吸综合征”。

袁国勇

裴伟士

图 1　2021 年未来科学大奖生命科学奖获奖人

应对这些疾病提供了证据和策略。

二、未来科学大奖物质科学奖

未来科学大奖物质科学奖获奖人中国科学院院士、上海交通大学教授、中国科学院物理研究所教授张杰（图 2）是开发利用太瓦（TW）到拍瓦（PW）激光束有效生成受控、高强度快电子束（约 100 keV 到 10 MeV）方法的先驱。

张杰

图 2　2021 年未来科学大奖物质科学奖获奖人

张杰领导的研究团队在快电子束方面取得了一系列重大突破，包括高效产生非热电子、用激光调节电子束能量、实现高定向电子发射，以及创时空分辨世界纪录的电子束成像。团队研发的可精确控制的高强度快电子束为一系列其他重要的科学探索提供了可能。例如，他们开发了 MeV 超快电子衍射和成像设备，并展示了亚埃空间分辨率和创纪录的 50 fs 时间分辨率；使用超快激光场成功地改变量子材料的维度，观察到了光诱导的新型相变；帮助实现了更紧凑、更高效的高能粒子加速器；通过超快电子衍射实现了单分子成像。

张杰对快电子束的研究最初是为了研究惯性约束聚变（ICF）。这个过程如果实现，可以为人类提供无限的能量供应。高强度超短脉冲快电子束为 ICF 提供了快速点火的关键工具，张杰是这一新物理过程最早探索者之一。快速点火方法将燃料点火与压缩分开，使这两个过程可以独立优化，同时避免不稳定性。

三、未来科学大奖数学与计算机科学奖

未来科学大奖数学与计算机科学奖获奖人是台湾“中研院”院士、美国国家工程院院士、中国工程院外籍院士、阳明交通大学电子工程学系暨电子研究所终身讲座教授施敏（图 3）。他为跨金属/半导体（金/半）载流子的传输理论和实践，做出了奠基性和开创性的贡献。本次数学与计算机科学奖所奖励的科学工作是他 1968～1969 年在新竹交通大学（今阳明交通大学）期间完成的。

施敏

图 3　2021 年未来科学大奖数学与计算机科学奖获奖人

施敏教授对于大范围掺杂（10^{14}～$10^{20}/cm^3$）和工作温度（硅：77～373 K；砷化镓：50～500 K）的金/半接触特性，通过跨金/半界面势垒的量子隧道穿越、热电子发射、镜像力降低和二维统计杂质变化的共同效应都做出了分析和实验。这些关于硅和砷化镓半导体的前沿贡献，不仅奠定了欧姆接触和肖特基接触的科学理论基础，并且开启了制造近代半导体器件的可扩展途径。在接下来的 50 年中，半导体器件被广泛用于计算、通信、传感、控制、成像和记忆之芯片电路的制造，对人类生活和文明有巨大贡献。

他还撰写了具有传奇色彩的著作《半导体器件物理学》[3]。这是一本全球半导体和集成电路研究人员“必学”之书，一直被研究生院教师/学生以及整个电子和光子行业的工程师使用和引用。

参考文献

[1] 未来科学大奖．2021未来科学大奖获奖名单公布：袁国勇、裴伟士、张杰、施敏获奖．http://www.futureprize.org/cn/nav/detail/1054.html[2023-01-06].

[2] The Lancet. Emerging stronger from the China crisis. The Lancet, 2003, 361(9366): 1311.

[3] S. M. Sze, Kwok K. Ng. Physics of Semiconductor Devices. New York: Wiley Interscience, 1969.

Introduction to the 2021 Future Science Prize Laureates

Ye Jing

The four laureates of 2021 were announced by the Future Science Prize Committee in Beijing on September 12, 2021. The Life Sciences Prize was awarded to Prof. Kwok-Yung Yuen and Prof. Joseph Sriyal Malik Peiris for their discovery of SARS-COV-1 as the causative agent of the global SARS outbreak in 2003 and its zoonotic origin, with impact on combating COVID-19 and emerging infectious the chain of transmission from animals to humans. The Physical Science Prize is awarded to Prof. Jie Zhang for his development of laser-based fast electron beam technologies and their applications in ultrafast time-resolved electron microscopy and fast ignition for research towards inertial confinement fusion. The Mathematics and Computer Sciences Prize is awarded to Prof. Simon Sze for his contributions to understanding carrier transports at the interface between metal and semiconductor, enabling Ohmic and Schottky-contact formations for scaling integrated circuits at the “Moore’s law” rate during the past five decades.

5.6　2022年诺贝尔自然科学奖简介

贾晓琪　王海霞

（中国科学院科技战略咨询研究院）

一、诺贝尔物理学奖

2022年10月4日，瑞典皇家科学院诺贝尔奖委员会宣布将2022年诺贝尔物理学奖授予阿兰·阿斯佩（Alain Aspect）、约翰·克劳泽（John Clauser）和安东·蔡林格（Anton Zeilinger），以表彰他们在“用纠缠光子进行实验、确立贝尔不等式的违背以及开创量子信息科学”方面所做的贡献[1]。

阿兰·阿斯佩教授1947年6月15日出生于法国阿让的，现就职于法国巴黎-萨克雷大学、巴黎综合理工学院。约翰·克劳泽1942年12月1日出生于美国加利福尼亚州帕萨迪纳市，现就职于美国加利福尼亚州核桃溪市的J. F. Clauser & Assoc. 公司。安东·蔡林格教授1945年5月20日出生于奥地利里德伊姆·因克瑞斯，现就职于奥地利维也纳大学、奥地利科学院量子光学和量子信息研究所。

阿兰·阿斯佩　　约翰·克劳泽　　安东·蔡林格

图1　2022年诺贝尔物理学奖获得者

量子力学最显著的特征之一是允许两个或多个粒子以所谓的纠缠态存在。纠缠对中的一个粒子发生什么决定了另一个粒子会发生什么，即便它们相距甚远。长期以来，人们对这种相关性是否是因为纠缠对中的粒子包含隐藏变量存有疑问。20 世纪 60 年代，英国物理学家约翰·斯图尔特·贝尔（John Stewart Bell）提出了著名的“贝尔不等式”（Bell's inequality）。贝尔不等式指出，如果存在隐藏变量，那么大量测量结果之间的相关性将永远不会超过某个特定值。然而，根据量子力学理论，某种类型的实验将违反贝尔不等式，从而导致更强的相关性。

约翰·克劳泽发展了贝尔的想法，并于 1972 年通过实际的实验进行测量，测量结果明显违反了贝尔不等式，从而支持了量子力学。这意味着，量子力学不能被使用隐变量的理论所取代。约翰·克劳泽的实验仍存在一些漏洞。1981～1982 年，阿兰·阿斯佩开发的实验设置弥补了一个重要漏洞。他在一个纠缠光子对离开光源后切换测量设置，使光子对被发射时的设置不会影响测量结果。1997～1998 年，安东·蔡林格利用精密工具和一系列实验，开始使用纠缠量子态。他的研究团队演示了量子隐形传态，使量子态从一个粒子远距离传递到另一个粒子成为可能。

这些实验证实了量子力学的正确性，并为量子计算、量子网络和量子保密通信铺平了道路。

二、诺贝尔化学奖

2022 年 10 月 5 日，瑞典皇家科学院诺贝尔奖委员会宣布将 2022 年诺贝尔化学奖授予卡罗琳·贝尔托齐（Carolyn R. Bertozzi）、莫滕·梅尔达尔（Morten Meldal）和卡尔·巴里·沙普利斯（K. Barry Sharpless），以表彰他们在“发展点击化学和生物正交化学”方面所做的贡献[2]。

卡罗琳·贝尔托齐教授 1966 年 10 月 10 日出生于美国马萨诸塞州波士顿，现就职于美国斯坦福大学、美国霍华德·休斯医学研究所。莫滕·梅尔达尔教授 1954 年出生于丹麦哥本哈根，现就职于丹麦哥本哈根大学。卡尔·巴里·沙普利斯教授 1941 年 4 月 28 日出生于美国宾夕法尼亚州费城，现就职于美国斯克利普斯研究所，曾获得 2001 年诺贝尔化学奖。

长期以来，化学家们致力于构建结构日益复杂的分子。在药物研究中，人工重建具有药用特性的天然分子导致了许多令人钦佩的分子结构，但这些结构的构建通常既耗时又非常昂贵。2000 年左右，沙普利斯创造了点击化学的概念，这是一种简单可靠的化学形式，反应迅速且避免生成不需要的副产物。

随后，梅尔达尔和沙普利斯各自独立地提出了点击化学皇冠上的明珠——铜催化

卡罗琳·贝尔托齐

莫滕·梅尔达尔

卡尔·巴里·沙普利斯

图2 2022年诺贝尔化学奖获得者

的叠氮化物—炔烃环加成反应，即CuAAC反应。该反应简单高效，现已被用于诸多领域，如药物开发、绘制DNA和开发更适用的材料等。

贝尔托齐教授将点击化学提升到了一个新的水平。为绘制细胞表面重要但难以捉摸的生物分子“聚糖”，她开发了在生物体内起作用的点击反应。这种生物正交反应在不破坏细胞正常化学的情况下发生，已被用于探索细胞和跟踪生物过程，以及提高癌症药物的靶向性，目前正在临床试验中进行测试。

点击化学和生物正交反应已将化学带入功能主义时代。

三、诺贝尔生理学或医学奖

2022年10月3日，瑞典皇家科学院诺贝尔奖委员会宣布将2022年诺贝尔生理学或医学奖授予斯万特·佩博（Svante Pääbo），以表彰他在“已灭绝古人类基因组和人类进化方面的发现”[3]。

斯万特·佩博教授1955年4月20日生于瑞典斯德哥尔摩。现就职于德国马克斯·普朗克演化人类学研究所、日本冲绳科学技术大学院大学。

在职业生涯的早期，斯万特·帕博就对利用现代遗传学方法研究尼安德特人DNA的可能性着迷。作为进化生物学领域的先驱艾伦·威尔逊（Allan Wilson）的博士后，帕博开始开发研究尼安德特人DNA的方法，这一努力持续了几十年。1990年，帕博被慕尼黑大学聘为教授，继续研究古DNA。凭借精湛的技术，他从一块40万年前的骨头中对尼安德特人线粒体DNA区域进行了测序。人类也因此第一次获得来自已灭绝亲戚的DNA序列。1999年，佩博在德国莱比锡建立了马克斯·普朗克演

斯万特·佩博

图 3 2022 年诺贝尔生理学或医学奖获得者

化人类学研究所，其团队稳步改进了从古骨遗骸中分离和分析 DNA 的方法，新技术使 DNA 测序变得非常高效。同时，他还与在群体遗传学和高级序列分析方面具有专业知识的重要合作者。2010 年，佩博发表了第一个尼安德特人基因组序列，比较分析表明尼安德特人和智人最近的共同祖先生活在大约 800 万年前。

佩博及其团队还发现了一种以前不为人知的古人类——丹尼索瓦人。他们发现丹尼索瓦人约在 7 万年前迁出非洲后向智人转移了基因。对于现代人类来说，这种古老的基因流动具有生理相关性，例如影响了我们的免疫系统对感染的反应。

帕博的开创性研究催生了一门全新的科学学科“古基因组学”。在最初的发现之后，他的团队已经完成了对灭绝古人类的几个额外基因组序列的分析。帕博的发现建立了一种独特的资源，科学界可以广泛利用这种资源去更好地了解人类的进化和迁徙。新的强大的序列分析方法表明，古人类也可能与非洲的智人发生基因混合。然而，由于热带气候中古 DNA 的加速降解，尚未对非洲已灭绝的古人类的基因组进行测序。通过揭示区分所有现存人类和已灭绝的人类的遗传差异，帕博的发现为探索是什么使我们成为独特的人类提供了基础。

参考文献

[1] The Nobel Prize. The Nobel Prize in Physics 2022. https://www.nobelprize.org/prizes/physics/2022/summary/[2022-11-20].

[2] The Nobel Prize. The Nobel Prize in Chemistry 2022. https://www.nobelprize.org/prizes/

chemistry/2022/summary/[2022-11-25].
[3] The Nobel Prize. The Nobel Prize in Physiology or Medicine 2022. https://www.nobelprize.org/prizes/medicine/2022/summary/[2022-11-25].

Introduction to the 2022 Nobel Prize in Natural Science

Jia Xiaoqi, Wang Haixia

The Nobel Prize in Physics 2022 was awarded jointly to Alain Aspect, John F. Clauser and Anton Zeilinger for experiments with entangled photons, establishing the violation of Bell inequalities and pioneering quantum information science. The Nobel Prize in Chemistry 2022 was awarded jointly to Carolyn R. Bertozzi, Morten Meldal and K. Barry Sharpless for the development of click chemistry and bioorthogonal chemistry. The Nobel Prize in Physiology or Medicine 2022 was awarded to Svante Pääbo for his discoveries concerning the genomes of extinct hominins and human evolution.

5.7　2022年度菲尔兹奖获奖者简介

赵　晶　聂吉冉

（中国科学院数学与系统科学研究院）

菲尔兹奖（Fields Medal）每四年颁发一次，在国际数学联盟（International Mathematical Union，IMU）主办的国际数学家大会上举行颁奖仪式，授予获得杰出成就并且有潜力的青年数学家。

菲尔兹奖委员会由国际数学联盟执行委员会选出，通常由国际数学联盟主席担任主席。委员会每次授予2～4名（尽量4名）菲尔兹奖获得者，并在选择时考虑到能足够代表数学领域的多样性。获奖者必须在获奖当年元旦前未满40岁。

2022年，菲尔兹奖授予4名科学家，分别是法国数学家米尼尔-柯平（Hugo Duminil-Copin）、韩裔美籍数学家许埈珥（June Huh）、英国数学家詹姆斯·梅纳德（James Maynard）与乌克兰数学家玛丽娜·维亚佐夫斯卡（Maryna Viazovska）。

1. 米尼尔-柯平

法国数学家度米尼尔-柯平，方向是概率论。菲尔兹奖表彰其在统计物理学相变的概率理论，特别是在三维和四维空间中长期存在的问题的解法方面做出的杰出贡献。

米尼尔-柯平改变了统计物理中相变的数学理论并解决了几个长期悬而未决的开放性问题，特别是在三维与四维的情形，以及二维的不可积情形。他的工作开辟了几个新的研究方向。这里，我们只描述他的众多结果中的一部分。

米尼尔-柯平最引人注目的结果是三维和四维中的伊辛（Ising）模型。他与合作者共同建立了三维相变的连续性和锐度，解决了自20世纪80年代开始的问题。在四维空间中，他与艾森曼（Aizenman）一起证明了Ising模型的平均场临界行为和四维欧几里得标量量子场论的平凡性。值得注意的是，后者是20世纪70年代以来物理学的公开猜想。

同样，在二维依赖的Fortuin-Kasteleyn（FK）渗流中，米尼尔-柯平与合作者已经证明了所有参数下转换是连续或不连续性，等边图上临界FK模型的普遍性。此外，通过证明对于大尺度临界FK模型的旋转不变性，他迈出了重要的一步，即建立它们的大规模共形不变性，这反过来又会弥补了严格连接到二维共形场世界的要素理论的缺失要素。

2. 许埈珥

菲尔兹奖表彰其将霍奇理论的思想引入组合学、对几何格上的道林-威尔逊（Dowling-Wilson）猜想和拟阵上的海伦-罗塔-威尔士（Heron-Rota-Welsh）猜想的证明、对洛伦兹多项式理论的发展，以及对强梅森猜想的证明。

许埈珥与他的合作者，运用霍奇理论、热带几何和奇点理论的方法，改变了组合几何学方向。

许埈珥和王博潼（Bottong Wang）使用代数几何和相交理论的工具来证明可实现拟阵的Dowling-Wilson猜想。

阿迪普拉西托（Karim Adiprasito）、许埈珥和卡茨（Eric Katz）发现了霍奇理论的组合类比理论，并证明了对任意拟阵的强莱夫谢茨（Lefschetz）定理和 Hodge-Riemann 关系。他们利用这些结果解决了 Heron-Rota-Welsh 关于拟阵特征多项式的对数凹性猜想。

Petter Brändén 和许埈珥发展了洛伦兹多项式理论，运用热带几何连接了连续和离散凸分析。他们证明了强梅森拟阵猜想，并且在不同数学方向进行应用，包括射影代数几何与统计力学中的波茨模型（Potts models）。

3. 詹姆斯・梅纳德

菲尔兹奖表彰其对解析数论做出了贡献，并极大促进了对素数结构和丢番图逼近的理解。

詹姆斯・梅纳德在解析数论方面做出了巨大贡献。他的工作非常出色，常常在看起来用现有技术不可能解决的重要问题取得惊人的突破。数论中一些最著名的问题与素数的分布相关，而质数的大规模分布由素数定理决定（更准确地说是黎曼假设），很多自然问题处理短（或稀疏）尺度。梅纳德在这个方向取得了许多了不起的成就。例如，即使素数的数列变得越来越稀疏，他依旧证明了对任意长度的 m，都存在无限多的“素数簇”，每个包含在一个有界的区间内（这个区间必然取决于 m）。这是对著名的张益唐研究的一个显著改进，$m=2$，这是张益唐在几个月前取得的最新成果。梅纳德的方法，以一种非常特别的方式突破了筛理论的界限。梅纳德接着证明了有时质数比平均水平更稀疏，这是众所周知的 Erdös 问题，近几十年这个问题没有取得任何实质性进展。

梅纳德也在丢番图近似中做出了重要贡献，他与 Koukoulopoulos 解决了 Duffin-

Schaeffer 猜想。这个在 1941 年提出的猜想可能被认为是 Khintchine 定理的最终推广，描述了有理数对典型实数的逼近程度。

4. 玛丽娜·维亚佐夫斯卡

菲尔兹奖表彰其证明了 E8 格提供了 8 维中相同球体的最密堆积，并进一步推进了傅里叶分析中的相关极值问题和插值问题的研究。

数学中一个由来已久的问题是如何在给定的维度中找到最密集的相同球体堆积方式。人们早已知道，圆的六边形堆积是二维中最密集的堆积；1998 年，匹兹堡大学的托马斯·黑尔斯（Thomas Hales）宣布用计算机辅助证明了开普勒猜想，即三维空间内球体最密堆积方式。直到 2016 年，维亚佐夫斯卡证明了 E8 格给出了 8 维空间的最密堆积。不久后，她又与 Cohn、Kumar、Miller 和 Radchenko 一起证明了 Leech 格给出了 24 维空间的最密堆积。维亚佐夫斯卡的方法建立在科恩和埃尔基斯的研究基础之上，科恩和埃尔基斯使用泊松求和公式给出了任意维度下球形堆积可能密度的上限。他们的研究表明，在 8 维和 24 维中，可能存在一个无穷的“辅助”函数序列，可以用来计算该维数中容许的球体堆积密度的上限。维亚佐夫斯卡根据模形式理论发明了一种全新的方法来生成此类函数。

维亚佐夫斯卡将这些想法发展到了其他方向。她与 Radchenko 一起证明了任何偶数 Schwartz 函数，只要它和它的傅里叶变换，在每个非负整数的平方根处消失，那么它一定是恒为 0 的。事实上，他们证明了对于某些特殊函数 an 和 bn 来说，任何偶数 Schwartz 函数都可以写成 $\sum_{m=0}^{\infty}(a_n(x)f(\sqrt{n})+b_n(x)\hat{f}(\sqrt{n}))$.

她与科恩（Cohn）、库马尔（Kumar）、米勒（Miller）和拉德琴科（Radchenko）一起证明，E8 和 Leech 格不仅给出了维数为 8 和 24 的最优球体堆积，而且它们能使关于距离平方的完全单调函数的每个势函数的能量最小化。

参考文献

[1] https://www.mathunion.org/imu-awards/fields-medal/fields-medals-2022
[2] https://www.mathunion.org/fileadmin/IMU/Prizes/Fields/2022/IMU_Fields22_Duminil-Copin_citation.pdf
[3] https://www.mathunion.org/fileadmin/IMU/Prizes/Fields/2022/IMU_Fields22_Huh_citation.pdf
[4] https://www.mathunion.org/fileadmin/IMU/Prizes/Fields/2022/IMU_Fields22_Maynard_citation.pdf
[5] https://www.mathunion.org/fileadmin/IMU/Prizes/Fields/2022/IMU_Fields22_Viazovska_citation.pdf
[6] https://doi.org/10.4007/annals.2003.157.689

Introduction to the 2022 Fields Medal

Zhao Jing, Nie Jiran

The Fields Medal is awarded to recognize outstanding mathematical achievement for existing work and for the promise of future achievement. In 2022, the Fields Medalwere awarded to four scientists: French mathematician Hugo Duminil-Copin, Korean mathematician June Huh, British mathematician James Maynard and Ukrainian mathematician Maryna Viazovska.

5.8 2022年度沃尔夫数学奖获奖者简介

赵 晶

（中国科学院数学与系统科学研究院）

2022年2月9日，沃尔夫基金会（Wolf Foundation）宣布2022年沃尔夫数学奖得主为麻省理工学院教授乔治·卢斯蒂格（George Lusztig），表彰其对表示论和相关领域的开创性贡献。

乔治·卢斯蒂格

乔治·卢斯蒂格是美籍罗马尼亚数学家，研究几何有限约化群的表示理论和代数群。卢斯蒂格的工作具有高度的独创性、包含宽泛的主题、卓越的精湛技术，并且深入到所涉及问题的核心。他开创性贡献标志着他是我们这个时代最伟大的数学家之一。

他对数学的热情始于小时候。事实上，正是在中小学的数学竞赛让他意识到自己在数学方面的天赋。10年级时，他代表罗马尼亚参加了1962年和1963年的国际数学奥林匹克竞赛：两次都获得银牌。卢斯蒂格于1968年毕业于布加勒斯特大学，并于1971年在阿蒂亚（Michael Atiyah）教授和William Browder教授的指导下获得从普林斯顿大学毕业，获得硕士和博士学位。在1974～1977年在华威大学被聘为教授，之后他于1978年加入麻省理工学院数学系。1999～2009年，他被任命为麻省理工学院诺伯特·维纳（Norbert Wiener）教授。

卢斯蒂格以在表示论方面的工作闻名数学界，特别是与代数群密切相关的研究，如有限约化群、Hecke 代数、P-进群、量子群和 Weyl 群。他基本上为现代表示论方向开创了先河。这包括一些基本的新概念，包括特征标层（character sheaves）、Deligne-Lusztig 簇和 Kazhdan-Lusztig 多项式。

1975 年左右，卢斯蒂格取得了第一个突破，与 Deligne 一同构造了 Deligne-Lusztig 表示。后来，他得到了有限域上约化群的不可约表示的完整描述。卢斯蒂格对有限约化群的特征标表的描述是 20 世纪由单独数学家做出的最非凡成就之一。为了实现他的目标，他发展了一整套至今仍被数百名数学家使用的技术。其亮点包括平展上同调的使用；对偶群的作用；相交上同调的使用，以及随后的特征标层理论，近似特征标（almost characters），以及非交换傅里叶变换。

1979 年，Kazhdan 和卢斯蒂格定义了 Coxeter 群的 Hecke 代数的 Kazhdan-Lusztig 基，并提出了 Kazhdan-Lusztig 猜想。Kazhdan-Lusztig 猜想直接导致了 Beilinson-Bernstein 局部化定理，40 年后，它仍然是我们理解约化李代数表示的最有力工具。Lusztig 与 Vogan 的合作随后引入了 Kazhdan-Lusztig 算法的变体，生成了 Lusztig-Vogan 多项式。这些多项式是我们理解实约化群及其酉表示的基础。

在 20 世纪 90 年代，卢斯蒂格对量子群论做出了开创性的贡献。他的贡献包括引入典范基、引入卢斯蒂格形式（允许单位根，与模表示建立联系）、量子 Frobenius 和小量子群，以及与仿射李代数的表示论的联系。卢斯蒂格的典范基理论（以及 Kashiwara 的晶体基的平行理论）在组合学和表示论方面取得了深刻的成果。最近通过“范畴化”在表示论和低维拓扑方面取得了重大进展，这项工作的根源可以追溯到卢斯蒂格通过箭图模上的偏曲层（perverse sheaves on quiver moduli）对量子群进行的几何分类。

参考文献

[1] https://wolffund.org.il/2022/02/08/george-lusztig/

Introduction to the 2022 Wolf Prize in Mathematics

Zhao Jing

The Wolf Foundation announced that the winner of the 2022 Wolf Prize in Mathematics is Romanian-American mathematician and MIT professor George Lusztig for his seminal contributions to representation theory and related fields.

5.9 2022年度图灵奖获奖者简介

黄 茹 唐 川

（中国科学院成都文献情报中心）

2023年3月22日，国际计算机协会（Association for Computing Machinery, ACM）宣布将2022年度图灵奖（Turing Award）授予鲍勃·梅特卡夫（Bob Metcalfe，图1），以表彰他在计算机和通信领域作出的杰出贡献，尤其是以太网的发明、标准化和商业化[1]。

鲍勃·梅特卡夫

图1 2022年图灵奖获得者

鲍勃·梅特卡夫于1946年出生在美国纽约，拥有麻省理工学院电气工程学士学位、硕士学位和哈佛大学工商管理硕士学位，在美国德克萨斯大学奥斯汀分校工作了11年，是该校电气和计算机工程系名誉教授，同时他还是3Com公司创始人，美国国家工程院院士，美国艺术与科学学院院士，麻省理工学院计算机科学与人工智能实验室计算工程研究成员。他曾被授予美国国家技术奖（the National Medal of Technology），IEEE荣誉勋章奖（IEEE Medal of Honor），马可尼奖（Marconi Prize），日本计算机与通信奖（Japan Computer & Communications Prize），IEEE亚历山大·格雷厄姆·贝尔奖章（IEEE Alexander Graham Bell Medal）等多项荣誉[2]。

以太网是一种计算机局域网技术，是目前应用最普遍的局域网技术。鲍勃·梅特卡夫被誉为“以太网之父”，1973 年，他在担任施乐帕洛阿尔托研究中心（PARC）计算机科学家时撰写了一份著名的备忘录，备忘录中描述了一个“广播通信网络”（broadcast communication network），用于在一栋建筑内连接个人计算机，第一个以太网雏形由此诞生。梅特卡夫随后邀请以太网的共同发明者大卫·博格斯（David Boggs）合作，建立了一个具有 100 个节点的 PARC 以太网。在此基础上，梅特卡夫领导团队开发出了一个10 Mbps 的以太网，为后续标准奠定了基础。1980 年，他联合数字设备公司（Digital Equipment Corporation）、英特尔（Intel）和施乐公司（Xerox）创建了10 Mbps 以太网标准——即 DIX 标准，同时还创建了 IEEE 802.3 标准，至今仍被广泛采用，从此，以太网成为一种开放标准。1979 年，梅特卡夫创建 3Com 公司，通过销售网络软件、以太网收发器以及小型计算机和工作站的以太网卡，来增强以太网的商业价值。当 IBM 公司推出其个人电脑时，3Com 公司为 IBM 个人电脑及相关产品提供了首批以太网接口。如今，以太网不仅是全球有线网络通信的主要渠道，还拥有一个巨大的市场[3]。

从 1980 年至 1990 年，梅特卡夫帮助普及了网络价值随着用户数量的增加而快速增长的观点，这一定律现在被称为“梅特卡夫定律”（Metcalfe's Law），对于理解网络效应和互联网经济发展有重要意义[4]

ACM 主席 Yannis Ioannidis 称赞梅特卡夫的发明极具影响力：以太网一直是计算机互联、与其他设备以及与互联网连接的主要方式，即使有 Wi-Fi 的出现，它仍是数据通信的主要模式，特别是在优先考虑安全性和可靠性的情况下。谷歌高级研究员、谷歌研究兼人工智能高级副总裁杰夫·迪恩（Jeff Dean）表示，以太网是互联网的基础技术，它支持 50 多亿用户，今天，全球估计有 70 亿个以太网端口。梅特卡夫的发明和他关于每台计算机都需要联网的理念促使了互联世界的高速发展[1]。

参考文献

[1] ACM. ACM Turing Award Honors Bob Metcalfe for Ethernet. https://amturing.acm.org/ [2023-08-27].

[2] Robert Metcalfe. https://www.aminer.cn/profile/robert-m-metcalfe/53f373cfdabfae4b349cac0a [2023-08-27].

[3] ACM. ACM A. M. Turing Award Honors Bob Metcalfe for Invention, Standardization, and Commercialization of Ethernet. https://awards.acm.org/about/2022-turing [2023-08-28].

[4] Ben Brubaker. Bob Metcalfe, Ethernet Pioneer, Wins Turing Award. https://www.quantamagazine.org/bob-metcalfe-ethernet-pioneer-wins-turing-award-20230322/ [2023-08-28].

Introduction to the 2022 Turing Award Winners

Huang Ru , Tang Chuan

ACM, the Association for Computing Machinery, has awarded the 2022 ACM A. M. Turing Award to Bob Metcalfe, who is Emeritus Professor of Electrical and Computer Engineering after 11 years at The University of Texas at Austin. He has recently become a researcher at the Massachusetts Institute of Technology Computer Science & Artificial Intelligence Laboratory. He received the award for his outstanding contributions in the fields of computer and communication, particularly the invention, standardization, and commercialization of Ethernet.

5.10　2022 年度泰勒环境成就奖获奖者简介

廖　琴

（中国科学院西北生态环境资源研究院）

2022 年度的泰勒环境成就奖（Tyler Prize for Environmental Achievement）授予安迪·海恩斯（Andy Haines）（图 1），以表彰其在理解气候变化对公共卫生的影响方面的贡献、在将公共卫生范围扩大到行星健康（Planetary Health）方面的领导作用，以及在 21 世纪预防性环境健康行动方面对下一代健康科学家和从业者的指导[1]。

安迪·海恩斯

图 1　2022 年泰勒环境成就奖获得者

海恩斯是英国伦敦卫生与热带医学学院气候变化与地球健康中心的环境变化与公共卫生教授。2001 年至 2010 年间，海恩斯担任伦敦卫生与热带医学学院院长。2005 年，海恩斯因对医学的贡献而被封为爵士。海恩斯是世界卫生组织卫生研究咨询委员会的成员，也是联合国政府间气候变化专门委员会（IPCC）第二次、第三次和第五次评估工作组成员。

海恩斯一直致力于理解和努力防止环境变化，特别是气候变化对人类健康的影响。20 世纪 90 年代，海恩斯与著名的流行病学家托尼·麦克迈克尔（Tony McMichael）教授共同撰写了关于气候变化和人类健康的重要早期评估[2]。自此，其工作重点是环境对健康的影响，包括气候变化的影响，以及低碳政策的健康协同效益。

海恩斯不仅是最早就气候变化对人类健康构成的危险发出警告的科学家之一，也是最早研究减少温室气体排放行动对健康益处的科学家之一。2008～2009 年，海恩斯主持了开创性的柳叶刀气候变化减缓和公共卫生工作组，评估了能源、交通、住宅、食品和农业方面的减缓战略。该工作组的报告标志着政策制定的关键转折点，使制定更具成本效益的政策解决方案成为可能。海恩斯研究了减缓气候变化行动的健康共同效益，例如步行、骑行和使用公共交通工具，转向更多的植物性饮食，以及转向清洁的可再生能源，这对重新构建气候话语，使其朝着积极和具有社会吸引力的叙事方向发展产生了深远的影响。2014—2015 年，海恩斯担任洛克菲勒/柳叶刀行星健康委员会主席，该委员会报告指出，人们已经抵押了子孙后代的健康，以实现当前的经济和发展收益，并提出了稳定地球关键生命支持系统的政策建议。

2019 年，海恩斯在《新英格兰医学杂志》（*New England Journal of Medicine*）上与人合著了《采取气候行动保护健康的必要性》[3]一文，详细说明了如果未来几十年不采取更多行动应对气候变化，预计“发病率和死亡率将大幅增加”，包括与高温、空气质量差、粮食不安全以及媒介传播疾病有关的疾病。2021 年，海恩斯与霍华德·弗鲁姆金（Howard Frumkin）合著了《行星健康：在人类世保护人类健康和环境》[4]一书。通过解决人类面临的环境挑战，海恩斯努力影响着政策和实践。

参考文献

[1] Tyler Prize. 2022 Laureate. https://tylerprize. org/laureates/ [2021-12-9]

[2] McMichael, Anthony J, Haines J A, et al. Climate change and human health : an assessment/prepared by a Task Group on behalf of the World Health Organization, the World Meteorological Association and the United Nations Environment Programme. World Health Organization. 1996.

[3] Haines A, Ebi K. The Imperative for Climate Action to Protect Health. N Engl J Med. 2019, 380 (3): 263-273.

[4] Andy Haines, Howard Frumkin. Planetary Health: Safeguarding Human Health and the Environment in the Anthropocene. Cambridge University Press. 2021.

Introduction to the 2022 Tyler Laureate

Liao Qin

The 2022 Tyler Prize for Environmental Achievement was awarded to Sir Andy Haines for his contributions in understanding the effects of climate change on public health, his leadership in expanding the scope of public health to one of Planetary Health, and for his mentorship of the next generation of health scientists and practitioners in preventive environmental health actions in the 21st century.

5.11 2022年未来科学大奖获奖者简介

贾晓琪

（中国科学院科技战略咨询研究院）

2022年8月21日，未来科学大奖科学委员会公布2022年获奖名单[1]。李文辉因其发现了乙型和丁型肝炎病毒感染人的受体为钠离子-牛磺胆酸共转运蛋白（NTCP），有助于开发更有效的治疗乙型和丁型肝炎的药物的成就，获得“生命科学奖”；杨学明因其研发新一代高分辨率和高灵敏度量子态分辨的交叉分子束科学仪器，揭示了化学反应中的量子共振现象和几何相位效应的成就，获得“物质科学奖”；莫毅明因其创立了极小有理切线簇（VMRT）理论并用以解决代数几何领域的一系列猜想，以及对志村簇上的Ax-Schanuel猜想的证明，获得“数学与计算机科学奖”。

一、未来科学大奖生命科学奖

未来科学大奖生命科学奖获奖人北京生命科学研究所资深研究员、清华大学生物医学交叉研究院教授李文辉（图1）发现了乙型和丁型肝炎病毒感染人的受体为钠离子-牛磺胆酸共转运蛋白（NTCP），有助于开发更有效的治疗乙型和丁型肝炎的药物。

李文辉

图1　2022年度未来科学大奖生命科学奖获奖人

乙型肝炎是人类健康的大敌，目前全球仍有超过两亿五千万人被乙型肝炎病毒感染，感染者会有高风险发展为肝硬化和肝癌。李文辉带领其实验室于 2012 年发现乙型和丁型肝炎病毒感染人的受体为钠离子-牛磺胆酸共转运蛋白（NTCP）。这一发现是乙肝病毒研究领域 30 年来里程碑式的突破，揭示了乙型和丁型肝炎病毒感染的分子机理，有助于开发更有效的治疗乙型和丁型肝炎的药物。

二、未来科学大奖物质科学奖

未来科学大奖物质科学奖获奖人南方科技大学教授、中国科学院大连化学物理研究所研究员杨学明（图 2）研发新一代高分辨率和高灵敏度量子态分辨的交叉分子束科学仪器，揭示了化学反应中的量子共振现象和几何相位效应。

杨学明

图 2 2022 年度未来科学大奖物质科学奖获奖人

自从 Eyring 和 Polanyi 在 20 世纪 30 年代提出化学反应过渡态理论以来，化学动力学研究取得了多个里程碑式的进展，并多次获得诺贝尔化学奖。

杨学明开发了新一代高分辨率和高灵敏度的交叉分子束科学仪器，在基元化学反应动力学研究领域，尤其是化学反应共振态、化学反应中的几何相位效应以及量子干涉等方面的研究取得了重大突破。他发展了量子态分辨的后向散射谱学技术，通过高分辨的散射实验与精确理论研究相结合，揭示了多类化学反应共振现象，大力推动了在量子水平上化学反应过渡态的研究。此外，他还发展了高分辨的交叉分子束反应成像技术，首次在实验上发现了化学反应中的几何相位效应以及自旋-轨道共振分波之间的量子干涉现象。

杨学明的科学研究和他研发的新一代分子束科学仪器为反应动力学领域进一步理解化学反应的量子特性提供了强有力的工具，他的新发现将化学动力学领域拓展到了前所未有的深度和广度。

三、未来科学大奖数学与计算机科学奖

未来科学大奖数学与计算机科学奖获奖人香港大学 Edmund and Peggy Tse 讲席教授莫毅明（图 3）创立了极小有理切线簇（VMRT）理论并用以解决代数几何领域的一系列猜想，以及对志村簇上的 Ax-Schanuel 猜想的证明。

莫毅明

图 3　2022 年度未来科学大奖数学与计算机科学奖获奖人

复几何是现代数学的一个核心研究方向，在理论物理和数学的其他分支都有重要作用。在与不同合作者的工作里，莫毅明在复几何及其应用有两项基本贡献。其一是他与 Jun-Muk Hwang 一起创造了代数几何领域中的极小有理切线簇（VMRT）。这个理论是基于他早期复几何的工作上发展起来的，并被应用于证明紧不可约厄米特对称空间（compact Hermitian symmetric spaces）在凯勒（Kähler）形变下的刚性，以及 Lazarsfeld 关于有理齐次空间上解析影射的一个猜想。

其二是他与 Jonathan Pila 和 Jacob Tsimerman 合作，证明了志村簇上的 Ax-Schanuel 猜想。经典的 Schanuel 猜想是数论中的主要猜想之一，志村簇上的 Ax-Schanuel 猜想是 Schanuel 猜想在双曲几何中的重要变种。莫毅明与合作者的定理已成为算术几

何中的重要工具。

参考文献

[1] 未来科学大奖 .2022 未来科学大奖获奖名单公布：李文辉、杨学明、莫毅明获奖 . http://www. futureprize. org/cn/nav/detail/1208. html[2023-08-28]

Introduction to the 2022 Future Science Prize Laureates

Jia Xiaoqi

The three laureates of 2022 were announced by the Future Science PrizeCommittee on August 21, 2022. The Life Science Prize is awarded to Prof. Wenhui Li for his discovery of the Hepatitis B and D virus receptor, sodium taurocholate cotransporting polypeptide (NTCP). This discovery facilitates the development of more effective treatment for hepatitis B and D. The Physical Science Prize is awarded to Prof. Xueming Yang for developing new-generation molecular beam techniques with high resolution and sensitivity for state-resolved reaction dynamics studies, revealing quantum resonances and geometric phase effects in chemical reactions. The Mathematics and Computer Science Prize is awarded to Prof. Ngaiming Mok for developing the theory of Varieties of Minimal Rational Tangents in algebraic geometry to solve several long-standing problems and proving Ax-Schanuel's conjecture for Shimura varieties.

第六章

中国科学发展建议

Suggestions on Science Development in China

6.1 发展自主、可持续的基础软件技术与产业的建议

（中国科学院学部咨询课题组①）

计算产业是数字经济的基石。核心芯片和基础软件是计算产业的基础。长期以来，中国信息产业“缺芯少魂”。在美国对我国进行精准施压的情况下，我国不仅要发展自主的核心芯片，还需要基于自主的基础软件，构建可持续的软硬件生态。

一、我国基础软件市场规模小且被国外厂商主导，国内产业发展散、小、弱

基础软件包括操作系统、数据库、中间件、编译器，以及基础工具库、算法库等通用目的软件，向下管理和操作多样化的计算硬件，向上支撑应用、促进产业生态繁荣发展。

目前国内市场上基础软件主要依赖美国厂商产品，我国信息产业发展、信息生态建设等受制于人的情况非常严峻。

从市场规模来看，我国基础软件市场规模较小，市场发育不良。根据 IDC 发布的中国服务器市场季度跟踪报告数据显示，2021 年我国服务器市场规模占全球 25%，网络设备占 14%，存储设备占 13%。但我国基础软件市场规模仅占全球 4.7%，操作系统仅占 0.8%，数据库占 6.9%，虚拟化仅占 0.5%，大数据仅占 0.5%，这与我国在全球信息技术产业中的市场地位极不相称。

从市场份额来看，主要基础软件市场都被国外厂商主导。电脑与手机操作系统几乎完全依赖国外厂商，美国微软、谷歌和苹果的 Windows、Android 和 iOS 已在全球电脑与智能手机操作系统领域占据垄断地位，牢牢把持着整个信息技术产业链的顶端。我国电脑终端所使用的操作系统中（2018 年 4 月至 2019 年 4 月）Windows 占比

---

① 咨询课题组组长为中国科学院院士、中国科学技术大学教授陈国良。

高达87.44%[1]，国内主流手机生产商均是在谷歌公司授权的Android系统上开发各自的操作系统。我国服务器虚拟化、数据库等重要基础软件同样大幅落后国外厂商。美国威睿（VMware）公司在我国服务器虚拟化市场中占据44.7%的份额，高居第一；我国厂商新华三公司和华为公司分别占16.2%和15.1%[2]。美国甲骨文（Oracle）公司在我国数据库市场的占有率超过50%[3]，而我国数据库企业的品牌和产品技术与之相比都存在显著差距。

国内基础软件产业分散，企业规模较小，生态能力不足。经过多年发展，国内已经出现了不少基础软件企业。据不完全统计，操作系统厂商超过20家，数据库厂商超过100家，但整体产业过于分散，营收规模都很小。据2019年工业和信息化部发布的《2019年中国软件业务收入前百家企业发展报告》，软件收入百强企业中没有一家基础软件企业。相比之下，全球软件企业前十位包括了微软、甲骨文、威睿等多家基础软件巨头。国内基础软件企业生态能力不足，难以承担产业链上下游的生态适配成本，产业很难聚力发展。过去我国基础软件产业主要面向桌面系统和办公需求，从未来发展角度看，要兼顾服务器、移动终端、可穿戴设备、工业互联网等多种应用场景需求，引入各行业用户和产业链各方，丰富产业生态。

二、全球基础软件技术创新和商业模式创新快速变革，带来新的竞争压力和挑战

从技术上看，开源数据库开发和开源社区正在成为基础软件技术创新和供给的源头。大量的基础软件厂商在大力投入开源社区，并基于开源软件开发商业版本。根据市场研究机构Strategy Analytics发布的数据，在智能终端操作系统方面，开源的安卓系统市场份额已超过85%；在桌面操作系统方面，根据Market Research Future的研究报告，预计到2023年，基于开源社区的Linux操作系统，市场规模将达到70.74亿美元，是2017年的2.6倍；在数据库方面，根据Gartner的预测，到2022年，超过70%的企业应用将基于开源数据库开发。

从商业模式上看，基础软件正在快速从卖产品向卖服务转变。基础软件的商业模式经历了从“许可证”到“许可证＋支持服务”，再到“云订阅服务”的发展历程。当前国际主要基础软件巨头厂商微软、甲骨文等，都把软件的订阅服务作为基础商业模式，并快速向云服务转变。对于用户来说，软件逐步从之前的固定资产投入（Capital Expenditure，Capex，资本性支出）转变为运营投入（Operating Expense，Opex，营运资本）。商业模式的变化既反映了行业对软件本身“知识价值”和“服务价值”的不断认识和强化，也使软件企业能够形成更加良性的盈利模式，进一步促进

其对研发和生态的持续投入。

以上两个方面快速变革的趋势，给国内软件产业发展带来新的竞争压力和挑战。

一是缺少自主的基础软件“根技术”开源社区，难以保证软件技术独立演进和持续发展。国内基础软件产品主要基于国外的开源社区版本进行二次开发，对国外开源社区存在较强的路径依赖。国内桌面操作系统厂商 90% 以上基于 CentOS、Ubuntu、Fedora 等国外开源社区软件开发；移动操作系统 98.5% 基于开源安卓系统；国产数据库基本上全部基于 MySQL、PostgreSQL 等国外开源社区开发。一旦上游开源社区被限制，国内绝大部分基础软件企业将面临技术断供风险。同时国内的代码托管平台、开源基金会等开源产业基础设施和环境还处在发展初期，与国外较为成熟的开源产业环境差距较大。

二是财政制度限制软件商业模式升级，不利于基础软件产业可持续发展。基于《财政部关于进一步规范和加强政府机关软件资产管理的意见》《政府采购品目分类目录》等文件，国内在财务制度上仍将软件视为与硬件相同的固定资产进行管理，忽视了软件中蕴含的持续改进、持续优化的“知识价值”和“服务价值”。软件的采购和支出仍以购买许可证为主，不利于基础软件企业形成良好的可持续盈利的商业模式，也就无法形成自我造血能力。

三、发展自主可控国内基础软件产业的若干建议

结合全球基础软件发展趋势，针对国内基础软件产业发展的现状和问题，提出以下建议。

一是在国家层面设立基础软件产业发展基金，支持龙头企业尽快发展壮大，形成自主可控产业生态。从国内集成电路产业发展的经验来看，国家集成电路产业投资基金发挥了很大的促进作用。当前国内基础软件产业散、小、弱，急需从国家层面设立产业发展基金，从专业角度重点支持龙头企业，提高基础软件产业的集聚化程度，提升自主可控生态建设能力。支持基于自主基础软件开发应用软件，提升国产软硬件兼容性与应用适配能力，促进国产基础软件做强做大。以计算产业场景化示范应用带动基础软件市场向国内基础软件倾斜、迁移，形成软硬件协同的自主生态体系以及提高技术迭代能力。

二是大力支持国内基础软件形成自主的“根技术”开源社区，形成自主可控的软件技术供给能力。自主开源社区要基于国内开源组织平台（开源基金会等）来构建。支持自主开源协议，通过自主技术标准等方式，牵引软硬件厂商在自主开源社区中快速形成合力和技术竞争力。在政策层面，适当放松对开源基金会、开源社区组织在成

立、运作方面的管理，促进自主开源社区、代码托管平台的发展。企业、高校等应大力鼓励开源文化，投入自主开源社区的建设和技术开发，国家科技计划支持的基础研究可以鼓励成果进行开源。

三是优化调整国内关于软件资产的财务政策，支持国内软件行业商业模式向“卖服务”优化。建议将购买软件拷贝/许可证的一次投资模式转变为获得软件服务的持续投入模式，将软件服务、维护等支出纳入运营投入。通过商业模式向“卖服务”优化，提升基础软件企业自我造血能力，提高软件在资产配置中的占比，建立国内基础软件商业正循环。加强软件知识产权保护，激励企业持续投入软件技术创新。

四是加强基础软件技术研发，加大面向自主基础软件的人才培养。在高校、研究机构中重点支持基础软件与硬件架构相结合的创新性技术研发，提出新体系，形成新赛道。在国内高等教育、职业培训、职称评定等环节加大面向自主基础软件的课程设置和方向引导，培养自主基础软件创新人才。

参考文献

[1] Netmarketshare. Operating System Market Share. https://netmarketshare.com/operating-system-market-share.aspx[2019-05-25].

[2] 计世资讯. 2016—2017年中国服务器虚拟化市场发展状况与趋势研究报告. https://www.csdn.net/article/a/2017-07-03/15929602[2018-08-01].

[3] IT168. 2015中国Oracle市场占有率达56%. http://www.sohu.com/a/74946310_374240[2018-08-01].

Developing Independent and Sustainable Basic Software Technology and Industry

Consultative Group of CAS Academic Division

This paper analyzes the current development of the basic software technology and industry in China, pointing out the small scale of China's basic software market, the market dominance of foreign firms and the scattered, small-scaled and weak structure of the domestic industry. Meanwhile, the following four measures are propesed: 1) to set up a national basic software industry development fund to support industry-leading enterprises to grow and expand their businesses as soon as possible, and thus an independent and controllable industrial ecology will come into being; 2) to provide strong support for the independent "root tech" open source community and the development of the supply capacity of independent and controllable software technology in the domestic basic software industry; 3) to optimize and adjust the domestic financial policy on software assets and support optimization of the business model of the domestic software industry to "selling service"; and 4) to strengthen the research and development of basic software technology and enhance the talent cultivation oriented to independent basic software industry.

6.2 关于协同水土资源与环境治理保障国家粮食安全的对策建议

中国科学院学部咨询课题组①

党的十九届五中全会要求坚持用系统观念开启更高质量、更有效率、更加公平、更可持续、更为安全的发展模式。如何在保障粮食安全下维持资源储备和消除环境污染是我国可持续发展面临的重大科学问题，需要切实破除“各管一摊、相互掣肘”的机制障碍，“全方位、全地域、全过程”开展生态文明建设[1]。

建立部门间和区域间内在统一、相互协调的科学目标是实现系统治理的基础。以全国粮食、水资源和水环境关系为例，《自然》发表的一项研究[2]显示：全国主要代表性水体的总氮（TN）浓度在1985年前就达到或超过了污染水平；若要在保持粮食不减产的条件下恢复水质环境，需要把当前的农田氮素利用效率提高近1倍，城乡全部有机肥还田率从当前的40%以下提高到86%以上，化肥氮总需求量减少55%～65%，才能将我国生态环境系统的氮输入总量控制在安全水平。而此前的“化肥施用零增长”“污水处理全覆盖”等行业政策无法实现这一科学管理目标[3]。在第二个百年奋斗目标的开局时期，亟须尽快调整发展模式和管理机制，将粮食安全、水土资源与污染防控纳入统一科学框架协同治理，实现“粮食安全”与“美丽中国”兼顾的远景目标。

一、当前水土资源、环境治理和国家粮食安全亟待解决的现实问题

尽管中央各部委、各地方政府都在各自的职能范围内为解决水安全和粮食安全问题开展了持续不懈的工作，但是仅从局部着眼难以在整体上理顺耕地资源、水资源、水环境和粮食安全的关系，缺乏系统的可持续协调发展模式。当前面临的亟待解决的

① 咨询课题组组长为中国科学院院士、武汉大学教授夏军。

现实问题主要有以下几个方面。

1. 粮食生产重心北移导致北方水资源短缺形势进一步加剧，亟待顶层协调

40 多年来，我国北方地区的灌溉面积持续增加，据统计[4]，2011 年的灌溉面积较 1980 年的灌溉面积扩大了 57%。在 2011 年中央水利工作会议后，2018 年的灌溉面积较 2011 年的灌溉面积进一步扩大了 11%。北方与南方粮食产量比值从 1990 年的 1.0，增加到 2011 年的 1.6 和 2018 年的 1.8[3]，粮食生产重心的持续北移加剧了我国农业与水资源环境的矛盾[5]。在 2012 年国家开始实行“最严格的水资源管理制度”后，用水效率得到明显提高，但并未根本改变南北粮食生产与资源不匹配、北方水资源难以可持续利用的宏观格局。

农业用水占北方地区供水量的 70%～75%[6]，是北方地区地下水储量快速消耗和水体自净功能减弱的主导因素。观测数据显示，北方不少地区的地下水埋深超过 50 m，如河北衡水一带的地下水埋深已经超过 100 m[7]。地下水的储量一旦枯竭，极端旱灾条件下北方粮食主产区产量可减产过半，受灾区供水量减少 2/3[8]，不仅会动摇我国现有的粮食安全根基，而且会危及社会稳定。

2. 化肥施用量过高、水体富营养化趋重和地下水污染加剧，亟待综合治理

已有研究表明，化肥的大量施用使得有机肥的还田率从 1980 年的 90% 左右下降到当前的 40% 以下[2]。我国由人为原因造成的氮排入淡水的速度为 1450 万 t/a，约为安全排放阈值估值（520 万 t/a）的 2.7 倍。全国污水处理系统的除氮量仅为全国总氮超排量的 16%[2]，污水处理厂排水口的平均总氮浓度[9,10]是劣五（V）类水质下限标准（2.0 mg/L）[11]的 7～10 倍。在除氮能力有限的情况下，环保政策会驱动污染物转移至监测缺失或难以监管的环境中去。尤其是，北方平原区的农田淋溶、养殖粪污和生活污水大多排放到难以监管的地下水环境。水利部 2016 年的监测数据表明，80.3% 的浅层地下水已达到或超过了Ⅳ类水质标准[12]。例如，2020 年 10 月央广网报道山东平邑暴发地下水污染危机[13]，急需采取系统化有效措施切断各种地下水污染链条。

3. 农业、水利和生态环境等部门政策协同性不足，亟待统筹协调

部门管理的不同目标会导致行业政策在地方执行过程中难以协调。例如，近 40 年，农业部门、卫生与健康部门在加快发展养殖业和推进饮食结构合理化方面做了大量工作，但客观上也产生了严重的养殖污染问题。环保部门为减少养殖污染影响，在畜禽养殖禁养区开展了大范围“禁养”执法，关闭了大量的养殖场，这被认为是自

2019年猪肉价格暴涨的重要因素之一。虽然国务院此后要求各地清理不合理“禁养”规定，但并没有从根本上解决农业生产和环境保护的矛盾。这凸显出中央各部委间的政策制定未能及时协调一致、政策出台前缺乏跨部门的系统性科学论证，给国家政策的科学性、连续性和权威性带来了一定的负面影响。

二、对策与建议

中国千百年来的传统循环农业曾被世界视为无污染、可持续农业发展的典范。而近40年来的农业生产模式和环境变化趋势表明，简单照搬西方的技术和标准体系，无法在中国这样的人口和资源压力下同时解决粮食安全和环境安全问题。有关氮污染、水资源和粮食安全方面的研究成果[2]表明，当前的科技水平已具备了打通气象、水利、农业、自然资源、生态环境、人口、城乡建设的跨行业、跨学科量化研究能力，可为我国根据自身条件探索符合国情的可持续发展道路提供科学支撑。为此，提出以下对策建议。

1. 建立省级农业生产责任制，彻底缓解北方地区水资源短缺的瓶颈

根据气候、水资源、耕地等生产条件评估各省级区域农业生产潜力和环境安全阈值，量化分配各省的农业生产任务[3]，确保全国粮食产量整体稳中有升。适当提升水资源相对丰富、经济相对发达地区的农业生产任务，倒逼这些地区积极改革农业管理模式、发展农业生产新技术，提高水资源利用效率、降低生产成本、减小耕地撂荒比例。加大南水北调中线和东线给地下水超采地区的调水量及节水灌溉和人畜饮水配套设施建设，加快南水北调西线建设，开发干旱半干旱地区农业生产潜力。探索海绵城市与洪水资源化利用模式，逐步恢复北方地区地下水储量和地表径流，增强水体自净能力，降低旱灾风险。继续推进节水措施，适量减少水资源匮乏和黑土地区的农业生产任务，增加生态用水供给，彻底解决粮食重心北移、北方地下水过度消耗等问题。

2. 全面重构城乡养分还田体系，发展水肥高效利用模式，消除环境富营养问题

改变当前以“污染治理”为主的环保方式，回归传统“水肥归田”的可持续生产模式。建立城乡建设与农业生产的联系，将城乡生活污水排放管网及处理系统与农田灌溉相连、与工业废水分流，在消除土壤污染和病毒风险后实现水肥还田[2]。减轻养殖业主的经济负担，由社会共同承担粪肥还田成本。建立养殖粪污监管体系，从土地配套、设施建设、利益分配、施肥技术、职业化服务、强化监管等方面进行系统化管

理，杜绝粪污直接排放。在农村家庭联产承包责任制基础上，建立农村规模化、岗位化服务体系[3]，为提高水肥利用效率、促进城乡有机肥循环利用、提升耕地有机碳储量、恢复撂荒地耕种，提供专业化服务。为增加农村就业、巩固脱贫攻坚成果建立长效机制。

3. 建立部委间统一的协调协商机制，统一编制国家环境安全与粮食安全保障规划

建议尽快成立跨行业、跨学科的科学技术指导委员会和工作组，在统一科学框架下制定能够促进各部委、各区域协同行动的国家环境安全与粮食安全保障规划，明确量化各部门、各地区管理目标和行动方案。针对当前行业管理条块分割的问题，建立部委间日常业务的协调机制，从任务、经费、考核等各个环节落实部门间、区域间的协同行动。重新核算和统筹分散在农业、水利、城乡建设、自然资源、生态环境、化工、食品安全等行业的资金，将水土环境保护与修复成本纳入经济核算体系，让农民成为农业生产、环境保护和生态恢复的主要力量[3]。

在坚持"高质量发展"指导思想的引领下，中国有能力从整体上解决耕地质量、水土资源、水土环境和粮食安全等重大综合问题，可为人类解决 21 世纪面临的全球人口、粮食和环境可持续发展难题树立典范。

参考文献

[1] 习近平．推动我国生态文明建设迈上新台阶．求是，2019.

[2] Yu C，Huang X，Chen H，et al. Managing nitrogen to restore water quality in China. Nature，2019，567(7749)：516-520.

[3] 喻朝庆．水-氮耦合机制下的中国粮食与环境安全. 中国科学：地球科学，2019，49(12)：2018-2036.

[4] 国家统计局．中国农村统计年鉴 1980—2022. 北京：中国统计出版社，2022.

[5] Yu C. China's water crisis needs more than words. Nature，2011，470(7334)：307.

[6] 中华人民共和国水利部．中国水资源公报 1997—2021. http：//www. mwr. gov. cn/sj/tjgb/szygb/，2022.

[7] 水利部．地下水动态月报 2010-2022. http：//xxzx. mwr. gov. cn/xxgk/gbjb/dxsdtyb/.

[8] Yu C，Huang X，Chen H，et al. Assessing the impacts of extreme agricultural droughts in china under climate and socioeconomic changes. Earth's Future，2018：10-1002.

[9] 宋连朋，魏连雨，赵乐军，等．我国城镇污水处理厂建设运行现状及存在问题分析．给水排水，2013，3：39-44.

[10] 赵银慧，李莉娜，景立新，等．污水处理厂氮排放特征．中国环境监测，2015，4：58-61.

[11] 国家环境保护总局，国家质量监督检验检疫总局. 中华人民共和国地表水环境质量标准. 北京：中

国环境科学出版社,2002.
[12] 中华人民共和国水利部. 地下水动态月报(2016年1月). http://xxzx.mwr.gov.cn/xxgk/gbjb/dxsdtyb/201711/t20171122_1014950.html.
[13] 山东广播电视台. 山东平邑:多处水井变质发臭,当地政府久拖不决. 央广网,2020:20201020.

Countermeasures and Suggestions on Safeguarding National Food Security by Coordinating Water and Land Resources and Environment Governance

Consultative Group of CAS Academic Division

This paper points out that at present a realistic problem needs to be urgently solved regarding water and land resources, environmental governance and food security, which is the lack of systematic and sustainable coordinated development mode. Meanwhile, the following suggestions are proposed: 1) to build the provincial-level agricultural production responsibility system to completely alleviate the bottleneck of water resource shortage in the northern of China; 2) to completely reconstruct the urban-rural nutrient return system, develop the effective use mode of water manure, and solve the problem of entrophication; 3) to form an inter-ministerial unified coordination and negotiation mechanism, and realize the realignment of programming support plans for the country's environmental safety and food security.

6.3 全球化新格局下全方位吸引国际一流科技人才的政策建议

（中国科学院学部咨询课题组[①]）

全方位培养、引进和用好国际一流的科技人才和创新团队，是贯彻落实好党中央决策要求、持续推动高质量发展的关键。2018 年以来，受中美贸易战、全球新冠疫情、新一轮科技革命等因素影响，全球化格局处在重构的关键时期，全球科技人才流动呈现新态势、新特点，亟须重新思考我国吸引国际一流科技人才面临的新机遇、新挑战，提出全方位吸引国际一流科技人才的新思路、新对策。

一、全球化新格局下我国引进全球科技人才面临新机遇、新挑战

科技人才的全球流动不是单个人和单个国家可以完全决定的，还受国际竞争合作关系、全球化格局的影响。

当前，三大因素推动全球化格局重构。一是美国试图构建“去中国化”的全球化新格局。2018 年以来，中美贸易战愈演愈烈，中美科技脱钩、人才脱钩甚至走向“新冷战”的风险加剧。民主党拜登政府上台后，这种状况并未有所缓解，美国已经形成的遏制中国发展、消解中国全球影响力、建立“去中国化”全球化新格局的跨党派共识没有改变。二是全球新冠疫情促进全球化格局重构。全球新冠疫情让美国“去中国化”的布局受挫、步伐变慢，我国在疫情防控中的出色表现用事实回应了美国等国“污名化”和孤立中国的错误行径，为我国构建和参与新型的、区域化的全球化格局提供了空间。三是新科技革命成为塑造全球化新格局的关键变量和重要的不确定因素。当前，新一轮科技革命和产业变革方兴未艾，包括我国在内的世界主要国家站在相近的起跑线上，增加了未来全球化格局多极化的可能。

① 咨询课题组组长为中国科学院院士、复旦大学教授杨玉良。

多样化、区域化的全球化新格局，使我国吸引全球科技人才面临新的挑战和机遇。一方面，美国阻挠中美科技合作、人才交流的政策取向不会有较大变化，在关键科技领域严控交流合作的做法会长期延续，但在应对气候变化等领域仍然存在较多的合作空间。另一方面，美国等少数国家的排外情绪、歧视行为，提高了华裔科技人才回国的倾向和意愿，助推了海外人才“回流”。同时，我国要增进与美国新政府及民间的科技合作交流布局，要通过共建“一带一路”倡议、上海合作组织（简称上合组织）、二十国集团（G20）、金砖国家、亚洲太平洋经济合作组织（APEC）等多边合作，拓展吸引全球科技人才的空间。

二、我国全方位吸引国际一流科技人才存在的突出问题

近年来，我国加大引进海外科技人才力度，一定程度上遏制了人才外流，出现人才回流趋向，但我国全方位吸引国际一流科技人才工作仍不能适应全球人才竞争新态势和全球化格局新变化，与以美国为主的西方国家存在较大差距。突出问题表现在以下几个方面。

一是战略规划缺乏总体设计和分工协同，碎片化、短期性、同质化等问题严重。海外人才规划缺少持续性政策、长期性设计和人才生态建设的统筹考虑，缺乏高水平全球人才流动监测的有效支撑。中央、部门、地方、用人单位职能分工不明确，不同层级竞相出台的引才计划同质化竞争严重。“引—留—用”法律保障和政策支撑体系不健全，引人用人政策碎片化。

二是人才流动集中在少数国家，多元、多边人才工作布局远未形成。我国年轻学者主要流向美国，引进高层次人才也以美国为主。我国的国际化战略和区域合作战略中，往往缺乏与之适应的多元、多边人才保障和交流政策。

三是以政府计划式、工程化引人方式为主，民间科技组织和用人单位的作用没有充分发挥。地方和部门多头实施人才计划、人才工程，存在一哄而上、一哄而散的短期行为，有重引进、轻服务、难管理的弊端，过度行政介入引起其他国家疑虑。非官方组织和机构参与度不高，国际人才中介机构发展不佳，科技类学会（协会）等民间机构参与度不够，高等院校、科研机构、企业等用人单位话语权不高、自主权不足，用人单位的主体作用难以充分发挥。

四是吸引全球人才的能力不强，适合科技人才发展的学术环境亟待完善。我国对全球科技人才的吸引力与我国科技大国地位不相匹配。清华大学关于全球科技人才吸引力指数的测算表明，我国经济发展速度、科研设施等位居世界前列，但科研文化和社会环境等方面的吸引力远低于发达国家，2019 年吸引力指数在 100 个国家中排名第

33 位。据中国科协创新战略研究院相关调查，科技评价导向偏差、非科研事务繁重、科研作风浮躁是我国科研环境中最严重的三个问题。科研环境缺乏吸引力已成为部分一流科学家决定回流到美国的主要原因。

五是对“党管人才”认识上存在误区，实践上存在偏差。对“党管人才”的理解存在政治化、意识形态化倾向，容易落入美国蓄意挑起的意识形态之争，增加“新冷战”风险。“党管人才”的制度优势在深化体制机制改革中的作用未能得到充分发挥。有些部门或机构存在用管理党政人员方式管理科技人才的倾向，违背了科技活动的特点和规律，不利于吸引国际一流科技人才。

三、全方位吸引国际一流科技人才的政策建议

1. 超前预判全球化格局新变化，制定长期可持续的海外人才整体战略

一是根据国家发展新主题、新格局，以及全球竞争合作新走向，适应吸引海外优秀人才工作特殊性，制定海外人才工作专项规划；二是明确中央、部门、地方和用人单位的职能分工，强化协作，避免无序竞争和短期行为；三是完善吸引海外优秀人才的制度体系，优化人才遴选程序和引进、使用机制，形成“引—留—用”相贯通的法律法规与政策，防止政策碎片化；四是建立全球科技人才流动观测体系，建立并不断完善关键人才、急需人才数据库，为抓住机遇精准引人用人提供支撑；五是加强海外人才战略与区域国际经济合作战略紧密结合，形成相互协同的战略合力；六是构建符合国际惯例的制度规范，推动国内人才市场与国际人才市场双循环相互促进并有效对接。

2. 尽快改变人才流动单极化状况，形成多边、多样、多向的海外人才工作布局

一是加强与共建“一带一路”合作国家和周边国家的科技合作交流；二是注重与欧盟、东盟等区域合作共同体、关键小国（如瑞士、芬兰、丹麦等）的科技合作与交流；三是注重人才引进类型的多元化，加大关键科技领域引进力度的同时，加强吸引和用好基础研究人才、青年科技人员，大幅度扩大国际博士后的比例，重视引进海外已退休高水平的“银发科学家”；四是注重科研合作交流方式的多样化，突出“不求所有、但求所用、重求所成”，重视“隐性流动、多向流动”的作用，充分发挥短期入境、研究合作、合作办学、国际会议等多种国际交流渠道的作用。

3. 大力提升我国科技社团、科研机构国际化水平，切实发挥民间组织与用人单位引人用人的主体作用

一是提高科技社团等民间组织的国际化水平，鼓励外籍科学家在我国科技学术组织任职，开放吸收外籍会员并建立相应的管理制度；二是逐步放开、加快审批在我国境内设立国际科技组织，聚焦绿色发展、清洁能源、气候变化、人类健康等共性问题，扩大全球科研合作，鼓励积极参与或牵头建立区域国际科技合作联盟；三是建立多渠道的国际科技和学术组织人才培养推送机制，强化国际组织任职人员储备和履职能力培训；四是扩大用人单位引进海外科技人才的自主权，充分发挥科技社团、企业、高校等用人单位引人用人的主体作用，有效运用减免税等市场机制激励用人单位吸引国际优秀人才，增强国际一流科技人才来华工作的吸引力。

4. 持续优化我国科技发展的环境与条件，夯实吸引国际一流科技人才的基础

一是大力推进科技评价制度改革，完善科研诚信和知识产权制度，充分发挥科学共同体在科研评价中的主体作用；二是构建具有创新活力、容错空间和可持续发展的科研环境，营造让科学家安心科研的学术氛围和创新文化，提高对海外优秀人才的吸引力和凝聚力；三是加大开放合作力度，试点更加开放的人才政策，倒逼科技人才体制改革，补齐国内人才市场结构不完整、制度不健全的短板，完善对接国际人才流动准则的法律法规；四是赋予一流科技人才组建科研团队、配置科研资源更大的自主权，形成“事业引人、事业留人”的制度环境；五是加强国际一流研究中心和研究机构建设，建立国际一流的学术交流、知识共享平台，打造吸引世界一流科技人才的创新高地；六是建立和扩大多样化的海外优秀人才基金，为吸引人才提供渠道更多、机制更活的经费支持。

5. 深化人才体制机制政策，形成吸引全球科技人才最广泛的“合作共同体”

一是在吸引全球人才的宏观规划、政策调整、综合协调、服务创新等方面，充分发挥“党管人才”的制度优势，推动深化我国人才体制机制改革；二是创新“党管人才”的方式方法，结合政府职能转变，加大向行业协会、科技社团、用人单位下放权力的力度，遵循科技人才成长和流动规律，适应国际人才流动趋势和特点，完善服务海内外优秀人才的工作机制；三是以促进科技进步和推动构建人类命运共同体为共同目标，以开放包容、互惠共享为基本准则，建立吸引全球科技人才最广泛的“合作共同体”。

Policy Suggestions on All-Round Recruitment of International Leading Sci-Tech Talent Under the New Pattern of Globalization

Consultative Group of CAS Academic Division

This paper reviews the new opportunities for recruiting top international leading sci-tech talent and discusses problems and challenges. The following suggestions are proposed: 1) to make advanced predictions about the new changes in the globalization pattern, and formulate the long-term and sustainable overall overseas talent strategy; 2) to change the unilateral polarization of talent flow as soon as possible and instead form the pattern of multilateral, diversified and multi-directional overseas talent recruitment patterns in our work; 3) to vigorously promote the level of internationalization of sci-tech societies and research institutes in China, and effectively play the role of non-governmental organizations and employers in talent introduction and employment; 4) to continuously optimize China's sci-tech development environment and conditions and lay a solid foundation for attracting top international sci-tech talents; 5) to strengthen the policies on talent system and mechanism and establish the most comprehensive "cooperation community" in the recruitment of sci-tech talents worldwide.

附　　录

Appendix

附录一　2021年中国与世界十大科技进展

一、2021年中国十大科技进展

1. 我国首次火星探测任务取得圆满成功

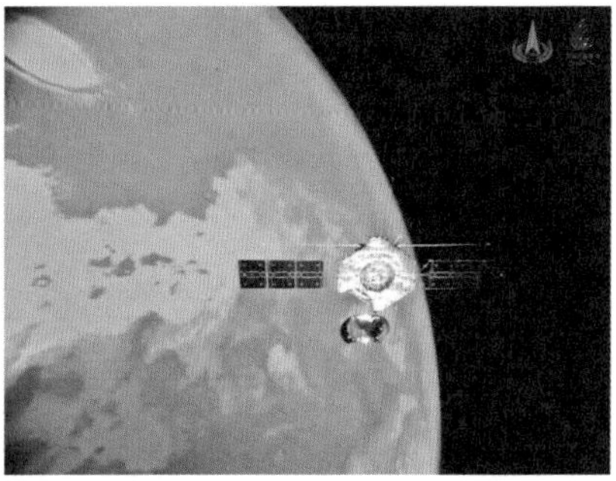

2021年6月11日，国家航天局在京举行“天问一号”探测器着陆火星首批科学影像图揭幕仪式，公布了由“祝融号”火星车拍摄的着陆点全景、火星地形地貌、“中国印迹”和“着巡合影”等影像图。首批科学影像图的发布，标志着我国首次火星探测任务取得圆满成功。我国首次火星探测任务于2013年全面启动论证，2016年1月批准立项。2020年7月23日“天问一号”探测器于海南文昌成功发射，历经地火转移、火星捕获、火星停泊、离轨着陆和科学探测等阶段，工程任务按计划顺利开展。

2. 中国空间站开启有人长期驻留时代

2021年6月17日和10月16日，“神舟十二号”“神舟十三号”载人飞船相继发射成功，顺利将航天员送入太空。“神舟十二号”与“天和”核心舱对接形成组合体，3名航天员进驻核心舱，进行了为期3个月的驻留，开展了一系列空间科学实验和技术试验，在轨验证了航天员长期驻留、再生生保、空间物资补给、出舱活动、舱外操作、在轨维修等空间站建造和运营关键技术。“神舟十三号”入轨后，与“天和”核心舱和“天舟二号”“天舟三号”组合体完成自主快速交会对接，3位航天员开启为期6个月的在轨驻留，其间开展了机械臂操作、出舱活动、舱段转位及空间科学实验与技术试验等工作，进一步验证航天员长期在轨驻留、再生生保等一系列关键技术，中国空间站有人长期驻留时代到来。

3. 我国实现二氧化碳到淀粉的从头合成

淀粉是粥、饭中最主要的碳水化合物，是面粉、大米、玉米等粮食的主要成分，也是重要的工业原料。其主要合成方式是由绿色植物通过光合作用固定二氧化碳来进行。长期以来，科研人员一直在努力改进光合作用这一生命过程，希望提高二氧化碳的转化速率和光能的利用效率，最终提升淀粉的生产效率。中国科学院天津工业生物技术研究所研究人员提出了一种颠覆性的淀粉制备方法，不依赖植物光合作用，以二氧化碳、电解产生的氢气为原料，成功生产出淀粉，在国际上首次实现了二氧化碳到淀粉的从头合成，使淀粉生产从传统农业种植模式向工业车间生产模式转变成为可能，取得原创性突破。相关研究成果于 2021 年 9 月 24 日在线发表于《科学》。

4. 我国团队凭打破“量子霸权”① 的超算应用摘得 2021 年度“戈登贝尔奖”

2021 年 11 月 18 日下午于美国密苏里州圣路易斯举行的全球超级计算大会（SC21）上，国际计算机协会（ACM）将 2021 年度“戈登·贝尔奖”授予中国超算应用团队。这支由之江实验室、国家超级计算无锡中心等单位研究人员组成的联合科研团队，基于新一代神威超级计算机的应用“超大规模量子随机电路实时模拟”（SWQSIM）获此殊荣。在这项工作中，研究人员引入了一个系统的设计过程，涵盖了模拟所需的算法、并行化和系统架构。使用新一代神威超级计算机，研究团队有效模拟了一个深度为 10×10（1+40+1）随机量子电路。与谷歌量子计算机“悬铃木”200 s 完成百万 0.2%保真度采样任务相比较，美国超级计算机“顶点”需要 1 万年完成同等复杂度的模拟，该团队 SWQSIM 应用则可在 304 s 以内得到百万更高保真度的关联样本，在一周内得到同样数量的无关联样本，一举打破“悬铃木”所宣称的“量子霸权”。

① 量子霸权，也称作量子优势，是指量子计算拥有的超越所有经典计算机的计算能力。

5. 1400 万亿电子伏特！我国科学家观测到迄今最高能量光子

中国科学院高能物理研究所牵头的国际合作组依托国家重大科技基础设施“高海拔宇宙线观测站（LHAASO）”，在银河系内发现 12 个超高能宇宙线加速器，并记录到能量达 1.4 PeV 的伽马光子，这是人类迄今观测到的最高能量光子，突破了人类对银河系粒子加速的传统认知，揭示了银河系内普遍存在能够把粒子加速到超过 1PeV 的宇宙线加速器，开启了“超高能伽马天文”观测时代。相关成果于 2021 年 5 月 17 日发表于《自然》。

6. “嫦娥五号”样品重要研究成果先后出炉

2021 年 10 月 19 日，中国科学院发布“嫦娥五号”月球科研样品最新研究成果。中国科学院地质与地球物理研究所和国家天文台主导，联合多家研究机构通过 3 篇《自然》论文和 1 篇《国家科学评论》论文，报道了围绕月球演化重要科学问题取得的突破性进展。在最新的研究中，科研人员利用超高空间分辨率铀-铅（U-Pb）定年技术，对“嫦娥五号”月球样品玄武岩岩屑中 50 余颗富铀矿物（斜锆石、钙钛锆石、静海石）进行分析，确定玄武岩形成年龄为（20.30 ± 0.04）亿年，表明月球直到 20 亿年前仍存在岩浆活动，比以往月球样品限定的岩浆活动延长了约 8 亿年。研究显示，“嫦娥五号”月球样品玄武岩初始熔融时并没有卷入富集钾、稀土元素、磷的克里普物质，嫦娥五号月球样品富集“克里普物质”的特征，是岩浆后期经过大量矿物结晶固化后，残余部分富集而来。这一结果排除了嫦娥五号着陆区岩石的初始岩浆熔融热源来自放射性生热元素的主流假说，揭示了月球晚期岩浆活动过程。据悉，此次研究采用的超高空间分辨率的定年和同位素分析技术处于国际领先水平，为珍贵地外样品年代学等研究提供了新的技术方法。

7. 异源四倍体野生稻快速从头驯化获得新突破

随着世界人口的快速增长，至2050年粮食产量或将增加50%才能完全满足需求。与此同时，近年来世界气候变化加剧，全球气候变暖、极端天气频发等都为粮食安全带来了巨大挑战。在此背景下，如何进一步提高作物单产成为亟待解决的严峻问题。中国科学院种子创新研究院/遗传与发育生物学研究所李家洋院士团队首次提出了异源四倍体野生稻快速从头驯化的新策略，旨在最终培育出新型多倍体水稻作物，从而大幅提升粮食产量并增加作物环境变化适应性。该项研究为未来应对粮食危机提出了一种新的可行策略，开辟了全新的作物育种方向。相关研究成果于2021年2月4日发表于《细胞》。

8. 我国研发成功－271℃超流氦大型低温制冷装备

2021年4月15日，由中国科学院理化技术研究所承担的国家重大科研装备研制项目“液氦到超流氦温区大型低温制冷系统研制”通过验收及成果鉴定，标志着我国具备了研制液氦温度（零下269℃）千瓦级和超流氦温度（－271℃）百瓦级大型低温制冷装备的能力，可满足大科学工程、航天工程、氦资源开发等国家战略高技术发展的迫切需要。该项目的成功实施，还带动了我国高端氦螺杆压缩机、低温换热器和低温阀门等行业的快速发展，提高了一批高科技制造企业的核心竞争力，使相关技术实现了从无到有、从低端到高端的提升，在我国初步形成了功能齐全、分工明确的低温产业群。

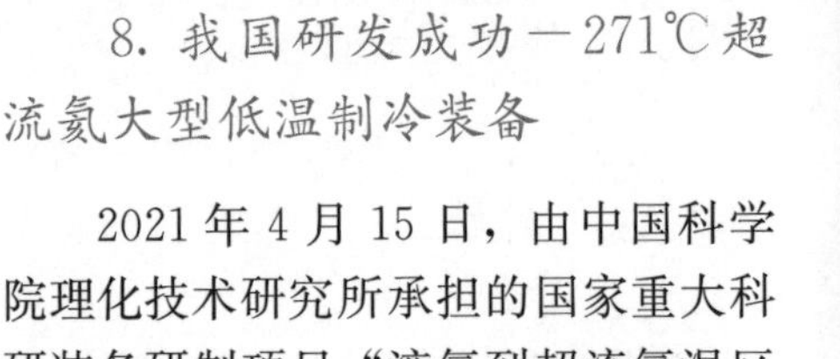

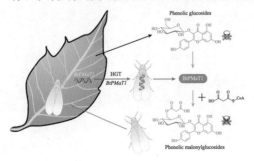

9. 植物到动物的功能基因转移首获证实

中国农业科学院蔬菜花卉研究所张友军团队经过20年追踪研究，发现被联合国粮农组织（FAO）认定的迄今唯一“超级害虫”烟粉虱，具有一种类似“以子之矛、攻子之盾”的本领：其从寄主植物那里获得了防御性基因。这是现代生物学诞生100多年来，首次研究证实植物和动物之间存在功能性基因水平转移现象。相关科研成果于2021年3月25日在线发表于《细胞》，并作为《细胞》封面文章于4月1日出版。这是我国农业害虫研究领域在《细胞》上发表的首篇论文，揭示了昆虫如何利用水平转移基因来克服宿主的防御，为探索昆虫适应性进化规律开辟了新的视角，也为新一代靶标基因导向的烟粉虱田间精准绿色防控技术研发提供全新思路。

10. 稀土离子实现多模式量子中继及1小时光存储

量子不可克隆定律赋予了量子通信基于物理学原理的安全性。而这一定律也决定了光子传输损耗不能使用传统的放大器来克服，使得远程量子通信成为当今量子信息科学的核心难题之一。量子中继和可移动量子存储是实现远程量子通信的两种可行方案，其共性需求是高性能的量子存储器。在量子中继方面，国际已有实验研究都聚焦于发射型存储器的架构，无法同时满足确定性发光和多模式复用这两个关键技术需求。可移动量子存储方面，国际上光存储的时间最长仅1分钟，无法满足可移动量子存储小时量级存储时间的需求。中国科学技术大学郭光灿院士团队李传锋、周宗权研究组基于稀土离子掺杂晶体研制出高性能的固态量子存储器，并在上述两条技术路线上取得了重要进展，实现了一种基于吸收型存储器的多模式量子中继，并成功将光存储时间提升至1小时。相关成果于2021年4月22日和6月2日分别发表于《自然-通讯》和《自然》。

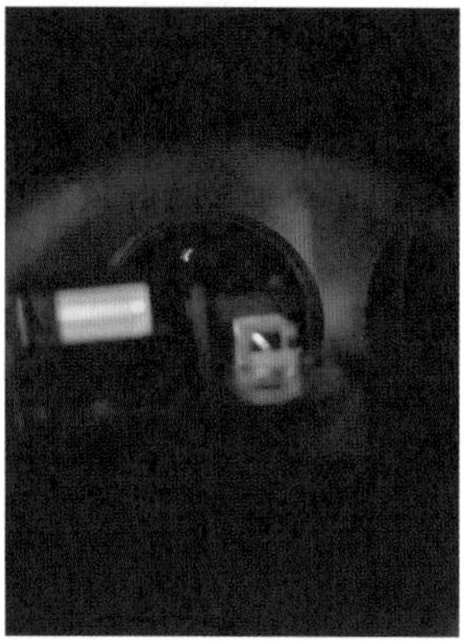

二、2021年世界十大科技进展

1. 全球首个“自我复制”的活体机器人诞生

美国佛蒙特大学、塔夫茨大学和哈佛大学威斯生物启发工程研究所的科学家发现了一种全新的生物繁殖方式，并利用其创造了有史以来第一个可进行自我复制多代的活体机器人——Xenobots 3.0。它仅有毫米大小，既不是传统的机器人，也不是已知的动物物种，而是一种从未在地球上出现过的、活的、可编程的全新有机体。据悉，该活体机器人或许可以有

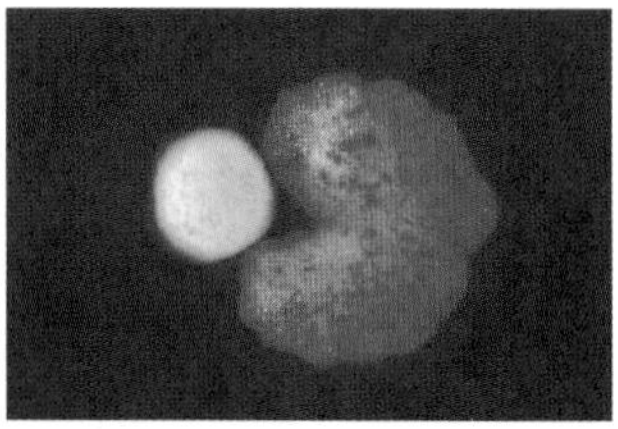

助于医学的全新突破——除了有望用于精准的药物递送之外，它的自我复制能力也使得再生医学有了新的帮手，或可为出生缺陷、对抗创伤、癌症与衰老提供开创性的解决思路。2021 年 11 月 29 日，相关研究成果发表于《美国国家科学院院刊》。

2. 核聚变向“点火”迈进一大步

我们在地球上之所以能看到阳光、感受到温暖，都是源自于发生在太阳核心的核聚变。核聚变指的是当原子合并在一起时，释放出巨大能量的过程，这个过程可以在

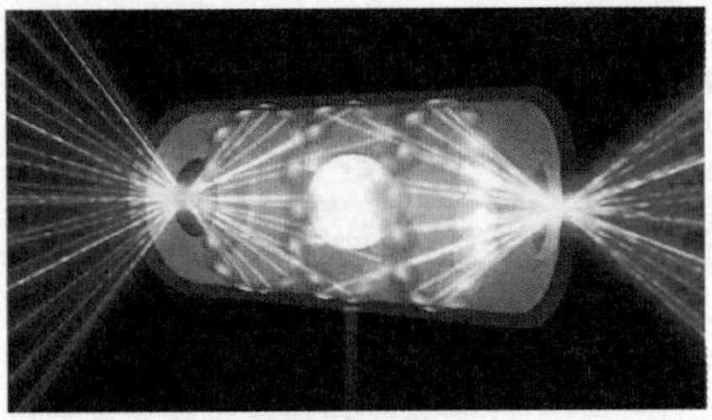

碳排放几乎为零的情况下，源源不断地提供绿色能源。但是，想在实验室里实现核聚变并非易事，一个重大的挑战就是“点火”（即聚变反应所产生的能量等于或超过输入能量的时刻）。2021 年 8 月 8 日，美国劳伦斯利弗莫尔国家实验室（LLNL）的国家点火装置（NIF）进行了一项新的实验。NIF 的科学家团队重现了存在于太阳核心的极端温度和压力，NIF 的强大的激光脉冲引发了燃料丸的核聚变爆炸，产生了 1.35 MJ 能量——大约相当于一辆时速 160 km 的汽车的动能。这一能量达到触发该过程的激光脉冲能量的 70%，意味着接近核聚变“点火”，即反应产生的能量足以使反应持续下去，在无限聚变能源的道路上迈出了一大步。

3. 科学家借助 AI 技术破解蛋白质结构预测难题

科学家们一直希望通过基因序列简单地预测蛋白质形状——如果能够成功，这将开启一个洞察生命运作机理的新世界。美国华盛顿大学和英国 DeepMind 公司分别公布了多年工作的成果：先进的建模程序，可以预测蛋白质和一些分子复合物的精确三维原子结构，并将这些结构放入公开的数据库免费供全球科研人员使用。据 DeepMind 公司报告显示，其人工智能程序 AlphaFold 预测出 98.5%的人类蛋白质结构，有助于深入理解一些关键生物学信息，从而更好开展药物研发。而美国华盛顿大学创建的高精确的蛋白质结构预测程序名叫 RoseTTAFold，基于深度学习，它不仅能预测蛋白质的结构，还能预测蛋白质之间的结合形式。仅需 10 min RoseTTAFold 就能用一台游戏电脑准确计算出蛋白质结构。相关论文于 2021

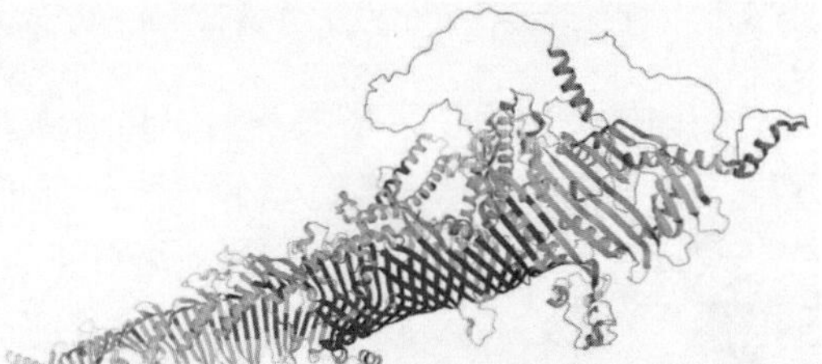

年 7 月 15 日分别刊登于《自然》和《科学》。

4.“基因剪刀”首次治疗遗传病

一直以来，人们若要使用被称为“基因剪刀”的 CRISPR 基因编辑技术治疗遗传疾病，需要清除一个巨大的障碍：将分子剪刀工具直接注射到受影响的细胞中，从而实现 DNA 切割。英国伦敦大学研究人员发现 CRISPR 技术能使一种突变基因失活。

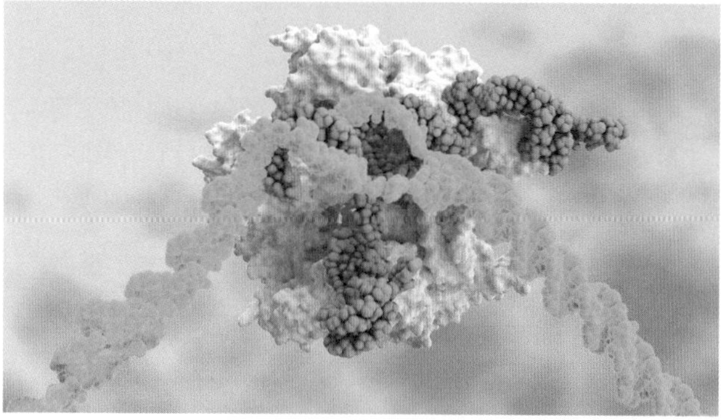

研究首次将 CRISPR 药物注射到一种罕见遗传病（转甲状腺素蛋白淀粉样变性病）患者的血液中，并发现其中 3 人的肝脏几乎停止产生有毒的蛋白质。虽然目前还不能确定 CRISPR 治疗是否能缓解该疾病的症状，但初步数据让人们对这种一次性治疗的效果感到兴奋。相关研究结果于 2021 年 5 月 28 日发表于《新英格兰医学杂志》。据悉，这项新工作在能够灭活、修复或替换身体任何部位的致病基因方面，迈出了关键的第一步。

5. 史上最冷反物质问世

加拿大国家粒子加速器中心的 Makoto Fujiwara 团队与合作者在瑞士日内瓦附近的欧洲核子研究组织粒子物理实验室进行了一项名为 ALPHA-2 的反氢捕获实验，演示了反氢原子的激光冷却，将样品冷却到了接近绝对零度。激光冷却经常被用来测量常规原子的能量跃迁——电子运动到不同能级。该团队开发了一种激光，它能以适当

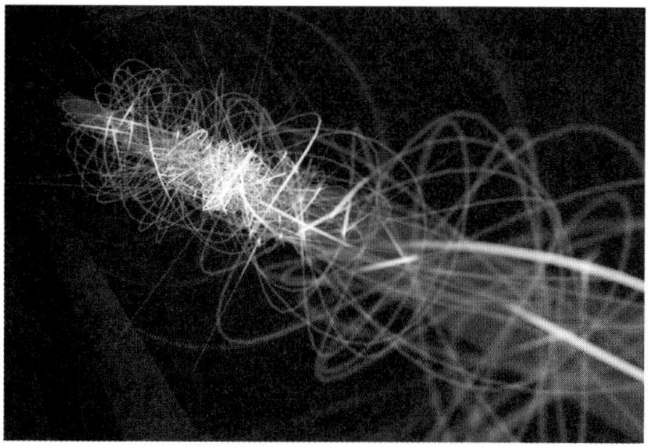

的波长发射被称为光子的光粒子，从而降低正在直接朝向激光移动的反原子的速度。研究人员将反原子的速度降低到 1/10 以下。对于冷却的反氢原子，该团队获得的测量精度几乎是未冷却的反原子的 3 倍。该研究产生了比以往任何时候都更冷的反物质，并使一种全新的实验成为可能，有助于科学家在未来更

多地了解反物质。相关研究成果于2021年3月31日刊登于《自然》。

6. “芝麻粒”大小心脏模型问世

奥地利科学院生物学家Sasha Mendjan和团队使用人类多能干细胞培养出芝麻大小的心脏模型，又称心脏线。它可以自发地进行组织，在不需要实验支架的情况下发展出一个中空的心房。Mendjan团队以特定的顺序激活所有参与胚胎心脏发育的6个已知信号通路，诱导干细胞自我组织。随着细胞分化，它们开始形成不同的层——类似心脏壁的结构。经过一周的发育，这些类器官自组织成一个有封闭腔的3D结构，几乎重现了人类心脏的自发生长轨迹。此外，研究小组还发现心脏壁状组织能有节奏地收缩，挤压腔内的液体。该团队还测试了心脏类器官对组织损伤的反应。他们用一根冷钢棒冷冻部分心脏类器官，并杀死该部位的许多细胞，研究发现，心脏成纤维细胞（一种负责伤口愈合的细胞）开始向损伤部位迁移，并产生修复损伤的蛋白质。相关研究于2021年5月20日发表于《细胞》，这项进展使得科学家能创造出一些迄今为止最真实的心脏类器官，为制药公司将更多药物引入临床试验提供了可能。

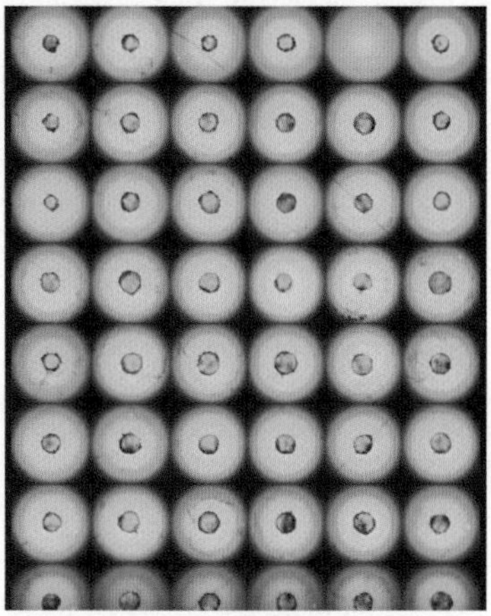

7. 科学家利用人工智能实现两项数学突破

纯数学研究工作的关键目标之一是发现数学对象间的规律，并利用这些联系形成猜想。从20世纪60年代开始，数学家开始使用计算机帮助发现规律和提出猜想，但人工智能系统尚未普遍应用于理论数学研究领域。2021年12月1日，一篇发表在《自然》上的论文显示，DeepMind公司研发出一个机器学习框架，能帮助数学家发现新的猜想和定理。此前，该框架已经帮助发现了不同纯数学领域的两个新猜想。研究人员将这一方法应用于两个纯数学领域，发现了拓扑学（对几何形状性质的研究）的一个新定理，和一个表示论（代数系统研究）的新猜想。研究人员表示，这是计算机科学家和数学家首次使用人工智能来帮助证明或提出复杂数学领域的新定理。

8. 科学家成功在实验室中构建人类早期胚胎样结构

美国得克萨斯大学达拉斯西南医学中心研究人员领衔的团队成功用人多能干细胞分化诱导出人类早期胚胎样结构。该结构与人囊胚期胚胎具有类似的结构，能正确表达相应的基因与蛋白，并且可在体外发育 2～4 d，形成类羊膜囊等结构。相关研究成果于 2021 年 3 月 17 日刊登于《自然》。据介绍，借助人类早期胚胎样结构，研究人员能深入研究胚胎的早期发育，更加了解人类早期重大疾病造成的流产、畸形儿、女性受孕障碍等现象，并为其寻找可行的解决方案。此外，研究人员还可以通过这项技术建立药物筛选模型，为进入临床应用的孕妇药品提供安全性模拟检测。

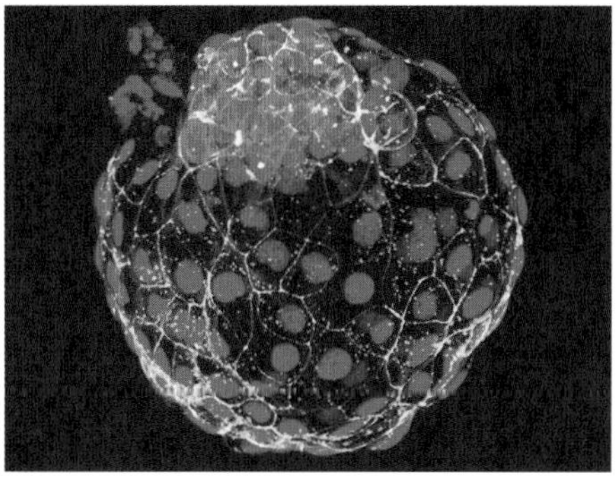

9. 激光传输稳定自如创世界纪录

澳大利亚国际射电天文学研究中心（ICRAR）和西澳大利亚大学（UWA）等机构的研究人员创造了在大气层中最稳定传输激光信号的世界纪录。该团队将相位稳定技术与先进的自导向光学终端相结合，实现了此次最稳定的激光传输。新技术有效地消除了大气湍流，允许激光信号从一个点发送到另一个点，而不会受到大气的干扰。这一结果是用一个通过大气传输的激光系统比较两个不同地点间时间流动的全球最精确的方法。相关论文于 2021 年 1 月 22 日发表于《自然-通讯》。据悉，这项研究有广阔的应用前景，可以用来精确地检验爱因斯坦的广义相对论，或者发现基本物理常数是否随着时间而变化。同时，这项技术的精确测量能力在地球科学和地球物理学中也有实际用途，可以改进有关地下水位如何随时间变化的卫星研究或寻找地下矿藏。此外，该技术在光通信领域的应用可以将卫星到地面的数据传输速率提高几个数量级，下一代大型数据收集卫星能更快地将关键信息传送到地面。

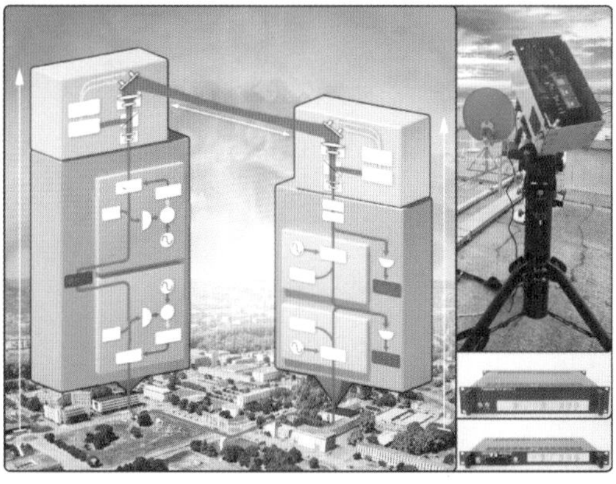

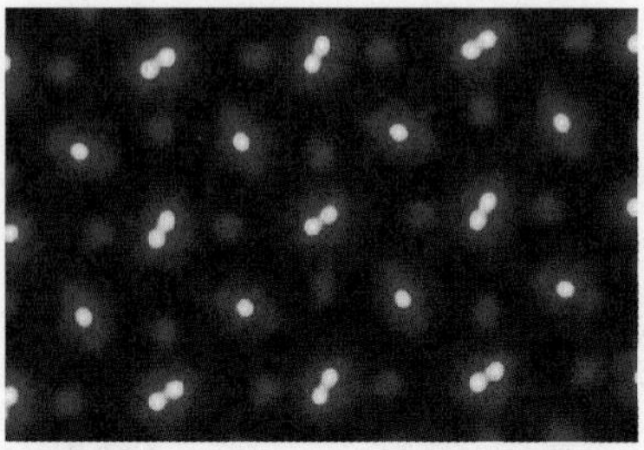

10. 科学家"绘制"最清晰原子"特写"

美国康奈尔大学的 Muller 团队捕捉到了迄今为止最高分辨率的原子图像，打破了其 2018 年所创下的纪录。据悉，Muller 团队使用叠层成像技术，用 X 射线照射钪酸镨晶体，然后利用散射电子的角度来计算散射它们的原子的形状。这些进步使得研究小组能够观察更稠密的原子样本，并获得更好的分辨率。据了解，这种最新形式的电子叠层成像分析技术使科学家可以在所有三个维度上定位单个原子。研究人员还将能够一次发现异常结构中的杂质原子，并对它们及其振动进行成像。相关论文于 2021 年 5 月 21 日刊登于《科学》。

附录二　2021年中国科学院、中国工程院新当选院士名单

一、2021年新当选中国科学院院士名单

（共65人，分学部按姓氏拼音为序）

数学物理学部（12人）

序号	姓名	年龄	专业	工作单位
1	陈松蹊	59	数学地球物理、数学	北京大学
2	封东来	48	实验凝聚态物理	中国科学技术大学
3	李　骏	60	数学	复旦大学
4	刘仓理	59	武器物理	中国工程物理研究院
5	马余强	56	软物质物理	南京大学
6	阮勇斌	58	基础数学	浙江大学
7	史生才	56	天体物理	中国科学院紫金山天文台
8	万宝年	59	磁约束等离子体物理	中国科学院合肥物质科学研究院
9	王玉鹏	56	凝聚态理论	中国科学院物理研究所
10	夏克青	63	流体力学	南方科技大学
11	张　平	51	数学	中国科学院数学与系统科学研究院
12	邹冰松	57	理论物理	中国科学院理论物理研究所

化学部（11人）

序号	姓名	年龄	专业	工作单位
1	卜显和	56	无机化学	南开大学
2	迟力峰（女）	63	物理化学	苏州大学
3	房　喻	64	物理化学	陕西师范大学
4	刘买利	62	分析化学	中国科学院精密测量科学与技术创新研究院
5	马光辉（女）	56	生物化工	中国科学院过程工程研究所
6	马於光	57	高分子化学与物理	华南理工大学

续表

序号	姓名	年龄	专业	工作单位
7	王梅祥	60	有机化学	清华大学
8	谢素原	53	无机化学	厦门大学
9	俞　飚	53	有机化学	中国科学院上海有机化学研究所
10	元英进	57	生物化工（合成生物学）	天津大学
11	周　翔	58	化学生物学	武汉大学

生命科学和医学学部（10人）

序号	姓名	年龄	专业	工作单位
1	窦科峰	65	器官移植	中国人民解放军空军军医大学
2	李劲松	49	干细胞与胚胎发育	中国科学院分子细胞科学卓越创新中心
3	林圣彩	57	生物化学与细胞生物学	厦门大学
4	宋保亮	46	生物化学	武汉大学
5	滕皋军	58	介入放射学	东南大学附属中大医院
6	王以政	63	神经生物学	中国人民解放军军事科学院
7	杨维才	57	植物发育生物学	中国科学院遗传与发育生物学研究所
8	杨正林	55	临床检验诊断学	四川省人民医院、电子科技大学
9	张克勤	62	植物保护、生物防治	云南大学
10	张　旭	58	泌尿外科学	中国人民解放军总医院

地学部（9人）

序号	姓名	年龄	专业	工作单位
1	邓　军	63	矿床学	中国地质大学（北京）
2	底青云（女）	56	应用地球物理学	中国科学院地质与地球物理研究所
3	胡瑞忠	62	矿床学	中国科学院地球化学研究所
4	黄建平	58	大气科学	兰州大学
5	朴世龙	45	自然地理学	北京大学
6	谈哲敏	56	气象学	南京大学
7	谢树成	53	地质微生物	中国地质大学（武汉）
8	朱　敏	55	古生物学	中国科学院古脊椎动物与古人类研究所
9	朱彤	58	环境科学（大气化学与环境健康）	北京大学

信息技术科学部（10人）

序号	姓名	年龄	专业	工作单位
1	丁赤飚	51	信号与信息处理	中国科学院空天信息创新研究院
2	黎　湘	53	信号与信息处理	中国人民解放军国防科技大学

续表

序号	姓名	年龄	专业	工作单位
3	李陟	59	导航、制导与控制	中国航天科工集团第二研究院
4	刘益春	58	光电功能材料与器件	东北师范大学
5	钱德沛	68	计算机系统结构	北京航空航天大学
6	乔　红（女）	56	机器人理论与应用	中国科学院自动化研究所
7	于登云	59	空间飞行器系统工程、动力学与控制	中国航天科技集团有限公司
8	郑婉华（女）	55	微电子学与固体电子学	中国科学院半导体研究所
9	朱鲁华	56	网络空间安全	中国人民解放军 61212 部队
10	祝宁华	61	微波光子学	雄安创新研究院、中国科学院半导体研究所

技术科学部（13 人）

序号	姓名	年龄	专业	工作单位
1	陈　光	59	金属材料加工工程	南京理工大学
2	范瑞祥	56	运载火箭总体设计	中国航天科技集团有限公司第一研究院
3	顾　宁	57	纳米医学材料	东南大学
4	贾金锋	55	量子材料	上海交通大学
5	江　涌	58	飞行器总体与制导	中国航天科工集团第二研究院
6	姜培学	56	热质传递理论与技术	清华大学
7	冷劲松	52	智能结构力学	哈尔滨工业大学
8	李　杰	63	土木工程（工程防灾）	同济大学
9	孙　军	62	金属材料	西安交通大学
10	唐志共	56	空气动力学	中国空气动力研究与发展中心
11	徐世烺	67	结构工程	浙江大学
12	张卫红	56	机械设计理论与制造技术	西北工业大学
13	周又和	64	非线性力学	兰州大学

二、中国工程院 2021 年当选院士名单

（共 84 人，分学部按姓氏笔画排列）

机械与运载工程学部（11 人）

序号	姓名	出生年月	工作单位
1	马玉山	1968 年 12 月	吴忠仪表有限责任公司

续表

序号	姓名	出生年月	工作单位
2	王云鹏	1966 年 12 月	北京航空航天大学
3	王向明	1962 年 12 月	中国航空工业集团公司沈阳飞机设计研究所
4	王国庆	1966 年 4 月	中国航天科技集团有限公司
5	王树新	1966 年 9 月	天津大学
6	付梦印	1964 年 11 月	南京理工大学
7	刘　宏	1966 年 12 月	哈尔滨工业大学
8	朱　坤	1966 年 2 月	中国航天科工集团第三研究院
9	李克强	1963 年 1 月	清华大学
10	邹汝平	1962 年 9 月	中国兵器工业第二□三研究所
11	郑津洋	1964 年 11 月	浙江大学

信息与电子工程学部（10 人）

序号	姓名	出生年月	工作单位
1	孔志印	1964 年 8 月	中央军委办公厅某研究所
2	龙　腾	1968 年 1 月	北京理工大学
3	江碧涛（女）	1967 年 8 月	中国人民解放军 61646 部队
4	吴剑旗	1966 年 7 月	中国电子科技集团公司第三十八研究所
5	张宏科	1957 年 9 月	北京交通大学
6	张宝东	1962 年 12 月	中国人民解放军 32057 部队
7	李得天	1966 年 6 月	兰州空间技术物理研究所
8	罗　毅	1960 年 2 月	清华大学
9	蓝羽石	1954 年 9 月	中国电子科技集团公司第二十八研究所
10	蒋昌俊	1962 年 5 月	同济大学

化工、冶金与材料工程学部（8 人）

序号	姓名	出生年月	工作单位
1	邓龙江	1966 年 11 月	电子科技大学
2	邢丽英（女）	1965 年 2 月	中国航空制造技术研究院
3	杨为民	1966 年 4 月	中国石油化工股份有限公司上海石油化工研究院
4	应汉杰	1969 年 7 月	南京工业大学
5	张立群	1969 年 12 月	北京化工大学
6	沈政昌	1960 年 5 月	矿冶科技集团有限公司
7	姜　涛	1963 年 10 月	中南大学
8	傅正义	1963 年 1 月	武汉理工大学

能源与矿业工程学部（9 人）

序号	姓名	出生年月	工作单位
1	王成山	1962 年 11 月	天津大学
2	孙友宏	1965 年 7 月	中国地质大学（北京）
3	孙焕泉	1965 年 1 月	中国石油化工集团有限公司
4	张来斌	1961 年 9 月	中国石油大学（北京）
5	周　刚	1964 年 6 月	中国人民解放军 63650 部队
6	饶　宏	1961 年 2 月	中国南方电网有限责任公司
7	胡晓棉（女）	1963 年 10 月	北京应用物理与计算数学研究所
8	葛世荣	1963 年 4 月	中国矿业大学（北京）
9	程杰成	1962 年 9 月	大庆油田有限责任公司

土木、水利与建筑工程学部（10 人）

序号	姓名	出生年月	工作单位
1	王明洋	1966 年 1 月	中国人民解放军陆军工程大学
2	刘加平	1967 年 1 月	东南大学
3	朱合华	1962 年 10 月	同济大学
4	许唯临	1963 年 10 月	四川大学
5	杜修力	1962 年 12 月	北京工业大学
6	张宗亮	1962 年 4 月	中国电建集团昆明勘测设计研究院有限公司
7	胡亚安	1965 年 2 月	水利部交通运输部国家能源局南京水利科学研究院
8	高宗余	1964 年 1 月	中铁大桥勘测设计院集团有限公司
9	唐洪武	1966 年 9 月	河海大学
10	梅洪元	1958 年 7 月	哈尔滨工业大学

环境与轻纺工程学部（8 人）

序号	姓名	出生年月	工作单位
1	王双飞	1963 年 9 月	广西大学
2	王军成	1952 年 12 月	山东省科学院海洋仪器仪表研究所
3	冯　起	1966 年 3 月	中国科学院西北生态环境资源研究院
4	吴明红（女）	1968 年 3 月	上海大学
5	单　杨	1963 年 2 月	湖南省农业科学院
6	徐卫林	1969 年 4 月	武汉纺织大学
7	高　翔	1968 年 10 月	浙江大学
8	谢明勇	1957 年 2 月	南昌大学

农业学部（10 人）

序号	姓名	出生年月	工作单位
1	尹飞虎	1954 年 12 月	新疆农垦科学院
2	许为钢	1958 年 10 月	河南省农业科学院
3	吴义强	1967 年 7 月	中南林业科技大学
4	陈松林	1960 年 10 月	中国水产科学研究院黄海水产研究所
5	沈其荣	1957 年 8 月	南京农业大学
6	周　卫	1966 年 8 月	中国农业科学院农业资源与农业区划研究所
7	侯水生	1959 年 10 月	中国农业科学院北京畜牧兽医研究所
8	柏连阳	1967 年 12 月	湖南省农业科学院
9	喻景权	1963 年 11 月	浙江大学
10	谯仕彦	1963 年 4 月	中国农业大学

医药卫生学部（11 人）

序号	姓名	出生年月	工作单位
1	田金洲	1956 年 12 月	北京中医药大学
2	朱兆云（女）	1954 年 3 月	云南白药集团股份有限公司
3	邬堂春	1965 年 7 月	华中科技大学
4	肖　伟	1959 年 10 月	江苏康缘药业股份有限公司
5	范先群	1964 年 6 月	上海交通大学医学院
6	赵铱民	1956 年 10 月	中国人民解放军空军军医大学
7	姜保国	1961 年 4 月	北京大学人民医院
8	高天明	1960 年 10 月	南方医科大学
9	徐兵河	1958 年 2 月	中国医学科学院肿瘤医院
10	蒋建东	1958 年 11 月	中国医学科学院医药生物技术研究所
11	蒋建新	1962 年 12 月	中国人民解放军陆军军医大学

工程管理学部（7 人）

序号	姓名	出生年月	工作单位
1	王自力	1964 年 10 月	北京航空航天大学
2	杨长风	1958 年 2 月	中国卫星导航系统管理办公室
3	杨　宏	1963 年 11 月	中国空间技术研究院
4	林　鸣	1957 年 10 月	中国交通建设股份有限公司
5	贾伟平（女）	1956 年 11 月	上海交通大学附属第六人民医院
6	黄殿中	1952 年 1 月	中国信息安全测评中心
7	谢玉洪	1961 年 2 月	中国海洋石油集团有限公司

附录三　2021年香山科学会议学术讨论会一览表

序号	会次	会议名称	执行主席	会期
1	S59	新污染物的健康风险及防控对策	陈宜瑜　江桂斌　赵进才　朱利中　朱永官　陶　澍	1月29～30日
2	Y5	面向睡眠健康的智能感知与计算	陈　炜　刘澄玉　王长明　张　远	3月25～26日
3	694	用于硼中子俘获肿瘤治疗的含硼药物	柴之芳　陈和生　罗志福　赵宇亮	3月30～31日
4	695	基于大数据的中医精准用药机制关键科学问题	康　乐　林国强　李振吉　孙晓波　谢雁鸣　王海南	4月8～9日
5	696	揭示生命领域三大科学问题，解析人体信息能量网络机制	丛　斌　陆朝阳　樊代明　谭光明　张世明	4月10～11日
6	S60	碳中和科技创新路径选择	刘燕华　杜祥琬　刘中民　聂祚仁	4月12～13日
7	S61	保护生物学研究与国家生态安全	张亚平　魏辅文　乔格侠　詹祥江	4月12～13日
8	698	车路协同自动驾驶关键科学问题及技术前沿	黄　卫　岑晏青　李德毅　李克强　陈山枝	4月14～15日
9	697	同一健康与人类健康	陈君石　高　福　陆家海　徐建国　赵国屏	4月21～22日
10	699	水的微观结构和超快动力学	曹则贤　胡　钧　江　雷　匡廷云　杨国桢	4月27～28日
11	700	大陆型强震孕育发生的物理机制及地震预测探索	陈晓非　邵志刚　石耀霖　吴忠良　张培震	5月12～13日
12	701	老年骨关节病发病机理与早期干预	关振鹏　张英泽　曹永平　胡宝洋	5月14～15日
13	702	新型孔材料与扩散传质工程	陈小明　何鸣元　刘中民　苏宝连　谢在库	6月3～4日
14	703	科学传播与科学教育	陈　进　郭传杰　汤书昆　刘嘉麒	6月21～22日

续表

序号	会次	会议名称	执行主席	会期
15	704	针灸面临的机遇与挑战：大科学与国际化的融合	景向红 刘保延 杨龙会 方剑乔	6月29～30日
16	S62	面向人民生命健康战略	裴 钢 张伯礼 丁 健 张学敏	7月6日
17	705	新阶段聚烯烃的困境与高端突破机制	董金勇 胡 杰 焦洪桥 王笃金 庄 毅	7月12～13日
18	S63	人工智能与结构生物学	饶子和 黄 卫 贺福初 蒋华良 许瑞明 许文青	8月22日
19	706	变化环境下自然资源及综合观测与模拟	周成虎 夏 军 陈 军	9月2～3日
20	707	宇宙缺失重子探寻的关键科学和技术问题	崔 伟 常 进 高 波 袁 峰	9月9～10日
21	708	作物表型组学与精准设计育种	陈 凡 张启发 杨维才 骆清铭	9月23～24日
22	709	强激光光学元件逼近体材料损伤阈值的超精密制造基础问题研究：机遇和挑战	郭东明 雒建斌 张维岩 邵建达	9月26～27日
23	710	杂交小麦育种与生产	邓兴旺 朱健康 种 康 付道林	9月27～28日
24	S64	定量合成生物学	赵国屏 杨焕明 田志刚 张先恩 刘陈立	9月29～30日
25	711	太阳物理前沿科学问题和立体探测中的关键技术	杨孟飞 汪景琇 王 赤 宗秋刚 张效信	10月11～12日
26	S65	植物高效碳汇	陈宜瑜 邓秀新 韩 斌 康绍忠	10月13～14日
27	712	中国西南山地生物多样性与生态安全：现状与未来	张亚平 傅伯杰 方精云 孙 航	10月18～19日
28	713	濒危药材代用品研究的科学基础	于德泉 黄璐琦 庾石山	10月20～21日
29	714	智慧农业与健康工程	冯玉林 匡廷云 庞国芳 戴小枫 赵春江	10月26～27日

附录四　2021年中国科学院学部“科学与技术前沿论坛”一览表

序号	会次	会议主题	执行主席	举办时间
1	110	声学	张仁和	3月28日
2	115	面向2035的分子科学前沿	朱道本　李玉良	4月12～13日
3	116	信息技术与法治建设	郭　雷　杨临萍　徐宗本 梅　宏　黄　维	5月16日
4	117	微生物组与大健康	邓子新　焦念志　岳建民	6月19～20日
5	118	早期大陆形成演化与环境资源效应	赵国春　郭敬辉	6月21日
6	119	变质地质学科学与技术	翟明国张立飞	6月22日
7	120	21世纪化学中的纳米科技前沿	朱道本　万立骏　赵宇亮	7月31日～8月1日
8	121	新时代的科技出版	朱作言　梅　宏	9月13日
9	122	深空探测与信息技术	包为民　杨元喜　吴一戎 王　赤　常　进	9月26日
10	123	文物保护与科技创新	成会明	10月12～14日
11	124	可循环高分子材料现状、机遇和挑战	唐勇　王玉忠　谢在库 陈学思	10月12～13日

附录五 2022年中国与世界十大科技进展

一、2022年中国十大科技进展

1. 中国天眼FAST取得系列重要进展

2022年1月6日，中国科学院国家天文台李菂研究员领导的团队，通过FAST平台，采用原创的中性氢窄线自吸收方法，首次获得原恒星核包层中具有高置信度的塞曼效应测量结果。3月18日，李菂领导的团队通过分析包括FAST、美国绿岸望远镜GBT在内的多项数据，首次提出了能够统一解释重复快速射电暴偏振频率演化的机制，为最终确定FRB起源提供了关键观测证据。6月9日，李菂领导的国际合作团队，在FAST的帮助下，发现了迄今为止唯一一例持续活跃的重复快速射电暴，并确认近源区域拥有目前已知的最大电子密度。9月21日，FAST快速射电暴优先和重大项目科学研究团队，利用FAST对一例位于银河系外的快速射电暴开展了深度观测，首次探测到距离快速射电暴中心仅1个天文单位（即太阳到地球的距离）的周边环境的磁场变化，向着揭示快速射电暴中心引擎机制迈出重要一步。10月19日，中国科学院国家天文台徐聪研究员领导的国际团队，利用FAST对致密星系群“斯蒂芬五重星系”及周围天区的氢原子气体进行了成像观测，发现了一个尺度大约为200万光年的巨大原子气体结构，比银河系大20倍，这是迄今为止在宇宙中探测到的最大的原子气体结构。上述5项重要成果均在《自然》《科学》上发表。

2. 中国空间站完成在轨建造并取得一系列重大进展

2022年11月29日23时08分，搭载神舟十五号载人飞船的长征二号F遥十五运载火箭在酒泉卫星发射中心发射成功。11月30日5时42分，神舟十五号载人飞船自主快速交会对接于空间站天和核心舱前向端口，加上问天、梦天实验舱，神舟十四号、天舟五号飞船，空间站由此形成“三舱三船”组合体，达到当前设计的最大构型，总重近百吨。神舟十五号航天员乘组于11月30日清晨入驻“天宫”，与神舟十四号航天员乘组相聚中国人的“太空家园”，开启中国空间站长期有人驻留时代。这是中国载人航天史上首次有两个航天员乘组在“太空会师”，也是中国航天员首次在空间站迎接神舟载人飞船来访。19个月内，中国载人航天密集实施11次发射、2次飞船返回、7次航天员出舱，4个飞行乘组12名航天员接续在轨驻留，空间站“T”字基本构型组装建造如期完成。展现了中国载人航天30年发展的厚重积淀与强大实力，跑出了新时代中国航天发展的加速度。

3. 我国科学家发现玉米和水稻增产关键基因

玉米、水稻和小麦是迄今驯化最为成功的三大农作物，为全人类提供了50%以上的能量摄入。由于它们的驯化地区、祖先各不相同，形态习性各异，其驯化过程是否遵循共同的遗传规律在科学界长期存在争论。2022年3月25日，《科学》杂志在线发表了中国农业大学教授杨小红/李建生与华中农业大学教授严建兵联合团队的研究论文。经过三代科学家18年研究发现，玉米基因KRN2和水稻基因OsKRN2受到趋同选择，并通过相似的途径调控玉米和水稻的产量。该团队进一步在全基因组层面阐明了趋同进化的遗传规律。据悉，这一成果不仅揭示了玉米与水稻的同源基因趋同进化从而增加玉米与水稻产量的机制，为育种提供了宝贵的遗传资源，而且为农艺性状关键控制基因的解析与育种应用，以及其它优异野生植物快速再驯化或从头驯化提供重要理论基础。

4. 科学家首次发现并证实玻色子奇异金属

电子科技大学电子薄膜与集成器件国家重点实验室主任李言荣院士团队与美国布朗大学教授James M. Valles Jr、北京大学物理学院/量子材料科学中心谢心澄院士等协同攻关，成功突破了费米子体系的限制，首次在玻色子体系中诱导出奇异金属态。

相关研究 2022 年 1 月 12 日发表于《自然》。宇宙中，基本粒子分为费米子与玻色子两种。其中，人类社会目前赖以生存的电子工业与器件发展几乎完全基于费米子体系，但该体系能耗高、损耗大，物理尺寸已近极限，面临性能持续提升的瓶颈，无法满足快速增长的信息传输需求。而以高温超导体为代表的玻色子器件，具有完美的零损耗能量传递特性，有望为电子信息工业带来革命性变化。据悉，该研究为理解凝聚态物理中奇异金属的物理规律、揭示奇异金属的普适性、完善量子相变理论奠定了科学基础，对揭示耗散效应对玻色子量子相干的定量影响、推动未来低能耗超导量子计算以及极高灵敏量子探测技术的发展具有重要的理论和实际意义。

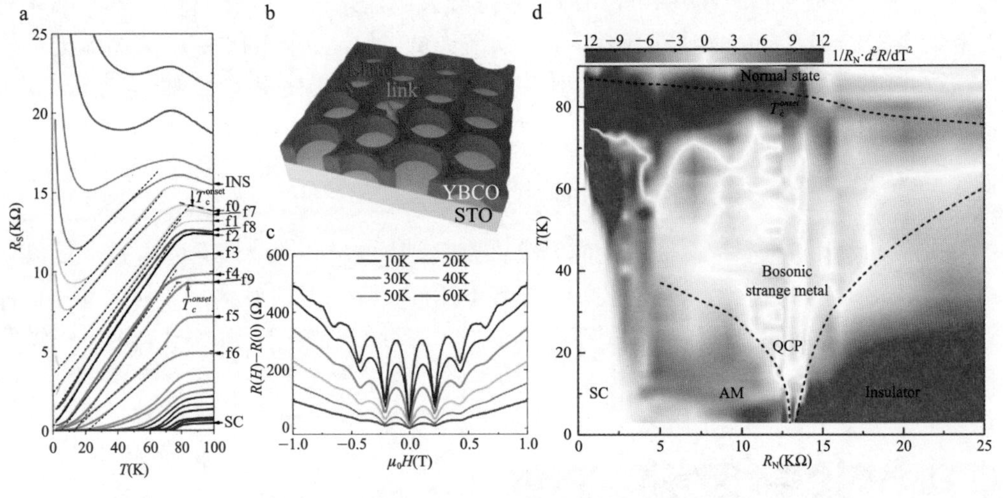

5. 我国科学家将二氧化碳人工合成葡萄糖和脂肪酸

将二氧化碳人工转化为高附加值化合物，“变废为宝”，是科技界持续攻关的重要领域。我国科学家此前在国际上首次实现了二氧化碳到淀粉的从头合成。2022 年，电子科技大学夏川课题组、中国科学院深圳先进技术研究院于涛课题组和中国科学技术大学曾杰课题组共同创建了一种二氧化碳转化新路径，通过电催化与生物合成相结合，成功以二氧化碳和水为原料合成了葡萄糖和脂肪酸，为人工和半人工合成“粮食”提供了新路径。该研究开辟了电化学结合活细胞催化制备葡萄糖等粮食产物的新策略，为进一步发展基于电力驱动的新型农业与生物制造业提

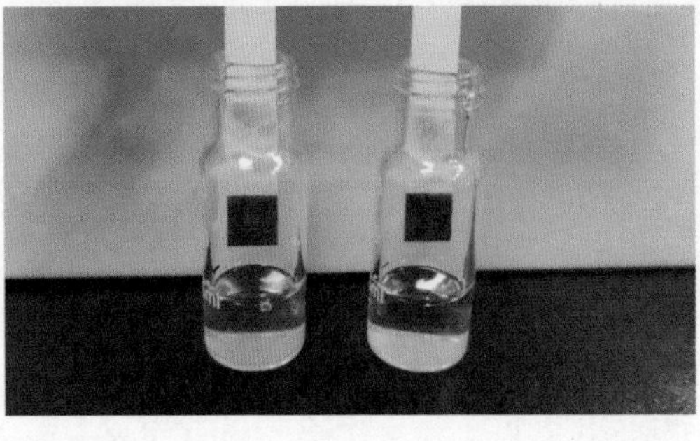

供了新范例，是二氧化碳利用的重要发展方向。该成果 4 月 28 日以封面文章形式在《自然—催化》发表。

6. 我国迄今运载能力最大固体运载火箭“力箭一号”首飞成功

2022 年 7 月 27 日 12 时 12 分，由中国科学院力学研究所抓总研制、中国迄今运载能力最大的固体运载火箭“力箭一号”（ZK-1A）在酒泉卫星发射中心成功发射，以“一箭六星”方式将六颗卫星送入预定轨道。“力箭一号”运载火箭首次飞行任务取得圆满成功，作为中小型卫星发射优先选择，丰富了中国固体运载火箭发射能力谱系。该款火箭是四级固体运载火箭，起飞重量 135 吨，起飞推力 200 吨，总长 30 米，芯级直径 2.65 米，首飞状态整流罩直径 2.65 米，500 公里太阳同步轨道运载能力 1500 公斤。据悉，“力箭一号”运载火箭由中国科学院“十四五”重大项目支持，其面向空间科学和空间技术发展需求，以“工程科学”思想为指导，以创新、先进、高效为设计思路，发展创新性、先进性、经济性运载火箭，对于推动中国运载技术和研制模式的变革和创新、推动空间科学发展具有重要意义。

7. “夸父一号”发射成功，并发布首批科学图像

我国综合性太阳探测专用卫星“夸父一号”首批科学图像于 2022 年 12 月 13 日在京正式对外发布。包括“夸父一号”自成功发射以来的 3 台有效载荷在轨运行两个月期间获取的若干对太阳的科学观测图像，这些成果实现多个国内外首次，在轨验证了“夸父一号”3 台有效载荷的观测能力和先进性。据了解，“夸父一号”卫星全称先进天基太阳天文台（ASO-S），是一颗综合性太阳探测专用卫星，由中国科学院国家空间科学中心负责工程大总体和地面支撑系统的研制建设，中国科学院微小卫星创新研究院、国家天文台、长春光学精密机械与物理研究所、紫金山天文台负责卫星平台

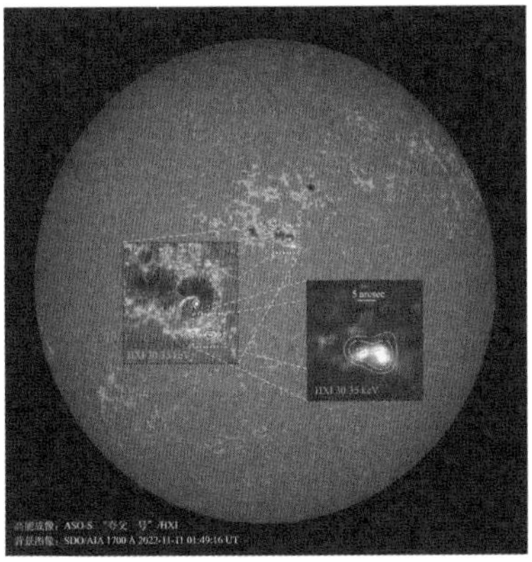

及有效载荷研制，科学应用系统由中国科学院紫金山天文台负责，测控系统由中国西安卫星测控中心负责实施，运载火箭由中国航天科技集团有限公司第八研究院研制生产。据悉，该卫星于 2022 年 10 月 9 日在酒泉卫星发射中心成功发射。卫星科学目标为“一磁两暴”，即同时观测太阳磁场及太阳上两类最剧烈的爆发现象——耀斑和日冕物质抛射，并研究它们的形成、演化、相互作用、关联等，同时为空间天气预报提供支持。

8. 新技术可在海水里原位直接电解制氢

由于淡水资源紧缺，向大海要水是未来氢能发展的重要方向。但复杂的海水成分（约 92 种化学元素）导致海水制氢面临诸多难题与挑战，先淡化后制氢工艺流程复杂且成本高昂。2022 年 11 月 30 日，中国工程院院士谢和平与他指导的深圳大学、四川大学博士生团队在《自然》发表论文，以物理力学与电化学相结合的全新思路，建立了相变迁移驱动的海水无淡化原位直接电解制氢全新原理与技术。该技术彻底隔绝了海水离子，实现了无淡化过程、无副反应、无额外能耗的高效海水原位直接电解制氢，即在海水里原位直接电解制氢。据悉，海水无淡化原位直接电解制氢技术未来有望与海上可再生能源相结合，构建无淡化、无额外催化剂工程、无海水输运、无污染处理的海水原位直接电解制氢工厂。

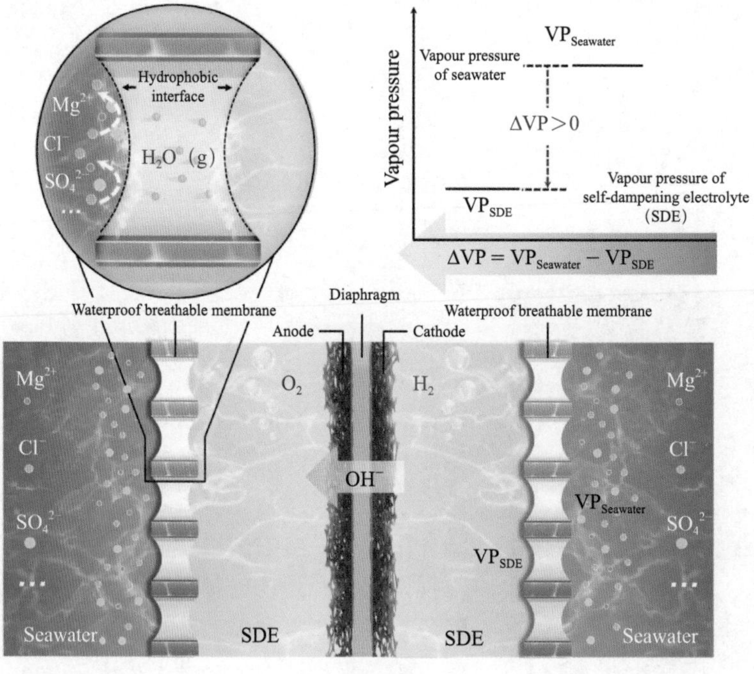

9. 国家重大科技基础设施“稳态强磁场实验装置”实现重大突破

2022 年 8 月 12 日，国家重大科技基础设施“稳态强磁场实验装置”实现重大突破，创造场强 45.22 万高斯的稳态强磁场，超越已保持了 23 年之久的 45 万高斯稳态强磁场世界纪录。国家稳态强磁场实验装置由中国科学院合肥物质科学研究院强磁场科学中心研制，是“十一五”期间国家发改委批准立项的重大科技基础设施，包括十台磁体——五台水冷磁体、四台超导磁体和一台混合磁体。此次国家稳态强磁场实验装置的混合磁体在 26.9 兆瓦的电源功率下产生 45.22 万高斯的稳态强磁场，达到国际领先水平，成为我国科学实验极端条件建设乃至世界强磁场技术发展的重要里程碑。据悉，稳态强磁场是物质科学研究需要的一种极端实验条件，是推动重大科学发现的“利器”。在强磁场实验环境下，物质特性会受到调控，有利于科学家发现物质新现象、探索物质新规律。

10. “巅峰使命”珠峰科考创造多项新纪录

2022 年 5 月 30 日，“巅峰使命”珠峰科考活动的主体任务完成，共有 5 个科考分队、16 支科考小组、270 多名科考队员参加。此次科考在西风-季风协同作用及影响、巅峰海拔的强烈升温、巅峰海拔的冰雪融化、高新技术平台观测的水汽和温室气体、珠峰地区的强大气氧化性过程、珠峰地区人体生理的特殊反应、珠峰地区变绿的生态过程等方面取得了众多亮点成果，创下多项科考新纪录。其中，“巅峰使命”珠峰科考首次建成了梯度联网的巅峰站并实现了数据实时传输，架设了世界上海拔最高的气象站（8830 米），建成了从 4276 米到 8830 米海拔梯度的观测网络，实现了观测数据实时传输；科考首次成功获取了海拔 6500 米、7028 米和 8848 米的冰雪样品；科考所使用的“极目一号”Ⅲ型系留浮

空艇长 55 米、高 19 米，体积达 9060 立方米，创造了海拔 9050 米浮空艇原位大气环境科学观测的纪录。此外，“巅峰使命”珠峰科考首次利用高精度雷达测量了珠峰顶部的冰雪厚度；首次采用多种先进技术获得地面至 39 公里高空大气臭氧浓度数据和三维风场；首次获得高原常驻和短居人群的高山生理适应数据等。

二、2022 年世界十大科技进展新闻

1. 首个完整人类基因组序列公布

由美国国家人类基因组研究所、加利福尼亚大学圣克鲁斯分校、华盛顿大学等机构研究人员领衔的国际科研团队 2022 年 3 月 31 日公布了首个完整、无间隙的人类基因组序列。与这项重大成果相关的 6 篇论文当天发表在美国《科学》杂志上。美国国家人类基因组研究所在一份公报中表示，人类基因组含有约 30 亿个 DNA（脱氧核糖核酸）碱基对，完成这些碱基对的完整、无间隙测序对于了解人类基因组变异全谱、掌握基因对某些疾病的影响至关重要。据悉，人类基因组测序项目的重要意义被视为与阿波罗登月计划相当。人类基因组蕴藏人类遗传信息，破译它能够为疾病诊断、新药研发、新疗法探索等带来革命性进步。早在 2001 年，由包括中国在内的 6 国科学家共同参与了国际“人类基因组计划”，并在英国《自然》杂志上发布了人类基因组草图及初步分析。但由于当时的测序技术所限，这份人类基因组草图中留有许多空白。

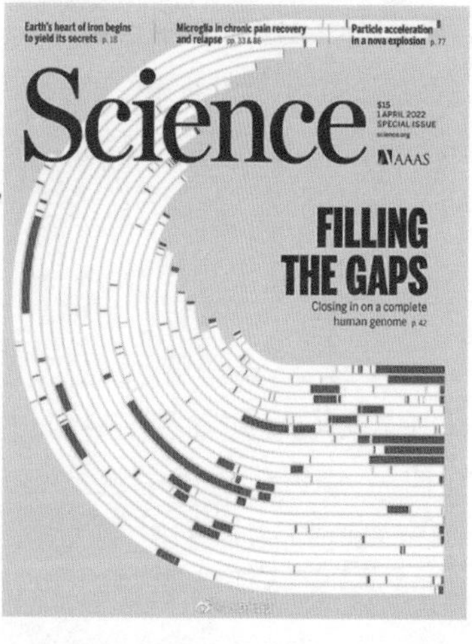

2. 人造心脏研究取得重要进展

为了从头开始构建人类心脏，研究人员需要复制构成心脏的独特结构。这包括重建螺旋几何形状——当心脏跳动时，螺旋几何形状会产生扭曲的运动。这种扭曲运动对大量泵血至关重要，但由于制造具有不同几何形状和排列的心脏难度较大，这项工作极具挑战性。如今，美国哈佛大学约翰·保尔森工程与应用科学学院（SEAS）生物工程师使用一种新的增材纺织品制造方法（FRJS），开发了

第一个具有螺旋排列跳动心脏细胞的人类心室生物杂交模型，并证明其肌肉排列确实会显著增加每次收缩时心室泵出的血液量。相关研究结果发表于2022年7月7日出版的《科学》杂志。研究的目标是建立一个模型，测试心脏的螺旋结构是否对达到大的射血分数（即每次收缩时心室泵送的血液百分比）至关重要，并研究心脏螺旋结构的相对重要性。这项工作是朝着器官生物制造迈出的重要一步，使人们更接近于建立用于移植的人体心脏的最终目标。

3. 银河系中心黑洞的首张照片面世

2022年5月12日，包括中国在内的全球多地天文学家同步公布了一个超大质量黑洞——人马座A*（Sgr A*）的照片。相关研究成果以特刊形式发表在《天体物理学杂志通讯》上。这是人类“看见”的第二个黑洞，也是银河系中心超大质量黑洞真实存在的首个直接视觉证据。这个超大质量黑洞距离太阳系约2.7万光年，质量超过太阳质量的400万倍。这张银河系中心黑洞的照片，与人类看到的第一张黑洞照片的拍摄者和拍摄时间均相同，都是由“事件视界望远镜”（EHT）合作组织在2017年通过分布在地球上由8个射电望远镜组成的一个等效于地球般口径大小的“虚拟望远镜”所拍摄。EHT研究团队花了五年时间，用超级计算机合成和分析数据，编纂了前所未有的黑洞模拟数据库，与观测结果进行严格比对，并提取出不同照片平均后的效果，最终得以将银河系中心这个超大质量黑洞的“真实容貌”第一次呈现出来。

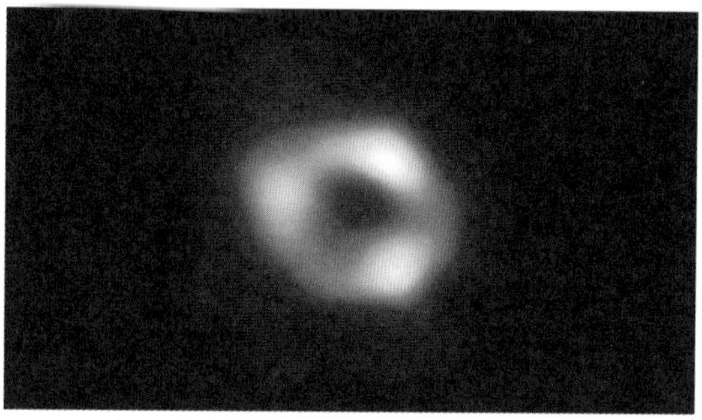

4. 人类首次成功改变小行星轨道

2022年9月26日，美国宇航局（NASA）利用双小行星重定向测试（DART）航天器，撞击了一颗近地双小行星系统中较小的小行星——Dimorphos，以期改变其运行轨道。这是世界上首个旨在防御地球免受小行星撞击威胁的测试任务。10月11日，NASA证实这次任务取得成功

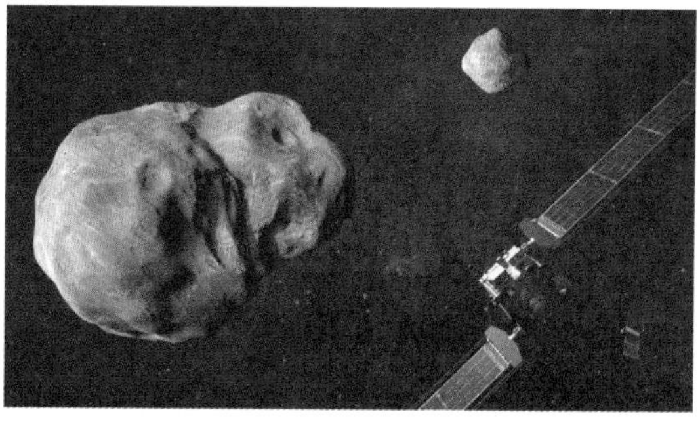

——DART 航天器的撞击，将 Dimorphos 推向其伴星 Didymos，并将前者近 12 小时的轨道周期缩短了 32 分钟。据悉，NASA 在撞击开始前表示，将轨道周期缩短 73 秒就代表任务成功。大多数天文学家则预测，撞击可能导致轨道周期缩短 10 分钟。但该撞击造成的偏斜程度远远大于预期。这也在一定程度上表明，动能撞击是行星防御的可行方法。

5. 美国首次成功在核聚变反应中实现“净能量增益”

2022 年 12 月 13 日，美国能源部（DOE）和能源部国家核安全管理局（NNSA）宣布，劳伦斯利弗莫尔国家实验室（LLNL）的美国国家点火装置（NIF）团队首次在可控核聚变实验中实现核聚变反应的净能量增益，即通过核聚变产生的能量比激发聚变所使用的能量更多，这项突破将为美国国防的发展和清洁能源的未来铺平道路。据悉，美国国家点火装置团队用 192 束激光束，向一个微型燃料颗粒输送了 205 万焦耳的激光能量，点燃核聚变燃料，最终产生了 315 万焦耳的聚变能量输出，实现净能量增益，首次证实了惯性核聚变能（IFE）的基本科学原理和可行性。

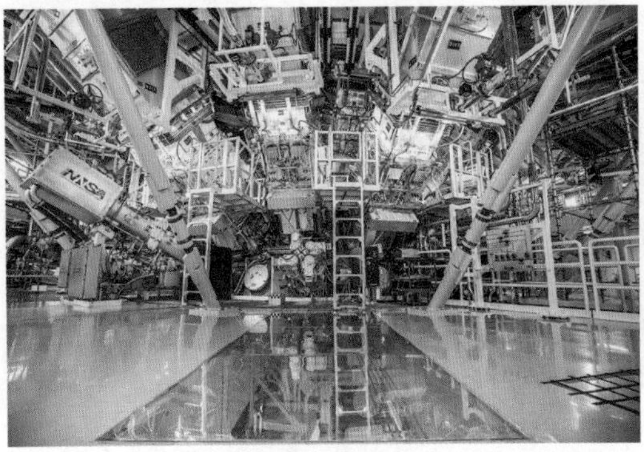

6. 詹姆斯·韦布空间望远镜顺利入轨首次传回照片

詹姆斯·韦布空间望远镜是由美国宇航局与欧洲空间局、加拿大航天局联合研究开发，是 NASA 建造的迄今最大、功能最强的空间望远镜，其主镜直径 6.5 米，由 18 片巨大六边形镜片构成；配有 5 层可展开的遮阳板，被认为是哈勃空间望远镜的“继任者”。该望远镜于 2021 年 12 月 25 日从法属圭亚那库鲁航天中心发射升空，2022 年 1 月 24 日顺利进入围绕日地系统第二拉格朗日点的运行轨道，并于 7 月 12 日正式公布了其拍摄的一批宇宙全彩色照片。此后，韦布空间望远镜还拍摄到距离地球约 280 亿光年的最遥远恒星的新图像并首次在系外行星上明确探测到二氧化碳。据悉，韦布空

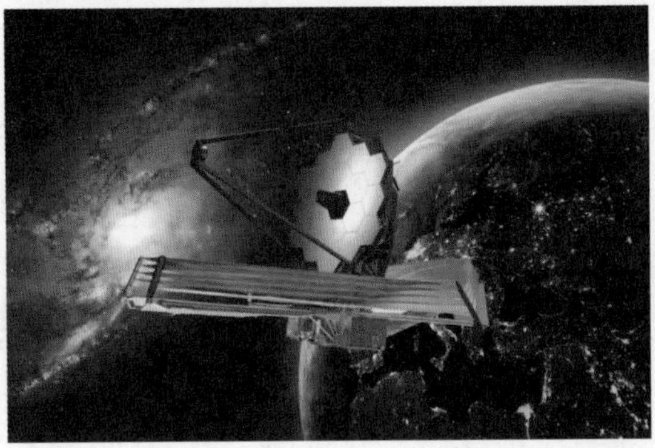

间望远镜任务目标主要有 4 个方面：寻找 135 亿多年前的宇宙中诞生的第一批星系；研究星系演化的各阶段；观察恒星及行星系统的形成；测定包括太阳系行星系统在内的行星系统的物理、化学性质，并研究其他行星系统存在生命的可能性。

7. 世界首台百亿亿次超级计算机打破速度纪录

2022 年 5 月 31 日，国际超算组织宣布，位于美国橡树岭国家实验室的超级计算机“前沿”在 2022 年国际超算 Top500 榜单中拔得头筹，成为现今世界上运行速度最快的超级计算机，算力高达每秒 1.1 百亿亿次，也是目前国际上公告的首台每秒能执行百亿亿次浮点运算的计算机。据悉，普通笔记本电脑每秒只能进行几万亿次运算，而“前沿”的运行速度是其 100 多万倍。百亿亿次超级计算机也被称为 E 级超级计算机，每秒计算次数超过 10^{18}，它的研制占据了国际高端信息技术创新和竞争的制高点，可用于对气候变化、核聚变模型进行精确建模，有助于新药的研发以及加密技术破解，因此也将成为国家安全的重要工具。

8. 猪蛋白角膜让人重见光明

长期以来，科学家一直在寻找可替代人类角膜的移植物。如今，瑞典林雪平大学和 LinkoCare Life Sciences 公司的研究人员通过提取猪胶原蛋白制成的人工角膜，成功使失明或视力受损的人恢复了视力，且手术两年后，患者没有严重并发症或副作用的报告。相关研究 2022 年 8 月 11 日发表于《自然-生物技术》。林雪平大学的 Mehrdad Rafat 和同事通过从猪皮中提取和纯化胶原蛋白，制造了一种柔韧有弹性的类似隐形眼镜的人工角膜。在相关实验成功后，研究小组开始在志愿者中对人工角膜进行测试。在接受人工角膜移植后，每个人的视力都有所提高，其中有 3 名失明患者术后视力恢复到正常人水平。该研究结果有助于开发出一种符合人类植入物标准、可以大规模生产并储存长达两年的生物材料，从而惠及更多有视力问题的人。

9. 人工智能加速“原创”新蛋白质设计

随着人工智能（AI）的巨大进步，美国西雅图华盛顿大学（UW）生物化学家 David Baker 领导的一个团队，只需几秒钟便可以设计出“原创”新蛋白质。相关研究发表于 2022 年 9 月 15 日出版的《科学》。最初，研究人员构想出一种新蛋白质的形状——通常是将其他蛋白质的片段拼凑在一起，然后由软件推导出与该形状对应的氨基酸序列。但在实验室中制作这些“草稿”蛋白质时很少能折叠成所需的形状，相反，它们最终被卡在不同的状态。而通过调整蛋白质结构预测软件 AlphaFold 和其他 AI 程序，这一耗时的步骤可以瞬间完成。在 Baker 团队开发的一种名为“幻觉”的方法中，研究人员将随机的氨基酸序列输入结构预测网络；根据网络的预测，改变其结构，使之变得更像蛋白质。

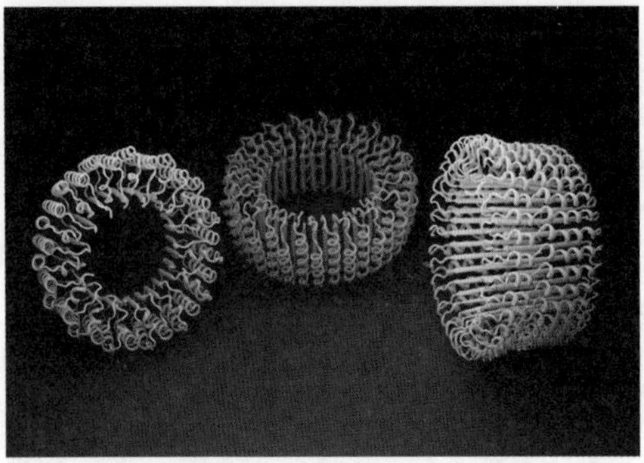

10. 科学家发现“四中子态”存在最明确证据

由数十个国家的科学家组成的联合团队发现了迄今“四中子态”（tetraneutron）奇异物质存在的最明确证据，相关论文 2022 年 6 月 22 日发表于《自然》。20 年前，科学家意外发现了一种奇异物质“四中子态”的存在迹象，该物质由 4 个中子组成。此次，国际联合团队找到了迄今“四中子态”存在的最明确证据。德国慕尼黑工业大学 Roman Gernhauser 等研究人员利用不同的粒子碰撞，制造出平常多出 4 个中子的氦原子，然后与质子碰撞，在碰撞后，只剩下四个中子，并且可以结合成一个“四中子态”。该实验旨在抑制可能干扰或被误认为是产生“四中子”的每一个反应，因此他们以无与伦比的精度测量了缺失的能量。通过追踪缺失的能量，他们推断出“四中子”形成的时间非常短暂，仅有 10^{-22} 秒钟。据悉，这一发现将有助于物理学家对核力本质的理论进行微调。

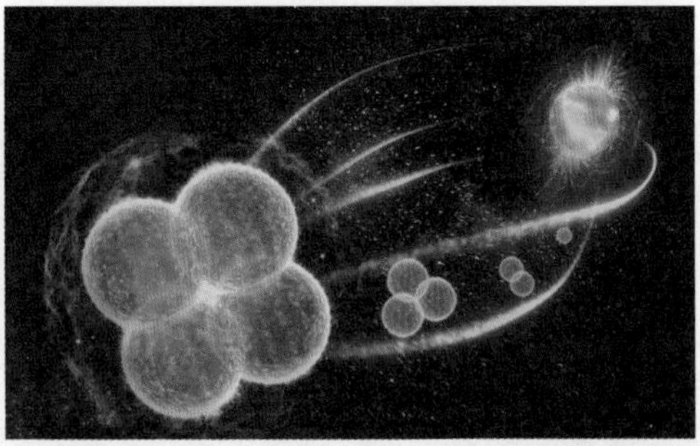

附录六　2022 年香山科学会议学术讨论会一览表

序号	会号	会议名称	执行主席	日期
1	S67	大科学装置前沿研究	常　进　马余刚　赵政国　高原宁　李　亮　陈和生　龚旗煌　李立毅　邰仁忠	7 月 13～15 日
2	715	低碳生物合成	赵国屏　邓子新　谭天伟　曾安平　马延和	7 月 21～22 日
3	716	单分子科学与技术	田中群　杨金龙　高　松　罗　毅　郭雪峰	8 月 2～3 日
4	717	远距离大容量连续无线功率传输的科学与技术问题	李　明　杨士中　段宝岩　李　军	8 月 15～16 日
5	718	中国居民营养素摄入量和慢性病防控	杨月欣　丁钢强　朱蓓薇　杨晓光　孙长颢　马爱国	8 月 18～19 日
6	S66	“宁静中国”与噪声治理关键技术	田　静　李风华　毛东兴　蒋伟康	8 月 22～23 日
7	719	食物系统转型的科学与创新	陈君石　樊胜根　唐华俊　张福锁　施小明	8 月 23～24 日
8	S68	地球系统与全球变化	朱日祥　戴民汉　金之钧　张人禾　马克平	8 月 24～25 日
9	720	数字眼科与全身疾病认知方法与关键技术	孙凝晖　黄　维　陈益强　魏文斌　卓业鸿	8 月 25～26 日
10	722	黑土地保护与利用	张福锁　康绍忠　于贵瑞　朱永官　张佳宝　沈其荣	8 月 30～31 日
11	723	面向人体系统调控的生命间质结构、功能与行为	韩　东　田中群　樊春海　顾　宁　谷晓红	8 月 30～31 日
12	724	大数据驱动中西医结合精准诊疗	陈香美　张伯礼　杨宝峰　刘　良	9 月 1～2 日
13	725	超快光电子成像技术及应用	祝世宁　沈保根　张　杰　龚旗煌	9 月 3～4 日
14	726	细胞治疗临床研究和监管	王福生　王广基　王军志　田志刚　张学敏	9 月 3～4 日

续表

序号	会号	会议名称	执行主席	日期
15	727	健康医学的理论体系和技术方法	刘德培 李振吉 陈东义 张启明	9月5～6日
16	728	中药资源的可持续开发与利用	黄璐琦 赵军宁 郝小江 林国强 岳建民 孙晓波 唐健元	9月8～9日
17	729	长寿命道路材料本构与结构力学行为理论	黄 卫 郑健龙 张肖宁 唐伯明 沙爱民 孙立军 王旭东	9月15～16日
18	730	激光加速器和深空探测：机遇和挑战	李儒新 欧阳晓平 蔡荣根 宾建辉	9月20～21日
19	732	火星探测中重要科学与技术问题	潘永信 王 赤 张荣桥 赵玉芬	9月22～23日
20	733	基于多学科视角的“脑-肠-微生态”前沿与技术	赵国屏 吴清平 魏 玮 荣培晶	9月24～25日
21	Y6	智能医学的基础理论与关键技术	刘 哲 聂 舟 孟 幻	9月26～27日
22	734	耕地内涵外延及其调查监测评价	张福锁 张佳宝 唐华俊 周成虎 朱永官	9月27～28日
23	735	新型精神健康诊疗技术的挑战与机遇	黄 维 陆 林 骆清铭 胡 斌	11月9～10日
24	736	实现建筑碳中和的挑战与应对技术路径	江 亿 刘加平 肖绪文 王元丰 肖建庄 徐润昌	11月14～15日
25	S69	科学数据治理与利用的前沿和热点	郭华东 刘德培 孙九林 邹自明 周国民 陈 刚	11月19～21日

附录七　2022年中国科学院学部“科学与技术前沿论坛”一览表

序号	会次	会议主题	执行主席	举办时间
1	125	缪子束加速和对撞技术及其应用	高原宁　张肇西　赵红卫　赵政国	4月16～17日
2	126	表观遗传学研究新进展	宋尔卫	5月21日
3	127	航空动力及航空航天交叉学科	李应红　闫楚良　唐志共	6月24日
4	128	先进装备技术发展	毛　明　芮筱亭　胡海岩	8月16～18日
5	129	环境、生态与健康	康　乐　陶　澍　陆　林	8月23～24日
6	130	数据驱动的新地学	陈发虎　徐宗本	10月29～30日
7	131	人工智能的基础研究	袁亚湘	10月29～30日
8	132	介催化科学与技术前沿	谢在库　苏宝连	11月1日
9	133	建设工程与交叉科学	吴硕贤　何满潮　杨　卫	11月13日